W9-AEC-293

# GENERAL TOPOLOGY

STEPHEN WILLARD
*University of Alberta*

# GENERAL TOPOLOGY

*Reading, Massachusetts • Menlo Park, California*
*London • Amsterdam • Don Mills, Ontario • Sydney*

ADDISON-WESLEY PUBLISHING COMPANY

This book is in the
**ADDISON-WESLEY SERIES IN MATHEMATICS**

*Consulting Editor:* LYNN H. LOOMIS

AMS 1968 Subject Classification 5401

 Printed in the United States of America. Published simultaneously in Canada. Library of Congress Catalog Card No. 74-100890

ISBN 0-201-08707-3
JKLMNOPQ-MA-89876

# Preface

This book is designed to develop the fundamental concepts of general topology which are the basic tools of working mathematicians in a variety of fields. The material here is sufficient for a variety of one- or two-semester courses, and presupposes a student who has successfully mastered the material of a rigorous course in advanced calculus or real analysis. Thus it is addressed primarily to the beginning graduate student and the good undergraduate.

A principal goal here has been to seek some sort of balance, in the treatment, between two broad areas into which general topology might (rather arbitrarily and, of course, inaccurately) be divided. The first, which could be called "continuous topology", centers on the results about compactness and metrization which are the indispensable tools of the modern analyst. This is what Kelley has labeled "what every young analyst should know", and is represented here by sections on convergence, compactness, metrization and complete metric spaces, uniform spaces and function spaces. The second area, which might be called "geometric topology", is primarily concerned with the connectivity properties of topological spaces and provides the cores of results from general topology which are necessary preparation for later courses in geometry and algebraic topology. This core is formed here by a series of nine sections on connectivity properties, topological characterization theorems and homotopy theory. By suitable surgical intervention, mixed audiences can be taught a mixture of the two approaches, using whatever recipe the instructor likes best. To aid in the concoction of such recipes this preface is followed by a table of some of the important topics in the book together with a list of the material which is prerequisite for each.

While trying to maintain the balance just described, I have also tried to keep in mind the potential uses of such a book both as a text and as a reference source. Thus, in a concession to pedagogy, I have paced the book rather more slowly at the beginning than at the end and have concentrated motivational comments at the beginning. I have also attempted to keep the pedagogical lines of force transparent by paring the material of each section down to what I believe is fundamental. At the same time, I have included a large selection of exercises (over 340, each containing several parts), which provide drill in the techniques developed in the text, develop limiting counterexamples and provide extensions of, and parallels to, the theory presented in the text. Some of the "theoretical" exercises are suitable for extended development and discussion in the classroom, and all should enhance the value of the book as a reference source. Worth particular mention are the exercises on normed linear spaces and topological groups, and many of the exercises in the sections on compactness, compactification, metrization and the Stone–Weierstrass theorem. To facilitate its use as a reference source, I have included at the end of the book a collection of background notes for each section, a large (but certainly not exhaustive) bibliography and an index as comprehensive as my patience would allow.

The primary organization of the book is into forty-four sections; chapter headings are provided, but not as a referencing device; they serve only to collect the sections into coherent groups. Within each section, the definitions, examples and theorems are further numbered consecutively so that Theorem 25.3 appears as the third item (not necessarily the third theorem) in Section 25. The one exception to this rule is Section 1, on set theory, where the material is somewhat condensed and the numbers 1.1, 1.2, . . . serve to designate subsections rather than specific results. One note of caution seems advisable. A reference to a theorem *number* only, omitting the word "theorem", should serve as a warning that the relevant observation may be made in the remarks following the proof of the theorem, rather than in the statement of the theorem itself. (This happens infrequently, however, and most references, even of this type, are to the numbered theorem itself.) Each section ends with a set of exercises, lettered consecutively; most exercises consist of several parts. A reference to 3E is a reference to the fifth exercise in Section 3; where more precision is needed, 3E.3 is used to designate the third part of this exercise.

A few notational and terminological conventions deserve special mention. Following the lead of Halmos and Kelley, we replace the cumbersome "if and only if" by "iff" and denote the end of a proof by ■. When discussing statements of the form "$P$ iff $Q$", we occasionally use "necessity" to mean "if $P$ then $Q$" and "sufficiency" to mean "$P$ if $Q$". Square brackets are used nonmathematically in two contexts in this book. At the end of an exercise, they enclose hints to the solution of that exercise, and placed at the end of an item in the bibliography, they enclose a reference to the review of that item in the *Mathematical Reviews* or (for items written between 1930 and 1940) the *Zentralblatt*.

Anyone who writes a book of this sort accumulates a sea of outstanding debts. My own personal sea has been fed by more rivers of kindness than I can count; many have no doubt achieved the status of underground streams and been forgotten. The one I cannot forget created the sea long before this project was conceived, and I here acknowledge my greatest debt to A. H. Stone. *J'en suis pas digne*.

The presentation here has been affected by countless conversations with friends and colleagues, who were not always aware they were speaking for posterity. I apologize, mentioning particularly Donald Plank, Melvin Henriksen, W. W. Comfort, Don Johnson, Ta Sun Wu, John Isbell, Anthony Hager and Phillip Nanzetta. A great many students deserve my thanks for stoically suffering through earlier versions of the manuscript: These include my own at Lehigh, Case Western Reserve and the University of Alberta, as well as those of Professor Johnson at New Mexico State University and Professor Comfort at Wesleyan University. Especially, parts of the manuscript were assiduously edited by Robert Shurtleff, and critically reviewed by the students in Professor Comfort's class. They will, I think, recognize their influence in the ultimate presentation.

If I mention the students who have suffered through one or another of the early versions of this manuscript, I cannot neglect my wife, Mary, who has suffered through every version, both as wife and as proof reader.

The typing was done by Elizabeth Roach and Rosemary Pappano. Virtually every mistake that survived their typing was my own and I am shaken to report that they caught several of my best and most subtle errors, mathematical and otherwise.

Case Western Reserve University deserves my thanks for making it possible for me to avoid dividing my time and myself between the classroom and preparation of this manuscript in the fall of 1968. Parts of the manuscript were prepared during my tenures on several grants from the National Science Foundation.

*Edmonton, Alberta*
*April 1970*

S.W.

| Topic | Prerequisite material (with 1.1–8.8 assumed) |
| --- | --- |
| Stone–Čech compactification (19.4–19.13) | 8.11–8.16, 11 or 12.1–12.14, 13, 14, 15.1–15.7, 17 |
| Urysohn's metrization theorem (23.1) | 13, 14.1–14.11, 15.1–15.7, 16, 22 |
| Uniform metrization theorem (23.4) | 13, 14.1–14.11, 20.1–20.8, 22, 23.1–23.3 |
| General metrization theorem (23.9) | 13, 14.1–14.11, 15.1–15.8, 20.1–20.13, 22, 23.1–23.2 |
| Banach's fixed-point theorem (24.16) | 22, 24.1–24.6 |
| Baire category theorem (25.4) | 13, 14, 15.1–15.5, 17, 18, 19.4–19.10, 22, 24 |
| Continuum characterization theorems (28.13, 28.14) | 13, 14, 15.1–15.5, 17, 26.1–26.10 |
| Cantor set characterization and mapping property (30.3, 30.7) | 13, 14, 15.1–15.8, 17, 22, 26–29, 30.1–30.2 |
| Hahn–Mazurkiewicz theorem (31.5) | 13, 14, 15.1–15.8, 17, 22, 26–30, 31.1–31.4 |
| Brouwer fixed-point theorem (34.6) | 26, 27, 32, 33, 34.1–34.5 |

For completeness, I will list also some of the special sections and their dependence on previous material.

# Table of Dependence

| Topic | Prerequisites |
| --- | --- |
| Products of normal spaces (21) | 13, 14.1–14.11, 15, 17, 19.4–19.13, 20.1–20.8 |
| Homotopy theory (32–34) | 26, 27 |
| Uniform spaces (35–39) | 8.11–8.16, 11, 13, 14.1–14.11, 15.1–15.5, 17, 20, 22, 23.1–23.4, 24.1–24.6 |
| Proximity spaces (40–41) | 8.11–8.16, 11, 13, 14.1–14.11, 15.1–15.5, 17, 19.4–19.13, 22, 24.1–24.6, 35, 36.1–36.15, 37, 39 |
| Function spaces (42–44) | 11, 13, 14.1–14.11, 15.1–15.5, 17, 18, 35, 36.1–36.15, 37, 39.1–39.6 |

# Contents

Chapter 1

# Set Theory and Metric Spaces

## 1 Set theory

The material of this section is introduced primarily to serve as a review for those with some background in set theory and as an introduction to our notational conventions and terminology. The reader entirely unfamiliar with any aspect of set theory should not be content with the intuitive discussion given here, but should consult one of the standard references on the subject (see the notes).

Most of the material in this book is accessible to anyone who understands 1.1 through 1.8 below. It is recommended that the remainder of this section be skipped on first reading and referred to later as needed.

**1.1 Sets.** A *set*, *family* or *collection* is an aggregate of things (for example, numbers or functions or desks or people), called the *elements* or *points* of the set. If $a$ is an element of the set $A$ we write $a \in A$ and if this is false we write $a \notin A$.

If $A$ is a set and $S$ is a statement which applies to some of the elements of $A$, the set of elements $a$ of $A$ for which $S(a)$ is true is denoted $\{a \in A \mid S(a)\}$. Thus if $\mathbf{N}$ is the set of positive integers, the positive divisors of 6 form the set $\{a \in \mathbf{N} \mid ab = 6$ for some $b \in \mathbf{N}\}$. In the case of small sets, such as this one, it is easy to describe the set by listing its elements in brackets. Thus the set just given is the set $\{1, 2, 3, 6\}$.

This discussion is rather naïve and leads to certain difficulties. Thus if $P$ is the set of all sets, we can apparently form the set $Q = \{A \in P \mid A \notin A\}$, leading to the contradictory $Q \in Q$ iff $Q \notin Q$. This is *Russell's paradox* (see Exercise 1A) and can be avoided (in our naïve discussion) by agreeing that *no aggregate shall be a set which would be an element of itself.*

**1.2 Elementary set calculus.** If $A$ and $B$ are sets and every element of $A$ is an element of $B$, we write $A \subset B$ or $B \supset A$ and say $A$ is a *subset* of $B$ or $B$ *contains* $A$. The collection $P(A)$ of all subsets of a given set $A$ is itself a set, called the *power set* of $A$.

We say sets $A$ and $B$ are *equal*, $A = B$, when both $A \subset B$ and $B \subset A$. Evidently, $A$ and $B$ are equal iff they have the same elements.

We write $A - B$ to denote the set $\{a \in A \mid a \notin B\}$ and (unlike some writers) use this notation even when $B$ is not a subset of $A$, i.e., even when $B \not\subset A$. When we do have $B \subset A$, $A - B$ is called the *complement* of $B$ in $A$.

The *empty set*, $\varnothing$, is the set having no elements. By the criterion for equality of sets, there is only one empty set and, by the criterion for containment, it is a subset of every other set.

Note that *element* and *subset* are different ideas. Thus, for example, $x \in A$ iff $\{x\} \subset A$.

A few sets will keep recurring and we will establish now a conventional notation for them.

$\mathbf{R}$: the set of real numbers,

$\mathbf{R}^n$: Euclidean $n$-space,

$\mathbf{N}$: the set of positive integers,

$\mathbf{I}$: the closed interval $[0, 1]$ in $\mathbf{R}$,

$\mathbf{Q}$: the set of rational numbers in $\mathbf{R}$,

$\mathbf{P}$: the set of irrational numbers in $\mathbf{R}$,

$\mathbf{S}^n$: the $n$-sphere, $\{x \in \mathbf{R}^{n+1} \mid |x| = 1\}$.

Eventually, each of these sets will be assumed to carry some "usual" structure (a metric, topology, uniformity or proximity) unless the contrary is noted. Additional less often used conventional notations will be introduced in the text. All can be found in the index.

**1.3 Union and intersection.** If $\Lambda$ is a set and, for each $\lambda \in \Lambda$, $A_\lambda$ is a set, the *union* of the sets $A_\lambda$ is the set $\bigcup_{\lambda \in \Lambda} A_\lambda$ of all elements which belong to at least one $A_\lambda$. When no confusion about the indexing can result, we will write the union of the sets $A_\lambda$ as simply $\bigcup A_\lambda$. The *intersection* of the sets $A_\lambda$ is the set $\bigcap_{\lambda \in \Lambda} A_\lambda$, or simply $\bigcap A_\lambda$, of all elements which belong to every $A_\lambda$. In case $\mathscr{A}$ is the collection $\{A_\lambda \mid \lambda \in \Lambda\}$, the union and intersection of the sets $A_\lambda$ are sometimes denoted $\bigcup \mathscr{A}$ and $\bigcap \mathscr{A}$, respectively.

When only finitely many sets $A_1, \ldots, A_n$ are involved, the alternative notations $A_1 \cup \cdots \cup A_n$ or $\bigcup_{k=1}^n A_k$ are sometimes used for the union of the $A_k$, while $A_1 \cap \cdots \cap A_n$ or $\bigcap_{k=1}^n A_k$ sometimes denotes their intersection. When denumerably many sets $A_1, A_2, \ldots$ are involved, their union will sometimes be denoted by $A_1 \cup A_2 \cup \cdots$ or $\bigcup_{k=1}^\infty A_k$, their intersection by $A_1 \cap A_2 \cap \cdots$ or $\bigcap_{k=1}^\infty A_k$.

We say *A meets B* iff $A \cap B \neq \varnothing$. Otherwise, $A$ and $B$ are *disjoint*. In general, a family $\mathscr{A}$ of sets is *pairwise disjoint* iff whenever $A, B \in \mathscr{A}$, $A \cap B = \varnothing$.

For those who wish to test themselves on the concepts just introduced, here are a few easily proved facts:

a) $A \subset B$ iff $A \cup B = B$,

b) $A \subset B$ iff $A \cap B = A$,

c) If $\mathscr{A}$ is the empty collection of subsets of $A$, then $\bigcup \mathscr{A} = \varnothing$ and $\bigcap \mathscr{A} = A$.

d) $A \cup B = A \cup (B - A)$.

e) $A \cap (B \cup C) = (A \cap B) \cup C$ iff $C \subset A$.

**1.4 Theorem.** *If $A$ is a set, $B_\lambda \subset A$ for each $\lambda \in \Lambda$ and $B \subset A$, then*

a) $A - (\bigcup_{\lambda\in\Lambda} B_\lambda) = \bigcap_{\lambda\in\Lambda} (A - B_\lambda),$
b) $A - (\bigcap_{\lambda\in\Lambda} B_\lambda) = \bigcup_{\lambda\in\Lambda} (A - B_\lambda),$ } *De Morgan's laws*

c) $B \cap (\bigcup_{\lambda\in\Lambda} B_\lambda) = \bigcup_{\lambda\in\Lambda} (B \cap B_\lambda),$
d) $B \cup (\bigcap_{\lambda\in\Lambda} B_\lambda) = \bigcap_{\lambda\in\Lambda} (B \cup B_\lambda).$ } *distributive laws*

*Proof.* a) If $x \in A - (\bigcup B_\lambda)$, then $x \in A$ and $x \notin B_\lambda$ for any $\lambda$, so $x \in A - B_\lambda$ for each $\lambda$; hence $x \in \bigcap (A - B_\lambda)$. Conversely if $x \in \bigcap (A - B_\lambda)$, then for each $\lambda$, $x \in A$ and $x \notin B_\lambda$; hence $x \in A - (\bigcup B_\lambda)$. Thus $x \in A - (\bigcup B_\lambda)$ iff $x \in \bigcap (A - B_\lambda)$, so that

$$A - (\bigcup B_\lambda) = \bigcap (A - B_\lambda).$$

b) Similar to (a). See Exercise 1B.

c) If $x \in B \cap (\bigcup B_\lambda)$, then $x \in B$ and $x \in \bigcup B_\lambda$; thus $x \in B$ and $x \in B_{\lambda_0}$ for some $\lambda_0$. Hence $x \in \bigcup (B \cap B_\lambda)$. Conversely, if $x \in \bigcup (B \cap B_\lambda)$, then $x \in B \cap B_{\lambda_0}$ for some $\lambda_0 \in \Lambda$; thus $x \in B$ and $x \in B_{\lambda_0}$, so that $x \in B$ and $x \in \bigcup B_\lambda$. Hence $x \in B \cap (\bigcup B_\lambda)$. We have shown $x \in B \cap (\bigcup B_\lambda)$ iff $x \in \bigcup (B \cap B_\lambda)$; it follows that $B \cap (\bigcup B_\lambda) = \bigcup (B \cap B_\lambda)$.

d) Similar to (c). See Exercise 1B. ■

**1.5 Small Cartesian products.** If $x_1$ and $y_2$ are distinct elements of some set, the two-element sets $\{x_1, x_2\}$ and $\{x_2, x_1\}$ are, by the criterion for set equality, the same. It is useful to have a device for reflecting priority as well as membership in this case, and it is provided by the notion of the *ordered pair* $(x_1, x_2)$. By definition, ordered pairs $(x_1, x_2)$ and $(y_1, y_2)$ are equal iff $x_1 = y_1$ and $x_2 = y_2$. For a somewhat more formal approach to ordered pairs, see Exercise 1C.

Now if $X_1$ and $X_2$ are sets, the *Cartesian product* $X_1 \times X_2$ of $X_1$ and $X_2$ is defined to be the set of all ordered pairs $(x_1, x_2)$ such that $x_1 \in X_1$ and $x_2 \in X_2$. This definition, for example, gives the plane as the set of all ordered pairs of real numbers. Other examples: $\mathbf{S}^1 \times \mathbf{I}$ is a cylinder, $\mathbf{S}^1 \times \mathbf{S}^1$ is a torus, $\mathbf{R} \times \mathbf{R}^n = \mathbf{R}^{n+1}$.

Once defined for two sets, Cartesian products of any finite number of sets can be defined by induction; thus, the last example in the previous paragraph could be taken as the definition of $\mathbf{R}^{n+1}$.

For more about finite Cartesian products, and for a bridge between the definition given here and the definition provided in Section 8 for products of infinitely many sets, see Exercise 1D.

**1.6 Functions.** A *function* (or *map*) $f$ from a set $A$ to a set $B$, written $f: A \to B$, is a subset of $A \times B$ with the properties:

a) For each $a \in A$, there is some $b \in B$ such that $(a, b) \in f$.

b) If $(a, b) \in f$ and $(a, c) \in f$, then $b = c$.

More informally, we are requiring that each $a \in A$ be paired with exactly one $b \in B$. The relationship $(a, b) \in f$ is customarily written $b = f(a)$ and functions are usually described by giving a rule for finding $f(a)$ if $a$ is known (rather than, for example, by giving some geometric or other description of the subset $f$ of $A \times B$). This reflects the common point of view, which is prone to regard a function not so much as a static subset of $A \times B$ as a "black box" which takes in elements of $A$ and spits out elements of $B$.

When regarded as a set in its own right, the collection of functions from $A$ to $B$ is denoted $B^A$.

If $f: A \to B$ and $C \subset A$, we define $f(C) = \{b \in B \mid b = f(a) \text{ for some } a \in A\}$. If $D \subset B$, we define $f^{-1}(D) = \{a \in A \mid f(a) \in D\}$. Hence every function $f: A \to B$ induces functions $f: P(A) \to P(B)$ and $f^{-1}: P(B) \to P(A)$ (and here we are following the unfortunate, but common, practice of denoting the elevation of $f$ from $A$ to $P(A)$ by $f$ also). The properties of these induced functions are investigated in Exercise 1H, which should be mandatory for anyone who cannot provide easily the answers to the questions it poses.

Note that if $f: A \to B$, then $f^{-1}(B) = A$ but it need not be true that $f(A) = B$. We call $f(A)$ the *image* of $f$ (or the image of $A$ under $f$), calling $B$ the *range* of $f$ and $A$ the *domain* of $f$. When $f(A) = B$, we say $f$ is *onto* $B$. Note also that, for $b \in B$, $f^{-1}(\{b\})$ [which is always abbreviated $f^{-1}(b)$] may consist of more than one point; in extreme cases, we may have $f^{-1}(b) = A$. When such behavior is proscribed, $f$ is called a *one–one* function. In addition to the usual requirements for a function, then, a one–one function $f: A \to B$ must evidently obey the rule: $a_1 \neq a_2 \Rightarrow f(a_1) \neq f(a_2)$. In words, such a function takes distinct elements of $A$ to distinct elements of $B$.

If $f: A \to B$ and $g: B \to C$, then $f$ and $g$ determine together a natural function, their *composition* $g \circ f: A \to C$, defined by

$$(g \circ f)(a) = g[f(a)], \qquad \text{for} \quad a \in A.$$

More formally, $(a, c) \in g \circ f$ iff for some $b \in B$, $(a, b) \in f$ and $(b, c) \in g$. Less formally, put two black boxes end to end.

**1.7 Special functions.** A function $f: \mathbf{N} \to A$ is called a *sequence* in $A$. It can be described by giving an indexed list $x_1, x_2, \ldots$ of its values at 1, 2, ... and this is often abbreviated $(x_n)_{n \in \mathbf{N}}$ or even simply $(x_n)$. Thus $f(n) = 1/n$, $(1/n)_{n \in \mathbf{N}}$ and $1, 1/2, \ldots, 1/n, \ldots$ describe the same sequence in $\mathbf{R}$.

A *real-valued function* on $A$ is a function on $A$ whose range is $\mathbf{R}$. The collection $\mathbf{R}^A$ of all real-valued functions on $A$ inherits an algebraic structure from $\mathbf{R}$ since we can define addition, multiplication and scalar multiplication in $\mathbf{R}^A$ as follows: given $a \in A$ and $r \in \mathbf{R}$,

$$(f + g)(a) = f(a) + g(a),$$
$$(fg)(a) = f(a)g(a),$$
$$(rf)(a) = r[f(a)].$$

For this and other reasons, the real-valued functions merit special attention in any branch of mathematics, and topology is no exception.

The *identity function* on any set $A$ is the function $i: A \to A$ defined by $i(a) = a$ for each $a \in A$. More generally, if $B \subset A$, the *inclusion* $j: B \to A$ is the function $j(b) = b$ for each $b \in B$.

**1.8 Relations.** A *relation* $R$ on a set $A$ is any subset of $A \times A$. (Thus every function from $A$ to $A$ is a relation on $A$, but not all relations on $A$ have the properties required of functions.) If $R$ is a relation on $A$, we usually denote the relationship $(a, b) \in R$ by $aRb$. For example, $\{(n_1, n_2) \in \mathbf{N} \times \mathbf{N} \mid n_1 < n_2\}$ is a relation on $\mathbf{N}$ and it would be typical to denote this relation by $<$, so that $(n_1, n_2) \in \, <$ iff $n_1 < n_2$.

A relation $R$ on $A$ is called *reflexive* iff $aRa$ for each $a \in A$, *symmetric* iff $aRb$ implies $bRa$ for all $a, b \in A$, *antisymmetric* iff $aRb$ and $bRa$ implies $a = b$ for all $a, b \in A$ and *transitive* iff $aRb$ and $bRc$ implies $aRc$ for all $a, b, c \in A$. For example, $<$ is a transitive relation on $\mathbf{R}$, $\leq$ is a reflexive, antisymmetric, transitive relation on $\mathbf{R}$, $\neq$ is a symmetric relation on $\mathbf{R}$.

An *equivalence relation* on $A$ is a reflexive, symmetric and transitive relation on $A$. As an example, let $f$ be any function from $A$ to $B$ and define a relation $R$ on $A$ by $xRy$ iff $f(x) = f(y)$. For other examples, see Exercise 1E.

If $R$ is an equivalence relation on $A$, the *equivalence class* (or *R-equivalence class* where confusion is possible) of $a \in A$ is the set $[a] = \{a' \in A \mid a'Ra\}$. If $a, b \in A$, note that either $[a] = [b]$ (and this happens precisely when $aRb$) or else $[a] \cap [b] = \emptyset$. Since $a \in [a]$ for each $a \in A$, the sets $[a]$, for $a \in A$, evidently form a partition of $A$, i.e., they are disjoint sets whose union is $A$. For example, if $R$ is the equivalence relation introduced in the preceding paragraph, the equivalence class of $a \in A$ is the set $f^{-1}[f(a)]$. Other examples can be found in 1E.

**1.9 Order relations.** A relation $R$ on $A$ is a *partial order* provided $R$ is reflexive, antisymmetric and transitive. Thus $\leq$ is a partial order on $\mathbf{R}$. It is the model partial order and thus it is customary to denote any partial order on any set by $\leq$. In this context, $\geq$ is defined by $a \geq b$ iff $b \leq a$.

Associated with any partial order $\leq$ on $A$ is a relation $<$ defined by $a < b$ iff $a \leq b$ and $a \neq b$. Note that $<$ is not reflexive or symmetric, but it is transitive and has the property that for any $a$ and $b$ in $A$, if $a < b$, then $b \nless a$. A transitive relation with this property will be called a *strict order*. Thus every partial order determines a strict order. Conversely, any strict order $<$ determines a partial order $\leq$ defined by $a \leq b$ iff $a < b$ or $a = b$. Moreover the passage from a partial order $\leq$ to its associated strict order $<$ to the partial order determined by $<$ returns us to $\leq$, and the assertion remains true with "strict order" and "partial order" interchanged. Thus, in dealing with a partially ordered set, the symbol "$<$" has a well-defined meaning.

A set $A$ is *linearly ordered* by a partial order $\leq$ provided that for any $a, b \in A$ exactly one of $a < b$, $b < a$ or $a = b$ holds. Then $\leq$ is called a *linear order*.

If $\leq$ is a partial order on $A$, the *smallest element* of $A$, if it exists, is the element $a_0$ such that $a_0 \leq a$ for each $a \in A$, and the *largest element* of $A$, if it exists, is the element $a_1$ such that $a \leq a_1$ for each $a \in A$. Smallest (largest) elements are unique, when they exist, by antisymmetry. They may not exist: $\mathbf{R}$ with the order $\leq$ has no smallest or largest element.

A set $A$ is *well-ordered* if it has a linear order $\leq$ such that every subset of $A$ has a smallest element (in the linear order induced on that subset by the linear order on $A$). The set $\mathbf{N}$ of positive integers is well-ordered by its usual order, the real line $\mathbf{R}$ is not.

**1.10 Minimal and maximal elements.** If $A$ is partially ordered by $\leq$, an element $b_0$ of $A$ is a *minimal element* of $A$ provided $b \leq b_0$ implies $b = b_0$ for each $b \in A$, and $b_1$ is a *maximal element* of $A$ provided $b_1 \leq b$ implies $b_1 = b$ for each $b \in A$. If a smallest (largest) element exists in $A$, then it is the *unique* minimal (maximal) element of $A$. In Fig. 1.1, where $x < y$ is represented by a rising line connecting $x$ to $y$, we find an example of a set with a unique maximal element $b$ which is not a largest element, so the converse fails.

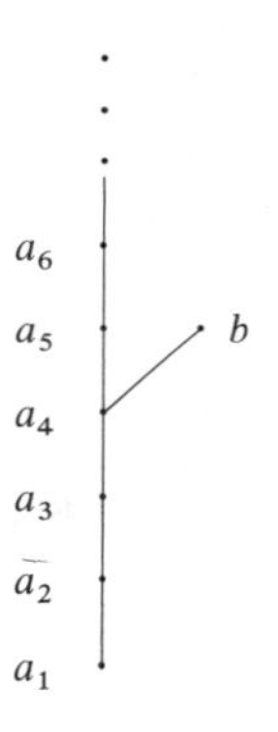


**Figure 1.1**

The reader is invited to draw a diagram illustrating that maximal elements need not be unique.

The *least upper bound* (lub) of a subset $B$ of a partially ordered set $A$ is the smallest element of the set $\{a \in A \mid b \leq a \text{ for each } b \in B\}$. It may or may not exist and, when it does, it may or may not belong to $B$. When it exists, it is unique. The *greatest lower bound* (glb) of $B$ is similarly defined.

**1.11 Lattices.** A partially ordered set $L$ is a *lattice* iff each two-element set $\{a, b\}$ in $L$ has a least upper bound $a \vee b$ and a greatest lower bound $a \wedge b$. If *every* nonempty subset of $L$ has a least upper bound and a greatest lower bound, $L$ is a *complete lattice.* Lattices having a least element 0 and a greatest element 1 are called *complemented* iff for each $a \in L$, there is some $a' \in L$ such that $a \vee a' = 1$, $a \wedge a' = 0$. A lattice is *distributive* iff for all $a, b, c \in L$,

$$a \vee (b \wedge c) = (a \vee b) \wedge (a \vee c)$$

and

$$a \wedge (b \vee c) = (a \wedge b) \vee (a \wedge c).$$

These rules are redundant since either can be deduced from the other.

A *Boolean lattice* is a lattice with 0 and 1 which is complemented and distributive.

The model lattice for most purposes is the set $P(A)$ of all subsets of a fixed set $A$. This becomes a complete Boolean lattice when partially ordered by the relation $B \leq C$ iff $B \subset C$. (See Exercise 1K.)

**1.12 Cardinality.** If $A$ and $B$ are sets, we say $A$ is *equipotent* with $B$ iff there is a one–one function $f$ from $A$ onto $B$. Intuitively, equipotent sets have the same number of elements. We now postulate the existence of sets, called *cardinal numbers*, so chosen that every set $A$ is equipotent with precisely one cardinal number, called the *cardinal number of* $A$ and denoted $|A|$.

If $C$ and $D$ are cardinal numbers, we say $C \leq D$ iff there is a one–one function $f: C \to D$. The result is a partial order on any family of cardinal numbers. Let us see what this says:

a) $\leq$ is reflexive: given a cardinal number $C$, there is a one–one function $f: C \to C$. The identity function will do nicely.

b) $\leq$ is antisymmetric: given cardinal numbers $C$ and $D$, if one–one functions $f: C \to D$ and $g: D \to C$ can be found, then $C = D$. This is the *Cantor–Bernstein theorem*, which in more general form says that if one–one functions $f: A \to B$ and $g: B \to A$ can be found, then there is a one–one function carrying $A$ onto $B$. (Existence of a one–one, onto function between cardinal numbers $C$ and $D$ ensures that $C = D$. Why?) A proof of the Cantor–Bernstein theorem is given in Exercise 1J.

c) $\leq$ is transitive: given cardinal numbers $C$, $D$ and $E$ and one–one functions $f: C \to D$ and $g: D \to E$, there is a one–one function $h: C \to E$. Here, the composition $g \circ f: C \to E$ will serve.

In fact, any set of cardinal numbers is well-ordered by the relation $\leq$, although we will not prove this, deferring to any of the standard references on set theory (see the notes).

Recalling that $|A|$ denotes the cardinal number of $A$, evidently

i) $|A| = |B|$ iff $A$ and $B$ are equipotent,
ii) $|A| \leq |B|$ iff $A$ is equipotent with some subset of $B$.

**1.13 Special cardinals.** We will distinguish notation for certain cardinal numbers. The empty set is the cardinal 0, and the cardinal number $n$ is the set $\{0, \ldots, n-1\}$. A set $A$ is *denumerable* iff $A$ is equipotent with $\mathbf{N}$ and, in this case, we write $|A| = \aleph_0$. A set $A$ is said to have the *cardinal of the continuum*, iff $A$ is equipotent with $\mathbf{R}$, and then we write $|A| = \mathfrak{c}$. A set $A$ is *countable* iff it is denumerable or has cardinal number $n$ for some $n = 0, 1, 2, \ldots$; otherwise, $A$ is *uncountable*.

The elements of a countable set $A$ can be listed in a (finite or infinite) sequence $a_1, a_2, \ldots$ and such a listing is called an *enumeration* of the elements of $A$.

**1.14 Facts about countability.** a) $n < \aleph_0 < \mathfrak{c}$,

b) The union of countably many countable sets is countable,

c) The product of two countable sets is countable,

d) The set $\mathbf{Q}$ of rational numbers is countable.

*Proof.* a) It is clear that $n < \aleph_0$ and, since $\mathbf{N}$ is equipotent with the subset $\{1/n \mid n = 1, 2, \ldots\}$ of $\mathbf{R}$, that $\aleph_0 \leq \mathfrak{c}$. To show $\aleph_0 \neq \mathfrak{c}$, it is enough to show that there is no one–one function from $\mathbf{N}$ onto $\mathbf{I}$. If such a function $f: \mathbf{N} \to \mathbf{I}$ exists, let the decimal expansion of $f(n)$ be $.a_{n_1}a_{n_2}a_{n_3}\cdots$. Define $.b_1b_2\cdots$ by taking $b_k$ to be 5 if $a_{kk} \neq 5$, $b_k$ to be 7 if $a_{kk} = 5$. Then $.b_1b_2\cdots$ is an element of $\mathbf{I}$ which can appear nowhere among the values of $f$, since it differs from $f(n)$ in the $n$th place, for each $n = 1, 2, \ldots$. This contradicts the assumption that $f$ is onto, showing no such function can exist, and completes the proof of (a).

b) Let $\{A_1, A_2, \ldots\}$ be a countable collection of countable sets. Set $B_1 = A_1$ and, for $n > 1$, $B_n = A_n - \bigcup_{k<n} A_k$. Then each $B_n$ is countable and

$$\bigcup_{n=1}^{\infty} B_n = \bigcup_{n=1}^{\infty} A_n.$$

Enumerate the elements of each $B_n$ as follows:

$$\begin{aligned} B_1 &= \{b_{11}, b_{12}, b_{13}, b_{14}, \ldots\} \\ B_2 &= \{b_{21}, b_{22}, b_{23}, b_{24}, \ldots\} \\ B_3 &= \{b_{31}, b_{32}, b_{33}, b_{34}, \ldots\} \\ B_4 &= \{b_{41}, b_{42}, b_{43}, b_{44}, \ldots\} \end{aligned}$$

and define $f: \mathbf{N} \to \bigcup_{n=1}^{\infty} B_n$ by $f(1) = b_{11}$, $f(2) = b_{21}$, $f(3) = b_{12}$, $f(4) = b_{13}$, $f(5) = b_{22}$, $f(6) = b_{31}, \ldots$ and so on, following the scheme indicated by the arrows. The result is a one–one function $f$ from $\mathbf{N}$ onto $\bigcup_{n=1}^{\infty} B_n = \bigcup_{n=1}^{\infty} A_n$, and the proof is complete.

c) If $A$ and $B$ are countable, enumerate the elements of $B$ as $b_1, b_2, \ldots$ and let $A_n = A \times \{b_n\} = \{(a, b_n) \mid a \in A\}$. Then $A_n$ is countable for each $n = 1, 2, \ldots$ and $A \times B = \bigcup_{n=1}^{\infty} A_n$; thus $A \times B$ is countable by part (b).

d) Write each element of $\mathbf{Q}$ in the form $m/n$, where $m$ and $n$ are integers in lowest terms. Then the function defined by $f(m/n) = (m, n)$ maps $\mathbf{Q}$ in one–one fashion onto a subset of $\mathbf{N} \times \mathbf{N}$. Since $\mathbf{N} \times \mathbf{N}$ is countable by part c), $\mathbf{Q}$ is countable. ■

**1.15 Cardinality and the power set.** It is possible to develop an arithmetic of cardinal numbers. We limit ourselves here to the definition of exponentiation. If $A$ and

$B$ are sets, $|A|^{|B|}$ is defined to be $|A^B|$ (recall $A^B$ denotes the set of all functions $f: B \to A$). The reader will verify in Exercise 1I that this definition gives the right answer if $|A| = n$ and $|B| = m$, where $n$ and $m$ are integers.

Let us pay particular attention to the cardinal number $2^{|A|}$ where $A$ is a set. Now $2 = \{0, 1\}$ and hence $2^{|A|} = |2^A| = |\{0, 1\}^A|$ is the cardinal number of the set of functions $f: A \to \{0, 1\}$. Such a function $f$ determines and is completely determined by the subset $B = \{a \in A \mid f(a) = 1\}$ of $A$ ($f$ is called the *characteristic function* of $B$) and hence $2^{|A|} = |P(A)|$.

By writing elements of **I** in binary form, it is not difficult to show that $2^{\aleph_0} = \mathfrak{c}$ (Exercise 1I). Hence, from 1.14(a), $\aleph_0 < 2^{\aleph_0} < 2^{\mathfrak{c}}$. It is generally true for any cardinal number $\alpha$ that $\alpha < 2^{\alpha}$; put another way, for any set $A$, $|A| < |P(A)|$. This is *Cantor s theorem* (Exercise 1I).

**1.16 The continuum hypothesis.** The *continuum hypothesis* states that there are no sets $A$ for which $\aleph_0 < |A| < 2^{\aleph_0}$. It has been proved independent of the other axioms needed to develop set theory (see notes); that is, either it or its negation can consistently be added to the other axioms. At present, intuition has provided us with little basis for preferring one assumption over the other (although in most contexts in which it arises, it, rather than its negation, is assumed) and it is definitely in order to attempt to eliminate from any proof any use of the continuum hypothesis. It follows, in the same vein, that whenever it or its negation *is* assumed, this should be explicitly pointed out.

**1.17 The axiom of choice.** The following axiom is assumed by most mathematicians when they need it, to the unremitting disgust of a few. We give it in two equivalent forms:

*Axiom of choice*

a) If $\{A_\lambda \mid \lambda \in \Lambda\}$ is a family of nonempty pairwise-disjoint sets, there is a set $B \subset \bigcup A_\lambda$ such that $B \cap A_\lambda$ has exactly one element, for each $\lambda \in \Lambda$.

b) If $\{A_\lambda \mid \lambda \in \Lambda\}$ is an indexed family of nonempty pairwise-disjoint sets, there is a function $f: \Lambda \to \bigcup A_\lambda$ such that $f(\lambda) \in A_\lambda$, for each $\lambda \in \Lambda$ ($f$ is called a *choice function*).

It is left to the reader to decide that these two statements both say the same thing. What they say is: given any collection of sets, however large, we can pick one element from each set in the collection. It bothers some people because it asserts the existence of a set (i.e., $B$ in part (a)) without giving enough information to determine that set uniquely (by applying a finite number of rules), and it is the *only* formal set-theoretic axiom which does this. For this reason it is customary to mention the axiom of choice whenever it is used. It need not be used if the number of sets is finite. In particular, if $A$ is a nonempty set, the statement "choose $a \in A$" need not be supported by an appeal to the axiom of choice.

The status of the axiom of choice bears some resemblance to that of the continuum hypothesis, with some differences. It, too, is known to be independent of the other axioms of set theory (that is, it or its negation can be consistently assumed), but it enjoys the status of an accepted part of the theory of sets in the minds of most modern mathematicians; that is, the intuition of almost all mathematicians now is that the axiom of choice should be assumed where needed without hesitation. Moreover, it is usually much clearer that, where it is used, it is needed, so that its presence does not usually provoke the same frenzy of attempt to eliminate it.

**1.18 Alternative forms of the axiom of choice.** We now provide some alternative, often-used forms of the axiom of choice. We say a family of sets is of *finite character* iff each finite subset of a member of the family is also a member, and each set belongs if each of its finite subsets belong.

***Theorem.*** *The following statements are all equivalent:*

a) (*Axiom of choice*)*: If* $\{A_\lambda \mid \lambda \in \Lambda\}$ *is an indexed family of nonempty pairwise disjoint sets, there is a set* $B \subset \bigcup A_\lambda$ *such that* $B \cap A_\lambda$ *is exactly one element for each* $\lambda \in \Lambda$.

b) (*Zorn's Lemma*)*: If each chain* (*linearly ordered set*) *in a nonempty partially ordered set* $A$ *has an upper bound, then* $A$ *has a maximal element.*

c) (*Zermelo's Theorem*)*: Every set can be well-ordered.*

d) (*Tukey's Lemma*)*: Each nonempty family of sets of finite character has a maximal element.*

As with the axiom of choice, it is customary to mention any one of these wherever it is used. The proof of equivalence will not be given here; it can be found in any standard reference.

**1.19 Ordinals.** For our purposes, it will be sufficient to postulate the existence of an uncountable well-ordered set $\mathbf{\Omega}$ with a largest element $\omega_1$, having the property that if $\alpha \in \mathbf{\Omega}$ with $\alpha < \omega_1$, then $\{\beta \in \mathbf{\Omega} \mid \beta \le \alpha\}$ is countable. Such a set $\mathbf{\Omega}$ exists if there exists any uncountable well-ordered set; see Exercise 1L. The elements of $\mathbf{\Omega}$ are *ordinals* with $\omega_1$ being the *first uncountable ordinal* and $\mathbf{\Omega}_0 = \mathbf{\Omega} - \{\omega_1\}$ being the set of *countable ordinals*.

If $\alpha$ and $\beta$ are ordinals with $\alpha < \beta$, we say $\alpha$ is a *predecessor* of $\beta$ and $\beta$ is a *successor* of $\alpha$. We call $\alpha$ an *immediate predecessor* of $\beta$, and $\beta$ an *immediate successor* of $\alpha$, if $\beta$ is the smallest ordinal larger than $\alpha$. Every ordinal $\alpha$ has an immediate successor, often denoted $\alpha + 1$; some ordinals, called the *limit ordinals* have predecessors without having an immediate predecessor ($\omega_1$, for example). The others are *nonlimit ordinals.*

To build a picture of $\mathbf{\Omega}$, observe that it has a least element, which we denote 1 for now. The immediate successor of 1 will be denoted 2, the immediate successor of 2 will be denoted 3, and so on, so that we can regard the first few elements of $\mathbf{\Omega}$ as being the positive integers 1, 2, 3, . . . . Since $\mathbf{\Omega}_0$ is well-ordered, there is a smallest ordinal larger than all of 1, 2, 3, . . . . It is called the *first infinite ordinal* $\omega_0$. It is still only a countable ordinal; it and its first few successors $\omega_0 + 1$, $\omega_0 + 2, \ldots$ evidently form another "copy" of **N** tacked on behind the first. The smallest ordinal larger than these is denoted $2\omega_0$, and we can apparently continue in this fashion through $3\omega_0, 4\omega_0, \ldots$ by adding denumerably many copies of **N** one after the other.

$$1, 2, 3, \ldots, \omega_0, \omega_0 + 1, \omega_0 + 2, \ldots, 2\omega_0, 2\omega_0 + 1, 2\omega_0 + 2, \ldots, \ldots.$$

The smallest ordinal larger than these is denoted $\omega_0^2$ and it is still only countable. Repeating the process obtained to reach $\omega_0^2$ denumerably many times leads us to $\omega_0^3$ and, repeating this over and over, we pass $\omega_0^4, \omega_0^5, \ldots$. The smallest ordinal larger than all these is still countable however, so the process continues. In fact, $\omega_1$ is unreachable by countable operations such as this, by the next theorem.

***1.20 Theorem.*** *If $A$ is a countable subset of $\mathbf{\Omega}$ not containing $\omega_1$, then* lub $A < \omega_1$.

*Proof.* For each $\alpha \in A$, $\{\beta \in \mathbf{\Omega} \mid \beta \leq \alpha\}$ is countable. Since $A$ is countable, the union of these sets, namely $B = \{\beta \in \mathbf{\Omega} \mid \beta \leq \alpha \text{ for some } \alpha \in A\}$, is also countable. Let $\gamma$ be the smallest element of $\mathbf{\Omega}$ not contained in $B$. Then $\beta \in B$ iff $\beta < \gamma$, so $\gamma$ has a countable number of predecessors, and hence $\gamma < \omega_1$. But $\gamma$ is an upper bound for $A$, so lub $A < \omega_1$. ■

**1.21 Induction.** The following theorem is a statement of the *principle of mathematical induction.* To prove it, we accept as obvious the fact that the positive integers **N** form a well-ordered set.

***Theorem.*** *Let $S(n)$ be a statement which is true or false, for $n = 1, 2, \ldots$. If*

a) *$S(1)$ is true,*

b) *$S(n)$ is true implies $S(n + 1)$ is true, for $n = 1, 2, \ldots$,*

*then $S(n)$ is true for all $n$.*

*Proof.* If the set $F$ of all integers $n$ for which $S(n)$ is false is nonempty, then it has a least element $n$, and $n \neq 1$ by (a). Since $n > 1$, $n - 1 \in \mathbf{N}$, and $n - 1 \notin F$, so $S(n - 1)$ is true. But then $S(n)$ is true, by (b); this contradiction establishes that $F = \emptyset$. ■

As an example, we prove that $1 + 2 + \cdots + n = n(n + 1)/2$ for any positive integer $n$. The formula certainly works for $n = 1$. Suppose it works for $n$. Then

$$1 + 2 + \cdots + (n + 1) = (1 + 2 + \cdots + n) + (n + 1)$$

$$= \frac{n(n + 1)}{2} + (n + 1) = \frac{n(n + 1) + 2(n + 1)}{2} = \frac{(n + 1)(n + 2)}{2},$$

which is the form the formula should take for $n + 1$. The "inductive step" is now established, so by the principle of mathematical induction, the formula applies to any $n$.

It is also instructive to point out an often used incorrect form of application of the principle of mathematical induction. A typical (wrong) argument would sound like this: "$\{1\}$ is a finite set, and, if $\{1, \ldots, n\}$ is a finite set, so is $\{1, \ldots, n + 1\}$. Therefore the positive integers form a finite set." This argument looks as absurd as it is, but uses of the principle of mathematical induction just as ridiculous logically are often submitted by those new to it.

**1.22 Transfinite induction.** A second method of induction, the *principle of transfinite induction*, can be applied to statements indexed by a well-ordered set of any sort. We will not need it in any form other than as stated here, however:

> ***Theorem.*** *For each ordinal* $\alpha \in \Omega_0$, *let* $S(\alpha)$ *be a statement which is true or false. If*
>
> a) $S(1)$ *is true,*
>
> b) $S(\beta)$ *is true for all* $\beta < \alpha$ *implies* $S(\alpha)$ *is true,*
>
> *then* $S(\alpha)$ *is true for each* $\alpha \in \Omega_0$.

The proof is in no essential way different from the proof of the principle of mathematical induction: one makes the same use of the well-ordering.

Both induction principles can be used as the basis for defining things. For example

$$f(1) = 1,$$
$$f(n + 1) = (n + 1)f(n)$$

is an inductive definition of the factorial function on $\mathbf{N}$. For an example of definition by transfinite induction, see 3I.

**1.23 Remarks.** The process which topology evolves from, outlined in the next section and the notes, is basic to any pure mathematical discipline. We wish to study a particular property enjoyed by some objects of interest (in this case, continuity of functions on some space) and the efficient way to proceed is to first clean the structure on the space down to the bare bones needed for introducing and developing the property we want. The passage to such abstraction has several well-documented advantages. Among them:

1. Since we have only what is essential, our proofs use only what is essential and thus clarify the nature of the object of study, and the logical dependence of the theorem in question.

2. Proofs become easier. Actually, this is a popular professional myth, with an element of truth. Occasionally, a proof really does get easier as a theorem gets more abstract, but this is offset by the need for more and more interpretive skill on the part of those who would use the theorem. What people really mean when

they say "proofs become easier" is something like this: "by establishing some notation and introducing the right definitions and conventions, we can draw together all the theorems about this subject and find common characteristics and even repetitions in their proofs, then prove lemmas which enable us to write large numbers of proofs more succinctly." If the subject matter is carefully chosen, the work done in abstracting the properties needed, establishing notation and proving those lemmas will be more than paid for by the gain in succinctness and clarity of the proofs later on, and by the acquisition of powerful methods for continued investigation of the original objects of study.

Such is the case with topology.

## Problems

### 1A. *Russell's Paradox*

The phenomenon to be presented here was first exhibited by Russell in 1901, and consequently is known as *Russell's Paradox.*

Suppose we allow as sets things $A$ for which $A \in A$. Let $P$ be the set of all sets. Then $P$ can be divided into two nonempty subsets, $P_1 = \{A \in P \mid A \notin A\}$ and $P_2 = \{A \in P \mid A \in A\}$. Show that this results in the contradiction: $P_1 \in P_1 \Leftrightarrow P_1 \notin P_1$. Does our (naïve) restriction on sets given in 1.1 eliminate the contradiction?

### 1B. *De Morgan's laws and the distributive laws*

1. $A - (\bigcap_{\lambda \in \Lambda} B_\lambda) = \bigcup_{\lambda \in \Lambda} (A - B_\lambda)$ [see 1.4a), b)].
2. $B \cup (\bigcap_{\lambda \in \Lambda} B_\lambda) = \bigcap_{\lambda \in \Lambda} (B \cup B_\lambda)$ [see 1.4c), d)].
3. If $A_{nm}$ is a subset of $A$ for $n = 1, 2, \ldots$ and $m = 1, 2, \ldots$ is it necessarily true that

$$\bigcup_{n=1}^{\infty}\left[\bigcap_{m=1}^{\infty} A_{nm}\right] = \bigcap_{m=1}^{\infty}\left[\bigcup_{n=1}^{\infty} A_{nm}\right]?$$

### 1C. *Ordered pairs*

Show that, if $(x_1, x_2)$ is *defined* to be $\{\{x_1\}, \{x_1, y_2\}\}$, then $(x_1, x_2) = (y_1, y_2)$ iff $x_1 = x_2$ and $y_1 = y_2$.

### 1D. *Cartesian products*

1. Provide an inductive definition of "the ordered $n$-tuple $(x_1, \ldots, x_n)$ of elements $x_1, \ldots, x_n$ of a set" so that $(x_1, \ldots, x_n)$ and $(y_1, \ldots, y_n)$ are equal iff their coordinates are equal in order, i.e., iff $x_1 = y_1, \ldots, x_n = y_n$.

2. Given sets $X_1, \ldots, X_n$ define the Cartesian product $X_1 \times \cdots \times X_n$
   a) by using the definition of ordered $n$-tuple you gave in part 1,
   b) inductively from the definition of the Cartesian product of two sets,

and show that the two approaches are the same.

3. Given sets $X_1, \ldots, X_n$ let $X = X_1 \times \cdots \times X_n$ and let $X^*$ be the set of all functions $f$ from $\{1, \ldots, n\}$ into $\bigcup_{k=1}^{n} X_k$ having the property that $f(k) \in X_k$ for each $k = 1, \ldots, n$. Show that $X^*$ is the "same" set as $X$, in the sense that there is a natural one–one mapping $F$ of $X^*$ onto $X$. [$F$ will take some value $F(f)$ in $X$ for each $f \in X^*$. What must such a value look like? Find a natural one.]

4. Use what you learned in part three to define the Cartesian product $X_1 \times X_2 \times \cdots$ of denumerably many sets as a collection of certain functions with domain **N**.

If you have completed part 4 successfully, the definition of Cartesian product given in Section 8 for infinitely many sets will give you little trouble.

### 1E. *Examples on equivalence relations*

Which of the following are equivalence relations on **R**? For each that is, describe the equivalence class $[x]$ of $x \in \mathbf{R}$.

1. $aRb$ iff $a - b$ is rational.
2. $aRb$ iff $a - b$ is irrational.
3. $aRb$ iff $a - b$ is an integer.
4. $aRb$ iff $|a - b| \leq 1$.

### 1F. *Cardinality*

1. $|\mathbf{P}| = \mathfrak{c}$.
2. $|\mathbf{I}| = \mathfrak{c}$.

See also 1I.

### 1G. *Well-ordering*

Assuming the axiom of choice, each of the following sets can be well-ordered. Try to think of a well-ordering for each (you may not be able to use the usual order).

1. **N**,
2. the rationals,
3. **R**.

When you have trouble, ask somebody who should know. Then think about the axiom of choice.

### 1H. *Inverses of functions are nice*

Let $f: A \to B$. Prove each of the following. For some, you will need to assume that $f$ is one–one; for others, that it is onto; some need neither. Precede your proof of each by a correct statement of what you are proving.

1. $f(\bigcup_{\lambda \in \Lambda} A_\lambda) = \bigcup_{\lambda \in \Lambda} f(A_\lambda)$,
2. $f(\bigcap_{\lambda \in \Lambda} A_\lambda) = \bigcap_{\lambda \in \Lambda} f(A_\lambda)$,
3. $f(A - A_0) = B - f(A_0)$,
4. $f^{-1}(\bigcup_{\lambda \in \Lambda} B_\lambda) = \bigcup_{\lambda \in \Lambda} f^{-1}(B_\lambda)$,
5. $f^{-1}(\bigcap_{\lambda \in \Lambda} B_\lambda) = \bigcap_{\lambda \in \Lambda} f^{-1}(B_\lambda)$,
6. $f^{-1}(B - B_0) = A - f^{-1}(B_0)$.

### 1I. *Cardinality revisited*

1. $|A| < |\mathscr{P}(A)|$. [This is proved by contradiction, in essentially the same way that Russell's paradox is established. First show $|A| \leq |\mathscr{P}(A)|$. Now if $|A| = |\mathscr{P}(A)|$, then there is a one–one mapping $f$ of $A$ onto $\mathscr{P}(A)$. For each $x \in A$, let $A_x$ be the image of $x$ under $f$. Then $\mathscr{P}(A) = \{A_x \mid x \in A\}$. Let $B = \{x \in A \mid x \notin A_x\}$. Then $B = A_y$ for some $y \in A$. Show that this leads to the contradiction: $y \in A_y \Leftrightarrow y \notin A_y$.] This is *Cantor's theorem.*

2. If $|A| = n$ and $|B| = m$, where $n$ and $m$ are integers, then $|A^B| = n^m$.

3. $2^{\aleph_0} = \mathfrak{c}$.

### 1J. *The Cantor–Bernstein theorem*

Let $A$, $B$ be sets.

1. Suppose that with each subset $C$ of $A$ there is associated a subset $C'$ of $A$ in such a way that $C \subset D$ implies $C' \subset D'$. Then $E = E'$ for some $E \subset A$. [Let $E = \bigcup \{C \in P(A) \mid C \subset C'\}$.]

2. If $f: A \to B$ and $g: B \to A$ are one–one functions, there is a one–one function $h$ of $A$ onto $B$. [For $C \subset A$, define $C' = A - g(B - f(C))$. Show that part 1 applies and, if $E$ is the resulting set, define $h$ to be $f$ on $E$ and $g^{-1}$ on $A - E$. Show that $h$ is one–one and onto from $A$ to $B$.]

### 1K. *Lattices*

1. Show that the power set $P(A)$ of a fixed set $A$, when partially ordered by $B \leq C$ iff $B \subset C$, becomes a complete Boolean lattice. Describe the largest and smallest elements of $P(A)$, the least upper bound and greatest lower bound operations in $P(A)$ and the lattice complement of $B \in P(A)$.

2. Exhibit a complemented lattice with an element $a$ having two distinct complements $b$ and $c$.

3. Show that in a complemented distributive lattice, complements are unique.

### 1L. *The ordinals*

We postulated the existence of the set $\mathbf{\Omega}$ (1.15). Show that such a set exists if there exists an uncountable well-ordered set. [There are two cases.]

## 2 Metric spaces

The concept of continuous function is central to the study of analysis and, as the functions in question are defined on more and more complicated spaces, the need for a notion of continuity which is as generally applicable as possible becomes acute. There were two steps in the development of general machinery for the definition of continuity for functions other than those defined on Euclidean spaces. Both came with (what was then) lightning speed on the heels of the development of a general theory of sets by Cantor, in the 1880's. The first step was taken by Frechet, in 1906, with the introduction of metric spaces, the second and conclusive step by Hausdorff, in 1914, with the introduction of topological spaces.

It is impossible now to give a faithful historical development of topology, but we can properly begin a book on topology with a brief motivational introduction to metric spaces. Thus, here we will define metric spaces, show that the abstract distance they provide is sufficient to define continuity, then conduct a brief and successful search for a way to define continuity for functions between metric spaces without mentioning the metrics. This will lead us naturally to the definition of topology in the next section.

**2.1 Definition.** A *metric space* is an ordered pair $(M, \rho)$ consisting of a set $M$ together with a function $\rho: M \times M \to \mathbf{R}$ satisfying, for $x, y, z \in M$:

$M$-a) $\rho(x, y) \geq 0$,

$M$-b) $\rho(x, x) = 0$;   $\rho(x, y) = 0$ implies $x = y$,

$M$-c) $\rho(x, y) = \rho(y, x)$,

$M$-d) $\rho(x, y) + \rho(y, z) \geq \rho(x, z)$ (*triangle inequality*).

The function $\rho$ is called the *metric* on $M$. If all axioms but the second part of $M$-b are satisfied, we call $(M, \rho)$ a *pseudometric space* and $\rho$ is then a *pseudometric.* Functions $\rho: M \times M \to \mathbf{R}$ (which are potentially metrics or pseudometrics but which have not yet been tested) are called *distance functions.* If a metric $\rho$ is fixed for a particular discussion, we may drop the ordered-pair notation and simply speak of "the metric space $M$."

Although all the material of this section will be developed for metric spaces, the basic results remain true for pseudometric spaces as well. In particular, the definitions of open set, closed set and continuous function given below for metric spaces can be applied to pseudometric spaces also (and now and then we will act as though they had).

**2.2 Examples.** a) The real line $\mathbf{R}$ with the distance function $\rho(x, y) = |x - y|$ is a metric space. More generally, $\mathbf{R}^n$ is a metric space when provided with the distance function

$$\rho((x_1, \ldots, x_n), (y_1, \ldots, y_n)) = \sqrt{\sum_{k=1}^{n} (x_k - y_k)^2},$$

called the *usual metric* on $\mathbf{R}^n$. The reader will verify that it is a metric in Exercise 2A.

b) The plane $\mathbf{R}^2$ with the distance function

$$\rho_1(x, y) = |x_1 - y_1| + |x_2 - y_2|$$

is a metric space; $\rho_1$ is called the *taxi-cab metric.*

c) The plane $\mathbf{R}^2$ with the distance function

$$\rho_2(x, y) = \max\{|x_1 - y_1|, |x_2 - y_2|\}$$

is a metric space.

d) If $(M, \rho)$ is a metric space and $A$ is a subset of $M$, then $A$ inherits a metric structure from $M$ in an obvious way, making $A$ a metric space. For example, **I**, **N** and **Q** all have "usual metrics," obtained by viewing them as subsets of **R** with its usual metric.

e) Let $X$ be any set and define $\rho$ on $X \times X$ by $\rho(x, x) = 0$ and $\rho(x, y) = 1$ if $x \neq y$. Then $\rho$ is a metric on $X$, called the *discrete metric*.

f) Let $X$ be any set and define $\rho$ on $X \times X$ by $\rho(x, y) = 0$ for all $x$ and $y$ in $X$. This is a pseudometric on $X$, called the *trivial pseudometric*. When is it a metric?

The distance functions available in metric spaces are precisely what we need to develop the notion of continuity in a more abstract setting, by mimicking the familiar definition for real-valued functions of a real variable. In fact, the following definition should look quite familiar when stated for **R** with its usual metric.

**2.3 Definition.** If $(M, \rho)$ and $(N, \sigma)$ are metric spaces, a function $f: M \to N$ is *continuous at* $x$ in $M$ iff for each $\epsilon > 0$, there is some $\delta > 0$ such that $\sigma(f(x), f(y)) < \epsilon$ whenever $\rho(x, y) < \delta$.

We turn now to the question: can we eliminate the dependence, in the previous definition, on the presence of distance functions? The answer is affirmative and depends on the development of the notion of an open set in a metric space.

**2.4 Definition.** Let $(M, \rho)$ be a metric space, $x$ a point of $M$. For $\epsilon > 0$, we define

$$U_\rho(x, \epsilon) = \{y \in M \mid \rho(x, y) < \epsilon\},$$

called the *$\epsilon$-disk* about $x$. If only one interpretation is possible, we will abbreviate $U_\rho(x, \epsilon)$ to $U(x, \epsilon)$.

If $E$ and $F$ are subsets of $M$, we define the *distance* between $E$ and $F$ to be

$$\rho(E, F) = \inf \{\rho(x, y) \mid x \in E, y \in F\}.$$

If $E$ has only one point, we usually write $\rho(x, F)$ rather than $\rho(\{x\}, F)$. Now we can extend the notation for $\epsilon$-disks to sets:

$$U_\rho(E, \epsilon) = \{y \in M \mid \rho(E, y) < \epsilon\}.$$

Using $\epsilon$-disks, we can reformulate the definition of continuity as follows: $f: (M, \rho) \to (N, \sigma)$ is continuous at $x$ in $M$ iff for each $\epsilon > 0$, there is some $\delta > 0$ such that $f(U_\rho(x, \delta)) \subset U_\sigma(f(x), \epsilon)$. This observation, together with the next definition, will make it possible to define continuity without mentioning the metrics involved at all.

**2.5 Definition.** A set $E$ in a metric space $(M, \rho)$ is *open* iff for each $x \in E$, there is an $\epsilon$-disk $U(x, \epsilon)$ about $x$ contained in $E$. A set is *closed* iff it is the complement of an open set. Evidently, a set $F$ is closed iff whenever every disk about $x$ meets $F$, then $x \in F$.

**2.6 Theorem.** *The open sets in a metric space* $(M, \rho)$ *have the following properties:*

a) *Any union of open sets is open.*

b) *Any finite intersection of open sets is open.*

c) $\varnothing$ *and* $M$ *are both open.*

*Proof.* a) If $A_\lambda$ is an open set for each $\lambda$ in $\Lambda$, and if $x$ is a point in $\bigcup A_\lambda$, then $x \in A_{\lambda_0}$ for some $\lambda_0$, so $A_{\lambda_0}$ contains some $\epsilon$-disk about $x$. Then $\bigcup A_\lambda$ will contain this same $\epsilon$-disk about $x$. It follows that $\bigcup A_\lambda$ is open. Arguments this simple will rarely be written out in such detail hereafter.

b) If $A_1, \ldots, A_n$ are open sets and $x \in \bigcap A_i$, then for each $i$, $x \in A_i$, so there is some disk $U(x, \epsilon_i)$ contained in $A_i$. Clearly, if $\epsilon$ is the minimum of $\epsilon_1, \ldots, \epsilon_n$, then the $\epsilon$-disk $U(x, \epsilon)$ is contained in $\bigcap A_i$.

c) $\varnothing$ contains a disk about each of its points since there are no points to worry about and $M$ contains a disk about each of its points because all disks are contained in $M$. Hence, $\varnothing$ and $M$ are open. ∎

**2.7 Examples.** a) Open sets in **R**. In the usual metric on **R**, the $\epsilon$-disk about a point $c$ is just the interval $(c - \epsilon, c + \epsilon)$. This makes it clear that each "open interval" in **R**, of the form $(a, b)$, is an open set. Hence every countable union of disjoint open intervals is an open set. We will prove the converse now; i.e., every open set in **R** is a countable union of disjoint open intervals. If $A$ is an open set in **R**, the relation $x \sim y$ iff there is some open interval $(a, b)$ with $\{x, y\} \subset (a, b) \subset A$ is an equivalence relation on $A$ and the resulting equivalence classes are disjoint open intervals whose union is $A$. The fact that there can be only countably many follows since each must contain a distinct rational.

b) Infinite intersections of open sets need not be open. In fact, the sets $A_n = (-1/n, 1/n)$ for $n = 1, 2, \ldots,$ are open in **R** with the usual metric, but $\bigcap_{n=1}^{\infty} A_n = \{0\}$ is not an open subset of **R**.

c) Disks are open. That is, in a metric space $X$, if $x \in X$ and $\delta > 0$, then $U(x, \delta)$ is an open set. This is left as a useful exercise on the triangle inequality, see Exercise 2D.

d) If $X$ is given the discrete metric, then for any point $x \in X$, the disk $U(x, 1)$ about $x$ is just the set $\{x\}$. Thus each one-point set in $X$ is open. But then, since any set is the union of its points, every set in $X$ is open.

e) One-point sets are always closed.

We can now rephrase the notion of a continuous function between metric spaces in terms of the open sets in these spaces, thus avoiding explicit mention of the metrics involved.

**2.8 Theorem.** *If $(M, \rho)$ and $(N, \sigma)$ are metric spaces, a function $f: M \to N$ is continuous at $x_0 \in M$ iff for each open set $V$ in $N$ containing $f(x_0)$, there is an open set $U$ in $M$ containing $x_0$ such that $f(U) \subset V$.*

*Proof.* If $f$ is continuous at $x$, and $V$ is an open set in $N$ containing $f(x)$, then $U_\sigma(f(x), \epsilon) \subset V$ for some $\epsilon > 0$, by the definition of open set. But, by continuity of $f$, there is a $\delta > 0$ such that $f(U_\rho(x, \delta)) \subset U_\sigma(f(x), \epsilon)$. Then $U = U_\rho(x, \delta)$ is an open set containing $x$ and $f(U) \subset V$.

Conversely, suppose for each open $V$ containing $f(x)$ there is an open $U$ containing $x$ such that $f(U) \subset V$. If $\epsilon > 0$ is given, then $U_\sigma(f(x), \epsilon) = V$ is an open set containing $f(x)$. Hence, there is an open $U$ containing $x$ such that $f(U) \subset V$. But since $x \in U$ and $U$ is open, $U_\rho(x, \delta) \subset U$ for some $\delta > 0$. Then $f(U_\rho(x, \delta)) \subset U_\sigma(f(x), \epsilon)$, so $f$ is continuous at $x$. ∎

Having Theorem 2.8, it is apparent that we can carry the notion of continuous function anywhere we can carry a reasonable notion of open set. "Reasonable" will simply be taken to mean "satisfying the properties (a), (b), and (c) of 2.6," and this, then, will be the basis of the definition of topological space, given in the next section.

Having given this brief motivational introduction, we will abandon the motivational approach now and develop topological spaces axiomatically. Thus, although topologies are introduced in the next section, continuous functions are not defined on general topological spaces until Section 7. However, the astute reader will see, in that definition, just a restatement of Theorem 2.8 (used there as the definition).

## Problems

2A. *Metrics on* $\mathbf{R}^n$

Verify that each of the following is a metric on $\mathbf{R}^n$:

1. $\rho(x, y) = \sqrt{\sum_{i=1}^n (x_i - y_i)^2}$
2. $\rho_1(x, y) = \sum_{i=1}^n |x_i - y_i|$
3. $\rho_2(x, y) = \max\{|x_1 - y_1|, \ldots, |x_n - y_n|\}$.

[For the first, make use of *Minkowski's inequality:* $\sqrt{\sum (a_n + b_n)^2} \le \sqrt{\sum a_n^2} + \sqrt{\sum b_n^2}$ for real numbers $a_n$, $b_n$ and $c_n$. The inequality is good for both finite and infinite sums.]

2B. *Metrics on* $C(\mathbf{I})$

Let $C(\mathbf{I})$ denote the set of all continuous real-valued functions on the unit interval $\mathbf{I}$ and let $x_0$ be a fixed point of $\mathbf{I}$.

1. $\rho(f, g) = \sup_{x \in \mathbf{I}} |f(x) - g(x)|$ is a metric on $C(\mathbf{I})$.
2. $\sigma(f, g) = \int_0^1 |f(x) - g(x)|\, dx$ is a metric on $C(\mathbf{I})$.
3. $\eta(f, g) = |f(x_0) - g(x_0)|$ is a pseudometric on $C(\mathbf{I})$.

These examples indicate that interesting and useful metrics can be defined on spaces other than the classical Euclidean spaces.

### 2C. *Pseudometrics*

1. Let $(M, \rho)$ be a pseudometric space. Define a relation $\sim$ on $M$ by $x \sim y$ iff $\rho(x, y) = 0$. Then $\sim$ is an equivalence relation.

2. If $M^*$ is the set of equivalence classes in $M$ under the equivalence relation $\sim$ and if $\rho^*$ is defined on $M^*$ by $\rho^*([x], [y]) = \rho(x, y)$, then $\rho^*$ is a well-defined metric on $M^*$. The metric space $(M^*, \rho^*)$ is called the *metric identification* of $(M, \rho)$.

3. If $h: M \to M^*$ is the mapping $h(x) = [x]$, then a set $A$ in $M$ is closed (open) iff $h(A)$ is closed (open) in $M^*$.

4. If $f$ is any real-valued function on a set $M$, then the distance function

$$\rho_f(x, y) = |f(x) - f(y)|$$

is a pseudometric on $M$.

5. If $(M, \rho)$ is any pseudometric space, then a function $f: M \to \mathbf{R}$ is continuous iff each set open in $(M, \rho_f)$ is open in $(M, \rho)$.

### 2D. *Disks are open*

For any subset $A$ of a metric space $M$ and any $\epsilon > 0$, the set $U(A, \epsilon)$ is open. (In particular, $U(x, \epsilon)$ is open for each $x \in M$.)

### 2E. *Bounded metrics*

A metric $\rho$ on $M$ is *bounded* iff for some constant $A$, $\rho(x, y) \leq A$ for all $x$ and $y$ in $M$.

1. If $\rho$ is any metric on $M$, the distance function $\rho^*(x, y) = \min\{\rho(x, y), 1\}$ is a metric also and is bounded.

2. A function $f$ is continuous on $(M, \rho)$ iff it is continuous on $(M, \rho^*)$. [It suffices to show that both $\rho$ and $\rho^*$ generate the same collection of open sets in $M$.]

### 2F. *The Hausdorff metric*

Let $\rho$ be a bounded metric on $M$; that is, for some constant $A$, $\rho(x, y) \leq A$ for all $x$ and $y$ in $M$.

1. Show that the elevation of $\rho$ to the power set $P(M)$ as defined in 2.4 is not necessarily a pseudometric on $P(M)$. (Take $M$ to be the unit disk $\{(x_1, x_2) \mid x_1^2 + x_2^2 \leq 1\}$ in the plane with the usual metric.)

2. Let $\mathscr{F}(M)$ be all nonempty closed subsets of $M$ and for $A, B \in \mathscr{F}(M)$ define

$$d_A(B) = \sup\{\rho(A, x) \mid x \in B\}$$
$$d(A, B) = \max\{d_A(B), d_B(A)\}.$$

Then $d$ is a metric on $\mathscr{F}(M)$ with the property that $d(\{x\}, \{y\}) = \rho(x, y)$. It is called the *Hausdorff metric* on $\mathscr{F}(M)$.

3. Prove that closed sets $A$ and $B$ are "close" in the Hausdorff metric iff they are "uniformly close"; that is, $d(A, B) < \epsilon$ iff $A \subset U_\rho(B, \epsilon)$ and $B \subset U_\rho(A, \epsilon)$.

The restriction in this problem to bounded metrics is, to a topologist, no problem at all, see 2E and 22.2. It is there so that $d_A(B)$, and hence $d(A, B)$, can never take the value $\infty$.

The Hausdorff metric is related to uniformities on the power set in Exercise 36E.

## 2G. *Isometry*

Metric spaces $(M, \rho)$ and $(N, \sigma)$ are *isometric* iff there is a one–one function $f$ from $M$ onto $N$ such that $\rho(x, y) = \sigma(f(x), f(y))$ for all $x$ and $y$ in $M$; $f$ is called an *isometry*.

1. If $f$ is an isometry from $M$ to $N$, then both $f$ and $f^{-1}$ are continuous functions.
2. $\mathbf{R}$ is not isometric to $\mathbf{R}^2$ (each with its usual metric).
3. $\mathbf{I}$ is isometric to any other closed interval in $\mathbf{R}$ of the same length.
4. Consider the pseudometric $\eta$ defined on $C(\mathbf{I})$ in 2B.3. What familiar space is the metric identification (2C.2) isometric to?

Isometric spaces are "metrically identical"; that is, there is nothing about their respective metrics which will serve to distinguish them.

## 2H. *Sequence spaces*

Let $\mathbf{m}$ denote the set of all bounded sequences $(x_n)_{n \in \mathbf{N}}$ of real numbers, $\mathbf{c}$ the set of all convergent sequences from $\mathbf{m}$, $\mathbf{c}_0$ the set of all sequences from $\mathbf{c}$ which converge to 0.

1. The distance function

$$\rho((x_n), (y_n)) = \sup \{|x_n - y_n| \mid n = 1, 2, \ldots\}$$

is a metric on $\mathbf{m}$ (and hence on each of the subspaces $\mathbf{c}$ and $\mathbf{c}_0$). On which of the three spaces is it bounded?

2. The distance function

$$\sigma((x_n), (y_n)) = \lim_{n \to \infty} |x_n - y_n|$$

is a pseudometric on $\mathbf{c}$. The metric identification (2C) of $(\mathbf{c}, \sigma)$ is isometric to the real line.

## 2I. *$l^p$-space*

For each $p > 0$, we denote by $l^p$ the set of all real sequences $(x_n)$ for which $\sum_{n=1}^{\infty} |x_n|^p < \infty$.

1. For $p \geq 1$, define a distance function $\rho$ on $l^p$ by

$$\rho((x_n), (y_n)) = \left( \sum_{n=1}^{\infty} |x_n - y_n|^p \right)^{1/p}.$$

This is a metric on $l^p$. [Use the *generalized Minkowski inequality:*

$$(\sum |a_n + b_n|^p)^{1/p} \leq (\sum |a_n|^p)^{1/p} + (\sum |b_n|^p)^{1/p},$$

for real sequences $(a_n)$, $(b_n)$ and $(c_n)$ and for $p \geq 1$.]

2. For $0 < p < 1$, define a distance function $\rho$ on $l^p$ as follows:

$$\rho((x_n), (y_n)) = \sum_{n=1}^{\infty} |x_n - y_n|^p.$$

Verify that this is a metric, using the inequality: $|a + b|^p \leq |a|^p + |b|^p$, for real numbers $a$ and $b$ and for $0 < p < 1$.

For $p = 2$, $l^p$ consists of all square-summable sequences, and as such, will be given its usual name and notation, (real) *Hilbert space* $\mathbf{H}$.

## 2J. *Normed linear spaces*

A *normed linear space* is a real linear space $X$ such that a number $\|x\|$, the *norm* of $x$, is associated with each $x \in X$, satisfying:

$NL$-a) $\|x\| \geq 0$ and $\|x\| = 0$ iff $x = 0$.

$NL$-b) $\|\alpha x\| = |\alpha| \cdot \|x\|$, for $\alpha \in \mathbf{R}$,

$NL$-c) $\|x + y\| \leq \|x\| + \|y\|$.

If ($NL$-a) is replaced by the weaker condition

$NL$-a)′ $\|x\| \geq 0$ and $\|0\| = 0$,

then $X$ is a *pseudonormed linear space*.

1. If $X$ is a pseudonormed linear space, the distance function $\rho(x, y) = \|x - y\|$ is a pseudometric on $X$. It is a metric iff $\|\cdot\|$ is a norm. We will call $\rho$ the *norm metric*, in case $\|\cdot\|$ is a norm.

2. If $\|\cdot\|_1$ and $\|\cdot\|_2$ are pseudonorms on the same linear space $X$, they give the same open sets (i.e., are *equivalent*) iff there are constants $C$ and $C'$ such that $\|x\|_1 \leq C \cdot \|x\|_2$ and $\|x\|_2 \leq C' \cdot \|x\|_1$, for all $x \in X$.

3. If $(X, \|\cdot\|)$ is a pseudonormed linear space and the metric identification procedure (2C) is applied to $X$ with its induced pseudometric $\rho$, producing a metric space $(X^*, \rho^*)$, then $X^*$ is a normed space with operations $[x] + [y] = [x + y]$ and $\alpha[x] = [\alpha x]$ and norm $\|[x]\|^* = \|x\|$, and furthermore the norm metric induced by $\|\cdot\|^*$ is $\rho^*$.

4. Let $X$ be any metric space, $C^*(X)$ the set of all bounded continuous functions from $X$ to $\mathbf{R}$. Then $C^*(X)$ is a normed linear space with the norm $\|f\| = \sup \{|f(x)| \mid x \in X\}$ and pointwise addition and scalar multiplication. This is the *sup norm* on $C^*(X)$. The associated metric was first introduced in 2B.1.

5. The collection $\mathscr{L}$ of all Riemann- (or, if you want, Lebesgue- ) integrable functions $f$ on $\mathbf{I}$ is a pseudonormed linear space with the pseudonorm $\|f\| = \int_0^1 |f(x)|\, dx$ and pointwise addition and scalar multiplication. But $\|\cdot\|$ is not a norm. (In fact, the set of all functions $f$ on $I$ such that $|f|^p$ is Lebesgue integrable is a pseudonormed space, with $\|f\| = [\int_0^1 |f|^p]^{1/p}$, for any $p$ with $1 \leq p < \infty$. It is called $\mathscr{L}^p(\mathbf{I})$ and the normed space resulting from part 3 above is $L^p(\mathbf{I})$. Verification of the axiom $NL$-c for the cases $p > 1$ requires the Hölder and Minkowski inequalities; see any reference on real analysis, e.g., Royden.)

6. On $\mathbf{R}^n$, with coordinatewise addition and scalar multiplication, each of the following is a norm:

a) $\|(x_1, \ldots, x_n)\| = (\sum_{k=1}^n x_k^2)^{1/2}$

b) $\|(x_1, \ldots, x_n)\|_1 = \sum_{k=1}^n |x_k|$

c) $\|(x_1, \ldots, x_n)\|_2 = \max \{|x_1|, \ldots, |x_m|\}$.

d) The norms $\|\cdot\|$, $\|\cdot\|_1$ and $\|\cdot\|_2$ have for their norm metrics the metrics $\rho$, $\rho_1$ and $\rho_2$ of 2A, respectively.

Chapter 2

# Topological Spaces

## 3 Fundamental concepts

As we pointed out in the previous section, open sets in metric spaces provide us with a way of phrasing the definition of continuous function without mentioning distance. Thus wherever we can carry a reasonable abstract notion of "open set," we can define continuous functions. The problem of what properties one should postulate as reasonable for our abstract open sets is, of course, a difficult one and any solution must ultimately live or die on the merits of the theory it produces. The "reasonableness" of the following definition, which is based on the observations made in Theorem 2.6, can thus be justified only by reading the forty-two sections which follow it.

**3.1 Definition.** A *topology* on a set $X$ is a collection $\tau$ of subsets of $X$, called the *open sets*, satisfying:

$G$-1) Any union of elements of $\tau$ belongs to $\tau$,

$G$-2) any finite intersection of elements of $\tau$ belongs to $\tau$,

$G$-3) $\varnothing$ and $X$ belong to $\tau$.

We say $(X, \tau)$ is a *topological space*, sometimes abbreviated "$X$ is a topological space" when no confusion can result about $\tau$.

Given two topologies $\tau_1$ and $\tau_2$ on the same set $X$, we say $\tau_1$ is *weaker* (*smaller*, *coarser*) than $\tau_2$, or $\tau_2$ is *stronger* (*larger*, *finer*) than $\tau_1$ iff $\tau_1 \subset \tau_2$.

**3.2 Examples.** a) Let $(M, \rho)$ be a metric space. Then, by Theorem 2.6, the open sets in $M$ defined by 2.5 form a topology on $M$, called the *metric topology* $\tau_\rho$. Whenever $(X, \tau)$ is a topological space whose topology $\tau$ is the metric topology $\tau_\rho$ for some metric $\rho$ on $X$, we call $(X, \tau)$ a *metrizable* topological space. Note the distinction: a "metrizable space" is a space with a topology which happens to have come from some metric, a "metric space" is a space with a metric. Every metric space $(X, \rho)$ determines a metrizable space $(X, \tau_\rho)$ and given a metrizable space $(X, \tau)$, one can always find many metrics $\rho$ on $X$ such that $\tau_\rho = \tau$ (for example, if $\tau_\rho = \tau$ then $\tau_{2\rho} = \tau$ also). The obvious modifications to the discussion above will define *pseudometrizable* topologies.

b) The metric topology generated by the usual metric on any subset of $\mathbf{R}^n$

will be called the *usual topology*. Hereafter, when a topology is used on a subset of $\mathbf{R}^n$ without mention it is assumed to be the usual topology.

c) Let $X$ be any set and let $\tau$ be the collection of *all* subsets of $X$. Then $\tau$ is clearly a topology for $X$; it is called the *discrete* topology. Moreover, it is metrizable, being the topology produced by the discrete metric on $X$, by part (d) of Example 2.7. It is finer than any other topology on $X$.

d) Let $X$ be any set and let $\tau = \{\varnothing, X\}$. Then $\tau$ is a topology for $X$, called the *trivial* (*indiscrete*) topology for $X$. It is pseudometrizable since it is the topology generated by the trivial pseudometric on $X$, by part (e) of Example 2.7. It is coarser than any other topology on $X$.

e) Let $X = \{a, b\}$ and let $\tau = \{\varnothing, \{a\}, X\}$. Then $\tau$ is a topology for $X$, and it is not even pseudometrizable. For suppose $\rho$ is a pseudometric on $X$ which produces $\tau$. Since $\{a\}$ is an open set, and $a \in \{a\}$, there must be an $\epsilon > 0$ such that $U(a, \epsilon) \subset \{a\}$; that is, $\rho(a, y) < \epsilon$ implies $y = a$. Hence, evidently $\rho(a, b) \geq \epsilon$. But then $U(b, \epsilon) = \{b\}$, so $\{b\}$ is an open set, contrary to the definition of $\tau$. Hence, no pseudometric $\rho$ can produce this topology on $X$. With this topology, $X$ is sometimes called the *Sierpinski space*.

The remainder of this section will be devoted to developing descriptive terminology which can be applied to subsets of a topological space. The notions of a closed set and of the closure, interior and frontier operations will be introduced and it will be observed that each of the first three completely describes the topology (the frontier operation does also, but this is not important).

**3.3 Definition.** If $X$ is a topological space and $E \subset X$, we say $E$ is *closed* iff $X - E$ is open.

The proof of the following theorem is an obvious application of De Morgan's laws in conjunction with the definition of a topology on $X$, and can be omitted.

**3.4 Theorem.** *If $\mathscr{F}$ is the collection of closed sets in a topological space $X$, then*

$F$-a) *Any intersection of members of $\mathscr{F}$ belongs to $\mathscr{F}$,*

$F$-b) *Any finite union of members of $\mathscr{F}$ belongs to $\mathscr{F}$,*

$F$-c) *$X$ and $\varnothing$ both belong to $\mathscr{F}$.*

*Conversely, given a set $X$ and any family $\mathscr{F}$ of subsets of $X$ satisfying* $F$-a, $F$-b *and* $F$-c, *the collection of complements of members of $\mathscr{F}$ is a topology on $X$ in which the family of closed sets is just $\mathscr{F}$.*

This theorem is a result of, and illustrates, the obvious duality between the notions of open set and closed set. More formally, *any result about the open sets in a topological space becomes a result about the closed sets upon replacing "open" by "closed" and interchanging $\bigcup$ and $\bigcap$.*

**3.5 Definition.** If $X$ is a topological space and $E \subset X$, the *closure* of $E$ in $X$ is the set

$$\bar{E} = \mathrm{Cl}\,(E) = \bigcap \{K \subset X \mid K \text{ is closed and } E \subset K\}.$$

Where confusion is possible as to what space the closure is to be taken in, we will write $\mathrm{Cl}_X(E)$. By property $F$-a for closed sets, $\bar{E}$ is closed. It is the smallest closed set containing $E$, in the sense that it is contained in every closed set containing $E$ (this is the precise meaning of "smallest" in 1.9 if the closed sets containing $E$ are ordered by $K_1 \leq K_2$ iff $K_1 \subset K_2$).

**3.6 Lemma.** *If $A \subset B$, then $\bar{A} \subset \bar{B}$.*

*Proof.* Since $B \subset \bar{B}$, if $A$ is contained in $B$, we have $A \subset \bar{B}$; since $\bar{B}$ is closed, we must then have $\bar{A} \subset \bar{B}$. ∎

**3.7 Theorem.** *The operation $A \to \bar{A}$ in a topological space $X$ has the following properties:*

$K$-a) $E \subset \bar{E}$,

$K$-b) $\overline{(\bar{E})} = \bar{E}$,

$K$-c) $\overline{A \cup B} = \bar{A} \cup \bar{B}$,

$K$-d) $\bar{\varnothing} = \varnothing$,

$K$-e) *$E$ is closed in $X$ iff $\bar{E} = E$.*

*Moreover, given a set $X$ and a mapping $A \to \bar{A}$ of $\mathscr{P}(X)$ into $\mathscr{P}(X)$ satisfying* $K$-a *through* $K$-d, *if we define closed sets in $X$ using* $K$-e, *the result is a topology on $X$ whose closure operation is just the operation $A \to \bar{A}$ we began with.*

*Proof.* First suppose $X$ is a topological space. We will show $K$-c holds, leaving the rest of $K$-a through $K$-e as an easy exercise. Since $\bar{A} \cup \bar{B}$ is closed and contains $A \cup B$, it contains $\overline{A \cup B}$. On the other hand, since $A \subset A \cup B$ and $B \subset A \cup B$ we have $\bar{A} \subset \overline{A \cup B}$ and $\bar{B} \subset \overline{A \cup B}$, by Lemma 3.6, and thus $\bar{A} \cup \bar{B} \subset \overline{A \cup B}$. This establishes $K$-c.

We proceed to the second part of the theorem. Let $X$ be any set and $A \to \bar{A}$ a mapping of $\mathscr{P}(X)$ into $\mathscr{P}(X)$ satisfying $K$-a through $K$-d. Let $\mathscr{F}$ be the collection of all sets $A$ such that $\bar{A} = A$. The assertion is that $\mathscr{F}$ satisfies $F$-a through $F$-c of Theorem 3.4.

First note that if $A \subset B$, then by $K$-c, $\bar{B} = \bar{A} \cup \overline{(B - A)}$ so that $\bar{A} \subset \bar{B}$ (why couldn't we just refer to Lemma 3.6?).

Now suppose $F_\lambda \in \mathscr{F}$ for each $\lambda \in \Lambda$. Then since $\bigcap F_\lambda$ is contained in $F_\lambda$, $\overline{\bigcap F_\lambda}$ is contained in $\bar{F}_\lambda$, for each $\lambda$, and hence $\overline{\bigcap F_\lambda} \subset \bigcap \bar{F}_\lambda = \bigcap F_\lambda$. But the reverse inclusion is given by $K$-a, so $\overline{\bigcap F_\lambda} = \bigcap F_\lambda$, that is, $\bigcap F_\lambda \in \mathscr{F}$. Thus $F$-a of Theorem 3.4 holds.

Next suppose $F_1, \ldots, F_n \in \mathscr{F}$. Then by $K$-c and induction,

$$\overline{F_1 \cup \cdots \cup F_n} = \overline{F_1} \cup \cdots \cup \overline{F_n} = F_1 \cup \cdots \cup F_n, \qquad \text{so} \quad F_1 \cup \cdots \cup F_n \in \mathscr{F}.$$

This establishes $F$-b of Theorem 3.4.

By $K$-d and $K$-a, it is clear that ø and $X$, respectively, belong to $\mathscr{F}$, so $F$-c of Theorem 3.4 is established.

Thus $\mathscr{F}$ is a collection of closed sets for $X$. It remains to show the resulting closure operation in $X$ is just the operation $A \to \bar{A}$ we began with; that is, that $\bar{A}$ is the smallest element of $\mathscr{F}$ containing $A$, for each $A \subset X$. Since $\overline{(\bar{A})} = \bar{A}$ by $K$-c, we know that $\bar{A} \in \mathscr{F}$, and from $K$-a, we know that $A \subset \bar{A}$. If $K$ is any element of $\mathscr{F}$ containing $A$, then $\bar{A} \subset \bar{K} = K$. Thus $\bar{A}$ is indeed the smallest element of $\mathscr{F}$ containing $A$. ■

An operation $A \to \bar{A}$ in a set $X$ which satisfies $K$-a through $K$-d is called a *Kuratowski closure operation* (which, incidentally, is the reason for the letter $K$ in the numeration). Thus every Kuratowski closure operation determines and is determined by some topology.

**3.8 Examples.** a) Let $X$ be an infinite set and for each $A \subset X$, define $\bar{A}$ as follows:

$$\bar{A} = A, \qquad \text{if } A \text{ is finite,}$$

$$\bar{A} = X, \qquad \text{if } A \text{ is infinite.}$$

The properties $K$-a through $K$-d can be verified for the resulting operation $A \to \bar{A}$, so we have a Kuratowski closure operation in $X$. The resulting topology on $X$, the *cofinite topology*, has for closed sets those sets $A$ for which $\bar{A} = A$. Apparently, then, the only closed sets are $X$, ø and all finite sets in $X$.

b) We always have $\overline{A \cup B} = \bar{A} \cup \bar{B}$. The corresponding statement for intersections is *not* true. Let $X$ be $\mathbf{R}$, $A$ the rationals in $\mathbf{R}$, $B$ the irrationals in $\mathbf{R}$, and give $X$ the usual topology. Check that $\bar{A} = \mathbf{R}$ and $\bar{B} = \mathbf{R}$. But $A \cap B = ø$, so $\overline{A \cap B} = ø$. Thus, $\overline{A \cap B} \neq \bar{A} \cap \bar{B}$. It *is* always true that $\overline{A \cap B} \subset \bar{A} \cap \bar{B}$.

c) As an exercise, you are asked to verify that if $(M, \rho)$ is a (pseudo)metric space, and $A \subset M$, then in the resulting (pseudo)metric topology on $M$,

$$\bar{A} = \{y \in M \mid \rho(y, A) = 0\}.$$

This provides a clue to the way the closure of a set is regarded in general. $\bar{A}$ is the set of points either in $A$ or sitting right next to $A$. (Further elucidation of this point of view will be found in Theorem 4.7.)

d) The closed disk $U(x, \bar{\epsilon}) = \{y \in M \mid \rho(x, y) \leq \epsilon\}$ in a metric space $(M, \rho)$ is a closed set in the metric topology but it need not be the closure of the disk $U(x, \epsilon)$. In Exercise 3E, you will verify that a counterexample exists. In $\mathbf{R}^n$ with the usual metric, the closure of $U(x, \epsilon)$ *is* $U(x, \bar{\epsilon})$.

e) The closure of a subset $A$ of a discrete space $X$ is $A$ itself.

f) The closure of any nonempty subset of a set $X$ with the trivial topology is $X$ (and, of course, the closure of $\varnothing$ is $\varnothing$).

**3.9 Definition.** If $X$ is a topological space and $E \subset X$, the *interior* of $E$ in $X$ is the set

$$E^\circ = \operatorname{Int}(E) = \bigcup \{G \subset X \mid G \text{ is open and } G \subset E\}.$$

Where confusion might otherwise result, we will write $\operatorname{Int}_X(E)$. Evidently, by property $G$-1 of open sets, $E^\circ$ is open. It is the largest open set contained in $E$, in the sense that it contains any other open set contained in $E$.

The notions of interior and closure are dual to each other, in much the same way that "open" and "closed" are. The strictly formal nature of this duality can be brought out in observing that

$$X - E^\circ = \overline{X - E}$$
$$X - \bar{E} = (X - E)^\circ.$$

Thus any theorem about closures in a topological space can be translated to a theorem about interiors. The next two results are, for example, the dual results to 3.6 and 3.7 about closures.

***3.10 Lemma.*** *If $A \subset B$, then $A^\circ \subset B^\circ$.*

*Proof.* It is clear that $A^\circ \subset A$, so if $A \subset B$, we have $A^\circ \subset B$. Thus $A^\circ$ is an open set contained in $B$, so $A^\circ \subset B^\circ$. ■

***3.11 Theorem.*** *The interior operation $A \to A^\circ$ in a topological space $X$ has the following properties:*

$I$-a) $A^\circ \subset A$.

$I$-b) $(A^\circ)^\circ = A^\circ$.

$I$-c) $(A \cap B)^\circ = A^\circ \cap B^\circ$.

$I$-d) $X^\circ = X$.

$I$-e) *$G$ is open iff $G^\circ = G$.*

*Conversely, given any map $A \to A^\circ$ of $\mathscr{P}(X)$ into $\mathscr{P}(X)$ in a set $X$, satisfying $I$-a through $I$-d, if open sets are defined in $X$ using $I$-e, the result is a topology on $X$ in which the interior of a set $A \subset X$ is just $A^\circ$.*

*Proof.* The proof can be done directly or by using the translation process on 3.7. Either way, it is easy and we will omit it. ■

**3.12 Examples.** a) In $\mathbf{R}$, with the usual topology, the interior of a closed interval $[a, b]$ is $(a, b)$. In $\mathbf{R}^2$ with the usual topology, the interior of the disk

$$\{(x_1, x_2) \mid x_1^2 + x_2^2 \leq 1\}$$

is the disk $\{(x_1, x_2) \mid x_1^2 + x_2^2 < 1\}$.

b) In $\mathbf{R}$, with the usual topology, if $A$ is the set of rationals, $B$ the set of irrationals, then $A^\circ = B^\circ = \varnothing$. But $(A \cup B)^\circ = \mathbf{R}^\circ = \mathbf{R}$. Hence,

$$(A \cup B)^\circ \neq A^\circ \cup B^\circ.$$

It is always true that $A^\circ \cup B^\circ \subset (A \cup B)^\circ$.

**3.13 Definition.** If $X$ is a topological space and $E \subset X$, the *frontier* of $E$ is the set

$$\mathrm{Fr}_X(E) = \bar{E} \cap \overline{(X - E)},$$

usually written $\mathrm{Fr}(E)$. Evidently, the frontier of $E$ is a closed set.

It is possible, but unrewarding, to characterize a topology completely by its frontier operation. We will be content to give the relationship between the frontier, closure and interior operations.

**3.14 Theorem.** *For any subset $E$ of a topological space $X$:*

a) $\bar{E} = E \cup \mathrm{Fr}(E)$

b) $E^\circ = E - \mathrm{Fr}(E)$

c) $X = E^\circ \cup \mathrm{Fr}(E) \cup (X - E)^\circ$.

*Proof.*

a)
$$\begin{aligned} E \cup \mathrm{Fr}(E) &= E \cup (\bar{E} \cap \overline{X - E}) \\ &= (E \cup \bar{E}) \cap (E \cup \overline{X - E}) \\ &= \bar{E} \cap X = \bar{E}. \end{aligned}$$

b)
$$\begin{aligned} E - \mathrm{Fr}(E) &= E - (\bar{E} \cap \overline{X - E}) \\ &= (E - \bar{E}) \cup (E - \overline{X - E}) \\ &= E - \overline{(X - E)} = E^\circ. \end{aligned}$$

c) Since $\mathrm{Fr}(E) \cup (X - E) = \overline{X - E}$ (as is easily verified) and since

$$X - E^\circ = \overline{X - E},$$

we have

$$X = E^\circ \cup \overline{X - E} = E^\circ \cup \mathrm{Fr}(E) \cup (X - E)^\circ. \blacksquare$$

**3.15 Examples.** a) The frontier of the closed interval $[a, b]$ in $\mathbf{R}$ is $\{a, b\}$, as is the frontier of any interval with the same endpoints. If $A$ denotes the set of rationals in $\mathbf{R}$, $\mathrm{Fr}_{\mathbf{R}}(A) = \mathbf{R}$.

b) For any space $X$, $\mathrm{Fr}_X(X) = \varnothing$.

c) If $D$ is the closed unit disk in the plane, and $X = \mathbf{R}^2$, $\mathrm{Fr}_X(D) = \mathbf{S}^1$, while $\mathrm{Fr}_D(D) = \varnothing$. In combinatorial topology, the word "boundary" would be used in such a way that the boundary of $D$ would always be $\mathbf{S}^1$. This prompts our use of the word "frontier."

## Problems

### 3A. *Examples of topologies*

1. If $\mathscr{F}$ is the collection of all closed, bounded subsets of $\mathbf{R}$ (in its usual topology), together with $\mathbf{R}$ itself, then $\mathscr{F}$ is the family of closed sets for a topology on $\mathbf{R}$ strictly weaker than the usual topology.

2. If $A \subset X$, show that the family of all subsets of $X$ which contain $A$, together with the empty set $\varnothing$, is a topology on $X$. Describe the closure and interior operations. What topology results when $A = \varnothing$? when $A = X$?

3. Let $B$ be a fixed subset of $X$ and for each nonempty $A \subset X$, let $\bar{A} = A \cup B$, with $\bar{\varnothing} = \varnothing$. Verify that $A \to \bar{A}$ is a closure operation. Describe the open sets in the resulting topology. What topology results when $B = \varnothing$? when $B = X$?

4. Call a subset of $\mathbf{R}^2$ *radially open* iff it contains an open line segment in each direction about each of its points. Show that the collection of radially open sets is a topology for $\mathbf{R}^2$. Compare this topology with the usual topology on $\mathbf{R}^2$ (i.e., is it weaker, stronger, the same or none of these?). The plane with this topology will be called the *radial plane.*

5. If $A \subset X$ and $\tau$ is any topology for $X$, then $\{U \cup (V \cap A) \mid U, V \in \tau\}$ is a topology for $X$. It is called the *simple extension* of $\tau$ over $A$.

### 3B. *Frontiers in the plane*

Any closed subset of the plane $\mathbf{R}^2$ is the frontier of some set in $\mathbf{R}^2$.

### 3C. *Complementation and closure*

If $A$ is any subset of a topological space, the largest possible number of different sets in the two sequences

$$A, A', A'^{-}, A'^{-\prime}, \ldots$$
$$A, A^{-}, A^{-\prime}, A^{-\prime -}, \ldots$$

(where $'$ denotes complementation and $^{-}$ denotes closure) is 14. There is a subset of $\mathbf{R}$ which gives 14. [For any open set $G$, $\mathrm{Cl}\,(\mathrm{Int}\,(\mathrm{Cl}\,G)) = \mathrm{Cl}\,G$.]

### 3D. *Regularly open and regularly closed sets*

An open subset $G$ in a topological space is *regularly open* iff $G$ is the interior of its closure. A closed subset is *regularly closed* iff it is the closure of its interior.

1. The complement of a regularly open set is regularly closed and vice versa.

2. There are open sets in $\mathbf{R}$ which are not regularly open.

3. If $A$ is any subset of a topological space, then $\mathrm{Int}\,(\mathrm{Cl}\,(A))$ is regularly open.

4. The intersection, but not necessarily the union, of two regularly open sets is regularly open. (Thus the same proposition, with "union" and "intersection" interchanged, holds for regularly closed sets.)

### 3E. *Metrizable spaces*

Let $X$ be a metrizable space whose topology is generated by a metric $\rho$.

1. The metric $2\rho$ defined by $2\rho(x, y) = 2 \cdot \rho(x, y)$ generates the same topology on $X$.

2. The closure of a set $E \subset X$ is given by $\bar{E} = \{y \in X \mid \rho(E, y) = 0\}$.

3. The closed disk $U(x, \bar{\epsilon}) = \{y \mid \rho(x, y) \leq \epsilon\}$ is closed in $X$, but may not be the closure of the open disk $U(x, \epsilon)$. [Consider $\epsilon = 1$ and the usual metric on

$$\{(x, y) \in \mathbf{R}^2 \mid x^2 + y^2 = 1\} \cup \{(x, 0) \in \mathbf{R}^2 \mid 0 \leq x \leq 1\}.]$$

### 3F. *Unions of closed sets*

1. Give an example of a sequence $B_1, B_2, \ldots$ of closed sets in a topological space $X$ whose union is not closed.

2. If $\rho$ generates the topology on a metrizable space $X$ and, for each $\lambda \in \Lambda$, $C_\lambda$ is a closed set in $X$ such that $\rho(C_{\lambda_1}, C_{\lambda_2}) \geq \epsilon$ for all $\lambda_1$ and $\lambda_2$, where $\epsilon$ is some fixed positive number, then $\bigcup C_\lambda$ is closed.

### 3G. *The lattice of topologies*

1. The intersection of any family of topologies on $X$ is a topology on $X$. [Note: intersect the topologies, not the sets which are elements of the topologies.]

2. The union of two topologies on $X$ need not be a topology on $X$. But for any family of topologies on $X$, there is a smallest topology larger than all of them.

Thus, the topologies on a fixed set $X$, when partially ordered by inclusion, form a complete lattice. The question of whether or not this lattice is complemented has only recently been answered (see notes).

### 3H. *$G_\delta$ and $F_\sigma$ sets*

A subset of a topological space $X$ is a $G_\delta$ iff it is a countable intersection of open sets and an $F_\sigma$ iff it is a countable union of closed sets.

1. The complement of a $G_\delta$ is an $F_\sigma$, and vice versa.

2. An $F_\sigma$ can be written as the union of an *increasing* sequence $F_1 \subset F_2 \subset \cdots$ of closed sets. (Hence, a $G_\delta$ can be written as a decreasing intersection.)

3. A closed set in a metric space is a $G_\delta$ (hence, an open set is an $F_\sigma$). [If $A$ is closed, let $A_n = \{y \mid \rho(A, y) < 1/n\}$ and see 2D.]

4. The rationals are an $F_\sigma$ in $\mathbf{R}$. (Much later, see 24.12 and 25A.4, it will be apparent that they cannot be a $G_\delta$.)

### 3I. *Borel sets*

The family of *Borel sets* in a topological space $X$ is the smallest family of sets $\mathscr{G}$ with the following properties:

a) $\mathscr{G}$ contains the open sets,

b) countable intersections of elements of $\mathscr{G}$ belong to $\mathscr{G}$.

c) complements of elements of $\mathscr{G}$ belong to $\mathscr{G}$.

1. In (a), "open" can be replaced by "closed"; in (b), "intersection" can be replaced by "union."

In any space, define the class $\mathscr{G}_\alpha$, $0 \leq \alpha < \omega_1$, by transfinite induction, as follows: the class $\mathscr{G}_0$ consists of the open sets, and for $\alpha > 0$, the class $\mathscr{G}_\alpha$ consists of the sets which are countable unions or countable intersections of sets of lower class. (Thus, for example, the class $\mathscr{G}_1$ will consist precisely of the $G_\delta$ sets (see 3H).)

2. In a metric space, $\bigcup \{\mathscr{G}_\alpha \mid 0 \leq \alpha < \omega_1\}$ is the family of Borel sets. [Show that $\bigcup \mathscr{G}_\alpha$ satisfies (a), (b) and (c). For (c), you will have to use transfinite induction and 3H.3.]

3. In a metric space, the family of Borel sets is the smallest family of sets satisfying:

a)′ $\mathscr{G}$ contains the open sets,

b)′ countable intersections of elements of $\mathscr{G}$ belong to $\mathscr{G}$.

c)′ countable unions of elements of $\mathscr{G}$ belong to $\mathscr{G}$.

"Open" can be replaced by "closed."

## 4 Neighborhoods

The means we have at hand so far for describing topologies (open sets, the closure operation, etc.) are not the most convenient, and for this reason are rarely used. In this and the next section, we present the two most popular ways to describe topologies.

Very often the topology we wish to present is quite "regular," in the sense that the open sets containing one point look no different from the open sets containing any other (this is true, for example, in the Euclidean spaces). In such cases, one can describe the topology by describing what it looks like "around" one point, or a few points, and then retiring from the field with the observation that around other points it is the same. Considerable saving of effort can result, and topologies will often be presented this way here, so we will present now a detailed discussion of the "local" description of topologies and topological concepts.

**4.1 Definition.** If $X$ is a topological space and $x \in X$, a *neighborhood* (hereafter abbreviated *nhood*) of $x$ is a set $U$ which contains an open set $V$ containing $x$. Thus, evidently, $U$ is a nhood of $x$ iff $x \in U^\circ$. The collection $\mathscr{U}_x$ of all nhoods of $x$ is the *nhood system* at $x$.

The next theorem is similar to Theorems 3.7 and 3.11 about closure and interior: it lists properties of the nhood system $\mathscr{U}_x$ at $x$ in a topological space, and provides a converse which says whenever nhoods have been assigned to each point of a set, satisfying these properties, one has a topology.

***4.2 Theorem.*** *The nhood system $\mathscr{U}_x$ at $x$ in a topological space $X$ has the following properties:*

N-a) *If $U \in \mathscr{U}_x$, then $x \in U$,*

N-b) *If $U, V \in \mathscr{U}_x$, then $U \cap V \in \mathscr{U}_x$,*

N-c) *If $U \in \mathscr{U}_x$, then there is a $V \in \mathscr{U}_x$, such that $U \in \mathscr{U}_y$ for each $y \in V$,*

N-d) *If $U \in \mathscr{U}_x$ and $U \subset V$, then $V \in \mathscr{U}_x$,* *and furthermore,*

*N-e) $G \subset X$ is open iff $G$ contains a nhood of each of its points.*

*Conversely, if in a set $X$ a nonempty collection $\mathcal{U}_x$ of subsets of $X$ is assigned to each $x \in X$ so as to satisfy N-a through N-d, and if N-e is used to define "open," the result is a topology on $X$, in which the nhood system at each $x \in X$ is precisely $\mathcal{U}_x$.*

*Proof.* $N$-a is obvious. For $N$-b: if $U, V \in \mathcal{U}_x$, then $x \in U^\circ$ and $x \in V^\circ$, so $x \in U^\circ \cap V^\circ = (U \cap V)^\circ$ and hence $U \cap V \in \mathcal{U}_x$. For $N$-c: let $U \in \mathcal{U}_x$ and pick $V = U^\circ$. Then for each $y \in V$, $y \in U^\circ$, so $U \in \mathcal{U}_y$. For $N$-d: if $U \in \mathcal{U}_x$, then $x \in U^\circ$. If $U \subset V$, then $U^\circ \subset V^\circ$, so $x \in V^\circ$. Hence $V \in \mathcal{U}_x$. Finally, to prove $N$-e, if $G$ is open, then $G = G^\circ$ and $G$ is a nhood of each of its points. On the other hand, if each $x \in G$ has a nhood $V_x \subset G$, then $G = \bigcup_{x \in G} V_x^\circ$ is a union of open sets and thus open.

The converse assertion is left to Exercise 4E. ∎

Neighborhoods provide us with an interesting description of what has happened in the passage from metric spaces to topological spaces. The linearly ordered "distances from $x$" have been replaced by the partially ordered "nhoods of $x$" (partially ordered by $U_1 \leq U_2$ iff $U_1 \supset U_2$), in describing closeness to $x$ of points nearby. Not only have we lost the linear order in our notion of closeness, we have lost the symmetry. If $y$ is close to $x$ in a metric space, then $x$ is close to $y$; but it can happen in a topological space that $y$ is in every nhood of $x$ while $x$ is in no nhood of $y$ (a very extreme example; this doesn't happen in useful topological spaces, although many useful spaces do lack symmetry in some degree).

Since supersets of nhoods are nhoods ($N$-d), it is not necessary to give all the nhoods of $x$ to describe the nhood system there. We can be content with a *nhood base.*

**4.3 Definition.** A *nhood base* at $x$ in the topological space $X$ is a subcollection $\mathcal{B}_x$ taken from the nhood system $\mathcal{U}_x$, having the property that each $U \in \mathcal{U}_x$ contains some $V \in \mathcal{B}_x$. That is, $\mathcal{U}_x$ must be determined by $\mathcal{B}_x$ as follows:

$$\mathcal{U}_x = \{U \subset X \mid V \subset U \text{ for some } V \in \mathcal{B}_x\}.$$

Once a nhood base at $x$ has been chosen (there are many to choose from, all producing the same nhood system at $x$) its elements are called *basic nhoods.*

Obviously, the nhood system at $x$ is itself always a nhood base at $x$. There are more interesting examples.

**4.4 Examples.** a) In any topological space, the open nhoods of $x$ form a nhood base at $x$, since for any nhood $U$ of $x$, $U^\circ$ is also a nhood of $x$. For this reason, it is the custom of a great many writers to use "nhood of $x$" to mean "open nhood of $x$" and to use the term "nhood" (without reference to a point $x$) to mean "nonempty open set." For us, nhoods will not necessarily be open, however, unless so described.

b) In any metrizable space, generated by a metric $\rho$ say, each open set containing $x$ contains some disk $U(x, \delta)$ about $x$; thus the disks $U(x, \delta)$ about $x$ form a nhood base at $x$. In fact, we need consider only the disks of rational radius to obtain a nhood base at $x$, so each point in a metric space has a countable nhood base. In particular, these comments apply to the usual topologies (and the usual metrics which generate them) on the spaces $\mathbf{R}^n$, $n = 1, 2, \ldots$ . A topological space in which every point has a countable nhood base is said to satisfy the *first axiom of countability* or to be *first countable*. Thus every metric space is first countable. We will meet the second axiom of countability in Exercise 5F; both axioms will be studied in greater detail in Section 16.

c) In $\mathbf{R}^2$, with the usual topology (and the usual metric), the set of all squares with sides parallel to the axes and centered at $x \in \mathbf{R}^2$ is a nhood base at $x$. Notice that this base at $x$ has no set in common with the nhood base described in (b), although they both describe the same topology. Thus, before one uses the term "basic nhood at $x$," one must fix for the discussion what nhood base at $x$ is being used. Sometimes context or general usage make this clear. It is customary, for example, to mean "disk about $x$" when one refers to a "basic nhood at $x$" in $\mathbf{R}^2$, or for that matter, in any metric space.

d) If $X$ is a discrete space, each point $x \in X$ has an acceptable nhood base consisting of a single set, namely $\{x\}$.

e) If $X$ is a trivial space, the only nhood base at $x \in X$ is the collection consisting of the single set $X$.

We turn now to the problem of specifying a topology by giving a collection of basic nhoods at each point of the space. Each of the properties $V$-a, $V$-b and $V$-c corresponds to the respective property $N$-a, $N$-b, $N$-c in Theorem 4.2. Note that $N$-d is dropped altogether.

The following theorem is used much more often than the corresponding Theorem 4.2 about nhood systems.

***4.5 Theorem.*** *Let $X$ be a topological space and for each $x \in X$, let $\mathscr{B}_x$ be a nhood base at $x$. Then*

$V$-a) *if $V \in \mathscr{B}_x$, then $x \in V$,*

$V$-b) *if $V_1, V_2 \in \mathscr{B}_x$, then there is some $V_3 \in \mathscr{B}_x$ such that $V_3 \subset V_1 \cap V_2$,*

$V$-c) *if $V \in \mathscr{B}_x$, there is some $V_0 \in \mathscr{B}_x$ such that if $y \in V_0$, then there is some $W \in \mathscr{B}_y$ with $W \subset V$,*

*and furthermore,*

$V$-d) *$G \subset X$ is open iff $G$ contains a basic nhood of each of its points.*

*Conversely, in a set $X$, if a collection $\mathscr{B}_x$ of subsets of $X$ is assigned to each $x \in X$ so as to satisfy $V$-a, $V$-b and $V$-c and if we define "open" using $V$-d, the result is a topology on $X$ in which $\mathscr{B}_x$ is a nhood base at $x$, for each $x \in X$.*

*Proof.* The properties $V$-a, $V$-b and $V$-c are easily verified for basic nhoods, by referring to the corresponding properties $U$-a, $U$-b, and $U$-c for nhoods. Similarly, $V$-d follows from $U$-e. We will proceed to the converse.

Suppose a collection $\mathscr{B}_x$ satisfying $V$-a, $V$-b and $V$-c has been prescribed at each $x \in X$ and define

$$\mathscr{U}_x = \{U \subset X \mid B \subset U \text{ for some } B \in \mathscr{B}_x\}$$

for each $x \in X$. The assertion is that $\mathscr{U}_x$ has the properties $N$-a through $N$-d of a nhood system at $x$.

Certainly each $U \in \mathscr{U}_x$ contains $x$, since each $B \in \mathscr{B}_x$ does, so $N$-a is clear. If $U_1, U_2 \in \mathscr{U}_x$, then for some $B_1, B_2, B_3 \in \mathscr{B}_x$ we have $B_1 \subset U_1$, $B_2 \subset U_2$ and (by $V$-b) $B_3 \subset B_1 \cap B_2 \subset U_1 \cap U_2$. Thus $U_1 \cap U_2 \in \mathscr{U}_x$, establishing $N$-b. For $N$-c, let $U \in \mathscr{U}_x$. Pick $B \in \mathscr{B}_x$ such that $B \subset U$. By $V$-c, there is some $B_0 \in \mathscr{B}_x$ such that each $y \in B_0$ has some $B_y \in \mathscr{B}_y$ contained in $B$. Thus $B \in \mathscr{U}_y$ for each $y \in B_0$. Hence $U \in \mathscr{U}_y$ for each $y \in B_0$, establishing $N$-c. Finally, the superset property $N$-d is clear from the definition of $\mathscr{U}_x$.

Thus $\mathscr{U}_x$ is a nhood system at $x$, for each $x \in X$. Moreover, it is clear that, at each $x$, $\mathscr{B}_x$ is a nhood base at $x$ in the resulting topology on $X$. ■

**4.6 Example.** There is a useful alternative to the usual topology on the real line which is best described in terms of basic nhoods. The *Sorgenfrey line*, **E**, is the real line with the topology in which basic nhoods of $x$ are the sets $[x, z)$ for $z > x$. Some of its basic properties will be studied in Exercise 4A, and we will find frequent occasion in later work to refer to it. It is named after the man who first produced it, in 1947.

Since nhood bases are important descriptive devices in dealing with topologies, it will be useful to have nhood characterizations of all the concepts so far introduced for topological spaces.

**4.7 Theorem.** *Let $X$ be a topological space and suppose a nhood base has been fixed at each $x \in X$. Then*

a) *$G \subset X$ is open iff $G$ contains a basic nhood of each of its points,*

b) *$F \subset X$ is closed iff each point $x \notin F$ has a basic nhood disjoint from $F$,*

c) *$\bar{E} = \{x \in X \mid$ each basic nhood of $x$ meets $E\}$,*

d) *$E^\circ = \{x \in X \mid$ some basic nhood of $x$ is contained in $E\}$,*

e) *$\mathrm{Fr}\,(E) = \{x \in X \mid$ each basic nhood of $x$ meets both $E$ and $X - E\}$.*

*Proof.* a) This is part of Theorem 4.5 and is recorded here for reference.

b) This follows directly from (a) together with the definition of a closed set as the complement of an open set.

c) Recall that $\bar{E} = \bigcap \{K \subset X \mid K \text{ is closed and } E \subset K\}$. If some nhood $U$ of $x$ does not meet $E$, then $x \in U^\circ$ and $E \subset X - U^\circ$. Since $X - U^\circ$ is closed, $\bar{E} \subset X - U^\circ$. Hence $x \notin \bar{E}$. Conversely, if $x \notin \bar{E}$, then $X - \bar{E}$ is an open set containing $x$, and hence containing a basic nhood of $x$, which does not meet $E$.

d) This follows from (c) by an application of De Morgan's laws.

e) Follows directly from (c) and the definition of Fr $(E)$ as $\bar{E} \cap \overline{(X - E)}$. ■

**4.8 Theorem.** (*Hausdorff criterion*) *For each $x \in X$, let $\mathscr{B}_x^1$ be a nhood base at $x$ for a topology $\tau_1$ on $X$, and let $\mathscr{B}_x^2$ be a nhood base at $x$ for a topology $\tau_2$ on $X$. Then $\tau_1 \subset \tau_2$ iff at each $x \in X$, given $B^1 \in \mathscr{B}_x^1$, there is some $B^2 \in \mathscr{B}_x^2$ such that $B^2 \subset B^1$.*

*Proof.* Suppose $\tau_1 \subset \tau_2$. Let $B^1 \in \mathscr{B}_x^1$. Then, since $B^1$ is a nhood of $x$ in $(X, \tau_1)$, $x$ is contained in some element $B$ of $\tau_1$ which is contained in $B^1$. But if $B \in \tau_1$, then $B \in \tau_2$ so $B$ is a nhood of $x$ in $(X, \tau_2)$. It follows that $B^2 \subset B$ for some $B^2 \in \mathscr{B}_x^2$, so $B^2 \subset B^1$.

Conversely, if $B \in \tau_1$, then $B$ contains some $B^1 \in \mathscr{B}_x^1$ for each $x \in B$; hence $B$ contains a corresponding element $B^2 \in \mathscr{B}_x^2$ for each $x \in B$. Thus $B \in \tau_2$. ■

The theorem above could be paraphrased: "small nhoods make large topologies." This is intuitively reasonable; the smaller the nhoods in a space are, the easier it is for a set to contain nhoods of all its points and the more open sets there will be.

We close this section by introducing a concept which depends for its definition on the use of nhoods.

**4.9 Definition.** An *accumulation point* (*cluster point*) of a set $A$ in a topological space $X$ is a point $x \in X$ such that each nhood (basic nhood, if you prefer) of $x$ contains some point of $A$, other than $x$. The set $A'$ of all cluster points of $A$ is called the *derived set* of $A$.

**4.10 Theorem.** $\bar{A} = A \cup A'$.

*Proof.* From 4.7, $A' \subset \bar{A}$, and since $A \subset \bar{A}$, we have $A \cup A' \subset \bar{A}$. On the other hand, if every nhood of $x$ meets $A$ (i.e., if $x \in \bar{A}$), then either $x \in A$ or every nhood of $x$ meets $A$ in a point different from $x$, so $x \in A \cup A'$. ■

## Problems

### 4A. *The Sorgenfrey line*

The following material concerns the Sorgenfrey line, **E**, introduced in 4.6.

1. Verify that the sets $[x, z)$, for $z > x$, do form a nhood base at $x$ for a topology on the real line.

2. Which intervals on the real line are open sets in the Sorgenfrey topology?

3. Describe the closure of each of the following subsets of the Sorgenfrey line: the rationals, the set $\{1/n \mid n = 1, 2, \ldots\}$, the set $\{-1/n \mid n = 1, 2, \ldots\}$, the integers.

### 4B. *The Moore plane*

Let $\mathbf{\Gamma}$ denote the closed upper half plane $\{(x, y) \mid y \geq 0\}$ in $\mathbf{R}^2$. For each point in the open upper half plane, basic nhoods will be the usual open disks (with the restriction, of course, that they be taken small enough to lie in $\mathbf{\Gamma}$). At the points $z$ on the $x$-axis, the basic nhoods will be the sets $\{z\} \cup A$, where $A$ is an open disk in the upper half plane, tangent to the $x$-axis at $z$.

1. Verify that this gives a topology on $\mathbf{\Gamma}$.

2. Compare the topology thus obtained with the usual topology on the closed upper half plane as a subspace of $\mathbf{R}^2$.

3. Describe the closure and interior operations in the space $\mathbf{\Gamma}$.

Hereafter, the symbol $\mathbf{\Gamma}$ will be reserved for the closed upper half plane with the topology described here. This space is often called the *Moore plane*. We will find consistent use for it as a counterexample.

### 4C. *The slotted plane*

At each point $z$ in the plane, the basic nhoods at $z$ are to be the sets $\{z\} \cup A$, where $A$ is a disk about $z$ with a finite number of straight lines through $z$ removed.

1. Verify that this gives a topology on the plane.
2. Compare this topology with the usual topology on the plane.
3. Can we re-replace "finite" in the definition of this space with "countable?"

This space will be called the *slotted plane*, $\mathbf{A}$.

### 4D. *The looped line*

At each point $x$ of the real line other than the origin, the basic nhoods of $x$ will be the usual open intervals centered at $x$. Basic nhoods of the origin will be the sets

$$(-\epsilon, \epsilon) \cup (-\infty, -n) \cup (n, \infty),$$

for all possible choices of $\epsilon > 0$ and $n \in \mathbf{N}$.

1. Verify that this gives a topology on the line.
2. Describe the closure operation in the resulting space.

This space is the *looped line*, $\mathbf{L}$.

### 4E. *Topologies from nhoods*

1. Show that if each point $x$ in a set $X$ has assigned a collection $\mathscr{U}_x$ of subsets of $X$ satisfying $N$-a through $N$-d of 4.2, then the collection

$$\tau = \{G \subset X \mid \text{for each } x \text{ in } G, x \in U \subset G \text{ for some } U \in \mathscr{U}_x\}$$

is a topology for $X$, in which the nhood system at each $x$ is just $\mathscr{U}_x$.

2. Show that, if $\mathscr{B}_x$ is a nhood base at $x$ for each $x$ in a topological space $X$, then $V$-a, $V$-b, $V$-c and $V$-d of 4.5 hold for elements of $\mathscr{B}_x$.

### 4F. *Spaces of functions*

Consider the set $\mathbf{R}^\mathbf{I}$ of all real-valued functions on the unit interval.

1. For each $f \in \mathbf{R}^\mathbf{I}$, each finite subset $F$ of $\mathbf{I}$ and each positive $\delta$, let

$$U(f, F, \delta) = \{g \in \mathbf{R}^\mathbf{I} \mid |g(x) - f(x)| < \delta, \text{ for each } x \in F\}.$$

Show that the sets $U(f, F, \delta)$ form a nhood base at $f$, making $\mathbf{R}^\mathbf{I}$ a topological space.

2. For each $f \in \mathbf{R}^\mathbf{I}$, the closure of the one-point set $\{f\}$ is just $\{f\}$. (This is not unusual. In fact, it is a situation to be desired; spaces without this property are difficult to deal with. See the discussion in Sections 13–15.)

3. For $f \in \mathbf{R}^\mathbf{I}$ and $\epsilon > 0$, let

$$V(f, \epsilon) = \{g \in \mathbf{R}^\mathbf{I} \mid |g(x) - f(x)| < \epsilon, \text{ for each } x \in \mathbf{I}\}.$$

Verify that the sets $V(f, \epsilon)$ form a nhood base at $f$, making $\mathbf{R}^\mathbf{I}$ a topological space.

4. Compare the topologies defined in 1 and 3.

5. If the definition in 3 is made to apply to continuous functions only, show that the resulting topology on $C(\mathbf{I})$ is the one induced by the metric defined in 2B.1.

We will return to the topology in 1, in a more general context, in Section 8 on product spaces. Both the topologies on $\mathbf{R}^\mathbf{I}$ introduced here are treated in the chapter on function spaces.

### 4G. *Nowhere dense sets*

A set $A$ in a topological space $X$ is *nowhere dense* in $X$ iff $\text{Cl}_X A$ contains no nonempty open set. A point $p$ is *isolated* iff the set $\{p\}$ is open and a set $D$ is *discrete in X* (or, *relatively discrete*) iff each $d \in D$ has a nhood $U$ in $X$ such that $U \cap D = \{d\}$.

1. In a metric space $X$ without isolated points, the closure of a discrete set in $X$ is nowhere dense in $X$.

2. In any space $X$, the frontier of an open set is closed and nowhere dense.

3. Conversely, every closed nowhere dense set is the frontier of an open set.

4. In a metric space $X$, the frontier of an open set is the set of accumulation points of a discrete set. [This requires the axiom of choice and is difficult.]

## 5 Bases and subbases

As we observed in the last section, we can specify the nhood system at a point $x$ of a topological space $X$ by giving a somewhat smaller collection of sets, a nhood base at $x$. In much the same way, the topology on all of $X$ can be specified, without describing each and every open set, by giving a *base for the topology*.

**5.1 Definition.** If $(X, \tau)$ is a topological space, a *base* for $\tau$ (sometimes we call it a *base for* $X$ when no confusion can result) is a collection $\mathscr{B} \subset \tau$ such that

$$\tau = \left\{ \bigcup_{B \in \mathscr{C}} B \,\middle|\, \mathscr{C} \subset \mathscr{B} \right\}.$$

That is, $\tau$ can be recovered from $\mathscr{B}$ by taking all possible unions of subcollections from $\mathscr{B}$. Evidently, $\mathscr{B}$ is a base for $X$ iff whenever $G$ is an open set in $X$ and $p \in G$, there is some $B \in \mathscr{B}$ such that $p \in B \subset G$.

**5.2 Examples.** a) In **R**, the collection $\mathscr{B}$ of all open intervals is a base for the usual topology. More generally, in any metric space $M$, the collection of all open disks about points of $M$ is a base for $M$.

b) The collection $\{\{x\} \mid x \in X\}$ is a base for the discrete topology on $X$.

The following theorem is similar to 3.7, 3.11, 4.2 and 4.5. That is, it lists a few properties that bases enjoy and provides the converse assertion: any structure on a set $X$ with these properties provides a topology on $X$. Note that no mention is made in this theorem of the topology. If you have a given topology $\tau$ and want to know whether a particular collection $\mathscr{B}$ of sets is a base for $\tau$, 5.3 can be used to show $\mathscr{B}$ is a base for *some* topology, but you must return to the Definition 5.1 to show the topology generated by $\mathscr{B}$ is $\tau$ (and here the form of the definition given in the last sentence of 5.1 is particularly useful).

***5.3 Theorem.*** *$\mathscr{B}$ is a base for a topology on $X$ iff*

a) $X = \bigcup_{B \in \mathscr{B}} B$

b) *whenever $B_1, B_2 \in \mathscr{B}$ with $p \in B_1 \cap B_2$, there is some $B_3 \in \mathscr{B}$ with*

$$p \in B_3 \subset B_1 \cap B_2.$$

*Proof.* If $\mathscr{B}$ is a base for a topology on $X$, the two properties are clear. Suppose, on the other hand, $X$ is a set and $\mathscr{B}$ a collection of subsets of $X$ with these properties. Let $\tau$ be all unions of subcollections from $\mathscr{B}$. Then any union of members of $\tau$ certainly belongs to $\tau$, so $\tau$ satisfies G-1 of 3.1. Moreover, if $\mathscr{B}_1 \subset \mathscr{B}$ and $\mathscr{B}_2 \subset \mathscr{B}$, so that $\bigcup_{B \in \mathscr{B}_1} B$ and $\bigcup_{C \in \mathscr{B}_2} C$ are elements of $\tau$, then

$$\left(\bigcup_{B \in \mathscr{B}_1} B\right) \cap \left(\bigcup_{C \in \mathscr{B}_2} C\right) = \bigcup_{B \in \mathscr{B}_1} \bigcup_{C \in \mathscr{B}_2} (B \cap C).$$

But by property (b), the intersection of two elements of $\mathscr{B}$ is a union of elements of $\mathscr{B}$, so

$$\left(\bigcup_{B \in \mathscr{B}_1} B\right) \cap \left(\bigcup_{C \in \mathscr{B}_2} C\right)$$

is a union of elements of $\mathscr{B}$, and hence belongs to $\tau$. Thus $\tau$ satisfies G-2 of 3.1. Finally, $X \in \tau$ by (a) and $\varnothing \in \tau$ since $\varnothing$ is the union of the empty subcollection from

$\mathscr{B}$. Hence $\tau$ satisfies $G$-3 of 3.1. This completes the proof that $\tau$ is a topology on $X$. ■

The reader might well suspect, especially after studying the examples given in 5.2, that more than a casual similarity exists between the idea of a nhood base at each point of $X$ on the one hand and the notion of a base for the topology of $X$ on the other. Indeed, as the next theorem makes clear, the only real difference between the two notions is that nhood bases need not consist of open sets.

**5.4 Theorem.** *If $\mathscr{B}$ is a collection of open sets in $X$, $\mathscr{B}$ is a base for $X$ iff for each $x \in X$, the collection $\mathscr{B}_x = \{B \in \mathscr{B} \mid x \in B\}$ is a nhood base at $x$.*

*Proof.* Suppose first that $\mathscr{B}$ is a base for $X$, $x \in X$, and $\mathscr{B}_x = \{B \in \mathscr{B} \mid x \in B\}$. The elements of $\mathscr{B}_x$ are clearly nhoods of $x$. Moreover, if $U$ is any nhood of $x$, then $x \in U^\circ$ and, since $U^\circ$ is a union of elements of $\mathscr{B}$, $x \in B \subset U^\circ$ for some $B \in \mathscr{B}$. Thus $B \in \mathscr{B}_x$ and $B \subset U$, so $\mathscr{B}_x$ is a nhood base at $x$.

Conversely, if $\mathscr{B}_x$ is an open nhood base at $x$, for each $x \in X$, and $\mathscr{B} = \bigcup_{x \in X} \mathscr{B}_x$, then for any open set $U$ in $X$, and each element $p$ of $U$, there is an element $B_p$ of $\mathscr{B}$ such that $p \in B_p \subset U$. Then $U = \bigcup \{B_p \mid p \in U\}$ is a union of elements of $\mathscr{B}$, so $\mathscr{B}$ is a base for $X$. ■

We can go one step further in reducing the size of the collection we must specify to describe a topology. The reduction from topology to base was accomplished essentially by dropping property $G$-1 of topologies. The further reduction to subbase is accomplished by dropping $G$-2 (see 3.1).

**5.5 Definition.** If $(X, \tau)$ is a topological space, a *subbase* for $\tau$ (or a subbase for $X$) is a collection $\mathscr{C} \subset \tau$ such that the collection of all finite intersections of elements from $\mathscr{C}$ forms a base for $\tau$.

**5.6 Theorem.** *Any collection of subsets of a set $X$ is a subbase for some topology on $X$.*

*Proof.* Exercise 5D. ■

## Problems

### 5A. *Examples of subbases*

1. The family of sets of the form $(-\infty, a)$ together with those of the form $(b, \infty)$ is a subbase for the usual topology on the real line.

2. Describe the topology on the plane for which the family of all straight lines is a subbase.

3. Describe the topology on the line for which the sets $(a, \infty)$, $a \in \mathbf{R}$, are a subbase. Describe the closure and interior operations in this topology.

### 5B. *Examples of bases*

1. The collection of all open rectangles is a base for a topology on the plane. Describe the topology in more familiar terms.

2. For each positive integer $n$, let $S_n = \{n, n + 1, \ldots\}$. The collection of all subsets of $\mathbf{N}$ which contain some $S_n$ is a base for a topology on $\mathbf{N}$. Describe the closure operation in this space.

3. The collection of all open intervals $(a, b)$ together with the one-point sets $\{n\}$ for all positive and negative integers $n$ is a base for a topology on the real line. Describe the interior operation in the resulting space.

### 5C. *The scattered line*

We introduce a new topology on the line as follows: a set is open iff it is of the form $U \cup V$ where $U$ is an open subset of the real line with its usual topology and $V$ is any subset of the irrationals. Call the resulting space $\mathbf{S}$, the *scattered line.*

1. With the definition of "open set" given, $\mathbf{S}$ is a topological space.

2. Describe an efficient nhood base at

   a) the rational points
   b) the irrational points

in $\mathbf{S}$. Put these together to describe a base for $\mathbf{S}$.

### 5D. *No axioms for subbase*

Any family of subsets of a set $X$ is a subbase for some topology on $X$ and the topology which results is the smallest topology containing the given collection of sets.

### 5E. *Bases for the closed sets*

A *base for the closed sets* in a topological space $X$ is any family of closed sets in $X$ such that every closed set is an intersection of some subfamily.

1. $\mathscr{F}$ is a base for the closed sets in $X$ iff the family of complements of members of $\mathscr{F}$ is a base for the open sets.

2. $\mathscr{F}$ is a base for the closed sets for some topology on $X$ iff (a) whenever $F_1$ and $F_2$ belong to $\mathscr{F}$, $F_1 \cup F_2$ is an intersection of elements of $\mathscr{F}$, and (b) $\bigcap_{F \in \mathscr{F}} F = \varnothing$.

### 5F. *Second countable and separable spaces*

A space $X$ is *second countable* iff $X$ has a countable base. $X$ is *separable* iff a countable subset $D$ of $X$ exists with $\mathrm{Cl}_X D = X$. (Such a set $D$ is said to be *dense* in $X$.)

1. A separable metric space is second countable. [The disks of rational radius about the points of a countable dense set form a countable base.]

2. Every second countable space is separable and first countable. [For separability, obtain a countable dense set by choosing one element from each member of a countable base. Note that this requires the axiom of choice.]

3. The Sorgenfrey line $\mathbf{E}$ (4.6) is first countable and separable; we will see later that it cannot be second countable.

Material on separable and second countable spaces will be developed in the text in Section 16.

Chapter 3

# New Spaces from Old

## 6 Subspaces

A subset of a topological space inherits a topology of its own, in an obvious way. This topology and some of its easily developed properties will be presented here.

**6.1 Definition.** If $(X, \tau)$ is a topological space and $A \subset X$, the collection $\tau' = \{G \cap A \mid G \in \tau\}$ is a topology for $A$, called the *relative topology* for $A$. The fact that a subset of $X$ is being given this topology is signified by referring to it as a *subspace* of $X$.

Any time a topology is used on a subset of a topological space without explicitly being described, it is assumed to be the relative topology. This natural and convenient convention has the result that any adjective which can be applied to topological spaces (e.g., "separable," see 5F) can be applied automatically to subsets of a topological space. We are *not* saying that if a space has a particular property, then every subspace of that space has the same property; see 6B.

**6.2 Examples.** a) The real line, regarded as the $x$-axis in $\mathbf{R}^2$, inherits its usual topology from $\mathbf{R}^2$. The integers, as a subspace of $\mathbf{R}$, inherit the discrete topology. Each of these examples is a special case of the general rule: *if $X$ is metrizable and $A \subset X$, then the relative topology on $A$ is generated by the restriction of any metric which generates the topology on $X$.* The proof of this will be made easy by the next theorem, so it is left to Exercise 6C.

b) By relativizing the usual topology on $\mathbf{R}^n$, we have a *usual topology* on any subset of $\mathbf{R}^n$. By part (a), the usual topology on $A$ is generated by the usual metric on $A$.

c) Any subspace of a discrete space is discrete and any subspace of a trivial space is trivial.

d) *A subspace of a subspace is a subspace.* That is, if $A_1 \subset A_2 \subset X$, then the relative topology induced on $A_1$ by the relative topology of $A_2$ in $X$ is just the relative topology of $A_1$ in $X$. The proof is easy.

The open sets in a subspace $A$ of $X$ are the intersections with $A$ of the open sets in $X$. Most, but not all, of the related topological notions are introduced into $A$ in the same way, by intersection, as the following theorem and example show.

**6.3 Theorem.** *If $A$ is a subspace of a topological space $X$, then:*

*a) $H \subset A$ is open in $A$ iff $H = G \cap A$, where $G$ is open in $X$,*

*b) $F \subset A$ is closed in $A$ iff $F = K \cap A$ where $K$ is closed in $X$,*

*c) if $E \subset A$, then $\mathrm{Cl}_A\, E = A \cap \mathrm{Cl}_X\, E$,*

*d) if $x \in A$, then $V$ is a nhood of $x$ in $A$ iff $V = U \cap A$, where $U$ is a nhood of $x$ in $X$,*

*e) if $x \in A$, and if $\mathscr{B}_x$ is a nhood base at $x$ in $X$, then $\{B \cap A \mid B \in \mathscr{B}_x\}$ is a nhood base at $x$ in $A$,*

*f) if $\mathscr{B}$ is a base for $X$, then $\{B \cap A \mid B \in \mathscr{B}\}$ is a base for $A$.*

*Proof.* a) is just the definition of the subspace topology on $A$, recorded here for reference.

b) follows directly from (a).

c) follows from (b) and the definition of the closure of $E$ as the intersection of all closed sets containing $E$.

d) follows from (a) and the definition of a nhood of $x$ as a set containing an open set containing $x$.

e) Each $B \cap A$ is a nhood of $x$ in $A$, by part (d). Further, if $V$ is any nhood of $x$ in $A$, then $V = U \cap Y$ where $U$ is a nhood of $x$ in $X$. Then $U \supset B$ for some $B \in \mathscr{B}_x$, so $V = U \cap A \supset B \cap A$. Thus the sets $B \cap A$ form a nhood base at $x$ in $A$.

f) follows from (e) and the theorem (5.4) on translation between bases and nhood bases. ■

The reader will notice that two concepts are missing from the list above; no mention is made of the interior operator or the frontier operator in subspaces. The following examples indicate why this is so.

**6.4 Examples.** a) Let $X$ be the plane with the usual topology while $A = E =$ the $x$-axis. Then $\mathrm{Int}_A\, E = A$ while $\mathrm{Int}_X\, E = \varnothing$, so that the former cannot be obtained by intersecting the latter with $A$. It *is* always true, however, that

$$\mathrm{Int}_A\, E \supset A \cap \mathrm{Int}_X\, E.$$

b) Using the same example, we have $\mathrm{Fr}_A\, E = \varnothing$ while $\mathrm{Fr}_X\, E = A$, so that, again, the former cannot be obtained by intersecting the latter with $A$. It *is* always true, however, that $\mathrm{Fr}_A\, E \subset A \cap \mathrm{Fr}_X\, E$.

## Problems

### 6A. *Examples of subspaces*

1. Recall that **A** denotes the slotted plane (4C). Any straight line in the plane has the discrete topology as a subspace of **A**. The topology on any circle in the plane as a subspace of **A** coincides with its usual topology.

2. We will let **B** denote the radial plane (3A). The relative topology induced on any straight line as a subspace of **B** is its usual topology. The relative topology on any circle in the plane as a subspace of **B** is the discrete topology.

3. Discuss the subspaces of the scattered line **S** (5C).

4. The rationals, as a subspace of **R**, do not have the discrete topology.

5. The topology on the nonnegative reals, regarded as the subspace $\{(0, y) \mid y \geq 0\}$ of the Moore plane **Γ** (4B) is the usual topology. The $x$-axis in the Moore plane inherits the discrete topology.

6. An open set in an open subspace of $X$ is open in $X$. This need not be true if the subspace is not open. A similar result holds for closed sets in closed subspaces.

7. If $\tau$ is the simple extension over $A$ (3A.5) of a topology $\tau'$ on $X$, then $A$ is open in $(X, \tau)$ and the topology $A$ inherits from $(X, \tau)$ is the same topology it inherits from $(X, \tau')$.

### 6B. *Subspaces of separable spaces*

1. The Moore plane **Γ** (4B) is separable (see 5F).

2. The $x$-axis in the Moore plane has for its relative topology the discrete topology. Thus, a subspace of a separable space need not be separable.

3. An open subset of a separable space is separable.

### 6C. *Subspaces of metrizable spaces*

If $M$ is metrizable and $N \subset M$, then the subspace $N$ is metrizable with the topology generated by the restriction of any metric which generates the topology on $M$.

### 6D. *Ordered spaces*

Let $X$ be linearly ordered by a relation $\leq$. Take as a subbase for a topology on $X$ all sets of the form $\{x \mid x < a\}$ and $\{x \mid x > a\}$, for $a \in X$. The resulting topology on $X$ is the *order topology* on $X$ and whenever we use the phrase *ordered space* we mean a linearly ordered set with its order topology. An *interval* in a linearly ordered space is any subset which contains all points between $x$ and $y$ whenever it contains $x$ and $y$.

1. If $a < b$ in $X$, the interval $\{x \in X \mid a < x < b\}$ is an open set in the order topology; but intervals of the form $\{x \in X \mid a \leq x \leq b\}$ may also be open.

2. The usual topology on the real line is the order topology given by the usual order.

3. In **I** × **I**, with the *lexicographic order*: $(x_1, x_2) < (y_1, y_2)$ iff either $x_1 < x_2$ or else $x_1 = x_2$ and $y_1 < y_2$, describe the nhoods of each of the following:

a) the points $(x, 0)$, with particular attention to $(0, 0)$,
b) the points $(x, 1)$, with particular attention to $(1, 1)$,
c) the points $(x, y)$, $0 < x < 1$, $0 < y < 1$.

4. A subset of an ordered space has a topology induced by the restricted order and a topology inherited from the order topology on the larger space. Show by an example that these two topologies on a subset need not be the same. [An example exists using for the large space the real line with its usual topology and order.] Find conditions on the subspace which will ensure that the two induced topologies agree.

## 7 Continuous functions

It is the purpose of this section to define continuous functions on a topological space and establish their elementary properties. The basis for our definition is Theorem 2.8, in which it was demonstrated that the notion of distance could be effectively suppressed in defining continuity of functions between metric spaces, by introduction of the use of open sets. In fact, the reader who restudies Theorem 2.8 at this point will see in the following definition just a rewording of that theorem, with the slight modification that here we use "nhood of $x_0$" instead of "open set containing $x_0$".

**7.1 Definition.** Let $X$ and $Y$ be topological spaces and let $f: X \to Y$. Then $f$ is *continuous* at $x_0 \in X$ iff for each nhood $V$ of $f(x_0)$ in $Y$, there is a nhood $U$ of $x_0$ in $X$ such that $f(U) \subset V$. We say $f$ is *continuous on* $X$ iff $f$ is continuous at each $x_0 \in X$.

It is left to the reader to verify that the effect of the definition is not altered if "nhood" is replaced by "basic nhood" throughout.

The next theorem provides an alternative, and somewhat surprising, set of characterizations of functions $f: X \to Y$ which are continuous on all of $X$. This theorem, in one or another of its forms, is more often used to check "global" continuity than the alternative, that is, checking continuity at each point of $X$ individually. The fourth characterization, although not often used as a test for continuity, is interesting. It provides us with a description of the continuous functions $f: X \to Y$ as precisely those functions which take the points close to a set $E$ in $X$ close to its image in $Y$.

**7.2 Theorem.** *If $X$ and $Y$ are topological spaces and $f: X \to Y$, then the following are all equivalent:*

a) *$f$ is continuous,*

b) *for each open set $H$ in $Y$, $f^{-1}(H)$ is open in $X$,*

c) *for each closed set $K$ in $Y$, $f^{-1}(K)$ is closed in $X$,*

d) *for each $E \subset X$, $f(\mathrm{Cl}_X E) \subset \mathrm{Cl}_Y f(E)$.*

*Proof.* a) $\Rightarrow$ b): If $H$ is open in $Y$, then for each $x \in f^{-1}(H)$, $H$ is a nhood of $f(x)$. Hence, by continuity of $f$, there is a nhood $V$ of $x$ such that $f(V) \subset H$; that is, $V \subset f^{-1}(H)$. Thus $f^{-1}(H)$ contains a nhood of each of its points and is therefore open.

b) $\Rightarrow$ c): If $K$ is closed in $Y$, then $f^{-1}(Y - K)$ is open in $X$, by part (b). Hence, since $f^{-1}(K) = X - f^{-1}(Y - K)$, $f^{-1}(K)$ is closed in $X$.

c) $\Rightarrow$ d): Let $K$ be any closed set in $Y$ containing $f(E)$. By part (c), $f^{-1}(K)$ is a closed set in $X$ containing $E$. Hence, $\mathrm{Cl}_X E \subset f^{-1}(K)$, and it follows that $f(\mathrm{Cl}_X E) \subset K$. Since this is true for any closed set $K$ containing $E$, we have

$$f(\mathrm{Cl}_X E) \subset \mathrm{Cl}_Y f(E).$$

d) $\Rightarrow$ a): Let $x \in X$ and let $V$ be an open nhood of $f(x)$. Set $E = X - f^{-1}(V)$ and let $U = X - \mathrm{Cl}_X E$. It is easy to verify that, since $f(\mathrm{Cl}_X E) \subset \mathrm{Cl}_Y f(E)$, we have $x \in U$. It is even clearer that $f(U) \subset V$. Hence, $f$ is continuous at $x$. ∎

The following theorem is intuitive, easily proved and surpassingly important.

**7.3 Theorem.** *If $X$, $Y$ and $Z$ are topological spaces and $f: X \to Y$ and $g: Y \to Z$ are continuous, then $g \circ f: X \to Z$ is continuous.*

*Proof.* If $H$ is open in $Z$, then $g^{-1}(H)$ is open in $Y$, by continuity of $g$. Hence, by continuity of $f$, $f^{-1}[g^{-1}(H)] = (g \circ f)^{-1}(H)$ is open in $X$. Thus $g \circ f$ is continuous. ∎

**7.4 Definition.** If $f: X \to Y$ and $A \subset X$, we will use $f \mid A$ (the *restriction* of $f$ to $A$) to denote the map of $A$ into $Y$ defined by $(f \mid A)(a) = f(a)$ for each $a \in A$.

**7.5 Theorem.** *If $A \subset X$ and $f: X \to Y$ is continuous, then $(f \mid A): A \to Y$ is continuous.*

*Proof.* If $H$ is open in $Y$, then $(f \mid A)^{-1}(H) = f^{-1}(H) \cap A$, and the latter is open in the relative topology on $A$. ∎

The theorem above has a sort of converse: if $f$ is continuous on each of a few properly fitting pieces of $X$, it is continuous on $X$. This is stated more precisely by the following theorem, and its generalizations in Exercise 7D.

**7.6 Theorem.** *If $X = A \cup B$, where $A$ and $B$ are both open (or both closed) in $X$, and if $f: X \to Y$ is a function such that both $f \mid A$ and $f \mid B$ are continuous, then $f$ is continuous.*

*Proof.* Suppose $A$ and $B$ are open. If $H$ is open in $Y$, then $f^{-1}(H)$ is open in $X$, since $f^{-1}(H) = (f \mid A)^{-1}(H) \cup (f \mid B)^{-1}(H)$ and each of the latter is open in an open subspace of $X$ and so open in $X$. The proof is similar if $A$ and $B$ are closed. ∎

If we write $f: X \to Y$, we have specified the domain of $f$ (as $X$), but the image of $f$ is not determined, except that it must be a subset of $Y$. The next theorem says, essentially, that it is not necessary to modify this procedure when dealing with continuous functions. The proof is left as Exercise 7E.

**7.7 Theorem.** *Suppose $Y \subset Z$ and $f: X \to Y$. Then $f$ is continuous as a map from $X$ to $Y$ iff it is continuous as a map from $X$ to $Z$.*

In the passage from $X$ to the image $Y$ of $X$ under a continuous map $f$, we lose information in two ways. The first is set-theoretical: $Y$ will have fewer (or, at least, no more) points than $X$. The second is topological: $Y$ will have fewer (or, at least, no more) open sets than $X$ in the sense that each open set $H$ in $Y$ is the image of an open set (for example, $f^{-1}(H)$) in $X$, but there may well be open sets $U$ in $X$ such that $f(U)$ is not open in $Y$.

The maps which preserve $X$ set-theoretically and topologically are called homeomorphisms.

**7.8 Definition.** If $X$ and $Y$ are topological spaces, a function $f$ from $X$ to $Y$ is a *homeomorphism* iff $f$ is one–one, onto and continuous and $f^{-1}$ is also continuous. In this case, we say $X$ and $Y$ are *homeomorphic*. If $f$ is everything but onto, we call it an *embedding* of $X$ into $Y$, and say that $X$ is *embedded* in $Y$ by $f$. Thus, $X$ is embedded in $Y$ by $f$ iff $f$ is a homeomorphism between $X$ and some subspace of $Y$.

Evidently, a continuous map $f: X \to Y$ is a homeomorphism iff there is a continuous map $g: Y \to X$ such that the compositions $g \circ f$ and $f \circ g$ are the identity maps on $X$ and $Y$ respectively. Various algebraic isomorphisms may be defined in the same formal way. The attempt to unify and systematize such notions has led to the development of *categorical algebra*.

The reader can easily verify the following theorem; it is a direct consequence of Theorem 7.2.

**7.9 Theorem.** *If $X$ and $Y$ are topological spaces and $f: X \to Y$ is one–one and onto, the following are all equivalent:*

a) *$f$ is a homeomorphism,*

b) *if $G \subset X$, then $f(G)$ is open in $Y$ iff $G$ is open in $X$,*

c) *if $F \subset X$, then $f(F)$ is closed in $Y$ iff $F$ is closed in $X$,*

d) *if $E \subset X$, then $f(\mathrm{Cl}_X E) = \mathrm{Cl}_Y f(E)$.*

Homeomorphic topological spaces are, for the purposes of a topologist, the same. That is, there is nothing about homeomorphic spaces $X$ and $Y$ having to do only with their respective topologies which we can use to distinguish them. Thus, for example, a "topological characterization" of the real line $\mathbf{R}$ would consist of a list of properties possessed by the real line which, if possessed by any other space $X$, ensure that $X$ is homeomorphic with $\mathbf{R}$.

If we denote "$X$ is homeomorphic with $Y$" by $X \sim Y$, then the relationship $\sim$ has the following properties:

a) $X \sim X$,

b) if $X \sim Y$, then $Y \sim X$,

c) if $X \sim Y$ and $Y \sim Z$, then $X \sim Z$.

Thus, the relation "is homeomorphic to" is an equivalence relation on any set of topological spaces. The reader might profit from thinking, at this point, about the question: is there a set of all topological spaces?

To prove two spaces are homeomorphic, one constructs a homeomorphism. To establish that two spaces are not homeomorphic, one must find a topological property possessed by one and not the other. The definition of "topological property" makes it clear why this works. A *topological property* is a property of topological spaces which, if possessed by $X$, is possessed by all spaces homeomorphic to $X$. First countability, second countability and separability are examples of topological properties which have already been introduced. We will introduce many more in sections to come.

**7.10 Examples.** a) The open interval $(a, b)$ in $\mathbf{R}$ is homeomorphic to $(0, 1)$, one homeomorphism being $f(x) = (x - a)/(b - a)$. Moreover, all intervals of the form $(a, \infty)$ are obviously homeomorphic by translation, and $(1, \infty)$ is homeomorphic to $(0, 1)$ under the map $f(x) = 1/x$. Also, the interval $(-\infty, -a)$ is homeomorphic to $(a, \infty)$ under the map $f(x) = -x$. Finally, $(-\infty, \infty)$ is homeomorphic to $(-\pi/2, \pi/2)$ under the map $f(x) = \arctan x$. The relations above can be summarized, using transitivity of the homeomorphism relation, as follows: *all open intervals in* $\mathbf{R}$, *including the unbounded intervals, are homeomorphic.* Verification of the details passed over here is left to Exercise 7G.

b) All bounded closed intervals in $\mathbf{R}$ which have more than one point are homeomorphic. In fact, $[a, b]$ is homeomorphic to $[0, 1]$ under the same map $f(x) = (x - a)/(b - a)$ used above. In 7G, we will see that we cannot include the unbounded intervals this time.

## Problems

### 7A. *Characterization of spaces using functions*

The *characteristic function* of a subset $A$ of a set $X$ is the function from $X$ to $\mathbf{R}$ which is 1 at points of $A$ and 0 at other points of $X$.

1. The characteristic function of $A$ is continuous iff $A$ is both open and closed in $X$.

2. $X$ has the discrete topology iff whenever $Y$ is a topological space and $f: X \to Y$, then $f$ is continuous.

3. $X$ has the trivial topology iff whenever $Y$ is a topological space and $f: Y \to X$, then $f$ is continuous.

### 7B. *No Cantor–Bernstein theorem for topological spaces*

Recall that the Cantor–Bernstein theorem states that if $A$ and $B$ are sets and if one–one functions $f: A \to B$ and $g: B \to A$ exist, then a one–one function of $A$ onto $B$ exists. The analog for topological spaces would be: whenever $X$ can be embedded in $Y$ and $Y$ can be embedded in $X$, then $X$ and $Y$ are homeomorphic. Find a counterexample. [See 7G.3].

### 7C. *Functions agreeing on a dense subset*

If $f$ and $g$ are continuous functions from $X$ to $\mathbf{R}$, the set of points $x$ for which $f(x) = g(x)$ is a closed subset of $X$. Thus two continuous maps on $X$ to $\mathbf{R}$ which agree on a dense subset (one whose closure is $X$) must agree on all of $X$. Rephrased: a real-valued continuous function is determined by its values on a dense set. [See also 13.14.]

### 7D. *Sufficient conditions for continuity*

There are useful extensions of Theorem 7.6. A family of subsets of a topological space is called *locally finite* iff each point of the space has a nhood meeting only finitely many elements of the family.

1. The union of any subfamily from a locally finite family of closed sets is closed.

2. If $\{A_\lambda \mid \lambda \in \Lambda\}$ is a locally finite collection of closed subsets of $X$ whose union is $X$, a function on $X$ is continuous iff its restriction to each $A_\lambda$ is continuous.

3. If $\{B_\lambda \mid \lambda \in \Lambda\}$ is any collection of open subsets of $X$ whose union is $X$, a function on $X$ is continuous iff its restriction to each $B_\lambda$ is continuous.

### 7E. *Range immaterial*

If $Y \subset Z$ and $f: X \to Y$, then $f$ is continuous as a map from $X$ to $Y$ iff $f$ is continuous as a map from $X$ to $Z$.

### 7F. *Functions to and from the plane*

The facts presented here for the plane will be proved in more generality for product spaces in Section 8.

If $f$ is a function on any space $X$ to the plane, associated with $f$ we have the coordinate functions $f_1$ and $f_2$, each mapping $X$ to $\mathbf{R}$. For each $x \in X$, $f_1(x)$ and $f_2(x)$ are the first and second coordinates, respectively, of $f(x)$.

On the other hand, if $g$ is a function from the plane to any space $Y$, for each fixed $x_0 \in \mathbf{R}$ we can define a function $g_{x_0}$ from $\mathbf{R}$ to $Y$ by $g_{x_0}(y) = g(x_0, y)$. Similarly, if $y_0 \in \mathbf{R}$ is fixed, $h_{y_0}(x) = g(x, y_0)$ defines a function $h_{y_0}$ from $\mathbf{R}$ to $Y$. We say $g$ is *continuous in* $x$ iff $h_{y_0}$ is continuous for each $y_0 \in \mathbf{R}$ and $g$ is *continuous in* $y$ iff $g_{x_0}$ is continuous for each $x_0 \in \mathbf{R}$.

1. A function $f: X \to \mathbf{R}^2$ is continuous iff both coordinate functions $f_1$ and $f_2$ are continuous.

2. If $g: \mathbf{R}^2 \to Y$ is continuous, then it is continuous in both $x$ and $y$.

3. The converse to part 2 fails. [Let $g(x, y) = xy/(x^2 + y^2)$, with $g(0, 0) = 0$.]

### 7G. *Homeomorphisms within the line*

1. Show that all open intervals in $\mathbf{R}$ are homeomorphic (see 7.10).

2. All bounded closed intervals in $\mathbf{R}$ are homeomorphic.

3. The property that every real-valued continuous function on $X$ assumes its maximum is a topological property. Thus $\mathbf{I}$ is not homeomorphic to $\mathbf{R}$.

### 7H. *Disjoint homeomorphisms*

Suppose $X$ and $Y$ are topological spaces such that $X = \bigcup X_n$ and $Y = \bigcup Y_n$, where $(X_n)$ and $(Y_n)$ are sequences of disjoint open sets in $X$ and $Y$ respectively. If $X_n$ and $Y_n$ are homeomorphic for each $n$, then $X$ and $Y$ are homeomorphic.

### 7I. *Topological properties*

Each of the following expresses a topological property of $X$:

1. $X$ has cardinal number $\aleph$,
2. the topology on $X$ has cardinal number $\aleph$,
3. the topology on $X$ has a base whose cardinal number is $\aleph$,
4. there is in $X$ a set of cardinal $\aleph$ whose closure is $X$,
5. $X$ is metrizable.

Each of the following expresses a property of $X$ which is not a topological property:

6. the topology on $X$ is generated by the metric $\rho$,
7. $X$ is a subset of $\mathbf{R}$.

### 7J. *Retracts*

A continuous function $r$ from a space $X$ onto a subspace $A$ of $X$ is called a *retraction* of $X$ onto $A$ iff $r \mid A$ is the identity map on $A$. When such a retraction exists, $A$ is called a *retract* of $X$.

1. If $A$ is a retract of $X$ and $B \subset X$, $A \cap B$ need not be a retract of $B$.
2. The unit disk is a retract of the plane.
3. If $A$ is a retract of $B$ and $B$ is a retract of $C$, then $A$ is a retract of $C$.

### 7K. *Semicontinuous functions*

A function $f: X \to \mathbf{R}$ is *lower semicontinuous* iff for each $a \in \mathbf{R}$, $f^{-1}(a, \infty)$ is open in $X$. We call $f$ *upper semicontinuous* iff for each $a \in \mathbf{R}$, $f^{-1}(-\infty, a)$ is open in $X$. Note that lower and upper semicontinuity bear no relation to continuity from the left or right for functions of a real variable; we are using the ordering of the range of our functions, not the domain. Most of the results below are stated for lower semicontinuous functions; they have obvious analogs for upper semicontinuous functions.

1. If $f_\alpha$ is a lower semicontinuous real-valued function on $X$ for each $\alpha \in A$, and if $\sup_\alpha f_\alpha(x)$ exists at each $x \in X$, then the function $f(x) = \sup_\alpha f_\alpha(x)$ is lower semicontinuous on $X$.

2. Every continuous function from $X$ to $\mathbf{R}$ is lower semicontinuous. Thus the supremum of a family of continuous functions, if it exists, is lower semicontinuous. Show by an example that "lower semicontinuous" cannot be replaced by "continuous" in the previous sentence.

3. The characteristic function (7A) of a set $A$ in $X$ is lower semicontinuous iff $A$ is open, upper semicontinuous iff $A$ is closed.

4. If $X$ is metrizable and $f$ is a lower semicontinuous function from $X$ to $\mathbf{I}$, then $f$ is the supremum of an increasing sequence of continuous functions on $X$ to $\mathbf{I}$. This provides a partial converse to part 2. [Given $f$, first find a sequence $h_n$ with $0 \le f(x) - h_n(x) \le 1/n$, where $h_n$ is a finite linear combination of characteristic functions of open sets. Then show that every characteristic function, hence each $h_n$, is the supremum of an increasing sequence of continuous functions. Finally, combine these two operations to obtain an increasing sequence of continuous functions whose supremum is $f$.]

5. Let $C^1(\mathbf{I})$ be the family of continuously differentiable real-valued functions on $\mathbf{I}$. For each $f \in C^1(\mathbf{I})$, define

$$L(f) = \int_0^1 \sqrt{1 + \left(\frac{df}{dx}\right)^2}\, dx.$$

Prove that $L$ is lower semicontinuous from $C^1(\mathbf{I})$ to $\mathbf{R}$, if $C^1(\mathbf{I})$ is given the topology of 4F.3.

## 7L. *Linear operators and linear functionals*

If $X$ and $Y$ are normed linear spaces (2J), a *linear operator* from $X$ to $Y$ is a function $\Gamma: X \to Y$ satisfying

a) $\Gamma(x_1 + x_2) = \Gamma(x_1) + \Gamma(x_2)$,
b) $\Gamma(ax) = a\Gamma(x)$,

for all $a$ in $\mathbf{R}$ and $x$, $x_1$, $x_2$ in $X$. A linear operator from $X$ to $\mathbf{R}$ is called a *linear functional.*

A linear operator $\Gamma$ from $X$ to $Y$ is *bounded* iff a constant $M$ exists such that $\|\Gamma(x)\| \le M\|x\|$, for all $x \in X$. (Here we indulge in the common bad habit of failing to use a distinguishing notation for the norms on $X$ and $Y$.) In case $\Gamma$ is a linear functional, the norm we use on $\mathbf{R}$ is $\|x\| = |x|$.

1. A linear operator is bounded iff $\sup\{\|\Gamma(x)\| \mid x \in X, \|x\| = 1\} < \infty$.

2. For a linear operator $\Gamma$ from $X$ to $Y$, the following are equivalent:

a) $\Gamma$ is continuous at some $x_0 \in X$,
b) $\Gamma$ is uniformly continuous on $X$,
c) $\Gamma$ is bounded.

3. Given normed linear spaces $X$ and $Y$, the collection $L(X, Y)$ of all bounded linear operators from $X$ to $Y$ is a linear space under pointwise addition and scalar multiplication $(\Gamma_1 + \Gamma_2)(x) = \Gamma_1(x) + \Gamma_2(x)$, $(a\Gamma)(x) = a \cdot \Gamma(x)$. It becomes a normed linear space if we define $\|\Gamma\| = \sup\{\|\Gamma(x)\| \mid \|x\| = 1\}$ (see part 1).

4. If $Y = \mathbf{R}$, the space $L(X, Y)$ given in part 3, consisting of all bounded linear functionals on $X$, is called the dual space of $X$, denoted $X^*$. Show that, in a natural way, $X \subset (X^*)^*$. [For each $x \in X$, define $F_x$ on $X^*$ by $F_x(\Gamma) = \Gamma(x)$. Show that the mapping $x \to F_x$ is a norm-preserving one–one map of $X$ into $(X^*)^*$.]

The spaces $X$ for which $(X^*)^* = X$ (that is, for which the mapping $x \to F_x$ given in part 4 is onto $(X^*)^*$) are called *reflexive*. In problem 24J, we will see that the norm metric on any dual space is complete, so that dual spaces are examples of "Banach spaces." Thus, only Banach spaces can be reflexive.

See Royden (*Real Analysis*) for a discussion of the representation of dual spaces of some familiar spaces; for example, the dual of $L^p(\mathbf{I})$ is $L^q(\mathbf{I})$, where $1/p + 1/q = 1$.

### 7M. $C(X)$ and $C^*(X)$

For topological spaces $X$ and $Y$, let $C(X, Y)$ denote the collection of all continuous functions from $X$ to $Y$. We will distinguish two special collections: $C(X)$ will be used to denote $C(X, \mathbf{R})$ and $C^*(X)$ will denote the set of all bounded functions from $C(X)$. We can define addition, multiplication and scalar multiplication of functions in $C(X)$ pointwise:

$$(f + g)(x) = f(x) + g(x),$$
$$(f \cdot g)(x) = f(x) \cdot g(x),$$
$$(a \cdot f)(x) = a \cdot f(x), \qquad \text{for} \quad a \in \mathbf{R}.$$

1. If $f$ and $g$ belong to $C(X)$, then so do $f + g$, $f \cdot g$ and $a \cdot f$, for $a \in \mathbf{R}$. If, in addition, $f$ and $g$ are bounded, then so are $f + g$, $f \cdot g$ and $a \cdot f$.

2. $C(X)$ and $C^*(X)$ are algebras over the real numbers. (Consult any book on abstract algebra for the definition of an algebra.)

3. $C^*(X)$ is a normed linear space (2J) with the operations of addition and scalar multiplication given above and the norm $\|f\| = \sup_{x \in X} |f(x)|$.

4. $C(X)$ and $C^*(X)$ are lattices when given the partial order $f \leq g$ iff $f(x) \leq g(x)$ for each $x \in X$. [If $f$, $g$ belong to $C(X)$, so do

$$m(x) = \min \{f(x), g(x)\} \qquad \text{and} \qquad M(x) = \max \{f(x), g(x)\}.]$$

Study of the interaction between the algebraic and lattice properties of $C(X)$ and $C^*(X)$ and the topological properties of $X$ is still actively being carried on. Some questions of importance in this direction are:

i) for what class of spaces is it true that $X$ and $Y$ are homeomorphic iff $C^*(X)$ and $C^*(Y)$ [or $C(X)$ and $C(Y)$] are isomorphic?

ii) how are topological properties of $X$ reflected in algebraic and lattice properties of $C^*(X)$ and $C(X)$?

iii) what properties of a ring $R$ (usually with a lattice structure) will ensure that $R$ is isomorphic with $C(X)$ for some topological space $X$?

An excellent introduction to the study of questions of this sort can be found in the book on rings of functions by Gillman and Jerison.

### 7N. *The group of homeomorphisms*

For any topological space $X$, let $H(X)$ denote the group of homeomorphisms of $X$ onto itself, with composition as the group operation. A central and obvious question is: if $\varphi$ is an isomorphism of $H(X)$ onto $H(Y)$, is there a homeomorphism $T$ of $X$ onto $Y$ such that $\varphi(h) = T \circ h \circ T^{-1}$, for each $h \in H(X)$?

1. $H(X)$ *is* a group, with composition as the operation.

2. Let $X = \mathbf{I}$ and $Y = (0, 1)$ and define $\varphi(h) = h \mid Y$ for each $h \in H(X)$. Then $\varphi$ is an isomorphism of $H(X)$ with $H(Y)$, but there is no homeomorphism of $X$ onto $Y$. [7G].

Part 2 effectively disposes of the question asked in the introduction for general spaces $X$ and $Y$. Affirmative answers are available, however, for suitably restricted classes of spaces. See the notes.

## 8 Product spaces, weak topologies

Our objective now is to define a topology on the Cartesian product of topological spaces, in some natural and useful way. First we extend the notion of Cartesian product to infinite collections of sets. The key to understanding the definition we are about to give is a careful study of Exercise 1D. There we show that the product of a finite collection of sets is, in a natural way, a collection of functions each defined on the indexing set.

**8.1 Definition.** Let $X_\alpha$ be a set, for each $\alpha \in A$. The *Cartesian product* of the sets $X_\alpha$ is the set

$$\prod_{\alpha \in A} X_\alpha = \left\{ x: A \to \bigcup_{\alpha \in A} X_\alpha \;\middle|\; x(\alpha) \in X_\alpha, \text{ for each } \alpha \in A \right\},$$

which we denote simply by $\prod X_\alpha$ if no confusion can result about the indexing set. Thus $\prod X_\alpha$ is a set of functions defined on the indexing set. In practice, the value of $x \in \prod X_\alpha$ at $\alpha$ is usually denoted $x_\alpha$, rather than $x(\alpha)$, and $x_\alpha$ is referred to as the *$\alpha$th coordinate* of $x$. The space $X_\alpha$ is the *$\alpha$th factor space.*

The map $\pi_\beta: \prod X_\alpha \to X_\beta$, defined by $\pi_\beta(x) = x_\beta$, is called the *projection map* of $\prod X_\alpha$ on $X_\beta$, or more simply, the *$\beta$th projection map.*

We need the axiom of choice (1.17) to ensure that the Cartesian product of a nonempty collection of nonempty sets is nonempty. This assertion is, in fact, equivalent to the axiom of choice; see Exercise 8F. If each $X_\alpha$ is nonempty and the axiom of choice *is* assumed, then the $\beta$th projection map carries $\prod X_\alpha$ *onto* $X_\beta$.

**8.2 Examples.** a) If the index set $A$ is finite, say $A = \{1, 2, \ldots, n\}$, it is customary to prescribe the function $x$ in $\prod_{k=1}^n X_k$ by listing its values as an ordered $n$-tuple, $x = (x_1, \ldots, x_n)$. Thus $\prod_{k=1}^n X_k = \{(x_1, \ldots, x_n) \mid x_k \in X_k, k = 1, \ldots, n\}$.

b) The notation in (a) is carried over to the case where $A = \mathbf{N}$. Thus $\prod_{k=1}^\infty X_k = \{(x_1, x_2, \ldots) \mid x_k \in X_k, k = 1, 2, \ldots\}$.

c) If $X_\alpha = X$ for each $\alpha \in A$, then $\prod_{\alpha \in A} X_\alpha$ is just the set $X^A$ of all functions from $A$ to $X$. (Finally, the reason for the notation $X^A$ is clear.) For example, $\mathbf{R}^\mathbf{R}$ is the set of all real-valued functions of a real variable.

d) If $X_\alpha \subset Y_\alpha$ for each $\alpha \in A$, then $\prod X_\alpha \subset \prod Y_\alpha$.

Now suppose $X_\alpha$ is a topological space, for each $\alpha \in A$. We want to define a topology on $\prod_{\alpha \in A} X_\alpha$ which is at the same time natural enough that, for example, the product topology on $\mathbf{R} \times \mathbf{R}$ will be the usual topology on $\mathbf{R}^2$, and tame enough that a number of theorems of the form "if each $X_\alpha$ has property $P$, then so does $\prod X_\alpha$" will remain true.

If naturality were the only requirement, the job would be easy. In fact, after recalling that the open squares in $\mathbf{R}^2$ form a base for the usual topology in $\mathbf{R}^2$, an obvious candidate for a topology on $\prod X_\alpha$ arises. Simply take as a base for such a topology all sets of the form $\prod U_\alpha$, where $U_\alpha$ is an open set in $X_\alpha$, for each $\alpha \in A$. In fact, this procedure gives a valid topology, called the *box topology*, on $\prod X_\alpha$. It satisfies our craving for naturality, but is not much used because it is not tame enough, having an over-abundance of open sets. The definition, given next, of the usual topology used on the product space rectifies this by sharply reducing the number of basis elements.

**8.3 Definition.** The *Tychonoff topology* (or *product topology*) on $\prod X_\alpha$ is obtained by taking as a base for the open sets, sets of the form $\prod U_\alpha$, where

$P$-a) $U_\alpha$ is open in $X_\alpha$, for each $\alpha \in A$,

$P$-b) For all but finitely many coordinates, $U_\alpha = X_\alpha$.

The reader will easily verify that $P$-a could have been replaced by

$P$-a)′ $U_\alpha \in \mathscr{B}_\alpha$, where for each $\alpha$, $\mathscr{B}_\alpha$ is a (fixed) base for the topology of $X_\alpha$.

Also, notice that the set $\prod U_\alpha$, where $U_\alpha = X_\alpha$ except for $\alpha = \alpha_1, \ldots, \alpha_n$, can be written

$$\prod U_\alpha = \pi_{\alpha_1}^{-1}(U_{\alpha_1}) \cap \cdots \cap \pi_{\alpha_n}^{-1}(U_{\alpha_n}).$$

Thus the product topology is precisely that topology which has for a subbase the collection $\{\pi_\alpha^{-1}(U_\alpha) \mid \alpha \in A,\ U_\alpha \text{ open in } X_\alpha\}$. Again, the sets $U_\alpha$ can be restricted to come from some fixed base (in fact, in this case, subbase) in $X_\alpha$.

In case only a finite number of spaces $X_1, \ldots, X_n$ is involved, the product topology on $\prod_{k=1}^n X_k$ coincides with the box topology, so in those cases where we have any intuition to begin with, the product topology will always seem "natural". In particular, $\mathbf{R} \times \mathbf{R} \times \cdots \times \mathbf{R}$ ($n$ times) with the product topology is homeomorphic to $\mathbf{R}^n$.

Hereafter, $\prod X_\alpha$ is always assumed to be endowed with the product (Tychonoff) topology if each $X_\alpha$ is a topological space.

**8.4 Examples.** a) Let $X = \mathbf{R}^{\mathbf{R}}$. Recall that $X$ is the set of all real-valued functions of a real variable. A basic nhood of $f \in X$ in the product topology is obtained by picking a finite subset $\{x_1, \ldots, x_n\}$ of the index set $\mathbf{R}$ and a corresponding set $\{\epsilon_1, \ldots, \epsilon_k\}$ of positive numbers, and letting

$$U(f; x_1, \ldots, x_n; \epsilon_1, \ldots, \epsilon_n) = \{g \in \mathbf{R}^{\mathbf{R}} \mid |g(x_k) - f(x_k)| < \epsilon_k, \text{ for } k = 1, \ldots, n\}.$$

We can obtain a somewhat simpler description of a base in $\mathbf{R}^{\mathbf{R}}$ by letting $F = \{x_1, \ldots, x_n\}$, $\epsilon = \min\{\epsilon_1, \ldots, \epsilon_n\}$ and noting that the nhood

$$U(f, F, \epsilon) = \{g \in \mathbf{R}^{\mathbf{R}} \mid |g(x) - f(x)| < \epsilon \text{ for } x \in F\}$$

is contained in $U(f; x_1, \ldots, x_n; \epsilon_1, \ldots, \epsilon_n)$, so that the sets $U(f, F, \epsilon)$, as $F$ ranges through all finite subsets of $\mathbf{R}$ and $\epsilon$ ranges through all positive numbers, form a nhood base at $f$ (see Fig. 8.1).

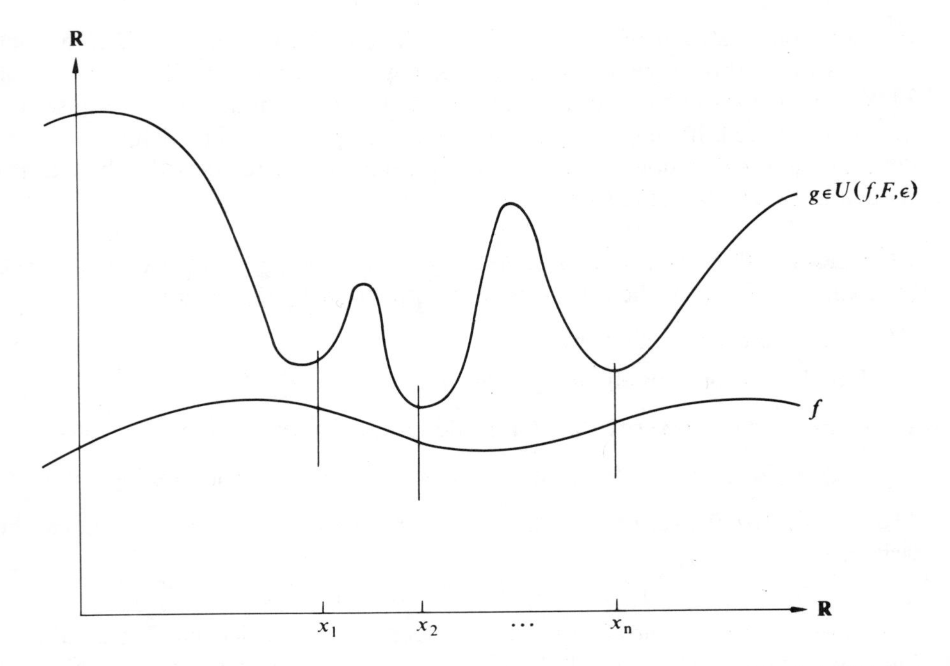


**Fig. 8.1** $U(f, F, \epsilon)$

b) For each $\alpha \in A$, let $X_\alpha$ be a discrete space of more than one point. Then $\prod X_\alpha$ will be a discrete space if and only if $A$ is finite.

c) If $\mathbf{S}^1$ is the unit circle in $\mathbf{R}^2$, then $\mathbf{S}^1 \times \mathbf{I}$ is a cylinder and $\mathbf{S}^1 \times \mathbf{S}^1$ is a *torus* (Fig. 8.2).

d) If $Y_\alpha \subset X_\alpha$ for each $\alpha \in A$, then the product topology on $\prod Y_\alpha$ coincides with its topology as a subspace of $\prod X_\alpha$.

**8.5 Definition.** If $X$ and $Y$ are topological spaces and $f: X \to Y$, we call $f$ an *open (closed) map* iff for each open (closed) set $A$ in $X$, $f(A)$ is an open (closed) set in $Y$.

If $f$ is one–one and onto, then $f$ is open iff $f$ is closed iff $f^{-1}$ is continuous. Thus a one–one onto map $f$ is a homeomorphism iff it is continuous and open iff it is continuous and closed.

In general, an open map need not be closed and vice versa; see 8A, 9C.

***8.6 Theorem.*** *The $\beta$th projection map $\pi_\beta: \prod X_\alpha \to X_\beta$ is continuous and open, but need not be closed.*

*Proof.* Left as Exercise 8A. ■

***8.7 Theorem.*** *The Tychonoff topology is the weakest topology on $\prod X_\alpha$ for which each projection $\pi_\beta$ is continuous.*

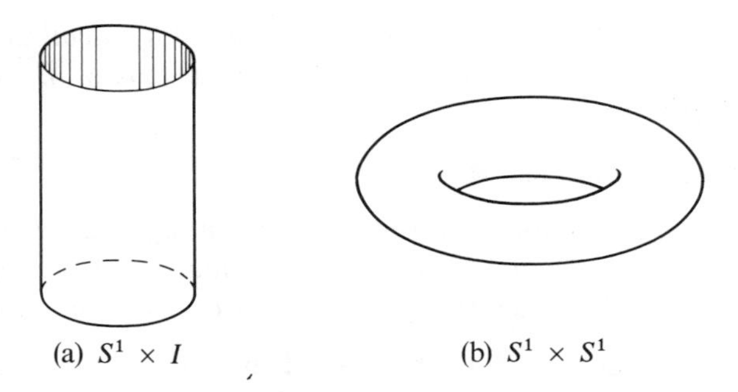

(a) $S^1 \times I$ (b) $S^1 \times S^1$

**Figure 8.2**

*Proof.* If $\tau$ is any topology on the product in which each projection is continuous, then for each $\beta$, if $U_\beta$ is open in $X_\beta$, $\pi_\beta^{-1}(U_\beta) \in \tau$. Consequently, the members of a subbase for the Tychonoff topology all belong to $\tau$, and hence the Tychonoff topology is contained in $\tau$. ∎

***8.8 Theorem.*** *A map $f: X \to \prod X_\alpha$ is continuous iff $\pi_\alpha \circ f$ is continuous for each $\alpha \in A$.*

*Proof.* Necessity of the composition condition is clear since the composition of continuous maps is continuous. Conversely, suppose $\pi_\alpha \circ f$ is continuous for each $\alpha \in A$. The sets of the form $\pi_\alpha^{-1}(U_\alpha)$, $\alpha \in A$ and $U_\alpha$ open in $X_\alpha$, form a subbase for the topology on $\prod X_\alpha$. But $f^{-1}(\pi_\alpha^{-1}(U_\alpha)) = (\pi_\alpha \circ f)^{-1}(U_\alpha)$. Thus the inverse images by $f$ of these subbasic open sets are open in $X$, by continuity of $\pi_\alpha \circ f$. This suffices to show $f$ is continuous. ∎

The previous two theorems form the penultimate justification for our choice of the Tychonoff topology on $\prod X_\alpha$ over the box topology. Directly or indirectly, these results lie at the heart of most useful investigations into the properties of product spaces. As is often the case, a theorem (in this case, 8.7) with a desirable conclusion becomes the basis for a definition.

**8.9 Definition.** Let $X$ be a set and $X_\alpha$ a topological space with $f_\alpha: X \to X_\alpha$, for each $\alpha \in A$. The *weak topology* induced on $X$ by the collection $\{f_\alpha \mid \alpha \in A\}$ of functions is the smallest topology on $X$ making each $f_\alpha$ continuous. It evidently is that topology on $X$ for which the sets $f_\alpha^{-1}(U_\alpha)$, for $\alpha \in A$ and $U_\alpha$ open in $X_\alpha$, form a subbase.

By Theorem 8.7 the product topology on $\prod_{\alpha \in A} X_\alpha$ is the weak topology induced by the collection $\{\pi_\alpha \mid \alpha \in A\}$ of projections. Moreover, Theorem 8.8 carries over to any weak topology, without essential change in the mechanics of the proof.

**8.10 Theorem.** *If $X$ has the weak topology induced by a collection $\{f_\alpha \mid \alpha \in A\}$ of functions $f_\alpha: X \to X_\alpha$, then $f: Y \to X$ is continuous iff $f_\alpha \circ f$ is continuous for each $\alpha \in A$.*

*Proof.* Mimic the proof of 8.8. ■

It is one of the remarkable and fruitful results in topology that, with a simple extra condition on the generating collection of maps, any space with a weak topology can be embedded as a subspace of the product of the range spaces.

**8.11 Definition.** If, for each $\alpha \in A$, $f_\alpha: X \to X_\alpha$, then the *evaluation map* $e: X \to \prod X_\alpha$ induced by the collection $\{f_\alpha \mid \alpha \in A\}$ is defined as follows: for each $x \in X$, $[e(x)]_\alpha = f_\alpha(x)$. That is, for $x \in X$, $e(x)$ is the point in $\prod X_\alpha$ whose $\alpha$th coordinate is $f_\alpha(x)$ for each $\alpha \in A$.

A collection $\{f_\alpha \mid \alpha \in A\}$ of functions on $X$ will be said to *separate points* in $X$ iff whenever $x \neq y$ in $X$, then for some $\alpha \in A$, $f_\alpha(x) \neq f_\alpha(y)$.

**8.12 Theorem.** *For each $\alpha \in A$, let $f_\alpha: X \to X_\alpha$. Then the evaluation map $e: X \to \prod X_\alpha$ is an embedding iff $X$ has the weak topology given by the functions $f_\alpha$ and the collection $\{f_\alpha \mid \alpha \in A\}$ separates points in $X$.*

*Proof.* The heart of the proof of this theorem lies in the observation that, for each $\alpha \in A$, $\pi_\alpha \circ e = f_\alpha$.

Now suppose $e$ is an embedding of $X$ into $\prod X_\alpha$. Then $e(X)$ has the weak topology induced by the restricted projections [see Exercise 8H]. Hence, since $e$ is a homeomorphism, it is clear that $X$ has the weak topology induced by the functions $\pi_\alpha \circ e = f_\alpha$. Moreover, if $x \neq y$, in $X$, then $e(x) \neq e(y)$ and hence $[e(x)]_\alpha \neq [e(y)]_\alpha$, that is, $f_\alpha(x) \neq f_\alpha(y)$, for some $\alpha \in A$. Thus the collection $\{f_\alpha \mid \alpha \in A\}$ separates points.

Now suppose the topology on $X$ is the weak topology induced by the functions $f_\alpha$ and that the collection $\{f_\alpha \mid \alpha \in A\}$ separates points in $X$. For each $\alpha \in A$, $\pi_\alpha \circ e = f_\alpha$ is continuous. Thus, by Theorem 8.8, $e$ is continuous. If $x \neq y$ in $X$, then for some $\alpha \in A$, $f_\alpha(x) \neq f_\alpha(y)$, i.e., $[e(x)]_\alpha \neq [e(y)]_\alpha$, and thus $e(x) \neq e(y)$. Hence $e$ is one–one. Finally, we will show $e$ is an open map; i.e., if $U$ is open in $X$, then $e(U)$ is open in $e(X)$. Since $e$ is one–one, it suffices to show $e(U)$ is open whenever $U$ is a subbasic open set. Hence we assume $U$ is of the form $f_\alpha^{-1}(V)$ for some $\alpha \in A$ and some open set $V$ in $X_\alpha$. But then

$$U = [(\pi_\alpha \mid e(X)) \circ e]^{-1}(V) = e^{-1}[(\pi_\alpha \mid e(X))^{-1}(V)]$$

and hence $e(U) = [\pi_\alpha \mid e(X)]^{-1}(V) = \pi_\alpha^{-1}(V) \cap e(X)$ which is an open set in $e(X)$, since $\pi_\alpha^{-1}(V)$ is open in $\prod X_\alpha$. (The last argument looks a lot nicer if you just carry the fact that $\pi_\alpha$ should be restricted to $e(X)$ in your head instead of writing it out.) ■

The following problem is, in various forms and with occasional modifications, one of the most important and often investigated questions in topology and related areas: given a space $X$ and a property $\mathscr{P}$ of spaces, can $X$ be embedded in a larger space $Y$ having property $\mathscr{P}$? The theorem just proved forms the essential core of a great many constructions intended to deal with such questions. The best known example, the Stone–Čech compactification $\beta X$ of a Tychonoff space $X$ (see Section 19) is typical of the use of 8.12 in this way.

In case $X$ already has a topology and we wish to know whether or not this topology is the weak topology given by a certain collection $\{f_\alpha \mid \alpha \in A\}$ of continuous functions on $X$, there is often a pleasant alternative to verifying that the sets $f_\alpha^{-1}(V)$, for $\alpha \in A$ and $V$ open in $X_\alpha$, form a subbase for the existing topology.

**8.13 Definition.** A collection $\{f_\alpha \mid \alpha \in A\}$ of functions on a space $X$ (to spaces $X_\alpha$) is said to *separate points from closed sets* iff whenever $B$ is closed in $X$ and $x \notin B$, then for some $\alpha \in A$, $f_\alpha(x) \notin \overline{f_\alpha(B)}$.

***8.14 Theorem.*** *A collection $\{f_\alpha \mid \alpha \in A\}$ of continuous functions on a topological space $X$ separates points from closed sets in $X$ iff the sets $f_\alpha^{-1}(V)$, for $\alpha \in A$ and $V$ open in $X_\alpha$, form a base for the topology on $X$.*

*Proof.* Exercise 8B. ■

***8.15 Corollary.*** *If $\{f_\alpha \mid \alpha \in A\}$ is a collection of continuous functions on a topological space $X$ which separates points from closed sets, then the topology on $X$ is the weak topology induced by the maps $f_\alpha$.*

Whenever one-point sets in $X$ are closed, a collection of functions which separates points from closed sets will separate points. A space is a *$T_1$-space* (see Section 13) iff one-point sets are closed.

***8.16 Theorem.*** *If $X$ is a $T_1$-space and $\{f_\alpha \mid \alpha \in A\}$ is a collection of continuous functions on $X$ (to spaces $X_\alpha$) which separates points from closed sets, then the evaluation $e: X \to \prod X_\alpha$ is an embedding.*

*Proof.* This is a direct consequence of 8.15, the remark preceding this theorem and 8.12. ■

## Problems

8A. *Projection maps*

1. The $\beta$th projection map $\pi_\beta$ is continuous and open. The projection $\pi_1: \mathbf{R}^2 \to \mathbf{R}$ is not closed.

2. Show that the projection of $\mathbf{I} \times \mathbf{R}$ onto $\mathbf{R}$ is a closed map.

8B. *Separating points from closed sets*

1. If $f_\alpha$ is a map of $X$ to $X_\alpha$ for each $\alpha \in A$, then $\{f_\alpha \mid \alpha \in A\}$ separates points from closed sets in $X$ iff $\{f_\alpha^{-1}(V) \mid \alpha \in A,\ V \text{ open in } X_\alpha\}$ is a base for the topology on $X$.

2. If $X$ has the weak topology induced by a collection of maps which separates points, this collection of maps need not separate points from closed sets.

## 8C. *Products are associative and commutative*

1. If $\{A_\lambda \mid \lambda \in \Lambda\}$ is a partition of the set $A$ (into disjoint subsets whose union is $A$), and $X_\alpha$ is a topological space for each $\alpha \in A$, then $\prod_{\lambda \in \Lambda} (\prod_{\alpha \in A_\lambda} X_\alpha)$ is homeomorphic to $\prod_{\alpha \in A} X_\alpha$.

2. If $\varphi$ is a one–one map of $A$ onto $B$ and for each $\alpha \in A$, $X_\alpha$ is homeomorphic to $Y_{\varphi(\alpha)}$, then $\prod_{\alpha \in A} X_\alpha$ is homeomorphic to $\prod_{\beta \in B} Y_\beta$.

## 8D. *Closure and interior in products*

Let $X$ and $Y$ be topological spaces containing subsets $A$ and $B$, respectively. In the product space $X \times Y$:

1. $(A \times B)^\circ = A^\circ \times B^\circ$.

2. $\overline{(A \times B)} = \bar{A} \times \bar{B}$.

3. Part 2 can be extended to infinite products, while part 1 can be extended only to finite products.

4. $\text{Fr}\,(A \times B) = [\bar{A} \times \text{Fr}\,(B)] \cup [\text{Fr}\,(A) \times \bar{B}]$.

5. If $X_\alpha$ is a nonempty topological space and $A_\alpha \subset X_\alpha$, for each $\alpha \in A$, then $\prod A_\alpha$ is dense (see 7C) in $\prod X_\alpha$ iff $A_\alpha$ is dense in $X_\alpha$, for each $\alpha$.

## 8E. *Miscellaneous facts about product spaces*

Let $X_\alpha$ be a nonempty topological space for each $\alpha \in A$, and let $X = \prod X_\alpha$.

1. If $V$ is a nonempty open set in $X$, then $\pi_\alpha(V) = X_\alpha$ for all but finitely many $\alpha \in A$.

2. If $b_\alpha$ is a fixed point in $X_\alpha$, for each $\alpha \in A$, then $X'_{\alpha_0} = \{x \in X \mid x_\alpha = b_\alpha$ whenever $\alpha \neq \alpha_0\}$ is homeomorphic to $X_{\alpha_0}$.

3. If $b_\alpha$ is a fixed point in $X_\alpha$, for each $\alpha \in A$, then $A = \{x \in X \mid x_\alpha = b_\alpha$ except for finitely many $\alpha \in A\}$ is a dense set in $X$; i.e., $\text{Cl}_X\, A = X$.

## 8F. *Products and the axiom of choice*

1. Show that the axiom of choice is equivalent to the assertion that the product of a nonempty collection of nonempty sets is nonempty.

2. Assuming the axiom of choice, show that each projection map is onto if each factor space is nonempty.

## 8G. *The box topology*

Let $X_\alpha$ be a topological space for each $\alpha \in A$.

1. In $\prod X_\alpha$, the sets of the form $\prod U_\alpha$, where $U_\alpha$ is open in $X_\alpha$ for each $\alpha \in A$, form a base for a topology.

2. What do nhoods of $f \in \mathbf{R}^{\mathbf{R}}$ look like in the box topology? [see 8.4(1)]. Compare with 4F3.

3. Work out formulas for the closure and interior of sets in a box product, similar to those given in 8D.

8H. *Weak topologies on subspaces*

Let $X$ have the weak topology induced by a collection of maps $f_\alpha: X \to X_\alpha$, for $\alpha \in A$.

1. If each $X_\alpha$ has the weak topology given by a collection of maps $g_{\alpha\lambda}: X_\alpha \to Y_{\alpha\lambda}$, for $\lambda \in \Lambda_\alpha$, then $X$ has the weak topology given by the maps $g_{\alpha\lambda} \circ f_\alpha: X \to Y_{\alpha\lambda}$, for $\alpha \in A$ and $\lambda \in \Lambda_\alpha$.

2. Any $B \subset X$ has the weak topology induced by the maps $f_\alpha | B$. [Any $B \subset X$ has the weak topology induced by the inclusion map $j: B \to X$.]

8I. *Weak topologies and the lattice of topologies*

Let $\{\tau_\alpha \mid \alpha \in A\}$ be a family of topologies on a fixed set $X$ and denote by $X_\alpha$ the space consisting of the set $X$ with the topology $\tau_\alpha$. The identity function from the set $X$ to the space $X_\alpha$ will be denoted $i_\alpha$.

1. The weak topology induced on $X$ by the maps $i_\alpha$ is the supremum $\tau$ of the topologies $\tau_\alpha$ (see 3G).

2. $(X, \tau)$ is homeomorphic to the diagonal $\Delta$ in the product space $\prod X_\alpha$. (Note: $\Delta = \{x \in \prod X_\alpha \mid x_\alpha = x_\beta \text{ for all } \alpha, \beta\}$.)

The corresponding theorems for the infimum of the topologies $\tau_\alpha$ are given in Exercise 9I.

8J. *Homeomorphic products*

Exhibit spaces $X$, $Y$ and $Z$ such that $X \times Y$ is homeomorphic to $X \times Z$, but $Y$ is not homeomorphic to $Z$.

It is also true that there are nonhomeomorphic spaces $X$ and $Y$ such that $X \times X$ and $Y \times Y$ are homeomorphic (see notes).

See also 30F.

## 9 Quotient spaces

Dual to the notion of the weak topology induced on $X$ by a collection of maps $f_\alpha: X \to X_\alpha$, which is the weakest topology making all these maps continuous, we have the notion of the *strong topology* induced on $Y$ by a collection of maps $g_\alpha: Y_\alpha \to Y$, which is the strongest topology on $Y$ making all these maps continuous. In the particular case when there is only one map $g: X \to Y$, the resulting strong topology on $Y$ is called the *quotient topology* induced on $Y$ by $g$. We will be solely concerned in this section with investigating three distinct but equivalent ways of viewing quotient spaces, leaving discussion of strong topologies to Exercise 9H (where we show that quotient spaces play a role for strong topologies similar to that played by product spaces relative to weak topologies).

**9.1 Definition.** If $X$ is a topological space, $Y$ is a set and $g: X \to Y$ is an onto mapping, then the collection $\tau_g$ of subsets of $Y$ defined by

$$\tau_g = \{G \subset Y \mid g^{-1}(G) \text{ is open in } X\}$$

is a topology on $Y$, called the *quotient topology* induced on $Y$ by $g$. When $Y$ is given some such quotient topology, it is called a *quotient space* of $X$, and the inducing map $g$ is called a *quotient map*.

It is clear that the quotient topology induced on $Y$ by $g$ is the largest topology on $Y$ making $g$ continuous. We should also note that the quotient topology can be completely described as follows: $F \subset Y$ is closed in the quotient topology induced by $g$ iff $g^{-1}(F)$ is closed in $X$.

The first and obvious question we must deal with is: under what conditions on $g$ will a preassigned topology $\tau$ on $Y$ be identical to the quotient topology $\tau_g$ induced by $g$? It is obvious that continuity of $g$ is necessary, to make $\tau \subset \tau_g$. Thus we search for additional conditions to force $\tau \supset \tau_g$. In fact, the conditions we need were given in Definition 8.5.

**9.2 Theorem.** *If $X$ and $Y$ are topological spaces and $f: X \to Y$ is continuous and either open or closed, then the topology $\tau$ on $Y$ is the quotient topology $\tau_f$.*

*Proof.* Suppose $f$ is continuous and open. Since $\tau_f$ is the largest topology making $f$ continuous, $\tau \subset \tau_f$. But if $U \in \tau_f$, then by definition of $\tau_f$, $f^{-1}(U)$ is open in $X$. Now $f$ is open as a map to $(Y, \tau)$, so $f[f^{-1}(U)] = U$ belongs to $\tau$. Thus $\tau_f \subset \tau$ and this establishes equality.

The reader can verify the theorem if $f$ is continuous and closed. ∎

**9.3 Example.** Let $X = [0, 2\pi]$ with its usual topology,

$$Y = \{(x, y) \in \mathbf{R}^2 \mid x^2 + y^2 = 1\}$$

with its usual topology, and define $f: X \to Y$ by $f(x) = (\cos x, \sin x)$. Then $f$ is continuous and closed, so the unit circle with its usual topology is a quotient space of $[0, 2\pi]$.

Just as 8.10 was the central useful fact about weak topologies, the following theorem states the fundamental result about quotient topologies.

**9.4 Theorem.** *Let $Y$ have the quotient topology induced by a map $f$ of $X$ onto $Y$. Then an arbitrary map $g: Y \to Z$ is continuous iff $g \circ f: X \to Z$ is continuous.*

$$\begin{array}{ccc} X & \xrightarrow{f} & Y \\ & {\scriptstyle g \circ f} \searrow \quad \swarrow {\scriptstyle g} & \\ & Z & \end{array}$$

*Proof.* Necessity is trivial, since the composition of continuous maps is continuous.

To prove sufficiency, suppose $g \circ f$ is continuous, and let $U$ be open in $Z$. Then $(g \circ f)^{-1}(U) = f^{-1}[g^{-1}(U)]$ is open in $X$, so by definition of the quotient topology on $Y$, $g^{-1}(U)$ is open in $Y$. Hence $g$ is continuous. ∎

There is another approach to quotient spaces which yields a great deal of insight. Essentially, we can regard any quotient space of $X$ as a certain collection of subsets of $X$ with a naturally defined topology. The best approach is to view the necessary construction abstractly, then show it can be used to describe quotient spaces.

**9.5 Definition.** Let $X$ be a topological space. A *decomposition* $\mathscr{D}$ of $X$ is a collection of disjoint subsets of $X$ whose union is $X$. If a decomposition $\mathscr{D}$ is endowed with the topology in which $\mathscr{F} \subset \mathscr{D}$ is open iff $\bigcup \{F \mid F \in \mathscr{F}\}$ is open in $X$, then $\mathscr{D}$ is referred to as a *decomposition space* of $X$. You are asked to show that this does give a topology on $\mathscr{D}$ in 9B.

Define a map $P$ of $X$ onto $\mathscr{D}$ by letting $P(x)$, for $x \in X$, be the element of $\mathscr{D}$ containing $x$. $P$ is called the *natural map* (or *decomposition map*) of $X$ onto $\mathscr{D}$.

The next theorem says that every decomposition space is a quotient space; the theorem following that says that every quotient space is (homeomorphic to) a decomposition space.

**9.6 Theorem.** *The topology on a decomposition space $\mathscr{D}$ of $X$ is the quotient topology induced by the natural map $P: X \to \mathscr{D}$.*

*Proof.* See Exercise 9B. ■

**9.7 Theorem.** *If $Y$ has the quotient topology induced by $f: X \to Y$, then $Y$ is homeomorphic to the decomposition space $\mathscr{D}$ whose elements are the sets $f^{-1}(y)$, $y \in Y$, under a homeomorphism $h: Y \to \mathscr{D}$ such that $h \circ f$ is the natural map $P$ of $X$ onto $\mathscr{D}$.* (We might paraphrase the situation by saying $f: X \to Y$ is "isomorphic" to $P: X \to \mathscr{D}$ under the isomorphism $h$.)

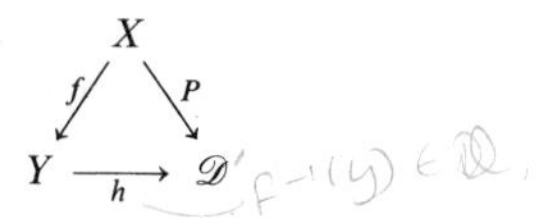


*Proof.* With the hint that $h$ is defined in the obvious way, that is, $h(y) = f^{-1}(y)$, we leave the details of this proof to Exercise 9B. ■

The natural map $P: X \to \mathscr{D}$ associated with a particular decomposition space $\mathscr{D}$ is, as noted in 9.6, a quotient map. It is often of interest, in investigations revolving around decomposition spaces, to know that $P$ is, in fact, closed. To state the basic result giving conditions on $\mathscr{D}$ which will make $P$ closed, we introduce the following definition.

**9.8 Definition.** An open set $V$ in a topological space $X$ is *saturated* relative to a given decomposition $\mathscr{D}$ of $X$ iff $V$ is a union of elements of $\mathscr{D}$ (i.e., iff $V = P^{-1}(W)$ for some open set $W$ in $\mathscr{D}$). A decomposition $\mathscr{D}$ is *upper semicontinuous* iff for each $F \in \mathscr{D}$ and each open set $U$ in $X$ containing $F$, there is some saturated open set $V$ in $X$ with $F \subset V \subset U$.

**9.9 Theorem.** *The natural map $P$ associated with a decomposition space $\mathscr{D}$ of $X$ is closed iff $\mathscr{D}$ is upper semicontinuous.*

*Proof.* Suppose $P$ is closed. Let $F \in \mathscr{D}$ and let $U$ be an open set in $X$ containing $F$. Then $P(X - U)$ is a closed set in $\mathscr{D}$, so $P^{-1}[P(X - U)]$ is a closed set in $X$ which is a union of elements of $\mathscr{D}$. Then clearly $V = X - P^{-1}[P(X - U)]$ is a saturated open set in $X$ and, without much effort, $F \subset V \subset U$.

Conversely, suppose $\mathscr{D}$ is upper semicontinuous, and let $K$ be a closed subset of $X$. To show $P(K)$ is closed, let $F \in \mathscr{D} - P(K)$. Then $F \subset X - K$, so there is a saturated open set $V$ with $F \subset V \subset X - K$, by upper semicontinuity. But then $P(V)$ is an open set and $F \in P(V) \subset \mathscr{D} - P(K)$, which establishes that $P(K)$ is closed in $\mathscr{D}$. ■

**9.10 Corollary.** *A quotient map $f: X \to Y$ is closed iff $\{f^{-1}(y) \mid y \in Y\}$ is an upper semicontinuous decomposition of $X$.*

Before moving on to some of the examples which typify the importance of quotient constructions in topology, it is convenient to introduce one last way of regarding quotient spaces. It requires nothing but a definition, but represents probably the most popular way of presenting quotient spaces.

**9.11 Definition.** If $\sim$ is an equivalence relation on the topological space $X$, then the *identification space* $X/\sim$ is defined to be the decomposition space $\mathscr{D}$ whose elements are the equivalence classes for $\sim$.

**9.12 Examples.** a) In 9.3, we saw that the unit circle is a quotient space of $[0, 2\pi]$. Viewed as a decomposition space, the appropriate elements of the decomposition are the one point sets $\{x\}$ for which $0 < x < 2\pi$ together with the set $\{0, 2\pi\}$. As an identification space, it is obtained through the equivalence relation $0 \sim 2\pi$ and otherwise $x \sim y$ iff $x = y$. Clearly, the last description is the neatest. We can (and do) simply say "the unit circle is obtained from $[0, 2\pi]$ by identifying endpoints."

b) Consider the square $[0, 2\pi] \times [0, 2\pi]$. If we identify each point $(0, x)$ with the point $(2\pi, x)$, the resulting identification space is homeomorphic to the cylinder $\mathbf{S}^1 \times [0, 2\pi]$ (Fig. 9.1).

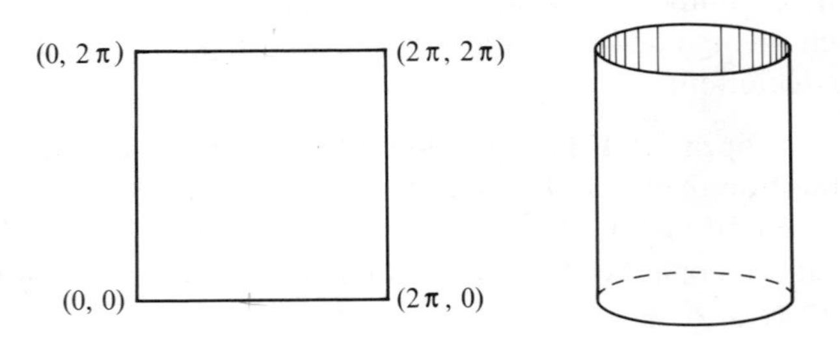


**Figure 9.1**

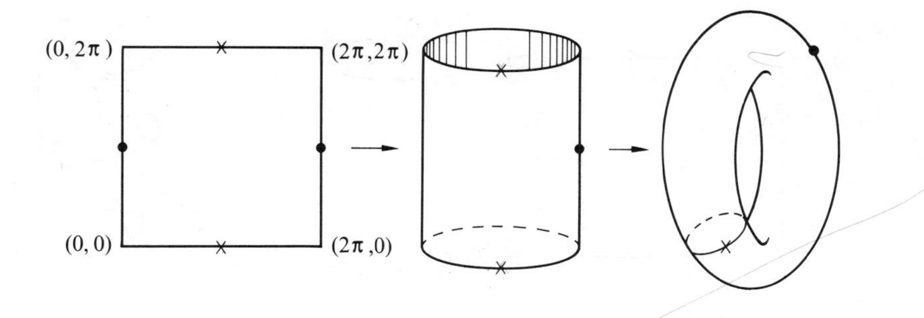


Figure 9.2

The corresponding quotient map of $[0, 2\pi] \times [0, 2\pi]$ which gives the cylinder $\mathbf{S}^1 \times [0, 2\pi]$ as a quotient space is $f(x, y) = ((\cos x, \sin x), y)$.

c) Again consider the square $[0, 2\pi] \times [0, 2\pi]$. This time, identify each point $(0, y)$ with the point $(2\pi, y)$ and *also* identify each point $(x, 0)$ with the point $(x, 2\pi)$. Intuitively, it is clear that the resulting identification space is what one obtains by first rolling the square to obtain a cylinder, as we did in (b), then matching the ends of the cylinder to obtain a torus (Fig. 9.2). More formally, the quotient map $f(x, y) = ((\cos x, \sin x), (\cos y, \sin y))$ gives the torus $\mathbf{S}^1 \times \mathbf{S}^1$ as a quotient space of $[0, 2\pi] \times [0, 2\pi]$.

We should mention here that it is clear that any square will produce a cylinder with one pair of sides identified and will give a torus with two pairs of sides identified, as above. The reason we chose $[0, 2\pi] \times [0, 2\pi]$ is obvious.

d) If we again consider $[0, 2\pi] \times [0, 2\pi]$, but now identify points $(x, 0)$ with points $(2\pi - x, 2\pi)$ the result is a twisted strip, called the *Moebius strip* (Fig. 9.3). It has several interesting properties most of which require combinatorial or algebraic methods to elucidate.

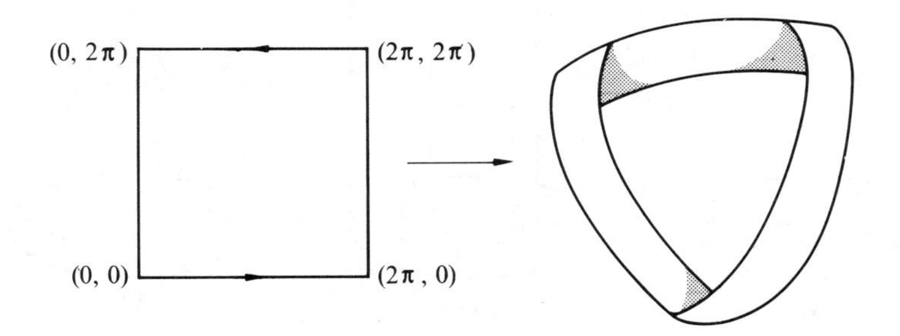


Figure 9.3

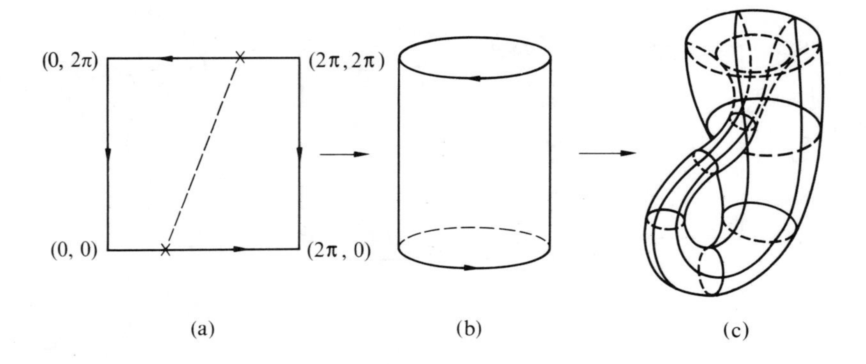


**Figure 9.4**

e) Once more we consider $[0, 2\pi] \times [0, 2\pi]$. Again the points $(0, y)$ are identified with the points $(2\pi, y)$; now, however, we identify each point $(x, 0)$ with the point $(2\pi - x, 2\pi)$. This can be conveniently represented by arrows, as in Fig. 9.4(a). The result, shown in Fig. 9.4(c), cannot be faithfully represented in 3-dimensional space without self-intersection. It is the so-called *Klein bottle*. It is a higher-dimensional relative of the Moebius strip.

f) Given any topological space $X$, we can describe two constructions. We obtain the *cone*, $\Lambda X$, over $X$ by identifying all the points $(x, 1)$ in $X \times \mathbf{I}$ with a single point (Fig. 9.5). The *suspension*, $\sum X$, of $X$ is obtained by identifying all the points $(x, 1)$ in $X \times [-1, 1]$ to a single point, and all the points $(x, -1)$ to another point (Fig. 9.6).

We conclude this section by providing two more methods for generating new spaces from old. The first is an obvious construction, based on the idea of "pulling apart" a collection of spaces to provide a topology on their union.

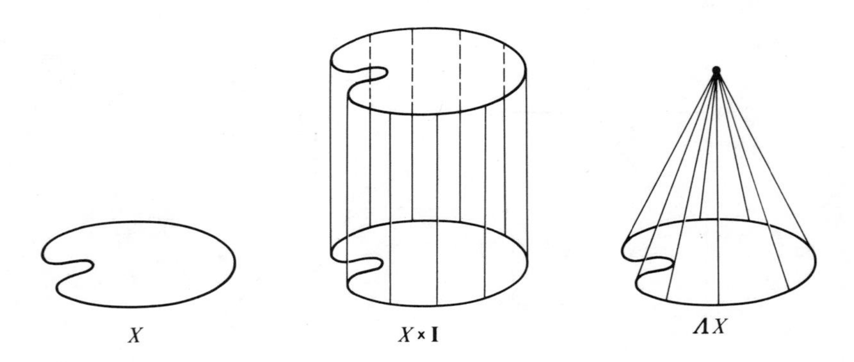


**Figure 9.5**

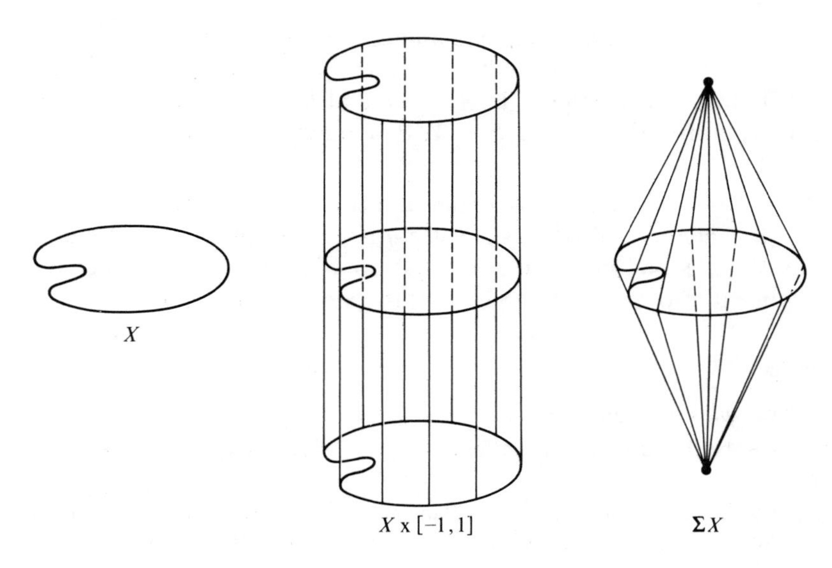


**Figure 9.6**

**9.13 Definition.** Let $X_\alpha$ be a topological space, for each $\alpha \in A$, and let

$$X_\alpha^* = \{(x, \alpha) \mid x \in X_\alpha\},$$

with the topology being defined on $X_\alpha^*$ in the obvious way, to make it homeomorphic to $X_\alpha$. The collection of spaces $X_\alpha^*$ is different from the collection of spaces $X_\alpha$, then, only in that $X_\alpha^* \cap X_\beta^* = \varnothing$ if $\alpha \neq \beta$.

Now define a topology on $X = \bigcup_{\alpha \in A} X_\alpha^*$ as follows: $U \subset X$ is open iff $U \cap X_\alpha^*$ is open for each $\alpha \in A$. The resulting space $X$ is called the *disjoint union* (or *free union*) of the spaces $X_\alpha$ and is denoted $\sum_{\alpha \in A} X_\alpha$, or just $\sum X_\alpha$. If only two spaces $X$ and $Y$ are involved, we write $X + Y$ for the disjoint union of $X$ and $Y$.

In practice, we almost always drop the distinction between $X_\alpha$ and $X_\alpha^*$, and treat $X_\alpha$ itself as a subset of the disjoint union. This will never cause any trouble; often, in fact, the spaces $X_\alpha$ will be disjoint to begin with.

We can now employ the construction just accomplished to provide one of the important and interesting ways of generating new spaces.

**9.14 Definition.** Let $X$ and $Y$ be disjoint topological spaces, with $f$ a continuous map of a closed subset $A$ of $X$ into $Y$. For each $p \in f(A)$, consider the set $A_p = \{p\} \cup f^{-1}(p)$ and form the quotient of $X + Y$ obtained by identifying the points of $A_p$ for each $p \in f(A)$. The resulting space is denoted by $X +_f Y$ and we say $X$ has been *attached to* $Y$ by $f$. The decomposition map of $X + Y$ onto $X +_f Y$ will be denoted $q$. For examples of attachings, see Exercise 9L.

***9.15 Theorem.*** a) *$q \mid Y$ is a homeomorphism and $q(Y)$ is closed in $X +_f Y$,*

b) *$q \mid (X - A)$ is a homeomorphism and $q(X - A)$ is open in $X +_f Y$.*

*Proof.* a) $q \mid Y$ is certainly one–one and continuous. Let $F$ be a closed subset of $Y$. Then $F$ is a closed subset of $X + Y$ and $F = q^{-1}[q(F)]$. Since $q$ is a quotient map, $q(F)$ must thus be closed in $X +_f Y$ and hence in $q(Y)$. Hence $q \mid Y$ is a homeomorphism. Also, letting $F = Y$, this argument shows that $q(Y)$ is closed in $X +_f Y$.

b) $q \mid (X - A)$ is certainly one–one and continuous. Let $G$ be an open subset of $X - A$. Then $q^{-1}[q(G)] = G$. Since $q$ is a quotient map, $q(G)$ must then be open in $X +_f Y$ and hence in $q(X - A)$, so $q \mid (X - A)$ is a homeomorphism. The argument also shows that $q(X - A)$ is open in $X +_f Y$. ■

## Problems

### 9A. *Examples of quotient spaces*

1. Let $\sim$ be the equivalence relation $(x_1, x_2) \sim (y_1, y_2)$ iff $x_2 = y_2$, on $\mathbf{R}^2$. Then $\mathbf{R}^2/\sim$ is homeomorphic to $\mathbf{R}$.

2. Let $\mathscr{D}$ be the decomposition of the plane into concentric circles about the origin. Prove that $\mathscr{D}$ is homeomorphic to $\{x \in \mathbf{R} \mid x \geq 0\}$; show directly that $\mathscr{D}$ is upper semicontinuous.

3. Let $\sim$ be the equivalence relation $x \sim y$ iff $x$ and $y$ are diametrically opposite, on $\mathbf{S}^1$. Then $\mathbf{S}^1/\sim$ is homeomorphic to $\mathbf{S}^1$. Is the corresponding result for $\mathbf{S}^2$ true?

### 9B. *Quotients versus decompositions*

1. The process given in 9.5 for forming the topology on a decomposition space does define a topology.

2. The topology on a decomposition space $\mathscr{D}$ of $X$ is the quotient topology induced by the natural map $P: X \to \mathscr{D}$. (See 9.6.)

### 9C. *Open and closed maps*

1. An open continuous map need not be closed, even if it is onto. [Consider the map $\pi_1$ of $\mathbf{R}^2$ onto $\mathbf{R}$ defined by $\pi_1(x_1, x_2) = x_1$.]

2. A closed continuous map need not be open, even if it is onto. [Consider the map of $[0, 2\pi]$ onto the unit circle given in 9.3.]

3. State and prove an analog to 9.9 for open maps, by appropriately defining "lower semicontinuous decomposition".

### 9D. *Quotients of subspaces and subspaces of quotients*

If $\mathscr{D}$ is a decomposition of $X$, then $\mathscr{D}$ induces an obvious decomposition $\mathscr{D}_A$ on any subset $A$ of $X$.

1. It is not, in general, true that $\mathscr{D}_A$ is homeomorphic to

$$\mathscr{D} \mid A = \{y \in \mathscr{D} \mid A \cap y \neq \varnothing \text{ in } X\}.$$

[Let $\mathscr{D}$ be the set of vertical lines in $\mathbf{R}^2$. For $A$ take the negative $x$-axis together with the point $(0, 1)$. Then $\mathscr{D}_A$ has an isolated point (4G), while $\mathscr{D} \mid A$ does not.]

2. If $A$ is a union of elements of $\mathscr{D}$, then $\mathscr{D}_A$ and $\mathscr{D} \mid A$ are homeomorphic.

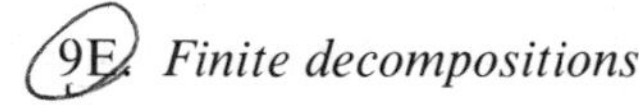

### 9E. *Finite decompositions*

A decomposition $\mathscr{D}$ of a space $X$ will be called *finite* iff only finitely many elements of $\mathscr{D}$ have more than one point. (Typically, $\mathscr{D}$ will contain only one element with more than one point.) Prove that a finite decomposition with closed elements is upper semicontinuous. Show that the restriction that the elements of $\mathscr{D}$ be closed is necessary.

### 9F. *Interpolation of quotient maps*

Let $f: X \to Y$ be continuous. Then there is a quotient map $q$ of $X$ onto a space $Z$ and a one–one continuous map $h$ of $Z$ into $Y$ such that $f = h \circ q$.

### 9G. *Quotient maps and product spaces*

The following conjecture is rather attractive: if $\mathscr{D}$ is a decomposition of $X$ into homeomorphic sets, say all homeomorphic to $Y$, then $X$ is homeomorphic to $\mathscr{D} \times Y$. Find a counterexample.

### 9H. *Strong topologies*

Here we develop the theory for strong topologies analogous to the theory for weak topologies given in 8.9 through 8.16.

Suppose $X_\alpha$ is a topological space and $f_\alpha$ is a map of $X_\alpha$ to a set $Y$, for each $\alpha \in A$. The *strong topology* coinduced by the maps $f_\alpha$ on $Y$ consists of all sets $U$ in $Y$ such that $f_\alpha^{-1}(U)$ is open in $X_\alpha$, for each $\alpha \in A$.

1. This is a topology on $Y$, the largest making each $f_\alpha$ continuous.

2. If $Y$ has the strong topology coinduced by the maps $f_\alpha$, for $\alpha \in A$, then a map $g: Y \to Z$ is continuous iff $g \circ f_\alpha$ is continuous for each $\alpha \in A$. (Compare with 8.10 and 9.4.)

The family of maps $f_\alpha$ will be said to *cover points* of $Y$ iff each $y \in Y$ is in the image of some $f_\alpha$. For families which cover points, the strong topology is just a quotient topology, according to what follows.

Let $X$ be the disjoint union of the spaces $X_\alpha$. If $x$ and $y$ are points of $X$, then (somewhat informally) $x \in X_\alpha$ and $y \in X_\beta$, for some choice of indices $\alpha$ and $\beta$. We define $x \sim y$ iff $f_\alpha(x) = f_\beta(y)$. This defines an equivalence relation on $X$, and we denote the resulting quotient space by $Z$.

3. If the maps $f_\alpha$ cover points of $Y$, then $Y$ has the strong topology coinduced by them iff $X$ is homeomorphic to the quotient space $Z$ constructed above, under the map $h$ which is defined as follows: for $y \in X$, pick $\alpha \in A$ and $x \in X$ so that $f_\alpha(x) = y$, and then define $h(y) = [x]$.

### 9I. *Strong topologies and the lattice of topologies*

Let $\{\tau_\alpha \mid \alpha \in A\}$ be a family of topologies on a fixed set $X$ and denote by $X_\alpha$ the space consisting of the set $X$ with the topology $\tau_\alpha$. The identity function from the space $X_\alpha$ to the set $X$ will be denoted $j_\alpha$.

1. The strong topology coinduced on $X$ by the maps $j_\alpha$ is the intersection (infimum) $\tau$ of the topologies $\tau_\alpha$.

2. $(X, \tau)$ is homeomorphic to the quotient space obtained by identifying points $x$ and $y$ in the disjoint union $\sum X_\alpha$ iff $j_\alpha(x) = j_\beta(y)$, where $x \in X_\alpha$ and $y \in X_\beta$.

These results compare with the results in 8I on weak topologies and suprema in the lattice of topologies.

### 9J. *Disjoint unions and products*

If $X_\alpha$ is homeomorphic to $X$, for each $\alpha \in A$, then the disjoint union $\sum X_\alpha$ is homeomorphic to $X \times A$, where $A$ is given the discrete topology.

### 9K. *Covering spaces*

Let $p$ be a continuous map of a space $\tilde{X}$ onto a space $X$. If each $x$ in $X$ has a nhood $U$ such that $p^{-1}(U)$ is a disjoint union of open sets $V$ each of which is homeomorphic to $U$ under the map $p \mid V$, then $p$ is called a *covering projection.* $X$ is called the *base space* and $\tilde{X}$ is the *covering space.*

A *local homeomorphism* from a space $X$ to a space $Y$ is a continuous map $f$ from $X$ to $Y$ such that each point $x$ in $X$ has an open nhood which is mapped homeomorphically by $f$ onto an open subset of $Y$.

1. The map $p(x) = (\cos x, \sin x)$ of $\mathbf{R}$ onto $\mathbf{S}^1$ is a covering projection.

2. Every covering projection is a local homeomorphism. The converse fails.

3. A local homeomorphism is an open map. Thus, under a covering projection, the base space is a quotient space of the covering space.

4. Give conditions under which $X \times Y$ is a covering space of $X$, with the usual projection map being the covering projection.

5. If $p: \tilde{X} \to X$ and $q: \tilde{Y} \to Y$ are covering projections, then the map $p \times q$ defined by $(p \times q)(x, y) = (p(x), q(y))$ is a covering projection from $\tilde{X} \times \tilde{Y}$ to $X \times Y$.

### 9L. *Attachings*

1. If $X$ is any space, $A$ is a closed subset of $X$, and $p \notin X$, the space $X +_f \{p\}$ resulting from the function $f$ which takes $A$ to $\{p\}$ is homeomorphic to the quotient space of $X$ obtained by identifying $A$ to a single point.

2. Let $X = \mathbf{I}$, $Y = [2, 3]$, $a = \{0, 1\}$, and let $f: A \to Y$ be defined by $f(0) = 2$, $f(1) = 3$. Then $X +_f Y$ is homeomorphic to $\mathbf{S}^1$.

### 9M. *Coherent topologies*

Let $\mathscr{A}$ be a collection of subsets of a topological space $X$. The topology on $X$ is said to be *coherent with* $\mathscr{A}$ provided a set $G$ is open in $X$ iff $G \cap A$ is open in $A$, for each $A \in \mathscr{A}$.

1. The topology on $X$ is coherent with $\mathscr{A}$ iff it is the strong topology (9H) coinduced by the inclusion maps $i_A: A \to X$, for $A \in \mathscr{A}$.

2. The topology on $X$ is coherent with $\mathscr{A}$ provided a set $F$ is closed in $X$ iff $F \cap A$ is closed in $A$, for each $A \in \mathscr{A}$.

3. If $\mathscr{A}$ is a collection of open sets whose union is $X$, then the topology on $X$ is coherent with $\mathscr{A}$.

 

4. If $\mathscr{A}$ is a locally finite collection (7D) of closed sets whose union is $X$, then the topology on $X$ is coherent with $\mathscr{A}$.

There is only one topology on $X$ coherent with any given collection $\mathscr{A}$ of subsets of $X$, of course. It is sometimes called the weak topology generated by the sets in $\mathscr{A}$, a term which we have already used to mean something quite different.

Coherent topologies are useful in the study of $k$-spaces; see Section 43.

Chapter 4

# Convergence

## 10 Inadequacy of sequences

The reader should be familiar with the fact that a function $f: \mathbf{R} \to \mathbf{R}$ is continuous at $x_0$ in $\mathbf{R}$ iff whenever $(x_n)$ is a sequence converging to $x_0$ in $\mathbf{R}$, then the sequence $(f(x_n))$ converges to $f(x_0)$. Since we introduced topologies for the purpose of providing a general setting for the study of continuous functions, this raises two obvious questions:

a) can we define sequential convergence in a general topological space?

b) if so, does the resulting notion describe the topology (as do the closure and interior operations, for example) and hence the continuous functions?

The answers (respectively, "yes" and "only for a limited class of spaces") are provided in this section. Succeeding sections constitute the successful search for a generally applicable and descriptive notion of convergence.

**10.1 Definition.** A sequence $(x_n)$ in a topological space $X$ is said to *converge* to $x \in X$, and we write $x_n \to x$, iff for each nhood $U$ of $x$, there is some positive integer $n_0$ such that $n \geq n_0$ implies $x_n \in U$. In this case, we say $(x_n)$ is *eventually* in $U$.

It is clear that we can replace "nhood" with "basic nhood" in this definition without altering its impact.

**10.2 Examples.** a) Let $\rho$ be a pseudometric on $X$. Then $x_n \to x$ in the topology generated by $\rho$ iff $\rho(x_n, x) \to 0$. This is clear, since $x_n \to x$ iff $(x_n)$ is eventually in each $\epsilon$-disk about $x$.

b) In the product space $\mathbf{R}^{\mathbf{R}}$ a sequence $f_n$ converges to $f$ iff $f_n(x) \to f(x)$ for each $x \in \mathbf{R}$. This is clear once it is remembered that basic nhoods of $f \in \mathbf{R}^{\mathbf{R}}$ have the form

$$U(f, F, \epsilon) = \{g \in \mathbf{R}^{\mathbf{R}} \mid |g(x) - f(x)| < \epsilon \text{ for each } x \in F\},$$

for $F$ a finite subset of $\mathbf{R}$ and $\epsilon > 0$. Thus $f_n \to f$ iff $f_n$ approaches $f$ on each finite set, which happens iff $f_n(x) \to f(x)$ for each $x \in \mathbf{R}$.

Sequential convergence will be able to describe only those topologies in which the minimum number of (basic) nhoods around each point is no greater than the number of terms in the sequences.

**10.3 Definition.** A topological space $X$ is *first countable* (or satisfies the *first axiom of countability*) iff each $x \in X$ has a countable nhood base.

Since the disks about $x$ of rational radius form a nhood base at $x$ in any pseudometric space, the pseudometrizable spaces are all first countable. They form the most important single class of first-countable spaces.

The first axiom of countability has been defined before, in 4.4(b), but you may have missed it. The second axiom was introduced in 5F. Both will be studied in detail in Section 16.

***10.4 Theorem.*** *If $X$ is a first countable space and $E \subset X$, then $x \in \bar{E}$ iff there is a sequence $(x_n)$ contained in $E$ which converges to $x$.*

*Proof.* If $x \in \bar{E}$, pick a countable nhood base $\{U_n \mid n = 1, 2, \ldots\}$ at $x$ in $X$. Replacing $U_n$ by $\bigcap_{k=1}^{n} U_k$ where necessary, we may assume that

$$U_1 \supset U_2 \supset \cdots .$$

Now $U_n \cap E \neq \varnothing$ for each $n$, so we can pick $x_n \in U_n \cap E$. The result is a sequence $(x_n)$ contained in $E$ which obviously converges to $x$.

Conversely, suppose $(x_n)$ is a sequence contained in $E$ and $x_n \to x$. Then each nhood of $x$ contains a tail of the sequence $(x_n)$ and thus meets $E$, so $x \in \bar{E}$. ■

***10.5 Corollary.*** *Let $X$ and $Y$ be first countable spaces. Then*

a) *$U \subset X$ is open iff whenever $x_n \to x \in U$, then $(x_n)$ is eventually in $U$,*

b) *$F \subset X$ is closed iff whenever $(x_n) \subset F$ and $x_n \to x$, then $x \in F$,*

c) *$f: X \to Y$ is continuous iff whenever $x_n \to x$ in $X$, then $f(x_n) \to f(x)$ in $Y$.*

*Proof.* This is left as Exercise 10C. ■

Thus sequential convergence describes the topology of any first countable space. A somewhat wider class of spaces can be described using sequences, in fact (see the notes), but the following examples show that the basic Theorem 10.4 fails in the general setting.

**10.6 Examples.** a) Consider $X = \mathbf{R}^{\mathbf{R}}$ with the product topology. Let

$$E = \{f \in \mathbf{R}^{\mathbf{R}} \mid f(x) = 0 \text{ or } 1 \text{ and } f(x) = 0 \text{ only finitely often}\},$$

and let $g \in \mathbf{R}^{\mathbf{R}}$ be the function which is 0 everywhere. Then if $U(g)$ is a basic nhood of $g$, we have

$$U(g) = \{h \in \mathbf{R}^{\mathbf{R}} \mid |h(y) - g(y)| < \epsilon \text{ if } y \in F\}$$

for some finite set $F \subset \mathbf{R}$ and some $\epsilon > 0$. But such a nhood $U(g)$ meets $E$ in the function $h$ which is 0 on elements of $F$ and 1 elsewhere. Hence, $g \in \mathrm{Cl}_X E$. On the other hand, if $(f_n)$ is a sequence in $E$, with each $f_n$ being 0 on the finite set $A_n$, then any function which is a limit of the sequence $(f_n)$ can be zero at most on the countable set $\bigcup_{n=1}^{\infty} A_n$. Since $g$ does not meet this requirement, no sequence in $E$ can converge to $g$.

Since sequences cannot describe the topology of $\mathbf{R}^{\mathbf{R}}$, the criterion for continuity given in Theorem 10.5 for first-countable spaces probably fails here. In Exercise 10B, you are asked to find a noncontinuous function $F: \mathbf{R}^{\mathbf{R}} \to \mathbf{R}$ with the property that whenever $f_n \to f$ in $\mathbf{R}^{\mathbf{R}}$, then $F(f_n) \to F(f)$.

b) Recall that $\mathbf{\Omega}$ denotes the set of ordinals $\leq \omega_1$, the first uncountable ordinal, and $\mathbf{\Omega}_0 = \mathbf{\Omega} - \{\omega_1\}$. Put the order topology (6D) on $\mathbf{\Omega}$, for which a subbase consists of all sets $[1, \alpha) = \{\gamma \mid 1 \leq \gamma < \alpha\}$, for $\alpha \in \mathbf{\Omega}$, together with all sets $(\beta, \omega_1] = \{\gamma \mid \beta < \gamma \leq \omega_1\}$, for $\beta \in \mathbf{\Omega}$. Note that if $\alpha$ is a nonlimit ordinal, $\{\alpha\}$ is a nhood of $\alpha$ in this topology, while if $\alpha$ is a limit ordinal, the nhoods $(\beta, \alpha]$, $\beta < \alpha$, form a nhood base at $\alpha$. Whenever $\mathbf{\Omega}$ is used as a topological space hereafter, this topology is assumed.

Now note that $\omega_1 \in \overline{\mathbf{\Omega}}_0$ in this topology. But if $(\alpha_n)$ were a sequence in $\mathbf{\Omega}_0$ with limit $\omega_1$, we would have $\omega_1 = \sup(\alpha_n)$, contradicting Theorem 1.20. Thus sequences fail to describe the topology on $\mathbf{\Omega}$.

## Problems

### 10A. *Sequential convergence in topological spaces*

For each of the following spaces, answer these questions:

a) Which sequences converge to which points?

b) Is $X$ first countable?

c) Does the result of Theorem 10.4 hold true for $X$?

(One of your answers should show that first countability is not necessary in Theorem 10.4.)

1. $X$ any uncountable set with the cofinite topology (in which the closed sets are $X$ and all finite subsets of $X$).

2. $X$ any uncountable set with the *cocountable topology*, in which the closed sets are $X$ and all countable subsets of $X$.

3. $X$ the real line with the topology in which the open sets are the sets of the form $(a, \infty)$, $a \in \mathbf{R}$.

4. $X$ the Sorgenfrey line $\mathbf{E}$ (4.6).

5. $X$ any discrete space.

6. $X$ any trivial space.

### 10B. *Sequential convergence and continuity*

Find spaces $X$ and $Y$ and a function $F: X \to Y$ which is not continuous, but which has the property that $F(x_n) \to F(x)$ in $Y$ whenever $x_n \to x$ in $X$.

### 10C. *Topology of first-countable spaces*

Let $X$ and $Y$ be first-countable spaces.

1. $U \subset X$ is open iff whenever $x_n \to x \in U$, then $(x_n)$ is eventually in $U$.
2. $F \subset X$ is closed iff whenever $(x_n)$ is contained in $F$ and $x_n \to x$, then $x \in F$.
3. $f: X \to Y$ is continuous iff whenever $x_n \to x$ in $X$, then $f(x_n) \to f(x)$ in $Y$.
4. Which of the properties above hold for an uncountable set $Z$ with the cofinite topology?

## 11 Nets

Formally, a sequence in $X$ is a mapping of $\mathbf{N}$ into $X$; in more informal terms, we are using the integers to order a collection of points in $X$. The key to successful generalization of the notion of sequence, for use in topological spaces, lies in retaining the idea of ordering a collection of points of $X$ by mapping some ordered set into $X$, while significantly relaxing the conditions on the ordered sets we will allow.

The linearity of the order on the integers can be dispensed with, provided we supply some other way of giving a definite "positive orientation" to our ordered sets. The following definition has stood the test of time.

**11.1 Definition.** A set $\Lambda$ is a *directed set* iff there is a relation $\leq$ on $\Lambda$ satisfying:

$\Lambda$-a) $\lambda \leq \lambda$, for each $\lambda \in \Lambda$,

$\Lambda$-b) if $\lambda_1 \leq \lambda_2$ and $\lambda_2 \leq \lambda_3$ then $\lambda_1 \leq \lambda_3$,

$\Lambda$-c) if $\lambda_1, \lambda_2 \in \Lambda$ then there is some $\lambda_3 \in \Lambda$ with $\lambda_1 \leq \lambda_3$, $\lambda_2 \leq \lambda_3$.

The relation $\leq$ is sometimes referred to as a *direction* on $\Lambda$, or is said to *direct* $\Lambda$.

The first two properties, $\Lambda$-a and $\Lambda$-b, are familiar requirements for an order relation . (Note, however, the lack of antisymmetry; a direction need not be a partial order.) $\Lambda$-c provides the positive orientation we were seeking for $\Lambda$. In fact, it models a property possessed by the set $\mathscr{U}_x$ of all nhoods of a point $x$ in a space $X$, when ordered by "reverse inclusion": $U_1 \leq U_2$ iff $U_2 \subset U_1$. Although directed sets were not first introduced (either historically or here) with this in mind, it is precisely this which makes them useful in describing convergence in general topological spaces.

The concept of a net, which generalizes the notion of a sequence, can now be introduced, using an arbitrary directed set to replace the integers.

**11.2 Definition.** A *net* in a set $X$ is a function $P: \Lambda \to X$, where $\Lambda$ is some directed set. The point $P(\lambda)$ is usually denoted $x_\lambda$, and we often speak of "the net $(x_\lambda)_{\lambda \in \Lambda}$" or "the net $(x_\lambda)$" if this can cause no confusion.

A *subnet* of a net $P: \Lambda \to X$ is the composition $P \circ \varphi$, where $\varphi: M \to \Lambda$ is an increasing cofinal function from a directed set $M$ to $\Lambda$. That is,

a) $\varphi(\mu_1) \leq \varphi(\mu_2)$ whenever $\mu_1 \leq \mu_2$ ($\varphi$ is *increasing*),

b) for each $\lambda \in \Lambda$, there is some $\mu \in M$ such that $\lambda \leq \varphi(\mu)$ ($\varphi$ is *cofinal* in $\Lambda$).

For $\mu \in M$, the point $P \circ \varphi(\mu)$ is often written $x_{\lambda_\mu}$, and we usually speak of "the subnet $(x_{\lambda_\mu})$ of $(x_\lambda)$".

If $(x_\lambda)$ is a net in $X$, a set of the form $\{x_\lambda \mid \lambda \geq \lambda_0\}$, for $\lambda_0 \in \Lambda$, is called a *tail* of $(x_\lambda)$.

The definition of net convergence is modeled after the definition of sequential convergence introduced in 10.1 and should provide no problems.

**11.3 Definition.** Let $(x_\lambda)$ be a net in a space $X$. Then $(x_\lambda)$ *converges* to $x \in X$ (written $x_\lambda \to x$) provided for each nhood $U$ of $x$, there is some $\lambda_0 \in \Lambda$ such that $\lambda \geq \lambda_0$ implies $x_\lambda \in U$. Thus $x_\lambda \to x$ iff each nhood of $x$ contains a tail of $(x_\lambda)$. This is sometimes said: $(x_\lambda)$ converges to $x$ provided it is *residually* (or *eventually*) in every nhood of $x$.

We say $(x_\lambda)$ has $x$ as a *cluster point* iff for each nhood $U$ of $x$ and for each $\lambda_0 \in \Lambda$ there is some $\lambda \geq \lambda_0$ such that $x_\lambda \in U$. This is sometimes said $(x_\lambda)$ has $x$ as a cluster point iff $(x_\lambda)$ is *cofinally* (or *frequently*) in each nhood of $x$.

Note that in both definitions above it is sufficient if we restrict attention to the nhoods in some fixed nhood base at $x$.

**11.4 Examples.** a) Let $X$ be a topological space, $x \in X$ and $\Lambda$ any fixed nhood base at $x$ in $X$. Then the order relation $U_1 \leq U_2$ iff $U_2 \subset U_1$ directs $\Lambda$. Hence if we pick $x_U \in U$ for each $U \in \Lambda$, the result is a net $(x_U)$ in $X$. Moreover, $x_U \to x$. For given any nhood $V$ of $x$, we have $U_0 \subset V$ for some $U_0 \in \Lambda$. Then $U \geq U_0$ implies $U \subset U_0$, so that $x_U \in U \subset V$. This example should be studied carefully; it is the model for most of the proofs, to be given later in this section, of the properties of nets in topological spaces.

b) The set $\mathbf{N}$ of positive integers is a directed set when given its usual order. Thus every sequence $(x_n)$ is a net. It is clear that the two definitions of convergence of $(x_n)$ (as a sequence in 10.1 and as a net in 11.3) coincide.

Note that every subsequence of a sequence $(x_n)$ is a subnet of $(x_n)$. The converse is *not* true; there is no guarantee that a subnet of $(x_n)$ is a subsequence, because there is no way of being sure that it is a sequence! This illustrates the (at first, strange) fact that a subnet can have a much richer index set than the original net.

c) The collection $\mathscr{P}$ of all finite partitions of the closed interval $[a, b]$ into closed subintervals is a directed set, when ordered by the relation $A_1 \leq A_2$ iff $A_2$ refines $A_1$. Thus, if $f$ is any real-valued function on $[a, b]$, we can define a net $P_L: \mathscr{P} \to \mathbf{R}$ by letting $P_L(A)$ be the lower Riemann sum of $f$ over the partition $A$; likewise, we can define $P_U: \mathscr{P} \to \mathbf{R}$ by letting $P_U(A)$ be the upper Riemann sum of $f$ over $A$. Convergence of both of these nets to the number $c$ simply means $\int_a^b f(x)\,dx = c$. This example is historically important; it is what first led Moore and Smith to the concept of a net. See the notes.

d) Let $(M, \rho)$ be a metric space, with $x_0 \in M$. Then $M - \{x_0\}$ becomes a directed set when ordered by the relation $x < y$ iff $\rho(y, x_0) < \rho(x, x_0)$. Hence if $f: M \to N$, where $N$ is a metric space, the restriction of $f$ to $M - \{x_0\}$ defines a net in $N$. The reader can check that this net converges to $z_0$ in $N$ iff $\lim_{x \to x_0} f(x) = z_0$ in the elementary calculus sense.

e) If $(x_\lambda)$ converges to $x$, every subnet of $(x_\lambda)$ converges to $x$.

f) If $x_\lambda = x$, for each $\lambda \in \Lambda$, then $x_\lambda \to x$.

**11.5 Theorem.** *A net has $y$ as a cluster point iff it has a subnet which converges to $y$.*

*Proof.* Let $y$ be a cluster point of $(x_\lambda)$. Define $M = \{(\lambda, U) \mid \lambda \in \Lambda, U$ a nhood of $y$ such that $x_\lambda \in U\}$, and order $M$ as follows: $(\lambda_1, U_1) \leq (\lambda_2, U_2)$ iff $\lambda_1 \leq \lambda_2$ and $U_2 \subset U_1$. This is easily verified to be a direction on $M$. Define $\varphi: M \to \Lambda$ by $\varphi(\lambda, U) = \lambda$. Then $\varphi$ is increasing and cofinal in $\Lambda$, so $\varphi$ defines a subnet of $(x_\lambda)$. Let $U_0$ be any nhood of $y$ and find $\lambda_0 \in \Lambda$ such that $x_{\lambda_0} \in U_0$. Then $(\lambda_0, U_0) \in M$, and moreover, $(\lambda, U) \geq (\lambda_0, U_0)$ implies $U \subset U_0$, so that $x_\lambda \in U \subset U_0$. It follows that the subnet defined by $\varphi$ converges to $y$.

Suppose $\varphi: M \to \Lambda$ defines a subnet of $(x_\lambda)$ which converges to $y$. Then for each nhood $U$ of $y$, there is some $u_U$ in $M$ such that $u \geq u_U$ implies $x_{\varphi(u)} \in U$. Suppose a nhood $U$ of $y$ and a point $\lambda_0$ in $\Lambda$ are given. Since $\varphi(M)$ is cofinal in $\Lambda$, there is some $u_0 \in M$ such that $\varphi(u_0) \geq \lambda_0$. But there is also some $u_U \in M$ such that $u \geq u_U$ implies $x_{\varphi(u)} \in U$. Pick $u^* \in M$ such that $u^* \geq u_0$ and $u^* \geq u_U$. Then $\varphi(u^*) = \lambda^* \geq \lambda_0$, since $\varphi(u^*) \geq \varphi(u_0)$, and $x_{\lambda^*} = x_{\varphi(u^*)} \in U$, since $u^* \geq u_U$. Thus for any nhood $U$ of $y$ and any $\lambda_0 \in \Lambda$, there is some $\lambda^* \geq \lambda_0$ with $x_{\lambda^*} \in U$. It follows that $y$ is a cluster point of $(x_\lambda)$. ■

**11.6 Corollary.** *If a subnet of $(x_\lambda)$ has $y$ as a cluster point, so does $(x_\lambda)$.*

*Proof.* A subnet of a subnet of $(x_\lambda)$ is a subnet of $(x_\lambda)$. ■

We turn now to the problem of showing that nets do indeed represent the correct way of approaching convergence questions in topological spaces.

**11.7 Theorem.** *If $E \subset X$, then $x \in \bar{E}$ iff there is a net $(x_\lambda)$ in $E$ with $x_\lambda \to x$.*

*Proof.* If $x \in \bar{E}$, then each nhood $U$ of $x$ meets $E$ in at least one point $x_U$. Then $(x_U)$ is a net contained in $E$ which converges to $x$. (See Example 11.4(a).)

Conversely, if $(x_\lambda)$ is a net contained in $E$ which converges to $x$, then each nhood of $y$ meets $E$ (in a tail of $(x_\lambda)$) and hence $x \in \bar{E}$. ■

**11.8 Theorem.** *Let $f: X \to Y$. Then $f$ is continuous at $x_0 \in X$ iff whenever $x_\lambda \to x_0$ in $X$, then $f(x_\lambda) \to f(x_0)$ in $Y$.*

*Proof.* Suppose $f$ is continuous at $x_0$ and $x_\lambda \to x_0$. Given a nhood $V$ of $f(x_0)$, $f^{-1}(V)$ is a nhood of $x_0$, so for some $\lambda_0$, $\lambda \geq \lambda_0$ implies $x_\lambda \in f^{-1}(V)$. Thus $\lambda \geq \lambda_0$ implies $f(x_\lambda) \in V$, showing that $f(x_\lambda) \to f(x_0)$.

On the other hand, if $f$ is not continuous at $x_0$, then for some nhood $V$ of $f(x_0)$, $f(U) \not\subset V$ for any nhood $U$ of $x_0$. Thus for each nhood $U$ of $x_0$, we can pick $x_U \in U$ such that $f(x_U) \notin V$. But then $(x_U)$ is a net in $X$ and $x_U \to x_0$, while $f(x_U) \nrightarrow f(x_0)$. ■

***11.9 Theorem.*** *A net $(x_\lambda)$ in a product $X = \prod_{\alpha \in A} X_\alpha$ converges to $x$ iff for each $\alpha \in A$, $\pi_\alpha(x_\lambda) \to \pi_\alpha(x)$ in $X_\alpha$.*

*Proof.* If $x_\lambda \to x$ in $\prod X_\alpha$, then since $\pi_\alpha$ is continuous, $\pi_\alpha(x_\lambda) \to \pi_\alpha(x)$, by the previous theorem, for each $\alpha$.

Suppose on the other hand that $\pi_\alpha(x_\lambda) \to \pi_\alpha(x)$ for each $\alpha \in A$. Let

$$\pi_{\alpha_1}^{-1}(U_{\alpha_1}) \cap \cdots \cap \pi_{\alpha_n}^{-1}(U_{\alpha_n})$$

be a basic nhood of $x$ in the product space. Then for each $i = 1, \ldots, n$ there is a $\lambda_i$ such that whenever $\lambda \geq \lambda_i$, $\pi_{\alpha_i}(x_\lambda) \in U_{\alpha_i}$. Thus if $\lambda_0$ is picked greater than all of $\lambda_1, \ldots, \lambda_n$, we have $\pi_{\alpha_i}(x_\lambda) \in U_{\alpha_i}$, $i = 1, \ldots, n$, for all $\lambda \geq \lambda_0$. It follows that for $\lambda \geq \lambda_0$, $x_\lambda \in \bigcap \pi_{\alpha_i}^{-1}(U_{\alpha_i})$, and hence that $x_\lambda \to x$ in the product. ■

In case all factor spaces are homeomorphic to $X$, the last theorem has a a pleasant re-interpretation. In the product topology on the set $X^A$ of all functions from $A$ to $X$, a net $f_\lambda$ converges to $f$ iff for each $\alpha \in A$, $f_\lambda(a) \to f(a)$. That is, convergence of functions in $X^A$ with this topology is just pointwise convergence. For sequences of functions, this was pointed out in 10.2. Thus if functions on a certain set $A$ to a space $X$ are to be studied with pointwise limits in mind, it is appropriate to consider them as elements in the product space $X^A$ with the Tychonoff topology. There are other kinds of functional convergence than pointwise, e.g., uniform convergence, and we will mention here that $X^A$ can be provided with appropriate structures to deal with these also. This is a topic which is deferred until the chapter on function spaces, where different convergence structures on $X^A$ and the interactions between them are studied.

**11.10 Definition.** A net $(x_\lambda)$ in a set $X$ is an *ultranet* (*universal net*) iff for each subset $E$ of $X$, $(x_\lambda)$ is either residually in $E$ or residually in $X - E$.

It follows from this definition that if an ultranet is frequently in $E$ then it is residually in $E$. In particular, an ultranet in a topological space must converge to each of its cluster points.

For any directed set $\Lambda$, the map $P\colon \Lambda \to X$, defined by $P(\lambda) = x$ for all $\lambda \in \Lambda$, gives an ultranet on $X$, called the *trivial ultranet*. Nontrivial ultranets can be proved to exist (relying on the axiom of choice; see 12D.5) but none has ever been explicitly constructed. Most facts about ultranets are best developed using filters and ultrafilters as a vehicle. We will do this in the next section.

***11.11 Theorem.*** *If $(x_\lambda)$ is an ultranet in $X$ and $f\colon X \to Y$, then $(f(x_\lambda))$ is an ultranet in $Y$.*

*Proof.* If $A \subset Y$, then $f^{-1}(A) = X - f^{-1}(Y - A)$, so $(x_\lambda)$ is eventually in either $f^{-1}(A)$ or $f^{-1}(Y - A)$, from which it follows that $(f(x_\lambda)$ is eventually in either $A$ or $Y - A$. Thus, $(f(x_\lambda))$ is an ultranet. ■

### Problems

#### 11A. *Examples of net convergence*

1. In $\mathbf{R}^{\mathbf{R}}$, let $E = \{f \in \mathbf{R}^{\mathbf{R}} \mid f(x) = 0 \text{ or } 1, \text{ and } f(x) = 0 \text{ only finitely often}\}$ and let $g$ be the function in $\mathbf{R}^{\mathbf{R}}$ which is identically 0. Then, in the product topology on $\mathbf{R}^{\mathbf{R}}$, $g \in \bar{E}$ (refer to Example 10.6). Find a net $(f_\lambda)$ in $E$ which converges to $g$.

2. In the ordinal space, recall that $\omega_1 \in \bar{\mathbf{\Omega}}_0$ (see Example 10.6). Find a net $(x_\lambda)$ in $\mathbf{\Omega}_0$ which converges to $\omega_1$ in $\mathbf{\Omega}$.

3. Let $M$ be any metric space. A mapping $P(\alpha) = x_\alpha$ of $\mathbf{\Omega}_0$ into $M$ will be a net. Show that $x_\alpha \to x$ in $M$ iff $x_\alpha$ is eventually equal to $x$.

4. Let $x \in \mathbf{R}^n$ and define $\leq$ on $\mathbf{R}^n$ by $y \geq z$ iff $|y - x| \leq |z - x|$. Then, with this order, $\Lambda = \mathbf{R}^n - \{x\}$ is a directed set. Thus any function $f: \mathbf{R}^n \to \mathbf{R}$ defines a net in $R$ (by restricting $f$ to $\mathbf{R}^n - \{x\}$). Show that this net converges to $L$ iff $\lim_{y \to x} f(y) = L$.

#### 11B. *Subnets and cluster points*

1. Every subnet of an ultranet is an ultranet.

2. Every net has a subnet which is an ultranet.

3. Exhibit a sequence $(x_n)$ on a set $X$ and a subnet of $(x_n)$ which is not a sequence.

4. If $(x_\lambda)$ is a net in a space $X$ and for each $\lambda_0$, $T_{\lambda_0} = \{x_\lambda \mid \lambda \geq \lambda_0\}$, then $y$ is a cluster point of $(x_\lambda)$ iff $y \in \overline{T_\lambda}$ for each $\lambda \in \Lambda$.

5. If an ultranet has $x$ as a cluster point, then it converges to $x$.

#### 11C. *Cluster points in products*

If $(x_\lambda)_{\lambda \in \Lambda}$ is a net in $\prod X_\alpha$ having $x$ as a cluster point then for each $\alpha$, $(\pi_\alpha(x_\lambda))_{\lambda \in \Lambda}$ has $\pi_\alpha(x)$ for a cluster point. The converse fails, even in $\mathbf{R} \times \mathbf{R}$.

#### 11D. *Nets describe topologies*

1. Nets have the following four properties (some have already been mentioned in the text):

   a) if $x_\lambda = x$ for each $\lambda \in \Lambda$, then $x_\lambda \to x$,
   b) if $x_\lambda \to x$, then every subnet of $(x_\lambda)$ converges to $x$,
   c) if every subnet of $(x_\lambda)$ has a subnet converging to $x$, then $(x_\lambda)$ converges to $x$,
   d) [*diagonal principal*] if $x_\lambda \to x$ and, for each $\lambda \in \Lambda$, a net $(x_\mu^\lambda)_{\mu \in M_\lambda}$ converges to $x_\lambda$, then there is a diagonal net converging to $x_\lambda$; i.e., the net $(x_\mu^\lambda)_{\lambda \in \Lambda, \mu \in M_\lambda}$, ordered lexicographically by $\Lambda$, then by $M_\lambda$, has a subnet which converges to $x$.

2. Conversely, suppose in a set $X$ a notion of net convergence has been specified (telling what nets converge to what points) satisfying a), b), c) and d) of part 1. If the closure of a subset $E$ of $X$ is *defined* by $\bar{E} = \{x \in X \mid x_\lambda \to x \text{ for some net } (x_\lambda) \text{ contained in } E\}$, the result is a topological space in which the notion of net convergence is as originally specified.

## 12 Filters

We have just seen that a good (i.e., topologically descriptive) notion of convergence can be obtained by simply using the nhoods of a single point as the model for an indexing set to replace the integers used for sequences. We now introduce a second way of describing convergence in a topological space in which we say, in effect, why not just treat the nhoods themselves as converging to the point? The result is the theory of filter convergence.

**12.1 Definition.** A *filter* $\mathscr{F}$ on a set $S$ is a nonempty collection of nonempty subsets of $S$ with the properties:

a) if $F_1, F_2 \in \mathscr{F}$ then $F_1 \cap F_2 \in \mathscr{F}$,

b) if $F \in \mathscr{F}$ and $F \subset F'$, then $F' \in \mathscr{F}$.

A subcollection $\mathscr{F}_0$ of $\mathscr{F}$ is a *filter base* for $\mathscr{F}$ iff each element of $\mathscr{F}$ contains some element of $\mathscr{F}_0$, that is, iff

$$\mathscr{F} = \{F \subset S \mid F_0 \subset F_0 \text{ for some } F_0 \in \mathscr{F}\}.$$

Evidently, a nonempty collection $\mathscr{C}$ of nonempty subsets of $S$ is a filter base for some filter on $S$ iff

a)′ if $C_1, C_2 \in \mathscr{C}$ then $C_3 \subset C_1 \cap C_2$ for some $C_3 \in \mathscr{C}$,

in which case the filter generated by $\mathscr{C}$ consists of all supersets of elements of $\mathscr{C}$.

If $\mathscr{F}_1$ and $\mathscr{F}_2$ are filters on $X$, we say $\mathscr{F}_1$ is *finer* than $\mathscr{F}_2$ (or $\mathscr{F}_2$ is coarser than $\mathscr{F}_1$) iff $\mathscr{F}_1 \supset \mathscr{F}_2$. A filter $\mathscr{F}$ on $X$ is *fixed* iff $\bigcap \mathscr{F} \neq \varnothing$ and *free* iff $\bigcap \mathscr{F} = \varnothing$.

**12.2 Examples.** a) Let $X$ be any set, $A \subset X$. Then $\{F \subset X \mid A \subset F\}$ is a filter on $X$ with a particularly simple filter base, the collection consisting of the single set $A$.

b) Let $X$ be any topological space, $A \subset X$. Then $\{U \subset X \mid A \subset U^\circ\}$ is a filter on $X$. In particular, the set $\mathscr{U}_x$ of all nhoods of $x \in X$ is a filter on $X$, and any nhood base at $x$ is a filter base for $\mathscr{U}_x$. This filter will sometimes be called the *nhood filter* at $x$.

c) Let $\mathscr{C} = \{(a, \infty) \mid a \in \mathbf{R}\}$. Then $\mathscr{C}$ is a filter base for a free filter on $\mathbf{R}$, which we will call the *Frechet filter* on $\mathbf{R}$.

**12.3 Definition.** A filter $\mathscr{F}$ on a topological space $X$ is said to *converge* to $x$ (written $\mathscr{F} \to x$) iff $\mathscr{U}_x \subset \mathscr{F}$, that is, iff $\mathscr{F}$ is finer than the nhood filter at $x$. We say $\mathscr{F}$ has $x$ as a *cluster point* (or, $\mathscr{F}$ *clusters* at $x$) iff each $F \in \mathscr{F}$ meets each $U \in \mathscr{U}_x$. Hence $\mathscr{F}$ has $x$ as a cluster point iff $x \in \bigcap \{\bar{F} \mid F \in \mathscr{F}\}$. Also, it is clear that if $\mathscr{F} \to x$, then $\mathscr{F}$ clusters at $x$.

It will be convenient to have the notions of convergence and clustering available for filter bases; they generalize easily and obviously. A filter base $\mathscr{C}$ *converges* to $x$ iff each $U \in \mathscr{U}_x$ contains some $C \in \mathscr{C}$ (iff the filter generated by $\mathscr{C}$ converges to $x$); $\mathscr{C}$ *clusters* at $x$ iff each $U \in \mathscr{U}_x$ meets each $C \in \mathscr{C}$ (iff the filter generated by $\mathscr{C}$ clusters at $x$).

**12.4 Examples.** a) Let $X$ be a topological space, $A \subset X$. The cluster points of the filter $\mathscr{F} = \{U \subset X \mid A \subset U\}$ include each point of $\bar{A}$. Under what conditions (on $A$ or on the topology) will $\mathscr{F}$ converge to some point?

b) The Frechet filter on **R** has no cluster points.

c) Let $\mathscr{F}$ be the filter on **R** generated by the filter base $\mathscr{C} = \{(0, \epsilon) \mid \epsilon > 0\}$. Then $\mathscr{F} \to 0$ (although 0 does not belong to every element of $\mathscr{F}$).

**12.5 Theorem.** *$\mathscr{F}$ has $x$ as a cluster point iff there is a filter $\mathscr{G}$ finer than $\mathscr{F}$ which converges to $x$.*

*Proof.* If $\mathscr{F}$ has $x$ as a cluster point, the collection $\mathscr{C} = \{U \cap F \mid U \in \mathscr{U}_x, F \in \mathscr{F}\}$ is a filter base for a filter $\mathscr{G}$ which is finer than $\mathscr{F}$ and converges to $x$.

Conversely, if $\mathscr{F} \subset \mathscr{G} \to x$, then each $F \in \mathscr{F}$ and each nhood $U$ of $x$ belong to $\mathscr{G}$ and hence meet, so $\mathscr{F}$ clusters at $x$. ■

According to the next three theorems, filter convergence is adequate to the task of describing topological concepts.

**12.6 Theorem.** *If $E \subset X$, then $x \in \bar{E}$ iff there is a filter $\mathscr{F}$ such that $E \in \mathscr{F}$ and $\mathscr{F} \to x$.*

*Proof.* If $y \in \bar{E}$, then $\mathscr{C} = \{U \cap E \mid U \in \mathscr{U}_y\}$ is a filter base. The resulting filter contains $E$ and converges to $y$.

Conversely, if $E \in \mathscr{F} \to y$, then $y$ is a cluster point of $\mathscr{F}$ and hence $y \in \bar{E}$. ■

**12.7 Definition.** If $\mathscr{F}$ is a filter on $X$ and $f: X \to Y$, then $f(\mathscr{F})$ is the filter on $Y$ having for a base the sets $f(F)$, $F \in \mathscr{F}$.

**12.8 Theorem.** *Let $f: X \to Y$. Then $f$ is continuous at $x_0 \in X$ iff whenever $\mathscr{F} \to x_0$ in $X$ then $f(\mathscr{F}) \to f(x_0)$ in $Y$.*

*Proof.* Suppose $f$ is continuous at $x_0$ and $\mathscr{F} \to x_0$. Let $V$ be any nhood of $f(x_0)$ in $Y$. Then for some nhood $U$ of $x_0$ in $X$, $f(U) \subset V$. Then since $U \in \mathscr{F}$, $V \in f(\mathscr{F})$.

Conversely, suppose whenever $\mathscr{F} \to x_0$ in $X$ then $f(\mathscr{F}) \to f(x_0)$ in $Y$. Let $\mathscr{F}$ be the filter of all nhoods of $x_0$ in $X$. Then each nhood $V$ of $f(x_0)$ belongs to $f(\mathscr{F})$, so for some nhood $U$ of $x_0$, $f(U) \subset V$. Thus $f$ is continuous at $x_0$. ■

**12.9 Theorem.** *A filter $\mathscr{F}$ converges to $x_0$ in $\prod X_\alpha$ iff $\pi_\alpha(\mathscr{F}) \to \pi_\alpha(x_0)$ in $X_\alpha$, for each $\alpha$.*

*Proof.* If $\mathscr{F} \to x_0$ in $\prod X_\alpha$, then $\pi_\alpha(\mathscr{F}) \to \pi_\alpha(x_0)$ in $X_\alpha$ because $\pi_\alpha$ is continuous.

Conversely, suppose $\pi_\alpha(\mathscr{F}) \to \pi_\alpha(x_0)$, for each $\alpha$. Let $\bigcap_{k=1}^n \pi_{\alpha_k}^{-1}(U_k)$ be a basic nhood of $x_0$ in $\prod X_\alpha$. Then $U_k$ is a nhood of $\pi_{\alpha_k}(x_0)$, for each $k$. So $U_k \in \pi_{\alpha_k}(\mathscr{F})$, for each $k$, and hence $\pi_{\alpha_k}(F_k) \subset U_k$ for some $F_k \in \mathscr{F}$. Then $\bigcap_{k=1}^n F_k \in \mathscr{F}$ and $\bigcap_{k=1}^n F_k \subset \bigcap_{k=1}^n \pi_{\alpha_k}^{-1}(U_k)$, so $\bigcap_{k=1}^n \pi_{\alpha_k}^{-1}(U_k) \in \mathscr{F}$. Thus $\mathscr{F} \to x_0$. ■

Many of the applications of filter convergence can be neatly done using only the ultrafilters.

**12.10 Definition.** A filter $\mathscr{F}$ is an *ultrafilter* iff there is no strictly finer filter $\mathscr{G}$ than $\mathscr{F}$. Thus the ultrafilters are the maximal filters.

The next theorem makes clear the analogy between ultrafilters and ultranets (11.10). In particular, it can be used to show that an ultrafilter must converge to each of its cluster points.

**12.11 Theorem.** *A filter $\mathscr{F}$ on $X$ is an ultrafilter iff for each $E \subset X$, either $E \in \mathscr{F}$ or $X - E \in \mathscr{F}$.*

*Proof.* Suppose $\mathscr{F}$ is an ultrafilter and $E \subset X$. Every element $F$ of $\mathscr{F}$ meets either $E$ or $X - E$ and hence (since no two elements of $\mathscr{F}$ have empty intersection) they must all meet one or the other, say $F \cap E \neq \varnothing$ for all $F \in \mathscr{F}$. Then

$$\{F \cap E \mid F \in \mathscr{F}\}$$

is a filter base for a filter $\mathscr{G}$ finer than $\mathscr{F}$ which contains $E$. Since $\mathscr{G}$ cannot be strictly finer than $\mathscr{F}$, we have $\mathscr{G} = \mathscr{F}$ and hence $E \in \mathscr{F}$.

Conversely, suppose $\mathscr{F}$ contains $E$ or $X - E$ for each $E \subset X$. If $\mathscr{G}$ is a strictly finer filter than $\mathscr{F}$, then for some $A \in \mathscr{G}$, $A \notin \mathscr{F}$. But then $X - A \in \mathscr{F}$, from the condition, and since $\mathscr{F} \subset \mathscr{G}$ we have the impossible situation that both $A$ and $X - A$ belong to $\mathscr{G}$. Thus $\mathscr{F}$ must be maximal. ■

**12.12 Theorem.** *Every filter $\mathscr{F}$ is contained in some ultrafilter.*

*Proof.* Let $\mathscr{S}$ be the collection of all filters finer than $\mathscr{F}$, partially ordered by $\mathscr{F}_1 \leq \mathscr{F}_2$ iff $\mathscr{F}_1 \subset \mathscr{F}_2$. Then a chain $\{\mathscr{F}_\alpha \mid \alpha \in A\}$ from $\mathscr{S}$ has $\bigcup \mathscr{F}_\alpha$ for an upper bound (that $\bigcup \mathscr{F}_\alpha$ is indeed a filter follows easily from the fact that if $F_1$ and $F_2$ belong to $\bigcup \mathscr{F}_\alpha$, then they both belong to some one $\mathscr{F}_\alpha$ by linearity of the inclusion order on $\{\mathscr{F}_\alpha \mid \alpha \in A\}$). Thus, by Zorn's lemma, $\mathscr{S}$ has a maximal element $\mathscr{G}$ and, obviously, $\mathscr{G}$ is an ultrafilter containing $\mathscr{F}$. ■

The proof of the last theorem, it should be noted, depends on the axiom of choice. Thus the following examples of free ultrafilters depend for the proof of their existence on the (nonconstructive) choice axiom. Explicit constructions of free ultrafilters have never been accomplished, although there are more free ultrafilters than fixed ultrafilters (that is, for a discrete space $X$, $|\beta X - X| > |X|$; see 19J and 19.13(d).

**12.13 Examples.** a) A filter $\mathscr{F}$ on $X$ is a fixed ultrafilter iff $\mathscr{F} = \{F \subset X \mid x \in F\}$ for some $x \in F$. By the criterion given in Theorem 12.11, each filter of this form is an ultrafilter. On the other hand, if $\mathscr{F}$ is a fixed ultrafilter, say $\bigcap \mathscr{F} = A \neq \varnothing$, then $\mathscr{F}$ must be the filter of all sets containing $A$ (since this is a filter containing $\mathscr{F}$) and $A$ must be a single point (since the filter of all sets containing $x \in A$ is finer than $\mathscr{F}$).

b) The Frechet filter $\mathscr{F}$ on $\mathbf{R}$ is, by Theorem 12.12, contained in some ultrafilter $\mathscr{G}$. Since $\mathscr{F}$ is free, $\mathscr{G}$ must be also be free.

c) The ultrafilter containing a given filter $\mathscr{F}$ need not be unique. For if $\mathscr{F}$ is the filter of all sets containing $A \subset X$, then for each $x \in A$, the filter of all sets containing $x$ is an ultrafilter containing $\mathscr{F}$. In fact, if a filter is contained in a unique ultrafilter, it is itself an ultrafilter; see Exercise 12C.

The following theorem is easily proved (using, for example, the criterion given in Theorem 12.11) and will be useful later.

***12.14 Theorem.*** *If $f$ maps $X$ into $Y$ and $\mathscr{F}$ is an ultrafilter on $X$, then $f(\mathscr{F})$ is an ultrafilter on $Y$.*

The similarities between net and filter convergence are manifest. Each describes the topology on a topological space with equal facility, "finer filters" provide a filter analog to "subnets" (by Theorem 12.5). In addition, there is more than a casual relationship between the ideas behind the two approaches. Thus the fact that a formal bridge can be built between the two notions should come as no surprise.

**12.15 Definition.** If $(x_\lambda)$ is a net in $X$, the filter generated by the filter base $\mathscr{C}$ consisting of the sets $B_{\lambda_0} = \{x_\lambda \mid \lambda \geq \lambda_0\}$, $\lambda_0 \in \Lambda$, is called the *filter generated by* $(x_\lambda)$.

**12.16 Definition.** If $\mathscr{F}$ is a filter on $X$, let $\Lambda_{\mathscr{F}} = \{(x, F) \mid x \in F \in \mathscr{F}\}$. Then $\Lambda_{\mathscr{F}}$ is directed by the relation $(x_1, F_1) \leq (x_2, F_2)$ iff $F_2 \subset F_1$, so the map $P: \Lambda_{\mathscr{F}} \to X$ defined by $P(x, F) = x$ is a net in $X$. It is called the *net based on* $\mathscr{F}$.

***12.17 Theorem.*** a) *A filter $\mathscr{F}$ converges to $x$ in $X$ iff the net based on $\mathscr{F}$ converges to $x$.*

b) *A net $(x_\lambda)$ converges to $x$ in $X$ iff the filter generated by $(x_\lambda)$ converges to $x$.*

*Proof.* a) Suppose $\mathscr{F} \to x$. If $U$ is a nhood of $x$, then $U \in \mathscr{F}$. Pick $p \in U$. Then $(p, U) \in \Lambda_{\mathscr{F}}$ and if $(q, F) \geq (p, U)$, then $q \in F \subset U$. Thus the net based on $\mathscr{F}$ converges to $x$.

Conversely, suppose the net based on $\mathscr{F}$ converges to $x$. Let $U$ be a nhood of $x$. Then for some $(p_0, F_0) \in \Lambda_{\mathscr{F}}$, we have $(p, F) \geq (p_0, F_0)$ implies $p \in U$. But then $F_0 \subset U$; otherwise, there is some $q \in F_0 - U$, and then $(q, F_0) \geq (p_0, F_0)$, but $q \notin U$. Hence $U \in \mathscr{F}$, so $\mathscr{F} \to x$.

b) The net $(x_\lambda)$ converges to $x$ iff each nhood of $x$ contains a tail of $(x_\lambda)$. Since the tails of $(x_\lambda)$ are a base for the filter generated by $(x_\lambda)$, the result follows. ■

Similar results are true of cluster points, the relationship between subnets and finer filters and the relationship between ultranets and ultrafilters. We will leave all these to Exercise 12D.

Filters are preferred to nets in dealing with convergence questions in topological spaces. The reason for this involves the difference that nets are, and must remain, essentially set-theoretic (or order-theoretic) in nature, and hence passive, while filters can, with the addition of topological restrictions on their sets, become intimately involved with the structure of the space itself. Examples of uses of filters which could hardly be duplicated with nets can be found in Exercises 17K, 17M, 19J, 19K and 19L. See also Exercise 12E in this section.

## Problems

### 12A. *Examples of filter convergence*

1. If the real line is given its topology as the looped line (4D), then the Frechet filter $\mathscr{F}$ converges to 0.

2. Which filters $\mathscr{F}$ will converge to $x$ in a discrete space $X$? In a trivial space $X$?

3. Let $X$ be an infinite set, $\mathscr{F}$ the filter on $X$ generated by the filter base consisting of all complements of finite sets. To which points does $\mathscr{F}$ converge if $X$ is given the cofinite topology?

4. Show that if a filter in a metric space converges, it must converge to a unique point.

### 12B. *Ultrafilters: lattices of filters*

1. The intersection of any number of filters on $X$ is a filter on $X$. But the set of all filters on $X$, ordered by $\mathscr{F}_1 \leq \mathscr{F}_2$ iff $\mathscr{F}_1 \subset \mathscr{F}_2$, is not a lattice because if $\mathscr{F}$ and $\mathscr{G}$ are different ultrafilters on $X$, then $\{\mathscr{F}, \mathscr{G}\}$ has no supremum.

2. The collection of all filters on $X$ contained in a given ultrafilter is a complete lattice with 0 and 1. Conversely, if a family of filters has a supremum, then the filters of the family are all contained in some single ultrafilter.

3. Under what condition is a filter the intersection of the ultrafilters containing it?

### 12C. *Ultrafilters: uniqueness*

If a filter $\mathscr{F}$ is contained in a unique ultrafilter $\mathscr{F}'$, then $\mathscr{F} = \mathscr{F}'$.

### 12D. *Nets and filters: the translation process*

1. A net $(x_\lambda)$ has $x$ as a cluster point iff the filter generated by $(x_\lambda)$ has $x$ as a cluster point.

2. A filter $\mathscr{F}$ has $x$ as a cluster point iff the net based on $\mathscr{F}$ has $x$ as a cluster point.

3. If $(x_{\lambda_\mu})$ is a subnet of $(x_\lambda)$, then the filter generated by $(x_{\lambda_\mu})$ is finer than the filter generated by $(x_\lambda)$.

4. The net based on an ultrafilter is an ultranet and the filter generated by an ultranet is an ultrafilter.

5. The net based on a free ultrafilter is a nontrivial ultranet. Hence, assuming the axiom of choice, there are nontrivial ultranets.

### 12E. *$\mathscr{P}$-filters*

Let $\mathscr{P}$ be a class of subsets of a topological space such that if $P_1$ and $P_2$ are sets from $\mathscr{P}$, then $P_1 \cap P_2$ and $P_1 \cup P_2$ belong to $\mathscr{P}$. A *$\mathscr{P}$-filter* on $X$ is a collection $\mathscr{F}$ of nonempty elements of $\mathscr{P}$ with the properties:

a) $P_1, P_2 \in \mathscr{F}$ implies $P_1 \cap P_2 \in \mathscr{F}$,

b) $P_1 \in \mathscr{F}$, $P_1 \subset P_2 \in \mathscr{P}$ implies $P_2 \in \mathscr{F}$.

A $\mathscr{P}$-*ultrafilter* is a maximal $\mathscr{P}$-filter.

A $\mathscr{P}$-filter $\mathscr{F}$ *converges* to $p \in X$ iff each nhood of $p$ contains an element of $\mathscr{F}$, and this definition is applied even when the $\mathscr{P}$-filter is defined on a dense subset of $X$ rather than on $X$ itself. A $\mathscr{P}$-filter $\mathscr{F}$ has $p$ as a *cluster point* iff $p$ belongs to the closure of each $P \in \mathscr{F}$. Then if $\mathscr{P}$ consists of closed sets, a $\mathscr{P}$-filter $\mathscr{F}$ has a cluster point iff $\bigcap \{P \mid P \in \mathscr{F}\} \neq \varnothing$.

The most important examples of $\mathscr{P}$-filters are obtained as follows:

a) $\mathscr{P}$ = all subsets of $X$; then the $\mathscr{P}$-filters are the filters in $X$, as defined in 12.1, and the theory we are about to outline reduces to the material of this section.

b) $\mathscr{P}$ = all open subsets of $X$; then the $\mathscr{P}$-filters are called *open filters*.

c) $\mathscr{P}$ = all closed subsets of $X$; then the $\mathscr{P}$-filters are called *closed filters*.

d) $\mathscr{P}$ = all zero sets in $X$ = all sets of the form $f^{-1}(0)$ for $f: X \to \mathbf{I}$ continuous; then the $\mathscr{P}$-filters are called *z-filters*.

Each of these collections $\mathscr{P}$, except the last, is known to satisfy the requirement set out at the beginning of this problem. Part 1 below takes care of the zero sets also.

1. If $Z_n$ is a zero set in $X$, for $n = 1, 2, \ldots$, then so is $\bigcap_{n=1}^{\infty} Z_n$. [Let $Z_n = f_n^{-1}(0)$. Prove $g(x) = \sum_{n=1}^{\infty} (|f_n(x)|/2^n)$ is continuous from $X$ to $\mathbf{I}$ and $g^{-1}(0) = \bigcap_{n=1}^{\infty} Z_n$.] Also, if $Z_1$ and $Z_2$ are zero sets, so is $Z_1 \cup Z_2$.

2. Every $\mathscr{P}$-filter is contained in a $\mathscr{P}$-ultrafilter.

3. For a $\mathscr{P}$-filter $\mathscr{F}$, the following are equivalent:

a) $\mathscr{F}$ is a $\mathscr{P}$-ultrafilter,

b) whenever $P \in \mathscr{P}$ and $P \cap F \neq \varnothing$ for each $F \in \mathscr{F}$, then $P \in \mathscr{F}$.

4. Suppose $p \in X$ has a base of nhoods with property $\mathscr{P}$. If a $\mathscr{P}$-ultrafilter has $p$ as a cluster point, then it converges to $p$.

5. Every $\mathscr{P}$-ultrafilter is *prime*; i.e., if $P_1$ and $P_2$ belong to $\mathscr{P}$ and $P_1 \cup P_2 \in \mathscr{F}$, then $P_1 \in \mathscr{F}$ or $P_2 \in \mathscr{F}$.

6. Every prime filter is an ultrafilter, but there are prime $z$-filters which are not $z$-ultrafilters. [In $\mathbf{R}$, let $J = \{1/n \mid n = 1, 2, \ldots\}$. The sets $J_m = \{1/n \mid n = m, m+1, \ldots\}$ form a filter base for a filter on $J$, and this filter is contained in some ultrafilter $\mathscr{U}$ on $J$. Define $\mathscr{F}$ to be the collection of all zero-sets $Z$ in $\mathbf{R}$ such that $Z \cap J \in \mathscr{U}$. Then $\mathscr{F}$ is a prime $z$-filter, but one can use part 4 above to show that $\mathscr{F}$ is not a $z$-ultrafilter.]

7. If $X$ is $T_3$, every prime $z$-filter is contained in a unique $z$-ultrafilter (compare with part 6 and 12C). But not every prime $\mathscr{P}$-filter is contained in a unique $\mathscr{P}$-ultrafilter. [Let $X = \{a, b, c\}$, $\mathscr{P} = \{\phi, \{a\}, \{b\}, \{a, b\}, X\}$.]

8. If $\mathscr{P}$ satisfies the condition that whenever $A, B \in \mathscr{P}$ and $A \cap B = \varnothing$, then there are $C, D \in \mathscr{P}$ such that $A \cap C = B \cap D = \varnothing$ and $C \cup D = X$, then every prime $\mathscr{P}$-filter is contained in a $\mathscr{P}$-ultrafilter. In particular, if $X$ is normal, every prime closed filter is contained in a closed ultrafilter.

## 12F. *Mappings of $\mathscr{P}$-filters*

For each topological space $X$, let $\mathscr{P}_X$ be a collection of subsets of $X$ such that if $f: X \to Y$ is continuous and $Q \in \mathscr{P}_Y$, then $f^{-1}(Q) \in \mathscr{P}_X$. (For example, each of the collections $\mathscr{P}$ described in the previous problem has this property.)

Let $f: X \to Y$ be continuous, $\mathscr{F}$ a $\mathscr{P}_X$-filter (12E) on $X$.

1. $f^{\#}(\mathscr{F}) = \{Q \in \mathscr{P}_Y \mid f^{-1}(Q) \in \mathscr{F}\}$ is a $\mathscr{P}_Y$-filter on $Y$.

2. Let $X$ and $Y$ be $T_3$. If $\mathscr{F}$ is a prime $z$-filter on $X$, $f^{\#}(\mathscr{F})$ is a prime $z$-filter on $Y$. In particular, if $\mathscr{F}$ is a $z$-ultrafilter on $X$, $f^{\#}(\mathscr{F})$ is contained in a unique $z$-ultrafilter on $Y$.

## 12G. *Open ultrafilters*

An open filter in a space $X$ is a $\mathscr{P}$-filter (12E) where $\mathscr{P}$ is the collection of open subsets of $X$. An open ultrafilter is a maximal open filter; by 12E.2, every open filter is contained in an open ultrafilter.

Show that the following are equivalent, for an open filter $\mathscr{F}$ on a topological space $X$:

a) $\mathscr{F}$ is an open ultrafilter,

b) if $G$ is any open set in $X$ and $G \cap H \neq \varnothing$ for each $H \in \mathscr{F}$, then $G \in \mathscr{F}$,

c) if $G$ is open and $G \notin \mathscr{F}$, then $X - \overline{G} \in \mathscr{F}$.

[If you did 12E, then in part 4 you showed (a) equivalent to (b).]

Chapter 5

# Separation and Countability

## 13 The separation axioms

Our definition of a topology admits structures which are, for most purposes, useless. The trivial topology on $X$, for example, makes $X$ look not much different from a single point, topologically. It would be much nicer if some of the set-theoretic structure of $X$ were reflected in its topology. What is needed, apparently, is a requirement that the topology on $X$ contain enough open sets to distinguish between the points of $X$, in some way. Increasing amounts of the sort of point separation needed can be introduced by requiring that $X$ satisfy one of the *separation axioms* (or, in German, *Trennungsaxiome*) $T_0$, $T_1$ or $T_2$.

**13.1 Definition.** A topological space $X$ is a $T_0$-*space* (or, the topology on $X$ is $T_0$) iff whenever $x$ and $y$ are distinct points in $X$, there is an open set containing one and not the other.

**13.2 Examples.** a) The trivial topology on a set $X$ of more than one point is not $T_0$.

b) The difference between pseudometrics and metrics is purely topological. In fact, *a pseudometric $\rho$ on $X$ is a metric iff the topology it generates is $T_0$*. For if the topology generated by $\rho$ is $T_0$, then whenever $x \neq y$ in $X$, there is some open set, and hence some $\epsilon$-disk, about one not containing the other. Then $\rho(x, y) \geq \epsilon > 0$, showing $\rho$ is a metric. Conversely, if $\rho$ is a metric then any two distinct points $x$ and $y$ are at some positive distance $\epsilon$ and hence the $\epsilon$-disk about $x$ is an open set containing $x$ and not $y$.

c) Let $X$ be any topological space and define $\sim$ on $X$ by $x \sim y$ iff $\overline{\{x\}} = \overline{\{y\}}$. Then $\sim$ is an equivalence relation on $X$ and the resulting quotient space $X/\sim$ is a $T_0$-space (the latter following easily from the observation that a space is $T_0$ iff whenever $x \neq y$ then $\overline{\{x\}} \neq \overline{\{y\}}$.) This procedure, and the space it produces, are referred to as the $T_0$-*identification of* $X$. You will prove the statements made here in Exercise 13C, as well as the additional fact that the $T_0$-identification and the metric identification are the same for any pseudometric space.

d) Subspaces and products of $T_0$ spaces are $T_0$; quotients need not be. See Exercise 13B.

**13.3 Definition.** A topological space $X$ is a $T_1$-*space* iff whenever $x$ and $y$ are distinct points in $X$, there is a nhood of each not containing the other.

Evidently, every $T_1$-space is $T_0$. But the set $X = \{a, b\}$ with the topology consisting of the open sets $\varnothing$, $\{a\}$ and $X$ is a $T_0$-space which is not $T_1$.

We can leave the proofs that subspaces and products of $T_1$-spaces are $T_1$, and the result on quotients of $T_1$-spaces, to Exercise 13B. The following theorem makes that exercise easy.

**13.4 Theorem.** *The following are equivalent, for a topological space $X$:*

a) *$X$ is $T_1$,*

b) *each one-point set in $X$ is closed,*

c) *each subset of $X$ is the intersection of the open sets containing it.*

*Proof.* a) $\Rightarrow$ b): If $X$ is $T_1$ and $x \in X$, then each $y \neq x$ has a nhood disjoint from $\{x\}$, so $X - \{x\}$ is an open set and thus $\{x\}$ is closed.

b) $\Rightarrow$ c): If $A \subset X$, then $A$ is the intersection of all sets of the form $X - \{x\}$, for $x \notin A$, and each of these is open, since one-point sets are closed.

c) $\Rightarrow$ a): If (c) holds, then $\{x\}$ is the intersection of its open nhoods and hence for any $y \neq x$, there is an open set containing $x$ and not $y$. ■

The real importance of $T_1$-spaces lies in the observation above: they are the spaces in which points are closed. The more restricted Hausdorff spaces about to be introduced will also have this property, however, and will have in addition an all-important unique-limits property. Thus the following separation property is the most important of those mentioned so far.

**13.5 Definition.** A space $X$ is a $T_2$-*space* (*Hausdorff space*) iff whenever $x$ and $y$ are distinct points of $X$, there are disjoint open sets $U$ and $V$ in $X$ with $x \in U$ and $y \in V$.

Evidently, every $T_2$-space is $T_1$.

**13.6 Examples.** a) Let $X$ be any infinite set with the cofinite topology (in which the closed sets are the finite sets and $X$). Since one-point sets are closed, $X$ is a $T_1$-space. But no two nonempty open sets are disjoint, so $X$ cannot be Hausdorff.

b) Every metric space is Hausdorff. If $x$ and $y$ are distinct points, then $\rho(x, y) = \epsilon > 0$, so the disks $U(x, \epsilon/2)$ and $U(y, \epsilon/2)$ are disjoint open sets containing $x$ and $y$ respectively.

**13.7 Theorem.** *The following are equivalent for a topological space $X$:*

a) *$X$ is Hausdorff,*

b) *limits in $X$ are unique (i.e., no net or filter in $X$ converges to more than one point),*

c) *the diagonal $\Delta = \{(x, x) \mid x \in X\}$ is closed in $X \times X$.*

*Proof.* First note that by the translation process between nets and filters, unique net limits imply unique filter limits and vice versa.

a) $\Rightarrow$ b): We will use filters. Suppose $X$ is Hausdorff and $\mathscr{F}$ is a filter on $X$ with $\mathscr{F} \to x$ and $\mathscr{F} \to y$. Then each nhood $U$ of $x$ and each nhood $V$ of $y$ belongs to $\mathscr{F}$, so $U \cap V \neq \varnothing$. But $X$ is $T_2$, so we must then have $x = y$.

b) $\Rightarrow$ c): We will use nets. If $\Delta$ is not closed, then for some $x \neq y$, a net $((x_\lambda, x_\lambda))$ in $\Delta$ converges to $(x, y)$. But then $(x_\lambda)$ is a net in $X$ converging to both $x$ and $y$, which is impossible.

c) $\Rightarrow$ a): Suppose $\Delta$ is closed. If $x \neq y$ in $X$, then $(x, y) \notin \Delta$, and hence there is a basic nhood $U \times V$ of $(x, y)$ in $X \times X$ which does not meet $\Delta$. But then $U$ and $V$ are disjoint nhoods of $x$ and $y$, respectively. Thus $X$ is Hausdorff. ■

Most of the literature in topology, including the monograph in which Hausdorff first introduced topological spaces, deals exclusively with Hausdorff spaces. The underlying reason for this is the existence of unique limits in Hausdorff spaces, which has pleasant consequences (for example, continuous functions with Hausdorff range are determined by their values on a dense set; see Theorem 13.14).

We will develop now the answers to some of the natural questions about products, subspaces and continuous images of Hausdorff spaces, following a pattern we will repeat with every important topological property we introduce.

**13.8 Theorem.** a) *Every subspace of a $T_2$-space is $T_2$.*

b) *A nonempty product space is $T_2$ iff each factor space is $T_2$.*

c) *Quotients of $T_2$-spaces need not be $T_2$.*

*Proof.* a) If $X$ is $T_2$ and $A$ is a subspace of $X$, distinct points $a$ and $b$ in $A$ have disjoint nhoods $U$ and $V$ in $X$ and then $U \cap A$ and $V \cap A$ are disjoint nhoods of $a$ and $b$ in $A$.

b) If $X_\alpha$ is a $T_2$-space, for each $\alpha \in A$, and $x \neq y$ in $\prod X_\alpha$, then for some coordinate $\alpha$, $x_\alpha \neq y_\alpha$, so disjoint nhoods $U_\alpha$ of $x_\alpha$ and $V_\alpha$ of $y_\alpha$ can be found in $X_\alpha$. Now $\pi_\alpha^{-1}(U_\alpha)$ and $\pi_\alpha^{-1}(V_\alpha)$ are disjoint nhoods of $x$ and $y$, respectively, in $\prod X_\alpha$.

Conversely, if $\prod X_\alpha$ is a nonempty $T_2$-space, pick a fixed point $b_\alpha \in X_\alpha$, for each $\alpha \in A$. Then the subspace $B_\alpha = \{x \in \prod X_\alpha \mid x_\beta = b_\beta \text{ unless } \beta = \alpha\}$ is $T_2$, by part (a), and is homeomorphic to $X_\alpha$ under the restriction to $B_\alpha$ of the projection map. Thus $X_\alpha$ is $T_2$, for each $\alpha$.

c) See the following examples. ■

**13.9 Examples.** a) The continuous closed image of a Hausdorff space need not be Hausdorff. Let $X$ be the real line, with nhoods as usual except that basic nhoods of 0 have the form $(-\epsilon, \epsilon) - A$, for $\epsilon > 0$, where $A = \{1/n \mid n \in \mathbf{N}\}$. Then $X$ is a Hausdorff space and $A$ is a closed subset of $X$ so the space $X/A$

ifying $A$ with a single point is a closed continuous image of $X$ is clearly upper semicontinuous). But $X/A$ is not Hausdorff, tion of $X$ onto $X/A$ then $p(0)$ and $p(A)$ are distinct points of annot be separated by open sets.

b) The continuous open image of a Hausdorff space need not be Hausdorff. Let $X$ be the union of the lines $y = 0$ and $y = 1$ in $\mathbf{R}^2$ and let $Y$ be the quotient of $X$ obtained by identifying each point $(x, 0)$, for $x \neq 0$, with the corresponding point $(x, 1)$. The resulting projection map $p: X \to Y$ is continuous and open, but $p(0, 0)$ and $p(0, 1)$ are distinct points of $Y$ which do not have disjoint nhoods.

The situation outlined in the examples above is quite unpleasant. Not only do continuous images of Hausdorff spaces fail, in general, to be Hausdorff, but even the best sorts of quotient maps may not preserve the $T_2$-axiom. This provokes the following series of results, giving various necessary conditions and sufficient conditions for image spaces to be Hausdorff, culminating with a characterization of the continuous open maps on any space $X$ which have Hausdorff range (13.12). The best available result on continuous closed images of Hausdorff spaces requires the prior development of compactness and is given in Exercise 17N.

**13.10 Theorem.** *If $f: X \to Y$ is continuous and $Y$ is Hausdorff, then*

$$\{(x_1, x_2) \mid f(x_1) = f(x_2)\}$$

*is a closed subset of $X \times X$.*

*Proof.* Let $A = \{(x_1, x_2) \mid f(x_1) = f(x_2)\}$. If $(x_1, x_2) \notin A$, then $f(x_1)$ and $f(x_2)$ are distinct and hence have disjoint nhoods $U$ and $V$ in $Y$. Then since $f$ is continuous, $f^{-1}(U)$ and $f^{-1}(V)$ are nhoods of $x_1$ and $x_2$ respectively, so $f^{-1}(U) \times f^{-1}(V)$ is a nhood of $(x_1, x_2)$. Obviously this nhood cannot meet $A$, so $A$ is closed. ■

**13.11 Theorem.** *If $f$ is an open map of $X$ onto $Y$ and the set*

$$\{(x_1, x_2) \mid f(x_1) = f(x_2)\}$$

*is closed in $X \times X$, then $Y$ is Hausdorff.*

*Proof.* Suppose $f(x_1)$ and $f(x_2)$ are distinct points of $Y$. Then

$$(x_1, x_2) \notin A = \{(x_1, x_2) \mid f(x_1) = f(x_2)\},$$

so there are open nhoods $U$ of $x_1$ and $V$ of $x_2$ such that $(U \times V) \cap A = \varnothing$. Then, since $f$ is open, $f(U)$ and $f(V)$ are nhoods of $f(x_1)$ and $f(x_2)$, respectively, and $f(U) \cap f(V) = \varnothing$ (otherwise $(U \times V) \cap A \neq \varnothing$). ■

**13.12 Theorem.** *If $f$ is a continuous open map of $X$ onto $Y$, then $Y$ is Hausdorff iff $\{(x_1, x_2) \mid f(x_1) = f(x_2)\}$ is a closed subset of $X \times X$.*

*Proof.* Simply combine 13.10 and 13.11. ■

We close this section with a result which implies that a continuous function which takes values in a Hausdorff space is determined once its values on a dense set are known. This result will have important ramifications later when we spend a great deal of time extending functions on subsets of $X$ to $X$ itself, since it implies that extensions of functions on dense subsets of $X$, when they exist, are unique.

***13.13 Theorem.*** *If $f, g: X \to Y$ are continuous and $Y$ is Hausdorff, then $\{x \mid f(x) = g(x)\}$ is closed in $X$.*

*Proof.* Let $A = \{x \mid f(x) = g(x)\}$. If $(x_\lambda)$ is a net in $A$ and $x_\lambda \to x$, then by continuity we have both $f(x_\lambda) \to f(x)$ and $g(x_\lambda) \to g(x)$ in $Y$. Since $f(x_\lambda) = g(x_\lambda)$ for each $\lambda$ and limits are unique in $Y$, we must have $f(x) = g(x)$. Thus $x \in A$ and $A$ is closed. ∎

***13.14 Corollary.*** *If $f, g: X \to Y$ are continuous, $Y$ is Hausdorff, and $f$ and $g$ agree on a dense set $D$ in $X$, then $f = g$.*

## Problems

### 13A. *Examples*

1. Let $B$ be a fixed subset of a set $X$ and for each nonempty $A \subset X$, define $\bar{A} = A \cup B$. This defines a topology on $X$ (according to 3A.3). Under what conditions on $B$ is the resulting space $T_0$? $T_1$? $T_2$?

2. If $\tau$ is a Hausdorff topology on $X$, any finer topology is also Hausdorff. The radial plane (3A.4), the Sorgenfrey line (4.6), the Moore plane (4B), the slotted plane (4C), the scattered line $\mathbf{S}$ (5C), and any simple extension (3A.5) of a Hausdorff topology are thus all Hausdorff.

3. The looped line (4D) is Hausdorff.

4. Recall that the sets $V(f, \epsilon)$ defined for $f \in \mathbf{R}^\mathbf{I}$ by

$$V(f, \epsilon) = \{g \in \mathbf{R}^\mathbf{I} \mid |g(x) - f(x)| < \epsilon, \text{ for each } x \in \mathbf{I}\}$$

form a nhood base at $f$, making $\mathbf{R}^\mathbf{I}$ a topological space (see 4F.3). Discuss the separation axioms for this space. (Note that the subspace of continuous functions on $\mathbf{I}$ is metrizable, by 4F.5, and thus has all the separation properties we could ask for.)

### 13B. *$T_0$- and $T_1$-spaces*

1. Any subspace of a $T_0$- or $T_1$-space is, respectively, $T_0$ or $T_1$.

2. Any nonempty product space is $T_0$ or $T_1$ iff each factor space is, respectively, $T_0$ or $T_1$.

3. Quotients of $T_1$-spaces need not be $T_0$, but the closed image of a $T_1$-space is $T_1$.

4. A quotient space of $X$ is $T_1$ iff each element of the corresponding decomposition is closed in $X$.

### 13C. *The $T_0$-identification*

For any topological space $X$, define $\sim$ by $x \sim y$ iff $\overline{\{x\}} = \overline{\{y\}}$.

1. $\sim$ is an equivalence relation on $X$.

2. The resulting quotient space $X/\sim\, = \tilde{X}$ is $T_0$.

3. The procedure above, when applied to a pseudometric space $(S, \rho)$ yields the metric identification $S^*$ of $S$ described in 2C.

## 13D. *The Zariski topology*

For a polynomial $P$ in $n$ real variables, let $Z(P) = \{(x_1, \ldots, x_n) \in \mathbf{R}^n \mid P(x_1, \ldots, x_n) = 0\}$. Let $\mathscr{P}$ be the collection of all such polynomials.

1. $\{Z(P) \mid P \in \mathscr{P}\}$ is a base for the closed sets of a topology (the *Zariski topology*) on $\mathbf{R}^n$.

2. The Zariski topology on $\mathbf{R}^n$ is $T_1$ but not $T_2$.

3. On $\mathbf{R}$, the Zariski topology coincides with the cofinite topology; in $\mathbf{R}^n$, $n > 1$, they are different.

## 13E. *Accumulation points and condensation points*

Recall that $a$ is an accumulation point of a set $A$ in a space $X$ iff each nhood of $a$ meets $A$ in some point other than $a$. We say $a$ is a *condensation point* of $A$ iff each nhood of $a$ meets $A$ in uncountably many points. Let $A'$ denote the set of accumulation points of $A$, $A^c$ the set of condensation points of $A$.

1. In a $T_1$-space, $a$ is an accumulation point of $A$ iff each nhood of $a$ meets $A$ in an infinite set.

2. For any set $A$, $A'$ and $A^c$ are closed sets, with $A^c \subset A'$.

3. Given a set $A$, let $A^1 = A'$, $A^2 = (A^1)'$, $A^3 = (A^2)'$ and so on. Then $A^1 \supset A^2 \supset \cdots$.

4. For any positive integer $n$, there is a set $A \subset \mathbf{R}$ such that $A, A^1, \ldots, A^{n-1}$ are nonempty, and $A^n = \varnothing$. (The result can be extended to countable ordinals. Let $A^\alpha = (A^{\alpha-1})'$ if $\alpha$ is a nonlimit ordinal, $A^\alpha = (\bigcap_{\beta<\alpha} A^\beta)'$ if $\alpha$ is a limit ordinal, and show that for any $\alpha < \omega_1$, a set $A$ can be found such that $A^\alpha = \varnothing$ and $A^\beta \neq \varnothing$ for $\beta < \alpha$).

5. Can the results 3 and 4 be proved for condensation points?

## 13F. *Hausdorffness and the lattice of topologies*

Let $\tau_1$ and $\tau_2$ be Hausdorff topologies on the same set $X$.

1. If $(X, \tau_1 \cap \tau_2)$ is Hausdorff, then the diagonal is closed in $(X, \tau_1) \times (X, \tau_2)$.

2. There are Hausdorff topologies $\tau_1$ and $\tau_2$ on a set $X$ such that the diagonal is closed in $(X, \tau_1) \times (X, \tau_2)$, but $(X, \tau_1 \cap \tau_2)$ is not Hausdorff. Thus the condition in 1 is necessary but not sufficient.

3. If disjoint $\tau_1$-open sets can be separated by disjoint $\tau_2$-open sets and vice versa, then $(X, \tau_1 \cap \tau_2)$ is Hausdorff.

4. The condition of 3 is not necessary.

The situation with suprema in the lattice of topologies is a bit more satisfactory. In fact, if $\{\tau_\alpha \mid \alpha \in A\}$ is any family of topologies on the same set $X$, then $(X, \sup(\tau_\alpha))$ is embedded in $\prod (X, \tau_\alpha)$, by 8I.2, so that any property preserved by products and subspaces will be preserved in passing to suprema. The $T_2$-axiom (as well as any other separation axiom from $T_0$ up through complete regularity) is thus inherited by suprema (by 13.8, 13B, 14.4 and 14.10).

## 13G. *Topological groups*

A *topological group* $G$ is a group with a Hausdorff topology satisfying the conditions:

a) multiplication is continuous; that is, the map $m: G \times G \to G$ defined by $m(x, y) = xy$ is continuous.

b) inversion is continuous; that is, the map $O: G \to G$ defined by $O(x) = x^{-1}$ is continuous.

The identity in $G$ is denoted $e$.

If $A \subset G$, $B \subset G$ and $x \in G$, the set $\{y \cdot z \mid y \in A, z \in B\}$ is denoted $AB$, and the sets $A^{-1}$, $xA$, $Ax$ are similarly defined. The set $A^2 = AA$, in particular, is the set $\{a \cdot a' \mid a, a' \in A\}$, *not* the set of all squares $a^2$ for $a \in A$.

1. The continuity conditions (a) and (b) can be expressed as follows:

a') for each nhood $W$ of $xy$ there are nhoods $U$ of $x$ and $V$ of $y$ such that $UV \subset W$,
b') for each nhood $W$ of $x^{-1}$ there is a nhood $U$ of $x$ such that $U^{-1} \subset W$.

2. The conditions (a) and (b) can be replaced by the single condition that the map $c: G \times G \to G$ defined by $c(x, y) = x \cdot y^{-1}$ be continuous.

3. **R**, with the usual topology and addition, is a topological group. Any group with the discrete topology is a topological group.

4. Let $a, b \in G$. Each of the maps

$$x \to x^{-1}$$
$$x \to ax$$
$$x \to xb$$
$$x \to axb$$

is a homeomorphism of $G$ onto $G$.

5. If $\{U \mid U \in \mathscr{U}\}$ is a nhood base at $e$, then for any $x \in G$, $\{xU \mid U \in \mathscr{U}\}$ is a nhood base at $x$, and so is $\{Ux \mid U \in \mathscr{U}\}$.

6. Let $\mathscr{U}$ be a nhood base of open sets at $e$ in $G$. Then

a) for each $U \in \mathscr{U}$, there exists $V \in \mathscr{U}$ with $V^2 \subset U$,
b) for each $U \in \mathscr{U}$, there exists $V \in \mathscr{U}$ with $V^{-1} \subset U$,
c) for each $U \in \mathscr{U}$ and $x \in U$, there exists $V \in \mathscr{U}$ with $xV \subset U$,
d) for each $U \in \mathscr{U}$ and $x \in G$, there exists $V \in \mathscr{U}$ with $xVx^{-1} \subset U$,
e) for each $U, V \in \mathscr{U}$, there exists $W \in \mathscr{U}$ with $W \subset U \cap V$,
f) $\{e\} = \bigcap \{U \mid U \in \mathscr{U}\}$.

Conversely, given any collection of sets satisfying (a)–(f) and using 5 to obtain a nhood base at each $x \in G$, the result is a topology on $G$ making $G$ a topological group.

7. The open symmetric nhoods of $e$ form a base. [If $U$ is open and a nhood of $e$, so is $U^{-1}$ and thus so is $U \cap U^{-1}$].

Look at

## 13H. *Open images of Hausdorff spaces*

1. Given any set $X$, there is a Hausdorff space $Y$ which is the union of a collection $\{Y_x \mid x \in X\}$ of disjoint subsets, each dense in $Y$.

2. If $X$ is any topological space and $Y$ is the space formed in part 1, let

$$Z = \{(x, y) \in X \times Y \mid y \in Y_x\}.$$

Then the restriction to $Z$ of the projection map from $X \times Y$ to $X$ is a continuous open map of $Z$ onto $X$. Thus *every topological space is the continuous open image of a Hausdorff space.*

## 14 Regularity and complete regularity

The separation axioms introduced in the previous section are rather weak and are added to the hypotheses of a theorem, if needed, without too much regret. Some theorems are simply not true for the trivial topology!

The properties to be introduced in this and the next section are rather more restrictive, although they are also defined in terms of separation. For one thing, we pass from a simple relationship in which the topology separates points to a more complex one in which the topology separates points from closed sets or closed sets from each other. Some pretty decent topologies are eliminated in the transition, so the concepts to be introduced now are not used in theorems without some attempt to justify their presence.

**14.1 Definition.** A topological space $X$ is a *regular space* iff whenever $A$ is closed in $X$ and $x \notin A$, then there are disjoint open sets $U$ and $V$ with $x \in U$ and $A \subset V$.

We have slipped backwards, in passing from Hausdorff to regular spaces, in the sense that the topology on a regular space $X$ may no longer reflect the set theoretic character of $X$. For example, a trivial space is always regular and thus a regular space need not be Hausdorff.

To remedy this deficiency, we note that separation of points from closed sets would imply separation of points if points were closed. Thus we define a *$T_3$-space* to be a regular $T_1$-space.

Clearly, then, every $T_3$-space is $T_2$.

**14.2 Example.** Not every $T_2$-space is $T_3$. Let $X$ be the real line with nhoods of any nonzero point being as in the usual topology, while nhoods of 0 will have the form $U - A$, where $U$ is a nhood of 0 in the usual topology and

$$A = \{1/n \mid n = 1, 2, \ldots\}.$$

Then $X$ is Hausdorff since this topology on the line is finer than the usual topology which is Hausdorff. But $A$ is closed in $X$ and cannot be separated from 0 by disjoint open sets, so $X$ is not $T_3$.

**14.3 Theorem.** *The following are equivalent for a topological space $X$:*

a) *$X$ is regular*

b) *if $U$ is open in $X$ and $x \in U$, then there is an open set $V$ containing $x$ such that $\overline{V} \subset U$.*

c) *each $x \in X$ has a nhood base consisting of closed sets.*

*Proof.* a) $\Rightarrow$ b): Suppose $X$ is regular, $U$ is open in $X$ and $x \in U$. Then $X - U$ is a closed set in $X$ not containing $x$, so disjoint open sets $V$ and $W$ can be found with $x \in V$ and $X - U \subset W$. Then $X - W$ is a closed set contained in $U$ and containing $V$, so $\overline{V} \subset U$.

b) $\Rightarrow$ c): If (b) applies, then every open set $U$ containing $x$ contains a closed nhood (namely $\overline{V}$) of $x$, so the closed nhoods of $x$ form a nhood base.

c) $\Rightarrow$ a): Suppose (c) applies and $A$ is a closed set in $X$ not containing $x$. Then $X - A$ is a nhood of $x$, so there is a closed nhood $B$ of $x$ with $B \subset X - A$. Then $B^\circ$ and $X - B$ are disjoint open sets containing $x$ and $A$, respectively. Thus $X$ is regular. ■

***14.4 Theorem.*** a) *Every subspace of a regular space ($T_3$-space) is regular ($T_3$).*

b) *A nonempty product space is regular ($T_3$) iff each factor space is regular ($T_3$).*

c) *Quotients of $T_3$-spaces need not be regular.*

*Proof.* It suffices to prove parts (a) and (b) for regular spaces; the assertions for $T_3$-spaces will then follow by combination with the corresponding results for $T_1$-spaces (13B).

a) If $X$ is regular, $Y \subset X$, and $A$ is a closed set in $Y$, then $A = B \cap Y$ where $B$ is closed in $X$. Now if $y$ is a point of $Y$ and $y \notin A$, then $y \notin B$, so there are disjoint open sets $U$ and $V$ in $X$ such that $y \in U$ and $B \subset V$. Then $U \cap Y$ and $V \cap Y$ are disjoint open sets in $Y$ containing $y$ and $A$, respectively.

b) If $\prod X_\alpha$ is regular and nonempty then each $X_\alpha$, since it is homeomorphic to a subspace of $\prod X_\alpha$, is regular. Conversely, suppose that each $X_\alpha$ is regular. Pick $x \in \prod X_\alpha$ and consider a basic nhood $\pi_{\alpha_1}^{-1}(U_1) \cap \cdots \cap \pi_{\alpha_n}^{-1}(U_n)$ of $x$ in $\prod X_\alpha$. Now $U_i$ is a nhood of $x_{\alpha_i}$ in $X_{\alpha_i}$, for $i = 1, \ldots, n$, and hence $U_i$ contains a closed nhood $C_i$ of $x_{\alpha_i}$. But then $\pi_{\alpha_1}^{-1}(C_1) \cap \cdots \cap \pi_{\alpha_n}^{-1}(C_n)$ is a closed nhood of $x$ contained in $\pi_{\alpha_1}^{-1}(U_1) \cap \cdots \cap \pi_{\alpha_n}^{-1}(U_n)$. Thus the closed nhoods of $x$ form a nhood base at $x$, showing that $\prod X_\alpha$ is regular.

c) See the following examples. ■

**14.5 Examples.** a) A closed continuous image of a $T_3$-space need not be $T_2$; if it is $T_2$, it need not be regular. Let $\mathbf{\Gamma}$ denote the closed upper half plane $\{(x, y) \mid y \geq 0\}$ in $\mathbf{R}^2$, with the topology specified as follows: nhoods of points $(x, y)$ with $y > 0$ will be as in the usual topology while basic nhoods of points $z$ on the $x$-axis in $\mathbf{\Gamma}$ will be sets of the form $\{z\} \cup A$, where $A$ is the interior of a circle in the upper half plane tangent to the $x$-axis at $z$. This space is the *Moore plane*. It was the object of study in Exercise 4B.

$\mathbf{\Gamma}$ is certainly Hausdorff. Since a base of closed nhoods can easily be constructed at each point of $\mathbf{\Gamma}$, it follows from 14.3 that $\mathbf{\Gamma}$ is $T_3$. Now let $D$ and $E$ be the sets of points on the $x$-axis in $\mathbf{\Gamma}$ whose first coordinates are rational and irrational, respectively. Then $D$ and $E$ are closed sets in $\mathbf{\Gamma}$ and we will see later

(25F) that $D$ and $E$ cannot be contained in disjoint open sets in $\mathbf{\Gamma}$. If $Y$ is the decomposition space of $\mathbf{\Gamma}$ whose elements are $D$, $E$ and the one-point sets in $\mathbf{\Gamma} - (D \cup E)$, then $Y$ is the image of $\mathbf{\Gamma}$ by a closed continuous map (9E), but $Y$ is not $T_2$ since $D$ and $E$ cannot be separated by disjoint open sets in $Y$. If $Z$ is obtained from $\mathbf{\Gamma}$ by identifying only the points of $D$, then $Z$ is a closed continuous image of $\mathbf{\Gamma}$ which is $T_2$ but not regular. ($Z$ is $T_2$ by 14.7 below, not regular because the point $D$ and the closed set $E$ cannot be separated by disjoint open sets.)

b) The open continuous image of a $T_3$-space need not be regular. In fact, in 13.9(b) we provided a space $X$ which is $T_3$ and a non-Hausdorff $T_1$-space $Y$ which is the image of $X$ under an open continuous map.

The following two theorems constitute a partial apology for the examples just given.

**14.6 Theorem.** *If $X$ is $T_3$ and $f$ is a continuous, open and closed map of $X$ onto $Y$, then $Y$ is $T_2$.*

*Proof.* By 13.11, it is sufficient to show that the set

$$A = \{(x_1, x_2) \in X \times X \mid f(x_1) = f(x_2)\}$$

is closed in $X \times X$. If $(x_1, x_2) \notin A$, then $x_1 \notin f^{-1}[f(x_2)]$ so that, since $X$ is regular, there are disjoint open sets $U$ and $V$ with $x \in U$ and $f^{-1}[f(x_2)] \subset V$. Since $f$ is closed, we can find a saturated open set in $X$ containing $f^{-1}[f(x_2)]$ and contained in $V$; that is, $f^{-1}[f(x_2)] \subset f^{-1}(W) \subset V$ for some open set $W$ in $Y$. Then $U \times f^{-1}(W)$ is a nhood of $(x_1, x_2)$ which cannot meet $A$, since $U \cap f^{-1}(W) = \emptyset$. ■

**14.7 Theorem.** *If $X$ is $T_3$ and $Y$ is obtained from $X$ by identifying a single closed set $A$ in $X$ with a point, then $Y$ is $T_2$.*

*Proof.* If $y_1$ and $y_2$ are distinct points of $Y$, then $f^{-1}(y_1)$ and $f^{-1}(y_2)$ are a point and a disjoint closed set (not necessarily in that order) in $X$ and hence there are disjoint open sets $U$ and $V$ in $X$ containing $f^{-1}(y_1)$ and $f^{-1}(y_2)$. Now $U$ and $V$ can be taken as saturated since $f$ is closed (9.8, 9.10). Then $U = f^{-1}(S)$ and $V = f^{-1}(T)$, where $S$ and $T$ are open sets in $Y$ which must contain $y_1$ and $y_2$, respectively. Since $U$ and $V$ are disjoint, so are $S$ and $T$. ■

The next axiom of separation which would seem natural would involve separating disjoint closed sets by disjoint open sets. We will set aside the study of this property, normality, until the next section, however, and take up a separation property intermediate between regularity and normality which has assumed a dominant role in the study of topology, primarily by virtue of Theorems 14.12 and 14.13.

**14.8 Definition.** A topological space $X$ is *completely regular* iff whenever $A$ is a closed set in $X$ and $x \notin A$, there is a continuous function $f: X \to \mathbf{I}$ such that

$f(x) = 0$ and $f(A) = 1$. It is clearly enough to find a continuous function $f: X \to \mathbf{R}$ such that $f(x) = b$ and $f(A) = a$, where $b \neq a$. Any such function $f$ will be said to *separate* $A$ and $x$. A completely regular $T_1$-space is called a *Tychonoff space*.

Completely regular spaces are regular. For suppose $A$ is closed, $x \notin A$, and $f: X \to \mathbf{I}$ is a continuous function with $f(x) = 0$ and $f(A) = 1$. Then $f^{-1}([0, \frac{1}{2}))$ and $f^{-1}((\frac{1}{2}, 1])$ are disjoint open sets in $X$ containing $x$ and $A$, respectively. But completely regular spaces need not be Hausdorff, as any trivial space of more than one point illustrates, and this is the reason Tychonoff spaces enjoy a separate identity. An early joke has somehow become semistandard, with some writers referring to Tychonoff spaces as $T_{3\frac{1}{2}}$-spaces.

A counterexample exists showing that not every regular space is completely regular. It is formidable and we have relegated it to Exercise 18G, where most people won't be bothered by it. There is an even more complicated example, also noted in 18G, of a $T_3$-space on which every continuous real-valued function is constant!

**14.9 Example.** Every metric space is Tychonoff. In fact, every pseudometrizable space is completely regular. For if $\rho$ is a pseudometric which gives the topology on $X$, $A$ is a closed subset of $X$ and $x \notin A$, then $f(y) = \rho(y, A)$ is a continuous function on $X$ to $\mathbf{R}$ such that $f(A) = 0$ and $f(y) \neq 0$.

We turn now to the basic questions about subspaces, products and quotients of Tychonoff spaces.

**14.10 Theorem.** a) *Every subspace of a completely regular (or Tychonoff) space is completely regular (respectively, Tychonoff).*

b) *A nonempty product space is completely regular (or Tychonoff) iff each factor space is completely regular (respectively, Tychonoff).*

c) *Quotients of Tychonoff spaces need not be completely regular or $T_2$.*

*Proof.* a) Suppose $X$ is completely regular and $Y \subset X$. If $A$ is closed in $Y$, then $A = B \cap Y$ where $B$ is closed in $X$. Given any $x \in Y - A$, $x \notin B$ so there is a continuous $f: X \to \mathbf{R}$ such that $f(x) = 1$, $f(B) = 0$. Then $f \mid Y$ separates $x$ and $A$ in $Y$, so $Y$ is completely regular. The result for Tychonoff spaces now follows from this together with the corresponding (easy) result for $T_1$-spaces.

b) If $\prod X_\alpha$ is nonempty, each $X_\alpha$ is homeomorphic to a subspace of $\prod X_\alpha$ and thus is completely regular if $\prod X_\alpha$ is.

Conversely, suppose $X_\alpha$ is completely regular, for each $\alpha$. Let $x \in \prod X_\alpha$ and let $A$ be a closed set in $\prod X_\alpha$ not containing $x$. Then some basic nhood $\pi_{\alpha_1}^{-1}(U_1) \cap \cdots \cap \pi_{\alpha_n}^{-1}(U_n)$ of $x$ does not meet $A$, where $U_k$ is an open set in $X_{\alpha_k}$. For $k = 1, \ldots, n$ there is a continuous $f_k: X_{\alpha_k} \to \mathbf{I}$ such that $f_k(x_{\alpha_k}) = 1$ and $f_k(X_{\alpha_k} - U_k) = 0$. Define $g: \prod X_\alpha \to \mathbf{I}$ by

$$g(y) = \min \{f_k(y_{\alpha_k}) \mid k = 1, \ldots, n\}.$$

Then $g$ is continuous (it is the infimum of the functions $f_k \circ \pi_{\alpha_k}$, $k = 1, \ldots, n$ and the infimum of finitely many continuous functions is continuous, by 7M.4) and $g(x) = 1$, $g(X - A) = 0$. Thus $\prod X_\alpha$ is completely regular.

Again, the result for Tychonoff spaces is easily derived from the result just given and the corresponding result for $T_1$-spaces.

c) See the following examples. ∎

**14.11 Examples.** a) The closed continuous image of a Tychonoff space need not be $T_2$; if it is $T_2$, it need not be Tychonoff. It is enough to show the Moore plane $\mathbf{\Gamma}$ is Tychonoff; the required closed continuous images are those constructed in 14.5(a). To show $\mathbf{\Gamma}$ is Tychonoff, let $p \in \mathbf{\Gamma}$ and let $V$ be a basic nhood of $p$ (so that $V$ is either a disk centered at $x$ or else $x$ together with a disk tangent to $x$, depending on the placement of $x$). Define $f: \mathbf{\Gamma} \to \mathbf{I}$ by setting $f(p) = 0$, setting $f(x) = 1$ for each $x \notin V$, and defining $f$ linearly along the straight-line segments between $x$ and the points on the boundary of $V$. Then $f$ is a continuous function on $\mathbf{\Gamma}$ such that $f(p) = 0$ and $f(X - V) = 1$. Since any closed set in $\mathbf{\Gamma}$ which does not contain $p$ is contained in $X - V$ for some basic nhood $V$ of $p$, it follows that $\mathbf{\Gamma}$ is Tychonoff.

b) In 13.9(b), we exhibited a space $X$ which is Tychonoff and a non-Hausdorff $T_1$-space $Y$ which is the image of $X$ by an open continuous map.

We close this section with the two theorems which embody much of the importance of completely regular and Tychonoff spaces.

The completely regular spaces are precisely the spaces having enough bounded continuous real-valued functions to determine their topology completely, according to the first of these results.

**14.12 Theorem.** *A topological space $X$ is completely regular iff it has the weak topology induced by its family $C^*(X)$ of bounded real-valued continuous functions.*

*Proof.* If $X$ is completely regular, then the functions in $C^*(X)$ separate points from closed sets so, by 8.15, $X$ has the weak topology induced by $C^*(X)$.

Conversely, suppose $X$ has the weak topology induced by $C^*(X)$. Suppose $U$ is open in $X$ and $x \in U$. There are functions $f_1, \ldots, f_n$ in $C^*(X)$ and subbasic open sets $V_1, \ldots, V_n$ in $\mathbf{R}$ such that

$$x \in f_1^{-1}(V_1) \cap \cdots \cap f_n^{-1}(V_n) \subset U.$$

Each $V_i$ is of the form $(a_i, \infty)$ or $(-\infty, a_i)$. But if $V_i = (-\infty, a_i)$, then

$$f_i^{-1}(V_i) = (-f_i)^{-1}(-a_i, \infty)$$

so that apparently, by occasionally replacing an $f_i$ by $-f_i$, we can assume each $V_i$ above has the form $(a_i, \infty)$. If we denote by $g_i$ the function defined by

$$g_i(x) = \sup\{f_i(x) - a_i, 0\},$$

then evidently $g_i$ is nonnegative and $g_i^{-1}(0, \infty) = f_i^{-1}(a_i, \infty)$. Hence, at this point, we have

$$x \in g_1^{-1}(0, \infty) \cap \cdots \cap g_n^{-1}(0, \infty) \subset U.$$

Finally, let $g = g_1 \cdot g_2 \cdot \cdots \cdot g_n$. Then $g(x) = g_1(x) \cdot \cdots \cdot g_n(x)$ is positive, so $x \in g^{-1}(0, \infty)$. Moreover, if $g(y) > 0$, then each $g_i(y) \neq 0$, so each $g_i(y) > 0$, and hence $y \in g_1^{-1}(0, \infty) \cap \cdots \cap g_n^{-1}(0, \infty) \subset U$. Thus

$$x \in g^{-1}(0, \infty) \subset U.$$

It follows that $g(x) \neq 0$ while $g(X - U) = 0$, so $X$ is completely regular. ■

Any product of closed bounded intervals will be called a *cube*. Thus a cube is (homeomorphic to) a product of copies of the unit interval **I**. We now can give the following elegant and all-important corollary to the previous theorem. (This result will be extended in the section on compactness; see 17.11.)

***14.13 Theorem.*** *A topological space $X$ is a Tychonoff space iff it is homeomorphic to some subspace of some cube.*

*Proof.* Every cube is a product of metric spaces and thus Tychonoff, and hence every subspace of a cube is Tychonoff.

Conversely, suppose $X$ is Tychonoff. Then $X$ is $T_1$ and, by the previous theorem, has the weak topology induced by the bounded continuous functions $f: X \to \mathbf{R}$. Each such function $f$ has a range contained in some closed bounded interval $I_f$ and thus can be regarded as a map of $X$ into $I_f$. Then the evaluation map $e: X \to \prod I_f$ defined by $[e(x)]_f = f(x)$ is a homeomorphism, by 8.16, so $X$ is homeomorphic to a subspace of the cube $\prod I_f$. ■

## Problems

### 14A. *Examples on regularity and complete regularity*

1. The family of all subsets of $X$ containing a fixed subset $A$, together with the empty set, is a topology for $X$ according to 3A.2. Under what conditions is it regular? completely regular?

2. Recall that if $\tau$ is a topology on $X$ and $A$ is a fixed subset of $X$, then the simple extension of $\tau$ over $A$ is the topology $\tau_A = \{U \cup (V \cap A) \mid U, V \in \tau\}$ on $X$. Show that if $\tau$ is regular or completely regular, and $A$ is closed in $X$, then $\tau_A$ has the same property. Find counterexamples if $A$ is not closed.

3. The slotted plane (4C) is $T_2$ but not $T_3$.

4. The looped line (4D) is Tychonoff.

### 14B. *The double of a topological space*

Let $X$ be any $T_1$-space and set $X_1 = X \times \{1\}$, $X_2 = X \times \{2\}$, $D(X) = X_1 \cup X_2$. For each $A \subset X$, let $A_1 = A \times \{1\}$ and $A_2 = A \times \{2\}$ be the corresponding subsets of $X_1$

and $X_2$ in $D(X)$ and, for each $x \in X$, let $x_1$ and $x_2$ be the corresponding points $(x, 1)$ and $(x, 2)$ in $D(X)$. Set $\mathscr{B} = \{U_1 \cup (U_2 - \{x\}) \mid U \text{ open in } X, x \in X\} \cup \{\{x_2\} \mid x \in X\}$.

1. $\mathscr{B}$ is a base for a topology on $D(X)$. With this topology, $D(X)$ will be called the *double* of $X$.

2. $X$ is homeomorphic to the closed subset $X_1$ of $D(X)$.

3. If $X$ is Hausdorff, regular or completely regular, then so is $D(X)$.

### 14C. *Zero sets in completely regular spaces*

A *zero-set* in a topological space $X$ is a set of the form $f^{-1}(0)$ for some continuous $f\colon X \to \mathbf{R}$.

1. If $f$ is a real-valued continuous function on $X$, then $\{x \mid f(x) \geq a\}$ and $\{x \mid f(x) \leq a\}$ are zero sets, for each $a \in \mathbf{R}$. [$g(x) = \max\{f(x) - a, 0\}$ is continuous.]

2. $X$ is completely regular iff the cozero-set nhoods of each point form a nhood base. (A cozero-set is the complement of a zero-set.)

3. $X$ is completely regular iff the zero sets form a base for the closed sets in $X$ (i.e., iff every closed set in $X$ is an intersection of zero sets).

The last two assertions provide handy ways of deciding whether or not a given space is completely regular.

### 14D. *Subsets of regular spaces*

If $X$ is $T_3$ and $A$ is an infinite subset of $X$, there is a sequence $U_1, U_2, \ldots$ of open subsets of $X$ such that $\overline{U}_n \cap \overline{U}_m = \emptyset$ if $n \neq m$ and $U_n \cap A \neq \emptyset$ for each $n$. [Use induction.]

### 14E. *Semiregular spaces*

A space is *semiregular* iff the regularly open sets (3D) form a base for the topology.

1. Every regular space is semiregular. Is the converse true?

2. A semiregular, $T_1$-space need not be Hausdorff.

3. Every space $X$ can be embedded in a semiregular space. [In the set $X \times \mathbf{I}$, define a topology as follows: nhoods of $(x, y)$ for $y \neq 0$ will be usual interval nhoods

$$\{(x, z) \mid y - \epsilon < z < y + \epsilon\} \qquad \text{in} \qquad I_x = \{x\} \times \mathbf{I},$$

for small positive $\epsilon$; nhoods of $(x, 0)$, $x \in X$, will have the form $\{(x', z) \mid x' \in U, 0 \leq z < \epsilon_{x'}\}$ where $U$ is a nhood of $x$ in $X$ and for each $x' \in U$, $\epsilon_{x'}$ is picked small and positive. The resulting space $Z$ is semiregular and $X$ is embedded in $Z$ as the closed, nowhere-dense subspace $\{(x, 0) \mid x \in X\}$.]

Thus subspaces of semiregular spaces need not be semiregular.

### 14F. *Urysohn spaces*

A space $X$ is a *Urysohn space* iff whenever $x \neq y$ in $X$, there are nhoods of $U$ of $x$ and $V$ of $y$ with $\overline{U} \cap \overline{V} = \emptyset$.

1. Every regular, $T_1$-space is Urysohn and every Urysohn space is Hausdorff.

2. Not every Urysohn space is semiregular (14E). Thus not every Urysohn space is regular.

3. Not every semiregular, Hausdorff space is Urysohn. [Give the real line the discrete topology and add the following points:

a) $+\infty$ whose nhoods have the form $\{+\infty\} \cup (a, \infty)$ for $a \in \mathbf{R}$,
b) $-\infty$ whose nhoods have the form $\{-\infty\} \cup (-\infty, a)$ for $a \in \mathbf{R}$,
c) $p_1, p_2, \ldots$ where the nhoods of $p_n$ have the form $\{p_n\} \cup$ all but finitely many points of $(-n - 1, -n) \cup (n, n + 1)$.

Verify that the resulting space $X$ has the required properties.]

### 14G. *Completely Hausdorff spaces*

A space $X$ is *completely Hausdorff* (*functionally Hausdorff*) iff whenever $x \neq y$ in $X$, there is a continuous $f: X \to \mathbf{I}$ with $f(x) = 0$, $f(y) = 1$.

1. Every completely Hausdorff space is Hausdorff. (The famous example of E. Hewitt of a regular $T_1$-space in which every continuous real-valued function is constant (see Exercise 18G) shows that not every regular $T_1$-space is completely Hausdorff.)

2. Discuss products and subspaces of completely Hausdorff spaces.

### 14H. *$C^*(X)$ for non-Tychonoff spaces*

Given any topological space $(X, \tau)$, there is a Tychonoff space $Y$ such that the rings $C^*(X)$ and $C^*(Y)$ of bounded continuous real-valued functions on $X$ and $Y$ are isomorphic. [Weaken the topology on $X$ to obtain a completely regular space with the same ring of functions. Then identify points to get a Tychonoff space.]

Thus $C^*(X)$ is studied only for Tychonoff spaces $X$.

## 15 Normal spaces

Regularity and complete regularity, as we have seen, constitute nontrivial restrictions on a topological space. Nonetheless, spaces with these properties behave decently with respect to the formation of products and subspaces. In the next (and obvious) step to normal spaces, we find ourselves confronted with the real bad boy among the separation axioms. So odd is the behavior of subspaces and products of normal spaces that their study is a separate subject. This is unfortunate, since as theorems late in this section will show, normal spaces possess many properties of paramount interest to topologists.

**15.1 Definition.** A topological space $X$ is *normal* iff whenever $A$ and $B$ are disjoint closed sets in $X$, there are disjoint open sets $U$ and $V$ with $A \subset U$ and $B \subset V$. A normal $T_1$-space will be called a *$T_4$-space*.

Now is the time to introduce a note of caution. The terminology in the literature with respect to the separation axioms beyond Hausdorff is more than a little confused. Some writers interchange our usage, using $T_3$, Tychonoff and $T_4$ for those spaces which need not be $T_1$ (and regular, completely regular and normal for those that are). Others use $T_3$ and regular to mean the same thing, which sometimes means it includes $T_1$ and sometimes not (and likewise for Tychonoff and completely regular, and $T_4$ and normal). Look before you leap.

The construction of examples of nonnormal spaces will be facilitated by the following lemma, due to F. B. Jones.

**15.2 Lemma.** *If $X$ contains a dense set $D$ and a closed, relatively discrete subspace $S$ with $|S| \geq 2^{|D|}$, then $X$ is not normal.*

*Proof.* If $X$ were normal then for each $T \subset S$, the sets $T$ and $S - T$ would be disjoint and closed in $X$ and hence would be contained in disjoint open sets $U(T)$ and $V(T)$. Now if $T_1 - T_2 \neq \varnothing$, then clearly $U(T_1) \cap V(T_2)$ is a nonempty open set in $X$, so $U(T_1) \cap V(T_2) \cap D$ is nonempty. But then $U(T_1) \cap V(T_2) \cap D$ is a subset of $U(T_1) \cap D$ and not a subset of $U(T_2) \cap D$. Thus if $T_1$ and $T_2$ are different subsets of $S$, then $U(T_1) \cap D$ and $U(T_2) \cap D$ are different subsets of $D$, so $|P(S)| \leq |P(D)|$. This is impossible if $|S| \geq 2^{|D|}$. ∎

**15.3 Examples.** a) A normal space need not be regular. If $X$ is the real line with the topology in which open sets are the sets $(a, \infty)$ for $a \in X$, then $X$ is normal since no two nonempty closed sets are disjoint, but $X$ is not regular since the point 1 cannot be separated from the closed set $(-\infty, 0]$ by disjoint open sets. Of course, every $T_4$-space is Tychonoff, but we need Urysohn's lemma (15.6) to prove this.

b) A Tychonoff space need not be $T_4$. As we have seen, the Moore plane $\Gamma$ is Tychonoff. But $\Gamma$ is not normal, by the lemma above, for the $x$-axis $T$ in $\Gamma$ is closed and relatively discrete, the set $K = \{(x, y) \in \Gamma \mid x \text{ and } y \text{ are rational}\}$ is dense in $\Gamma$, and $|T| \geq 2^{|K|}$. (Note that $K$ is countable and $|T| = \mathfrak{c}$.)

c) *Every metrizable space is $T_4$.* In fact, every pseudometrizable space is normal. For suppose $\rho$ is a pseudometric which gives the topology on $X$ and let $A$ and $B$ be disjoint closed sets in $X$. For each $x \in A$, pick $\delta_x > 0$ such that $U(x, \delta_x)$ does not meet $B$ and for each $y \in B$, pick $\epsilon_y > 0$ such that $U(y, \epsilon_y)$ does not meet $A$. Let

$$U = \bigcup_{x \in A} U\left(x, \frac{\delta_x}{3}\right), \qquad V = \bigcup_{y \in B} U\left(y, \frac{\epsilon_y}{3}\right).$$

Then $U$ and $V$ are open sets in $X$ containing $A$ and $B$ respectively. Suppose $z \in U \cap V$. Then $\rho(x, z) < \delta_x/3$ and $\rho(z, y) < \epsilon_y/3$ so $\rho(x, y) < \delta_x/3 + \epsilon_y/3 < \delta_x$, assuming $\delta_x = \max\{\delta_x, \epsilon_y\}$. But then $y \in U(x, \delta_x)$, which is impossible. Thus $U$ and $V$ must be disjoint, showing that $X$ is normal.

**15.4 Theorem.** a) *Closed subspaces of normal (or $T_4$) spaces are normal (respectively, $T_4$).*

b) *Products of normal spaces need not be normal.*

c) *The closed continuous image of a normal (or $T_4$) space is normal (respectively, $T_4$).*

*Proof.* a) If $Y$ is closed in $X$ and $A$ and $B$ are disjoint closed sets in $Y$, then $A$ and $B$ are disjoint closed sets in $X$, and hence are contained in disjoint open sets $U$ and $V$ in $X$. Then $U \cap Y$ and $V \cap Y$ are disjoint open subsets of $Y$ containing $A$ and $B$. Thus $Y$ is normal. The assertion for $T_4$-spaces follows now from the fact that every subspace of a $T_1$-space is $T_1$.

b) See Example 15.5(b).

c) Suppose $X$ is normal and $f$ is a closed continuous map of $X$ onto $Y$. If $A$ and $B$ are disjoint closed sets in $Y$, then $f^{-1}(A)$ and $f^{-1}(B)$ are disjoint closed sets in $X$ and hence we can find disjoint open sets $U_1$ and $U_2$ in $X$ containing $f^{-1}(A)$ and $f^{-1}(B)$. Since $f$ is closed, the sets $V_1 = Y - f(X - U_1)$ and $V_2 = Y - f(X - U_2)$ are open in $Y$. It is easily checked that $V_1$ and $V_2$ are disjoint and contain $A$ and $B$, respectively. Thus $Y$ is normal. The assertion for $T_4$-spaces follows, since the image of a $T_1$-space under a closed continuous map is $T_1$. ■

**15.5 Examples.** a) Arbitrary subspaces of $T_4$-spaces need not be $T_4$. In fact, the nonnormal Tychonoff space $\mathbf{\Gamma}$ (15.3(b)) can be embedded in some cube, by 14.13. But in 17.10, we note that every cube is normal. If every subspace of a space $X$ is normal, $X$ is said to be *completely normal*; see Exercise 15B.

b) Products of $T_4$-spaces need not be $T_4$. It is noted, in 16D, that the Sorgenfrey line $\mathbf{E}$ is $T_4$ (in fact, that $\mathbf{E}$ has even stronger properties). But $\mathbf{E} \times \mathbf{E}$ is not normal. (The proof makes an easy exercise in the use of Lemma 15.2; see 15A.2.) Normality in product spaces is studied in some detail in Section 21.

c) Arbitrary quotients of $T_4$-spaces need not be $T_4$. See 13.9(b), which provides an open continuous map from a $T_4$-space $X$ onto a non-Hausdorff $T_1$-space $Y$.

The remainder of this section will be devoted to giving some useful properties of normal spaces. The theorems which follow deal, in order, with separation of sets by continuous functions, with existence of extensions of continuous functions and with existence of certain kinds of open coverings. Each of the properties thus exhibited for normal spaces is, in fact, characteristic of normal spaces. Since separation, extension and covering are among the most important topics in topology, any one of them would be enough to overcome the stigma we originally attach to normal spaces because of the trouble we get into when forming subspaces and (especially) products. That all three should be true for the same kind of space certainly ranks with other wonders of the world (e.g., the embedding of Tychonoff spaces in cubes).

The first of these results has as an immediate consequence that every $T_4$-space is Tychonoff. Intrinsically, it ranks among the greatest theorems in topology, since it provides, starting from scratch, a bare-hands construction of a continuous function where none was assumed to exist.

**15.6 Urysohn's Lemma.** *A space $X$ is normal iff whenever $A$ and $B$ are disjoint closed sets in $X$, there is a continuous function $f: X \to [0, 1]$ with $f(A) = 0$ and $f(B) = 1$.*

*Proof.* Suppose $X$ is normal and $A$, $B$ are disjoint closed sets in $X$. By normality, there is an open set $U_{1/2}$ such that $A \subset U_{1/2}$ and $\overline{U}_{1/2} \cap B = \emptyset$. But now $A$ and $X - U_{1/2}$ are disjoint and closed and so are $\overline{U}_{1/2}$ and $B$. Hence open sets $U_{1/4}$ and $U_{3/4}$ exist such that

$$A \subset U_{1/4}, \overline{U}_{1/4} \subset U_{1/2}, \overline{U}_{1/2} \subset U_{3/4}, \overline{U}_{3/4} \cap B = \emptyset.$$

Now suppose sets $U_{k/2^n}$, $k = 1, \ldots, 2^n - 1$ have been defined in such a way that

$$A \subset U_{1/2^n}, \ldots, \overline{U}_{k-1/2^n} \subset U_{k/2^n}, \ldots, \overline{U}_{(2^n-1)/2^n} \cap B = \emptyset;$$

then the process can, by normality, be continued so as to provide sets $U_{k/2^{n+1}}$, $k = 1, \ldots, 2^{n+1} - 1$ with the same properties. By induction, then, we have for each "dyadic rational" $r$ (i.e., each rational of the form $r = k/2^n$ for some $n > 0$ and $k = 1, \ldots, 2^n - 1$) an open set $U_r$ subject to the conditions:

a) $A \subset U_r$ and $\overline{U}_r \cap B = \emptyset$ for each dyadic $r$,

b) $\overline{U}_r \subset U_s$ whenever $r < s$.

Now define $f: X \to [0, 1]$ as follows:

$$f(x) = \begin{cases} 1 & \text{if } x \text{ belongs to no } U_r, \\ \inf \{r \mid x \in U_r\} & \text{otherwise.} \end{cases}$$

It is apparent that $f(A) = 0$ and $f(B) = 1$, so we have the function we want provided $f$ is continuous. But continuity of $f$ follows easily from facts like these:

a) if $x \notin \overline{U}_r$, then $f(x) \geq r$ (continuity at points $x$ where $f(x) = 1$),

b) if $x \in U_r$, then $f(x) \leq r$ (continuity at points where $f(x) = 0$),

c) if $x \in U_r - \overline{U}_s$, where $s < r$, then $s \leq f(x) \leq r$ (continuity at all other points).

This proves necessity in the theorem.

Conversely, suppose $A$ and $B$ are disjoint closed sets in $X$ and $f: X \to [0, 1]$ is a continuous function such that $f(A) = 0$, $f(B) = 1$. Then apparently $f^{-1}[0, \frac{1}{2})$ and $f^{-1}(\frac{1}{2}, 1]$ will be disjoint open sets in $X$ containing $A$ and $B$, respectively, so the condition of the theorem is sufficient for normality. ∎

**15.7 Corollary.** *Every $T_4$-space is Tychonoff.*

As a convenience for later use, we point out that 0 and 1 in the statement of Urysohn's Lemma can obviously be replaced by any pair of real numbers $a$ and $b$ with $a \neq b$.

If $A$ and $B$ are disjoint closed sets in a normal space, a function of the type whose existence is guaranteed by Urysohn's Lemma is called a *Urysohn function* for $A$ and $B$.

It is *not* in general true that, given disjoint closed sets $A$ and $B$ in a normal space, there will be a Urysohn function such that $A = f^{-1}(0)$, $B = f^{-1}(1)$. The spaces with this property are called *perfectly normal*; see Exercise 15C.

Now we turn to the next in our series of three characterizations of normal spaces. Its importance cannot be overstated. It provides for the existence of extensions of continuous functions and some of the best and hardest work being done today in topology (in particular, by algebraic topologists) deals with variations on the extension question; if $A \subset X$, when can a continuous function $f: A \to Y$ be extended to a continuous function $F: X \to Y$?

**15.8 Tietze's extension theorem.** *$X$ is normal iff whenever $A$ is a closed subset of $X$ and $f: A \to \mathbf{R}$ is continuous, there is an extension of $f$ to all of $X$; i.e., there is a continuous map $F: X \to \mathbf{R}$ such that $F \mid A = f$.*

*Proof.* $\Rightarrow$: First suppose $f: A \to [-1, 1]$. Let

$$A_1 = \{x \in A \mid f(x) \geq \tfrac{1}{3}\}, \qquad B_1 = \{x \in A \mid f(x) \leq -\tfrac{1}{3}\}.$$

Now $A_1$ and $B_1$ are disjoint closed sets in $A$, and therefore in $X$, so by Urysohn's Lemma, there is a continuous $f_1 : X \to [-\frac{1}{3}, \frac{1}{3}]$ such that $f_1(A_1) = \frac{1}{3}$, $f_1(B_1) = -\frac{1}{3}$. Evidently, for each $x$ in $A$, $|f(x) - f_1(x)| \leq \frac{2}{3}$, so that $f - f_1$ is a mapping of $A$ into $[-\frac{2}{3}, \frac{2}{3}]$.

Now we repeat the process with $f - f_1 = g_1$. That is, divide $[-\frac{2}{3}, \frac{2}{3}]$ into thirds (at $-\frac{2}{9}$ and $\frac{2}{9}$) and let $A_2 = \{x \in A \mid g_1(x) \geq \frac{2}{9}\}$, $B_2 = \{x \in A \mid g_1(x) \leq -\frac{2}{9}\}$. Then there is a Urysohn function $f_2: X \to [-\frac{2}{9}, \frac{2}{9}]$ such that $f_2(A_2) = \frac{2}{9}$, $f_2(B_2) = -\frac{2}{9}$. Evidently, $|(f - f_1) - f_2| \leq (\frac{2}{3})^2$ on $A$.

Continuing the process, we obtain a sequence $f_1, f_2, \ldots$ of continuous functions on $A$ such that

$$\left| f - \sum_{k=1}^{n} f_k \right| \leq (\tfrac{2}{3})^n.$$

Define $F(x) = \sum_{i=1}^{\infty} f_i(x)$, for each $x \in X$. Certainly $F(x) = f(x)$ for each $x \in A$, so it remains only to show $F$ continuous.

Let $x \in X$ and $\epsilon > 0$ be given. Pick $N > 0$ so that $\sum_{n=N+1}^{\infty} (\frac{2}{3})^n < \epsilon/2$. Since each $f_i$ is continuous for $i = 1, \ldots, N$, pick open $U_i$ containing $x$ such that

$$y \in U_i \Rightarrow |f_i(x) - f_i(y)| < \epsilon/2N.$$

Then $U = U_1 \cap \cdots \cap U_N$ is open in $X$, and

$$\begin{aligned} y \in U \Rightarrow |F(x) - F(y)| &\leq \sum_{i=1}^{N} |f_i(x) - f_i(y)| + \sum_{i=N+1}^{\infty} (\tfrac{2}{3})^n \\ &< N \cdot \frac{\epsilon}{2N} + \frac{\epsilon}{2} = \epsilon, \end{aligned}$$

so that $F$ is continuous at $x$. This completes the proof in the case where $f$ maps $A$ into $[-1, 1]$.

Since $(-1, 1)$ is homeomorphic to $\mathbf{R}$, we can prove the general case by considering a map $f: A \to (-1, 1)$. Since we can regard $f$ as mapping $A$ into $[-1, 1]$, we can find an extension $F': X \to [-1, 1]$. Let $A_0 = \{x \in X \mid |F'(x)| = 1\}$. Then $A$ and $A_0$ are disjoint closed sets in $X$, so there is a Urysohn function $g: X \to [0, 1]$ such that $g(A_0) = 0$ and $g(A) = 1$. Define $F: X \to (-1, 1)$ by $F(x) = g(x) \cdot F'(x)$. Then $F$ is continuous, and if $x \in A$, $F(x) = g(x) \cdot F'(x) = 1 \cdot f(x) = f(x)$, so $F$ is the desired extension of $f$.

$\Leftarrow$: Suppose the condition holds. If $A$ and $B$ are disjoint closed sets in $X$, then $A \cup B$ is closed in $X$ and the function $f: A \cup B \to [0, 1]$ defined by $f(A) = 0$ and $f(B) = 1$ is continuous on $A \cup B$. The extension of $f$ to all of $X$ will be a Urysohn function for $A$ and $B$. Thus, by 15.6, $X$ is normal. ■

It is worth mentioning that implicit in the proof of the Tietze theorem is the proof that if the function $f$ carries $A$ to $[a, b]$, then the extension $F$ can be made to have the same property.

The last property characteristic of normal spaces will play an important role in later work both on paracompactness and Dowker's conjecture. The terminology associated with this theorem is unusually descriptive.

**15.9 Definition.** A *cover* (or *covering*) of a space $X$ is a collection $\mathscr{A}$ of subsets of $X$ whose union is all of $X$. A *subcover* of a cover $\mathscr{A}$ is a subcollection $\mathscr{A}'$ of $\mathscr{A}$ which is a cover. An *open cover* of $X$ is a cover consisting of open sets, and other adjectives applying to subsets of $X$ apply similarly to covers.

An open cover $\mathscr{U} = \{U_\alpha \mid \alpha \in A\}$ of $X$ is *shrinkable* provided an open cover $\mathscr{V} = \{V_\alpha \mid \alpha \in A\}$ exists with the property that $\overline{V}_\alpha \subset U_\alpha$ for each $\alpha \in A$. Of course, $\mathscr{V}$ is called a *shrinking* of $\mathscr{U}$.

A covering $\mathscr{U}$ is *point finite* provided each $x \in X$ belongs to only finitely many elements of $\mathscr{U}$.

***15.10 Theorem.*** *$X$ is normal iff every point-finite open cover is shrinkable.*

*Proof.* Suppose $X$ is normal and $\mathscr{U} = \{U_\alpha \mid \alpha \in A\}$ is a point-finite open cover of $X$. Well-order the set $A$; for convenience, then, suppose $A = \{1, 2, \ldots, \alpha, \ldots\}$. Now construct $\{V_\alpha \mid \alpha \in A\}$ by transfinite induction as follows: let

$$F_1 = X - \bigcup_{\alpha > 1} U_\alpha.$$

Then $F_1 \subset U_1$ so there is an open set $V_1$ such that $F_1 \subset V_1$ and $\overline{V}_1 \subset U_1$ by normality. Suppose $V_\beta$ has been defined for each $\beta < \alpha$ now, and let $F_\alpha = X - [(\bigcup_{\beta<\alpha} V_\beta) \cup (\bigcup_{\gamma>\alpha} U_\gamma)]$. Then $F_\alpha$ is closed and $F_\alpha \subset U_\alpha$, so we let $V_\alpha$ be any open set such that $F_\alpha \subset V_\alpha$ and $\overline{V}_\alpha \subset U_\alpha$. Now $\mathscr{V} = \{V_\alpha \mid \alpha \in A\}$ is a shrinking of $\mathscr{U}$ provided it is a cover. But if $x \in X$, then $x$ belongs to only finitely many elements of $\mathscr{U}$, say $U_{\alpha_1}, \ldots, U_{\alpha_n}$. Let $\alpha = \max(\alpha_1, \ldots, \alpha_n)$. Now $x \notin U_\gamma$

for any $\gamma > \alpha$ and hence, if $x \notin V_\beta$ for any $\beta < \alpha$, then $x \in F_\alpha \subset V_\alpha$. Hence, in any case, $x \in V_\beta$ for *some* $\beta \leq \alpha$. This completes the proof that $\mathscr{V}$ is a shrinking of $\mathscr{U}$.

For the converse, let $A$ and $B$ be disjoint closed subsets of $X$. Then $\{X - A, X - B\}$ is a point-finite open cover of $X$. But any shrinking $\{U, V\}$ of $\{X - A, X - B\}$ induces a separation $X - \overline{U}, X - \overline{V}$ of $A$ and $B$. ■

## Problems

### 15A. *Examples on normality*

1. Let $A$ be a fixed subset of $X$, $\tau$ the topology for $X$ consisting of ø and all supersets of $A$. Discuss normality of $(X, \tau)$.

2. Recall that $\mathbf{E}$ denotes the Sorgenfrey line (4.6). Show that $\mathbf{E} \times \mathbf{E}$ is not normal.

3. The radial plane (3A.4) is not normal.

4. The scattered line (5C) is $T_4$.

5. Suppose $(X, \tau)$ is normal and $A$ is closed in $X$. Show that $(X, \tau_A)$ is normal iff $X - A$ is a normal subspace of $(X, \tau)$, where $\tau_A$ denotes the simple extension (3A.5) of $\tau$ over $A$.

### 15B. *Completely normal spaces*

A space $X$ is *completely normal* iff every subspace of $X$ is normal.

1. $X$ is completely normal iff whenever $A$ and $B$ are subsets of $X$ with $A \cap \overline{B} = \overline{A} \cap B = ø$, then there are disjoint open sets $U \supset A$ and $V \supset B$. [To do necessity, consider the subspace $X - (\overline{A} \cap \overline{B})$, which contains both $A$ and $B$, and in which $A$ and $B$ have disjoint closures. Sufficiency is easy.]

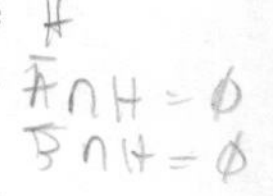

2. Why can't the method used to show every subspace of a regular space is regular be carried over to give a proof that every subspace of a normal space is normal?

3. Every metric space is completely normal.

### 15C. *Perfectly normal spaces*

A $T_1$-space $X$ is called *perfectly normal* iff for each pair of disjoint closed sets $A$ and $B$ in $X$, there is a continuous function $f: X \to \mathbf{I}$ such that $A = f^{-1}(0)$ and $B = f^{-1}(1)$. Recall that a *$G_\delta$-set* in a topological space is a countable intersection of open sets.

1. A space $X$ is perfectly normal iff it is $T_4$ and each closed set in $X$ is a $G_\delta$-set. [For sufficiency, if $A$ is a closed set and $A = \bigcap G_n$ where each $G_n$ is open, then a Urysohn function $f_n$ exists such that $f_n(A) = 0$ and $f_n(X - G_n) = 1$, for each $n$. Set $f_A(x) = \sum (f_n(x)/2^n)$. If $A$ and $B$ are now disjoint closed sets, set

$$f(x) = \frac{f_A(x)}{f_A(x) + f_B(x)}.$$

Then $f$ is continuous and $f^{-1}(0) = A$, $f^{-1}(1) = B$.]

2. Every metric space is perfectly normal.

3. It is not sufficient for perfect normality that $X$ be $T_4$ and every point in $X$ be a $G_\delta$-set.

## 15D. *Retraction and extension*

A continuous function $r$ from $X$ onto a subspace $A$ of $X$ is a *retraction* iff $r \mid A$ is the identity on $A$. The subspace $A$ of $X$ is then called a *retract* of $X$. Questions about existence of extensions can be phrased in terms of existence of retractions, according to part 2 below. This is the way algebraic topologists like to view extension questions.

1. A retract in a Hausdorff space is a closed set.

2. A subset $A$ of $X$ is a retract of $X$ iff every continuous function $f: A \to Z$ has an extension to a continuous function $F: X \to Z$. [If $r$ is a retraction, consider $f \circ r$.]

Related to retracts are the absolute retracts. A space $R$ is an *absolute retract* iff given any $T_4$-space $X$, any closed subset $A$ of $X$, and a continuous $f: A \to R$, then $f$ has an extension to all of $X$. The reason for the name is given in 3 below.

3. A $T_4$-space is an absolute retract iff it is a retract of every $T_4$-space in which it can be embedded as a closed subset.

4. **R** is an absolute retract; any closed interval in **R** is an absolute retract.

5. Any product of absolute retracts is an absolute retract.

Related to absolute retracts are the absolute nhood retracts. A space $Y$ is an *absolute nhood retract* (ANR) iff whenever $A$ is a closed subset of a normal space $X$ and $f: A \to Y$ is continuous, then $f$ can be extended over some open set $U$ containing $A$ in $X$.

6. A normal space is an ANR iff whenever it is embedded as a closed subset of a normal space, it is a retract of some open set in that space.

7. $\mathbf{S}^1$ is an ANR, but not an absolute retract. [The second statement follows from the *no-retraction theorem*: there is no retraction from the unit disk onto $\mathbf{S}^1$, which you may assume for now. It is proved in the text in Section 34.]

8. The product of finitely many ANRs is an ANR.

## 15E. *C*-embedding: Urysohn's extension theorem*

A subspace $A$ of $X$ is *C-embedded* (*C*-embedded*) in $X$ iff every continuous function $f: A \to \mathbf{R}$ ($f: A \to \mathbf{I}$) can be extended to a continuous function $F: X \to \mathbf{R}$ ($F: X \to \mathbf{I}$). Subsets $B$ and $C$ of a topological space $X$ are called *completely separated* iff there is a continuous $g: X \to \mathbf{I}$ such that $g(B) = 0$ and $g(C) = 1$. Show that a subspace $A$ of $X$ is $C^*$-embedded in $X$ iff every pair of completely separated sets in $A$ is completely separated in $X$ (this is *Urysohn's extension theorem*).

## 15F. *Order topologies*

Every ordered space (6D) is $T_4$.

## 15G. *Extremally disconnected spaces*

A topological space $X$ is *extremally disconnected* iff the closure of every open set in $X$ is open.

1. For any space $X$, the following are equivalent:

   a) $X$ is extremally disconnected,
   b) every two disjoint open sets in $X$ have disjoint closures,

c) every two disjoint open sets in $X$ are completely separated (15E),

d) every open subspace of $X$ is $C^*$-embedded (15E),

[Prove that (a) $\Rightarrow$ (c) $\Rightarrow$ (b) $\Rightarrow$ (a) and (c) $\Rightarrow$ (d) $\Rightarrow$ (b), using the Urysohn extension Theorem 15E for (c) $\Rightarrow$ (d).]

2. Every dense subspace and every open subspace of an extremally disconnected space is extremally disconnected. (Closed subspaces need not be; see the book by Gillman and Jerison.) By Exercise 19I, products of extremally disconnected spaces need not be extremally disconnected.

3. The only convergent sequences in an extremally disconnected $T_2$-space are those which are ultimately constant. [Suppose $x_n \to p$, but $(x_n)$ is not ultimately constant. Construct a sequence $U_1, U_2, \ldots$ of disjoint open sets in $X$ such that $x_{n_k} \in U_k$ for some subsequence $(x_{n_k})$ of $(x_n)$, and such that $p \in \overline{U}_k$, for each $k$. Let $G = \bigcup_{k=1}^{\infty} U_{2k}$. Then $\bar{G}$ is an open set containing $p$, but $x_{n_k} \notin \bar{G}$ for any odd $k$.]

Part 3 shows that sequential convergence cannot be used to describe the topology of any nondiscrete extremally disconnected space. In particular, such spaces cannot be first countable.

Extremally disconnected spaces are important in studying the Stone–Čech compactification of a product space (19I.2) as well as, more generally, in the study of the Stone space of any complete Boolean algebra. They also crop up in investigations of the reducibility of mappings of compact spaces (17P), and the extremally disconnected compact spaces are precisely the compact-projective spaces (17Q).

## 15H. *Hahn–Banach theorem*

In the presence of algebraic structure on a space $X$, e.g., if $X$ is a normed linear space, one can ask whether a function $f$ on a subset $A$ of $X$ which is continuous and has certain algebraic properties can be extended to all of $X$ in such a way that continuity and these algebraic properties are preserved. The answer, if $A$ is a subspace of $X$ and $f$ is a continuous linear functional on $X$, is yes. This follows (see part 2) as an intermediate corollary to the Hahn–Banach theorem (part 1) below. Prerequisite to the understanding of this material is a careful study of Problems 2J and 7L.

1. (*Hahn–Banach theorem*) Let $X$ be any linear space, $p\colon X \to \mathbf{R}$ a function such that $p(x + y) \leq p(x) + p(y)$ and $p(\alpha x) = \alpha \cdot p(x)$ for $\alpha \geq 0$. If $A$ is a linear subspace of $X$ and $f$ is a linear functional on $A$ such that $f(x) \leq p(x)$ for all $x \in A$, then $f$ can be extended to a linear functional $F$ on $X$ such that $F(x) \leq p(x)$ for all $x \in X$. [First note that if $A'$ is a subspace of $X$ with $A \subset A'$ and $F'$ is an extension of $f$ to $A'$ which is less than or equal to $p$ on $A'$, then for any $y \notin A'$, a further extension of $f$ to the subspace $\{x + \lambda y \mid x \in A', \lambda \in R\}$ can be found which is less than or equal to $p$ there. Next, use Zorn's lemma to conclude that there exists a maximal extension of $f$, when extensions of $f$ are ordered by $g_1 \leq g_2$ iff

$$\operatorname{dom} g_1 \subset \operatorname{dom} g_2 \qquad \text{and} \qquad g_1 = g_2 \mid \operatorname{dom} g_1,$$

which is less than or equal to $p$. Finally, combine these two results to conclude that the domain of this maximal extension must be all of $X$.]

2. If $X$ is a normed linear space and $f$ is a bounded linear functional on a subspace $A$

of $X$, then $f$ can be extended to a bounded linear functional $f$ on $X$ with $||F|| = ||f||$. [Use the Hahn–Banach theorem with $p(x) = ||f|| \cdot ||x||$.]

3. If $X$ is a normed linear space, $A$ is a subspace of $X$ and $y \notin \bar{A}$, then there is a bounded linear functional $F$ on $X$ such that

a) $F(A) = 0$,
b) $F(y) =$ the distance from $A$ to $y$,
c) $||F|| = 1$.

[Define $f$ on $\{a + \lambda \cdot y \mid a \in A,\ \lambda \in \mathbf{R}\}$ by $f(a + \lambda y) = \lambda \cdot \delta$ and use the Hahn–Banach theorem to conclude $f$ can be extended to a functional $F$ on all of $X$ with $|F(x)| \leq ||x||$ at each $x \in X$.]

Part 3 can be regarded as giving a form of complete regularity on the space $X$, in which subspaces can be separated from points by linear continuous maps. Part 2 could be called a Tietze extension theorem for normed linear spaces.

15I. *Jones' lemma*

Prove Jones' lemma (15.2) by comparing the number of continuous functions on $D$ with the number on $X$ and using the Tietze extension theorem.

## 16 Countability properties

We will introduce three topological properties in this section and investigate the relationships between them, as well as the basic combinatorial questions (about subspaces, products and quotients) for each individually.

Recall that the first axiom of countability, providing for countable nhood bases, was introduced in Section 4. We are ready now for the second axiom.

**16.1 Definition.** $X$ is *second countable* (or, satisfies the *second axiom of countability*) iff its topology has a countable base.

Every second-countable space is first countable. On the other hand, any uncountable discrete space is first countable without being second countable.

***16.2 Theorem.*** a) *The continuous open image of a second countable space is second countable.*

b) *Subspaces of second countable spaces are second countable.*

c) *A product of Hausdorff spaces is second countable iff each factor is second countable and all but countably many factors are one-point spaces.*

*Proof.* a) Let $f$ be a continuous open map of $X$ onto $Y$. It is sufficient to check that if $\mathscr{B}$ is a base for $X$, then $f(\mathscr{B}) = \{f(B) \mid B \in \mathscr{B}\}$ is a base for $Y$. For this purpose, let $V$ be an open set in $Y$, $p \in V$. Then $f^{-1}(V)$ is open in $X$, and if we pick $q \in f^{-1}(p) \subset f^{-1}(V)$, then for some basic open set $B$, $q \in B \subset f^{-1}(V)$. It follows that $p \in f(B) \subset V$, and thus that the sets $f(B)$ do form a base for $Y$ (where did we use openness of $f$?).

b) The restriction of a base for $X$ to a subspace $A$ of $X$ is a base for $A$.

c) Suppose $X = \prod X_\alpha$ is second countable. By (a), each $X_\alpha$ is second countable and by Exercise 16A, since $\prod X_\alpha$ is first countable, there are at most countably many nontrivial factors.

Conversely, suppose $\{B_{\alpha n} \mid n = 1, 2, \ldots\}$ is a base for $X_\alpha$, for each $\alpha \in A$. Then the sets of the form

$$B_{\alpha_1 n_1} \times \cdots \times B_{\alpha_k n_k} \times \prod \{X_\alpha \mid \alpha \neq \alpha_1, \ldots, \alpha_k\}$$

form a base for the product space. It is easily verified that, since $A$ is countable, there are only countably many sets of this form. ■

An example showing that the requirement that $f$ be open in 16.2(a) is not frivolous, is given to be worked out in 16B.1.

**16.3 Definition.** A topological space $X$ is *separable* iff $X$ has a countable dense subset. (A set $D$ is *dense* in $X$ iff $\text{Cl}_X D = X$.)

The real line is separable, since the rationals are dense. A discrete space is separable iff it is countable.

***16.4 Theorem.*** a) *The continuous image of a separable space is separable.*

b) *Subspaces of separable spaces need not be separable. However, an open subspace of a separable space is separable.*

c) *A product of Hausdorff spaces, each with at least two points, is separable iff each factor is separable and there are $\leq \mathfrak{c}$ factors.*

*Proof.* a) A continuous map of $X$ onto $Y$ carries a dense subset of $X$ to a dense subset of $Y$.

b) The Moore plane $\Gamma$ is separable, while the $x$-axis $T$ in $\Gamma$ is not; see Exercise 6B. The assertion for open subspaces is an easy exercise.

c) $\Rightarrow$: Since projection is continuous, each $X_\alpha$ is separable if $\prod X_\alpha$ is, by part (a). We proceed to show $|A| \leq \mathfrak{c}$. For each $\alpha \in A$, let $U_\alpha$ and $V_\alpha$ be disjoint nonempty open sets in $X_\alpha$ (using the fact that each $X_\alpha$ is Hausdorff and has at least two points). Let $D$ be a countable dense set in $\prod_{\alpha \in A} X_\alpha$ and, for each $\alpha \in A$, let $D_\alpha = D \cap \pi_\alpha^{-1}(U_\alpha)$. Then $D_\alpha \neq \varnothing$ for each $\alpha$, and for distinct $\alpha$ and $\beta$, $D_\alpha \neq D_\beta$ since points in $\pi_\alpha^{-1}(U_\alpha) \cap \pi_\beta^{-1}(V_\beta)$ which belong to $D$ will belong to $D_\alpha$ and not $D_\beta$. Thus the map $F: A \to P(D)$ defined by $F(\alpha) = D_\alpha$ is one–one and therefore

$$|A| \leq |P(D)| = 2^{\aleph_0} = \mathfrak{c}.$$

$\Leftarrow$: In $X_\alpha$, let $\{d_{\alpha 1}, d_{\alpha 2}, \ldots\} = D_\alpha$ be a countable dense subset. If we suppose $|A| \leq \mathfrak{c}$, then we can regard $A$ as a subset of the unit interval **I**. For each sequence $J_1, \ldots, J_k$ of disjoint closed intervals with rational endpoints and each sequence $n_1, \ldots, n_k$ of positive integers, define a point $p(J_1, \ldots, J_k; n_1, \ldots, n_k)$ as follows:

$$p_\alpha = d_{\alpha n_i} \quad \text{if} \quad \alpha \in J_i,$$

$$p_\alpha = d_{\alpha 1} \quad \text{otherwise.}$$

The set $D$ of points $p$ so defined is countable. Moreover, it is dense. For a (basic) open set in $\prod X_\alpha$ has the form

$$B = \pi_{\alpha 1}^{-1}(U_{\alpha 1}) \cap \cdots \cap \pi_{\alpha m}^{-1}(U_{\alpha m})$$

where $U_{\alpha_i}$ is open in $X_{\alpha_i}$, $i = 1, \ldots, m$. Then $U_{\alpha_i}$ contains a point $d_{\alpha_i n_i}$ of $D_{\alpha_i}$, for each $i$, and there are disjoint closed rational intervals $J_1, \ldots, J_m$ containing the points $\alpha_1, \ldots, \alpha_m$, respectively. The point $p(J_1, \ldots, J_m; n_1, \ldots, n_m)$ belongs to $B$ since $p_{\alpha_i} = d_{\alpha_i n_i}$, $i = 1, \ldots, m$. Hence, the set $D$ is dense. ■

**16.5 Definition.** $X$ is *Lindelöf* iff every open cover of $X$ has a countable subcover.

***16.6 Theorem.*** a) *The continuous image of a Lindelöf space is Lindelöf.*

b) *Closed subspaces of Lindelöf spaces are Lindelöf; arbitrary subspaces of Lindelöf spaces need not be Lindelöf.*

c) *Products of (even two) Lindelöf spaces need not be Lindelöf.*

*Proof.* a) Suppose $f\colon X \to Y$ is continuous and onto and $X$ is Lindelöf. Let $\{U_\alpha \mid \alpha \in A\}$ be an open cover of $Y$. Then $\{f^{-1}(U_\alpha) \mid \alpha \in A\}$ is an open cover of $X$ from which we can choose a countable subcover $\{f^{-1}(U_{\alpha_i}) \mid i = 1, 2, \ldots\}$. $\{U_{\alpha_i} \mid i = 1, 2, \ldots\}$ will be the desired countable subcover from $\{U_\alpha \mid \alpha \in A\}$.

b) Suppose $F$ is closed in $X$ and $X$ is Lindelöf. If $\{U_\alpha \mid \alpha \in A\}$ is an open cover of $F$, find for each $\alpha$ an open set $V_\alpha$ in $X$ with $V_\alpha \cap F = U_\alpha$. Then $X - F$ and the sets $V_\alpha$ form an open cover of $X$, for which there will be a countable subcover, $\{X - F, V_{\alpha 1}, V_{\alpha 2}, \ldots\}$. Then the corresponding $U_{\alpha_i}$, $i = 1, 2, \ldots$, cover $F$, so $\{U_\alpha \mid \alpha \in A\}$ has a countable subcover.

For the remaining assertions of the theorem, see the examples below. ■

**16.7 Examples.** a) Arbitrary subspaces of Lindelöf spaces need not be Lindelöf. Recall $\mathbf{\Omega}$ denotes the set of ordinals which are less than or equal to the first uncountable ordinal $\omega_1$ (as described in 1.19). Since $\mathbf{\Omega}$ is a totally ordered space, it can be provided with its order topology; recall that a basic nhood of $\alpha \in \mathbf{\Omega}$ is then of the form $(\alpha_1, \alpha_2) = \{\beta \in \mathbf{\Omega} \mid \alpha_1 < \beta < \alpha_2\}$, where $\alpha_1 < \alpha < \alpha_2$, with the modification that nhoods of $\omega_1$ have the form $(\gamma, \omega_1] = \{\beta \in \mathbf{\Omega} \mid \gamma < \beta \leq \omega_1\}$, for $\gamma < \omega_1$.

Now $\mathbf{\Omega}$ is a Lindelöf space. In fact, given any open cover of $\mathbf{\Omega}$, find one element $U$ which contains $\omega_1$. Then $U$ contains an interval $(\gamma, \omega_1]$ for some $\gamma < \omega_1$. But this leaves at most the set $[1, \gamma]$ to be covered, and this set is countable, so at most countably many more elements of the cover will be needed to cover $\mathbf{\Omega}$.

The subspace $\mathbf{\Omega}_0 = \mathbf{\Omega} - \{\omega_1\}$, however, is not Lindelöf. If for each $\alpha \in \mathbf{\Omega}_0$, we set $U_\alpha = [1, \alpha)$, then $\{U_\alpha \mid \alpha \in \mathbf{\Omega}_0\}$ is an open cover of $\mathbf{\Omega}_0$ which has no countable subcover. For if $\{U_{\alpha 1}, U_{\alpha 2}, \ldots\}$ covers $\mathbf{\Omega}_0$, then $\sup\{\alpha_1, \alpha_2, \ldots\} = \omega_1$, which is impossible, by Theorem 1.20.

b) The product of two Lindelöf spaces need not be Lindelöf. Consider the Sorgenfrey line $\mathbf{E}$ which is the real line with the topology in which basic open sets have the form $[a, b)$, $a < b$. In Exercise 16D you will prove this space is Lindelöf. Now $\mathbf{E} \times \mathbf{E}$ is not normal, as we pointed out in Example 15.2, but it is regular, since $\mathbf{E}$ is. But a regular Lindelöf space is normal according to the next theorem, so $\mathbf{E} \times \mathbf{E}$ cannot be Lindelöf.

***16.8 Theorem.*** *A regular, Lindelöf space is normal.*

*Proof.* Let $A$ and $B$ be disjoint closed sets in a regular Lindelöf space $X$. For each $a \in A$, let $U_a$ be an open set containing $a$ such that $\overline{U}_a \cap B = \varnothing$, by regularity. Similarly, find a set $V_b$ for each $b \in B$ separating $b$ from $A$. Since $A$ and $B$ are Lindelöf subspaces of $X$, apparently a countable number of the sets $U_a$ cover $A$, say $A \subset U_1 \cup U_2 \cup \cdots$; similarly, $B \subset V_1 \cup V_2 \cup \cdots$. Now construct open sets $S_n$ and $T_n$ inductively as follows:

$$\begin{aligned} S_1 &= U_1 & T_1 &= V_1 - \overline{S}_1 \\ S_2 &= U_2 - \overline{T}_1 & T_2 &= V_2 - \overline{(S_1 \cup S_2)} \\ S_3 &= U_3 - \overline{(T_1 \cup T_2)} & T_3 &= V_3 - \overline{(S_1 \cup S_2 \cup S_3)} \\ &\ \ \vdots & &\ \ \vdots \end{aligned}$$

Then it is easily seen that $S = \bigcup S_n$ and $T = \bigcup T_n$ are disjoint open sets containing $A$ and $B$, respectively. ■

***16.9 Theorem.*** *If $X$ is second countable, then $X$ is*

a) *Lindelöf,*

b) *Separable.*

*Proof.* a) Let $\mathscr{B}$ be a countable base for $X$. Suppose $\mathscr{U}$ is any open cover of $X$. For each $U \in \mathscr{U}$ and $x \in U$, there is some $B_{x,U} \in \mathscr{B}$ such that $x \in B_{x,U} \subset U$. Now $\mathscr{B}' = \{B_{x,U} \mid x \in U,\ U \in \mathscr{U}\}$ is really a countable set, since $\mathscr{B}' \subset \mathscr{B}$. Say $\{B_{x,U} \mid x \in U, U \in \mathscr{U}\} = \{B_{x_1,U_1}, B_{x_2,U_2}, \ldots\}$. Then $U_1, U_2, \ldots$ is a countable subcover from $\mathscr{U}$.

b) You did this as Exercise 5F.2. Simply pick one point from each element of a countable base and verify that the resulting countable set is dense. ■

The next examples show that, in general, no other implications between the properties in Theorem 16.9 will hold.

**16.10 Examples.** a) A separable space not Lindelöf. The space $\mathbf{E}$ is separable, hence so is $\mathbf{E}^2 = \mathbf{E} \times \mathbf{E}$. But $\mathbf{E}^2$ is not Lindelöf (otherwise it would be normal by Theorem 16.8).

b) A Lindelöf space not separable. Let $X$ be uncountable and discrete. Adjoin an extra point $x^*$ to $X$ and specify that its nhoods will be $\{x^*\} \cup A$, where

$A$ is the complement of a finite set in $X$, while nhoods of points in $X$ remain the same. Then the resulting space $X^*$ is Lindelöf (in fact, every open cover has a finite subcover) but not separable, since there are uncountably many points $x \in X$ and each is open in $X^*$.

**16.11 Theorem.** *For a (pseudo)metric space X, the following are equivalent:*

a) *X is second countable,*

b) *X is Lindelöf,*

c) *X is separable.*

*Proof.* By 16.9 it suffices to show (b) implies (a) and (c) implies (a) for a pseudometric space. Thus, let $(X, \rho)$ be a pseudometric space.

b) $\Rightarrow$ a): Suppose $X$ is Lindelöf. Let $\mathcal{U}_n = \{U(x, 1/n) \mid x \in X\}$. For each $n$, $\mathcal{U}_n$ is an open cover of $X$ and hence has a countable subcover $\mathcal{U}_n^*$. Then $\mathcal{U} = \mathcal{U}_1^* \cup \mathcal{U}_2^* \cup \cdots$ is a countable collection of open sets in $S$. Let $W$ be a nonempty open set in $X$, and $x \in W$. Then $U(x, 1/m) \subset W$ for some $m$. Now since $\mathcal{U}_{2m}^*$ covers $X$, there is some $y \in X$ such that $x \in U(y, 1/2m)$. Then

$$U(y, 1/(2m)) \subset U(x, 1/m) \subset W;$$

i.e., $U(y, 1/2m)$ is an element of $\mathcal{U}$ containing $x$ and contained in $W$. Thus $\mathcal{U}$ is a countable base for $X$, so $X$ is second countable.

c) $\Rightarrow$ a): Let $\{d_1, d_2, \ldots\}$ be a countable dense subset of $X$ and let

$$\mathcal{U}_{nm} = U(d_n, 1/m), \; n = 1, 2, \ldots, m = 1, 2, \ldots.$$

Then $\{U_{nm} \mid n = 1, 2, \ldots, m = 1, 2, \ldots\}$ is countable. We claim it is a base. Let $x \in W$, $W$ a nonempty open set in $X$. Then $U(x, 1/m) \subset W$ for some $m$. But some $d_n \in U(x, 1/2m)$ and then $U(d_n, 1/2m) \subset U(x, 1/m)$ so

$$x \in U_{n2m} = U(d_n, 1/2m) \subset W.$$

Thus, $\{U_{nm}\}$ is a base as advertised. ∎

**16.12 Example.** Experience indicates the necessity of pointing out that a separable, first-countable space need not be second countable. **E** provides an easy counterexample. It is separable since the rationals are dense, and first countable since the sets $[x, x + 1/n)$ form a nhood base at $x$, but not second countable. For if **E** were second countable, then **E** $\times$ **E** would (in two easy steps) be normal, which is not true.

## Problems

### 16A. *First countable spaces*

1. Every subspace of a first-countable space is first countable.

2. A product $\prod X_\alpha$ of first-countable spaces is first countable iff each $X_\alpha$ is first countable, and all but countably many of the $X_\alpha$ are trivial spaces.

3. The continuous image of a first-countable space need not be first countable [discrete spaces are first countable]; but the continuous open image of a first-countable space is first countable. (See also part 5 below and 23K.)

4. For a space to be first countable, it is *not* sufficient that each point be a $G_\delta$. [Construct a space $X$ by adjoining to the real line (whose topology is unchanged) a single point $p$ whose nhoods are all sets of the form $(a, \infty) - C$, where $C$ is a countable subset of $(a, \infty)$ with no cluster points. Verify that $p$ is a $G_\delta$ but has no countable nhood base. Why the condition that $C$ be "scattered?"]

The condition that each point be a $G_\delta$ *is* sufficient for first countability of a compact Hausdorff space. (See 17F.7.)

5. For each $n \in \mathbf{N}$, let $X_n$ be a copy of the subspace $\{0\} \cup \{1/m \mid m = 1, 2, \ldots\}$ of $\mathbf{R}$. Let $X$ be the disjoint union of the $X_n$. Is the quotient $Y$ of $X$ obtained by identifying all accumulation points of $X$ first countable?

## 16B. *Second countable spaces*

1. A quotient of a second-countable space need not be second countable. [For each $n \in \mathbf{N}$, let $I_n$ be a copy of $[0, 1]$ and let $X$ be the disjoint union of the spaces $I_n$. Now identify the left-hand endpoints of all the intervals $I_n$. The resulting space $Z$ is not first countable at the distinguished point, and hence is not second countable, although $X$ is second countable.]

2. Any base for the open sets in a second countable space has a countable subfamily which is a base.

3. Any increasing chain of real numbers which is well ordered by the usual order must be countable.

## 16C. *The countable chain condition*

Let $\aleph$ be any cardinal number. A space $X$ has *caliber* $\aleph$ iff whenever $\mathscr{U}$ is a family of open subsets of $X$ with $|\mathscr{U}| = \aleph$, a subfamily $\mathscr{V}$ of $\mathscr{U}$ exists with $|\mathscr{V}| = \aleph$ and $\bigcap \{V \mid V \in \mathscr{V}\} \neq \varnothing$. We say $X$ satisfies the *countable chain condition* iff every family of disjoint open subsets of $X$ is countable.

1. Every separable space has caliber $\aleph_1$.
2. Every product of separable spaces has caliber $\aleph_1$.
3. If $X$ has caliber $\aleph_1$, then $X$ satisfies the countable chain condition.
4. Investigate the three properties mentioned in 1 and 2 for a space $X$ with $\aleph_1$ elements and the "co-countable" topology, in which the open sets are $\varnothing$ and all complements of countable sets.

It is an open question whether the product of two spaces, each with the countable chain condition, has the countable chain condition. Also, this condition plays a key role in the enunciation of the *Souslin hypothesis* with which we will be concerned in Section 21.

## 16D. *Lindelöf spaces*

A subset $A$ of a space $X$ is *$G_\delta$-closed* in $X$ iff each point $p \notin A$ is contained in a $G_\delta$ disjoint from $A$.

1. The Sorgenfrey line $\mathbf{E}$ is Lindelöf. Conclude that $\mathbf{E}$ is a $T_4$-space.
2. If $X$ is Lindelöf, every uncountable subset of $X$ has an accumulation point.

3. A regular space is Lindelöf iff each open cover has a countable subcollection whose closures cover (i.e., has a *countable dense subsystem*).

4. Any space is Lindelöf iff each closed filter $\mathscr{F}$ with the countable intersection property (whenever $F_1, F_2, \ldots \in \mathscr{F}$, then $\bigcap F_n \neq \varnothing$) has a nonempty intersection. [The complements of the sets in an open cover with no countable subcover generate a base for a closed filter with the countable intersection property.]

5. A regular space is Lindelöf iff each open filter with the countable intersection property has a cluster point [the complements of the closures of the sets in an open cover having no countable dense subsystem form an open filter with the countable intersection property].

6. A regular space is Lindelöf iff whenever it is embedded in a Hausdorff space, it is $G_\delta$-closed. [A non-Lindelöf space $X$ has an open filter with the countable intersection property but no cluster point. Add a point $p$ to the space whose nhoods are $\{p\} \cup U$, where $U$ is any element of this filter. This provides a Hausdorff space in which $X$ is not $G_\delta$-closed. The reverse is easier.]

## 16E. *Hereditarily Lindelöf spaces*

A space $X$ is *hereditarily Lindelöf* iff every subspace of $X$ is Lindelöf.

1. Every second-countable space is hereditarily Lindelöf.

2. Any space $X$ can be embedded as a dense subset of a Lindelöf space. [Adjoin a point $p$ to $X$ whose nhoods are the sets $\{p) \cup E$, where $E$ is a subset of $X$ whose complement in $X$ is Lindelöf.

3. If $X$ is hereditarily Lindelöf and $E \subset X$, the set $E^*$ of points of $E$ which are not accumulation points of $E$ is countable.

## 16F. *Cardinality and the countability axioms*

1. A separable first-countable space has cardinal $\leq \mathfrak{c}$ [$\mathfrak{c} = \aleph_0^{\aleph_0}$].

2. If $X$ is separable and $C(X)$ denotes all continuous functions $f: X \to \mathbf{R}$, then $|C(X)| \leq \mathfrak{c}$. [A continuous function is determined by its values on a dense set.]

3. If $(X, \tau)$ is second countable, then $|\tau| \leq \mathfrak{c}$.

## 16G. *Separable spaces*

1. Every subspace of a separable metric space is separable.

2. Prove the irrationals are separable directly by finding a countable dense subset.

3. The set $\Omega_0$ of ordinals less than the first uncountable ordinal is not separable.

4. Give an example of a regular, separable space which is not normal. (Compare with 16.8.)

## 16H. *Examples on countability properties*

1. The plane with slotted disks (4C) is separable, but neither first countable nor Lindelöf (hence not second countable).

2. The plane with the topology given by radially open sets (3A.4) is separable, but neither first countable nor Lindelöf.

3. The Moore plane **Γ** is separable and first countable, but not Lindelöf.

4. The sequence space **m** (2H) is not separable. [An uncountable subset $A$ of **m** can be found such that $\rho(a, b) = 1$ whenever $a, b \in A$.]

5. The sequence spaces **c** and $\mathbf{c}_0$ (2H) are separable. [Consider sequences with rational terms which are ultimately constant.]

166. 4
E = Sorgenfrey line
E×E (regular)

Chapter 6

# Compactness

## 17 Compact spaces

Many of the most important theorems in a course in classical analysis are proved for closed bounded intervals (e.g., a continuous function on a closed bounded interval assumes its maximum). The basis for the proof of such theorems is almost without exception the Heine–Borel theorem, that a cover of a closed bounded interval by open sets has a finite subcover. It is not surprising, then, that the (topological) property of closed bounded intervals thus expressed has been made the subject of a definition in topology, the definition of compactness.

This section is long, but falls naturally into three parts. In the first (17.1 through 17.4) we study compactness and equivalent conditions for compactness, in the second (17.5 through 17.9) we give the basic theorems and examples about subspaces, products and continuous images of compact spaces; in the third (17.10 through 17.14) we study some of the properties of compact spaces which are the reasons this section is so long.

**17.1 Definition.** A space $X$ is *compact* iff each open cover of $X$ has a finite subcover. $X$ is *countably compact* iff each countable open cover of $X$ has a finite subcover.

Evidently, $X$ is compact iff $X$ is countably compact and Lindelöf. Countable compactness played an important role in the early stages of topology, because for the spaces then considered (usually metric spaces) it is equivalent to compactness (see 17F.6). It is still important in certain restricted directions. Another variation of compactness, sequential compactness, is introduced in Exercise 17G. It, too, was once more important than it now is.

**17.2 Examples.** a) $\mathbf{R}$ *is not compact.* In fact, the cover of $\mathbf{R}$ by the open sets $(-n, n)$, for $n \in \mathbf{N}$, can have no finite subcover.

b) $\mathbf{I}$ *is compact.* Let $\mathscr{U}$ be any open cover of $\mathbf{I}$ and let $K$ be the set of all points $c$ in $\mathbf{I}$ such that some finite subcollection from $\mathscr{U}$ covers $[0, c]$. Clearly $0 \in K$. Also, if $c \in K$ and $b \leq c$, then $b \in K$. Thus $K$ is a subinterval of $\mathbf{I}$ containing 0. Moreover, if $c \in K$, then any finite subcollection from $\mathscr{U}$ which covers $[0, c]$ also covers $[0, c + \epsilon]$ for some $\epsilon > 0$ (unless $c = 1$; in which case we have finished). Thus $K$ is an open set in $\mathbf{I}$. Finally, if $k$ is the right-hand endpoint of $K$, then $k \in K$. For pick $U \in \mathscr{U}$ such that $k \in U$. Then $(k - \epsilon, k] \subset U$ for some $\epsilon > 0$ so that, by adding $U$ to a finite subcollection from $\mathscr{U}$ which covers $[0, k - \epsilon]$, we obtain a finite subcollection from $\mathscr{U}$ which covers $[0, k]$. Now $K$ is a closed subinterval of $\mathbf{I}$ which contains 0 and is an open set in $\mathbf{I}$. Thus $K = \mathbf{I}$. This proves that $\mathbf{I}$ is compact.

c) *The ordinal space* $\mathbf{\Omega}$ *is compact.* Let $\mathscr{U}$ be any open cover of $\mathbf{\Omega}$. Let $\alpha_1$ be the least element of $\mathbf{\Omega}$ such that $(\alpha_1, \omega_1]$ is contained in some element $U_1$ of $\mathscr{U}$. If $\alpha_1 \neq 1$, let $\alpha_2$ be the least element of $\mathbf{\Omega}$ such that $(\alpha_2, \alpha_1]$ is contained in some element $U_2$ of $\mathscr{U}$. Continue this process. Then for some $n$, $\alpha_n = 1$, since otherwise we would have a sequence $\alpha_1 > \alpha_2 > \cdots$, which would contradict the well-ordering of $\mathbf{\Omega}$. Then $\{U_1, \ldots, U_n\}$ is a subcollection from $\mathscr{U}$ which covers all of $\mathbf{\Omega}$ except possibly 1, so an $(n + 1)$-element subcollection from $\mathscr{U}$ covers $\mathbf{\Omega}$. Note that each of the closed subspaces $[1, \alpha]$ of $\mathbf{\Omega}$ is now compact, by 17.5.

Some of the properties of the subspace $\mathbf{\Omega}_0$ of $\mathbf{\Omega}$ will be of interest. First note that $\mathbf{\Omega}_0$ *is countably compact.* For let $\mathscr{U} = \{U_1, U_2, \ldots\}$ be a countable open cover of $\mathbf{\Omega}_0$. If no finite subcover of $\mathbf{\Omega}_0$ exists, then for each $n$, pick $\alpha_n \notin U_1 \cup \cdots \cup U_n$. If $\alpha = \sup\{\alpha_1, \alpha_2, \ldots\}$, then $\alpha \in \mathbf{\Omega}_0$ and no finite subcollection from $\mathscr{U}$ covers the compact set $[1, \alpha]$, which is impossible. Next note that $\mathbf{\Omega}_0$ *is not compact*, since the cover of $\mathbf{\Omega}_0$ by the sets $[1, \alpha)$, for $\alpha \in \mathbf{\Omega}_0$, can have no finite subcover. Also, letting $\Omega(\alpha)$ denote the set of all ordinals $\leq \alpha$, $\mathbf{\Omega}_0 - \Omega(\alpha)$ *is homeomorphic to* $\mathbf{\Omega}_0$, for each $\alpha \in \mathbf{\Omega}_0$. The homeomorphism is easily constructed; it takes the least element of $\mathbf{\Omega}_0 - \Omega(\alpha)$ to 1, the next element to 2, and so on by transfinite induction. Finally, *every continuous real-valued function on* $\mathbf{\Omega}_0$ *is constant on some tail.* To see this, let $f: \mathbf{\Omega}_0 \to \mathbf{R}$ be continuous. Then $f(\mathbf{\Omega}_0)$ is countably compact, by 17F.5, and Lindelöf. Hence $f(\mathbf{\Omega}_0)$ is compact. By the next theorem, the net $(f(\alpha))_{\alpha \in \mathbf{\Omega}_0}$ must then have a cluster point in $f(\mathbf{\Omega}_0)$. This cluster point $y$ is unique. For suppose $z$ is another cluster point of the same net. Then we can find an increasing sequence $\alpha_1, \alpha_2, \ldots$ of countable ordinals such that $|f(\alpha_{2n-1}) - y| < 1/n$ and $|f(\alpha_{2n}) - z| < 1/n$, for $n = 1, 2, \ldots$. Thus if $\alpha = \sup\{\alpha_1, \alpha_2, \ldots\}$, we have $f(\alpha) = y$ and $f(\alpha) = z$, so that $y = z$. Next we claim the net $(f(\alpha))$ converges to this unique cluster point $y$. If not, then for some open nhood $U$ of $y$, $\mathbf{\Omega}_0 - f^{-1}(U)$ contains a cofinal subset of $\mathbf{\Omega}_0$. But $\mathbf{\Omega}_0 - f^{-1}(U)$ is a closed subset of $\mathbf{\Omega}_0$ and thus countably compact (17F.5), and the argument above can be re-applied to yield a cluster point of $(f(\alpha))$ other than $y$. Since this is impossible, $(f(\alpha))$ must converge to $y$. Now for $n = 1, 2, \ldots$, pick $\alpha_n \in \mathbf{\Omega}_0$ such that $\alpha \geq \alpha_n$ implies $|f(\alpha) - y| < 1/n$. Let $\alpha_0 = \sup\{\alpha_1, \alpha_2, \ldots\}$. Then $\alpha \geq \alpha_0$ implies $f(\alpha) = y$, so $f$ is constant on the tail $\{\alpha \in \mathbf{\Omega}_0 \mid \alpha \geq \alpha_0\}$ of $\mathbf{\Omega}_0$.

This last property of $\mathbf{\Omega}_0$ yields an extension theorem: *every continuous real-valued function on* $\mathbf{\Omega}_0$ *can be extended to a continuous function on* $\mathbf{\Omega}$.

**17.3 Definition.** A family $\mathscr{E}$ of subsets of $X$ has the *finite intersection property* iff the intersection of any finite subcollection from $\mathscr{E}$ is nonempty.

Families with the finite intersection property are somewhat like filters; in fact, if $\mathscr{E}$ is such a family and $\mathscr{F}$ is the collection of all possible finite intersections from $\mathscr{E}$ then $\mathscr{F}$ is a filter base, so every family $\mathscr{E}$ with the finite intersection property generates a filter. Conversely, every filter *is* a family with the finite intersection property. Some of the implications in the following theorem will now be clear.

**17.4 Theorem.** *For a topological space $X$, the following are equivalent:*

a) *$X$ is compact,*

b) *each family $\mathscr{E}$ of closed subsets of $X$ with the finite intersection property has nonempty intersection,*

c) *each filter in $X$ has a cluster point,*

d) *each net in $X$ has a cluster point,*

e) *each ultranet in $X$ converges,*

f) *each ultrafilter in $X$ converges.*

*Proof.* a) $\Rightarrow$ b): If $\{E_\alpha \mid \alpha \in A\}$ is a family of closed sets in $X$ having empty intersection, then $\{X - E_\alpha \mid \alpha \in A\}$ is an open cover of $X$. By compactness, there is a finite subcover $\{X - E_{\alpha_1}, \ldots, X - E_{\alpha_n}\}$ and then $\bigcap_{k=1}^n E_{\alpha_k} = \varnothing$, so $\{E_\alpha \mid \alpha \in A\}$ does not have the finite intersection property.

b) $\Rightarrow$ c): If $\mathscr{F}$ is a filter on $X$, then $\{\bar{F} \mid F \in \mathscr{F}\}$ is a family of closed sets with the finite intersection property, so there is a point $x$ in $\bigcap \{\bar{F} \mid F \in \mathscr{F}\}$. Then $\mathscr{F}$ has $x$ for a cluster point.

c) $\Rightarrow$ d): This is an easy exercise in the use of the standard translation process from filters to nets. See 12.15–12.17 and 12D.

d) $\Rightarrow$ e): If an ultranet has a cluster point, it converges to that point.

e) $\Rightarrow$ f): Let $\mathscr{F}$ be an ultrafilter on $X$. The net based on $\mathscr{F}$ is then an ultranet (12D.4) and hence converges. Then $\mathscr{F}$ converges (12.17).

f) $\Rightarrow$ a): Suppose $\mathscr{U}$ is an open cover of $X$ with no finite subcover. Then $X - (U_1 \cup \cdots \cup U_n) \neq \varnothing$ for each finite collection $\{U_1, \ldots, U_n\}$ from $\mathscr{U}$. The sets of the form $X - (U_1 \cup \cdots \cup U_n)$ then form a filter base (since the intersection of two such sets has again the same form), generating a filter $\mathscr{F}$. Now $\mathscr{F}$ is contained in some ultrafilter $\mathscr{F}^*$ and, by ($f$), $\mathscr{F}^*$ converges, say to $x$. Now $x \in U$ for some $U \in \mathscr{U}$. Since $U$ is a nhood of $x$, $U \in \mathscr{F}^*$. But, by construction, $X - U \in \mathscr{F} \subset \mathscr{F}^*$. Since it is impossible for both $U$ and $X - U$ to belong to $\mathscr{F}^*$, we have a contradiction. Thus $\mathscr{U}$ must have a finite subcover. ■

The previous theorem gives a hint of one of the lines from topology to more "applied" branches of mathematics. Compactness can be used by "existential" (as opposed to "constructive") analysts, in the following way. Given a differential equation, it may be possible to topologize some set of functions (among which are the solutions, if any, of that equation) in such a way that convergence of an appropriate net or sequence of functions to the limit $f$ implies that $f$ is a solution of the original differential equation. Thus the study of compactness (every net has a convergent subnet), countable compactness (every sequence has a convergent subnet; see 17F) and sequential compactness (every sequence has a convergent subsequence; see 17G) in spaces of functions is germane to the study of *existence* of solutions to differential equations.

We turn now to investigation of the basic structural questions about subspaces, continuous images and products of compact spaces. That the answers are as pleasing as they are is one of the primary reasons for the importance of compactness. In particular, we will have more to say about the Tychonoff Theorem (which is about products).

We begin with subsets.

**17.5 Theorem.** a) *Every closed subset of a compact space is compact.*

b) *A compact subset of a Hausdorff space is closed.*

*Proof.* a) If $A$ is closed in the compact space $X$ and $\mathscr{U}$ is any open cover of $A$, then for each $U \in \mathscr{U}$ we can find an open set $V_U$ in $X$ such that $V_U \cap A = U$. Now $\{X - A\} \cup \{V_U \mid U \in \mathscr{U}\}$ is an open cover of $X$ which, by compactness, has a finite subcover. The intersections with $A$ of this finite cover form a finite subcover of $A$ from $\mathscr{U}$.

b) Suppose $A$ is a compact subset of the Hausdorff space $X$. If $a \in \bar{A}$, then a net $(x_\lambda)$ exists in $A$ with $x_\lambda \to a$ in $X$. But since $A$ is compact, $(x_\lambda)$ has a cluster point $b$ in $A$ and thus a subnet which converges to $b$. Since this subnet converges to $a$ also and limits in $X$ are unique, we must have $a = b$. Thus $a \in A$, showing that $A$ is closed. ■

For non-Hausdorff spaces, the second part of the theorem above may fail; see Exercise 17B.4. Before turning to continuous images and products, we note that compact subsets of a topological space "behave like points" in a sense made more precise by the following theorem. The proof is left to Exercise 17B.

**17.6 Theorem.** a) *Disjoint compact subsets of a Hausdorff space can be separated by disjoint open sets.*

b) *A compact set and a disjoint closed set in a regular space can be separated by disjoint open sets.*

c) *If $A \times B$ is a compact subset of a product $X \times Y$ contained in an open set $W$ in $X \times Y$, then open sets $U$ in $X$ and $V$ in $Y$ can be found such that*

$$A \times B \subset U \times V \subset W.$$

**17.7 Theorem.** *The continuous image of a compact space is compact.*

*Proof.* Suppose $X$ is compact and $f$ is a continuous map of $X$ onto $Y$. If $\mathscr{U}$ is an open cover of $Y$, then $\{f^{-1}(U) \mid U \in \mathscr{U}\}$ is an open cover of $X$ and, by compactness, a finite subcover exists, say $\{f^{-1}(U_1), \ldots, f^{-1}(U_n)\}$. Then, since $f$ is onto, the sets $U_1, \ldots, U_n$ cover $Y$. Thus $Y$ is compact. ■

This theorem has a nice consequence. If $f$ is a continuous mapping from a compact space $X$ to a Hausdorff space $Y$, then each closed subset $E$ of $X$ is compact, so $f(E)$ is compact and thus closed in $Y$. Hence *every continuous map from a compact space to a Hausdorff space is a closed map* (*and thus a quotient map*). One consequence of this is given in 17.14 at the end of this section.

For use in the next theorem, we recall that an onto mapping takes ultranets to ultranets (11.11). A proof similar to the one given here can easily be constructed using ultrafilters.

***17.8 Theorem (Tychonoff).*** *A nonempty product space is compact iff each factor space is compact.*

*Proof.* $\Rightarrow$: If the product space is nonempty, then the projection maps are all continuous and onto, so the result here follows from 17.7.

$\Leftarrow$: Let $(x_\lambda)_{\lambda \in \Lambda}$ be an ultranet in $\prod_{\alpha \in A} X_\alpha$. Then for each fixed $\alpha$, $(\pi_\alpha(x_\lambda))_{\lambda \in \Lambda}$ is an ultranet in $X_\alpha$ and hence converges, since $X_\alpha$ is compact. By 11.9 it follows that $(x_\lambda)$ converges. Thus the product space is compact. ∎

The theorem just proved can lay good claim to being the most important theorem in general (nongeometric) topology. It plays a central role in the development of a wealth of theorems within topology and applications of topology to other fields. To mention but a few examples: the construction of the Stone–Čech compactification $\beta X$ of any Tychonoff space $X$ is based on it (see 19.4), Ascoli's theorem on compactness of function spaces (see 43.15) relies on it (and Ascoli's theorem can, in turn, be used to provide existence theorems for various differential equations), the proof of compactness of the maximal ideal space of a Banach algebra requires it and hence it is central to the development of the Gelfand representation theorem.

It is worth mentioning that the proof of 17.8, as we have given it, is deceptively simple; it hides a good deal of muscle. Tychonoff did not have available for his proof the powerful convergence theorems which roam around in ours. Some idea of the strength of his theorem can be had by studying Exercise 17O, in which you show that 17.8 implies the axiom of choice. (Thus the axiom of choice must be used somehow in our proof, since it cannot be derived from the other axioms of set theory.)

We can use the Tychonoff theorem to provide a number of important examples of compact spaces.

**17.9 Examples.** a) *A subset of* $\mathbf{R}^n$ *is compact iff it is closed and bounded.* For if $A$ is compact, it is closed. Moreover the sets $U(x, 1)$ for $x \in A$ form an open cover of $A$ which, by compactness, has a finite subcover. A routine calculation shows that $A$ is thus bounded.

Conversely, each closed interval $[a, b]$ in $\mathbf{R}$ is homeomorphic to $\mathbf{I}$ and thus is compact. But a closed, bounded subset of $\mathbf{R}^n$ will be a closed subset of an $n$-fold product $[-c, c] \times \cdots \times [-c, c]$ of such intervals and thus will be compact.

b) *Every cube is compact.* This follows directly from Tychonoff's theorem, since a cube is just a product of closed bounded intervals. Of particular interest is the *Hilbert cube*, which is the product $\mathbf{I}^{\aleph_0}$ of countably many copies of $\mathbf{I}$. To us it makes no difference, but a metric geometer working with Hilbert space $\mathbf{H}$ (18.7) would rather think of the Hilbert cube as the product

$$[0, 1] \times [0, \tfrac{1}{2}] \times [0, \tfrac{1}{3}] \times \cdots$$

(since then it is isometric, rather than just homeomorphic, to a subspace of $\mathbf{H}$).

c) *The Cantor set.* Beginning with the unit interval $\mathbf{I}$, define closed subsets $A_1 \supset A_2 \supset \cdots$ in $\mathbf{I}$ as follows. We obtain $A_1$ by removing the interval $(\frac{1}{3}, \frac{2}{3})$ from $\mathbf{I}$. $A_2$ is then obtained by removing from $A_1$ the open intervals $(\frac{1}{9}, \frac{2}{9})$ and $(\frac{7}{9}, \frac{8}{9})$. In general, having $A_{n-1}$, $A_n$ is obtained by removing the open middle thirds from each of the $2^{n-1}$ closed intervals that make up $A_{n-1}$. The *Cantor set* is the subspace $\mathbf{C} = \bigcap A_n$ of $\mathbf{I}$. It is a nonempty compact metric space.

We can develop an interesting alternative description of the Cantor set. Each $x \in \mathbf{I}$ has an expansion $(x_1, x_2, \ldots)$ in ternary form (that is, each $x_i$ is 0, 1 or 2) obtained by writing $x = \sum x_i/3^i$. These expressions are unique, except that any number but 1 expressible in a ternary expansion ending in a sequence of 2's can be re-expressed in an expansion ending in a sequence of 0's (for example, $\frac{1}{3}$ can be written as $(1, 0, 0, \ldots)$ or as $(0, 2, 2, \ldots)$). Then *the Cantor set* $\mathbf{C}$ *is precisely the set of points in* $\mathbf{I}$ *having a ternary expansion without* 1's. For this reason, $\mathbf{C}$ is sometimes referred to as the *Cantor ternary set.*

Using the ternary representation, it is possible to show that $\mathbf{C}$ *is homeomorphic to a product of denumerably many copies of the two-point discrete space.* In fact, by writing the discrete space as $D = \{0, 2\}$, the ternary correspondence $x \to (x_1, x_2, \ldots)$ becomes a homeomorphism. The proof is left as an exercise. You should do it if you think you can't, since it will teach you a lot about product spaces. Later, in the section devoted to the Cantor set, we will see that the product of denumerably many nontrivial finite discrete spaces is homeomorphic to the Cantor set. For this reason, (possibly nondenumerable) products of finite discrete spaces are called *Cantor spaces.* The Cantor spaces occupy a special place in topology. Compactness and discreteness are, in a sense, dual properties, and only the Cantor spaces carry the banners of both.

We close this section with a study of some of the (nonstructural) properties of compact spaces which make them important. In particular, we will develop the relationship between compact Hausdorff spaces, Tychonoff spaces and normal spaces.

***17.10 Theorem.*** *A compact Hausdorff space $X$ is a $T_4$-space.*

*Proof.* It suffices to prove regularity since a regular Lindelöf space is normal. Let $A$ be closed in $X$, $x \notin A$. For each $a \in A$, pick disjoint open sets $U_a$ containing $x$ and $V_a$ containing $a$. The sets $V_a$, $a \in A$, cover $A$ and, by compactness of $A$, some finite collection $V_{a_1}, \ldots, V_{a_n}$ is sufficient. Let $V = \bigcup_{i=1}^n V_{a_i}$ and

$U = \bigcap_{i=1}^{n} U_{a_i}$. Then $U$ and $V$ are disjoint open sets containing $x$ and $A$, respectively. ■

One importance of this theorem can be brought into focus by recalling that normal spaces enjoy very nice separation, extension and covering properties, but that products of normal spaces need not be normal. By combining the above theorem with Tychonoff's theorem, we obtain the only result which asserts normality for a large class of product spaces: *every product of compact, Hausdorff spaces is* $T_4$. One of the immediate consequences of this is a result we have already mentioned without proof: *every cube is* $T_4$. The search for theorems which assert normality for various product spaces has occupied the time of some very good mathematicians; we will return to this topic in Section 21.

Another (related) consequence of Theorem 17.10 provides the important relationship between compact Hausdorff spaces, Tychonoff spaces and normal spaces.

***17.11* Corollary.** *The following are equivalent, for a topological space X:*

a) *X is Tychonoff,*

b) *X is homeomorphic to a subspace of some cube,*

c) *X is homeomorphic to a subspace of some compact Hausdorff space,*

d) *X is homeomorphic to a subspace of some $T_4$-space.*

*Proof.* a) ⇒ b): We have already shown (a) equivalent to (b) in 14.13.

b) ⇒ c): Every cube is a compact Hausdorff space and thus is normal.

c) ⇒ d): Every compact Hausdorff space is a $T_4$-space.

d) ⇒ a): Every subspace of a $T_4$-space is a Tychonoff space. ■

In studying the interplay between compactness and the strong-side separation axioms (normality and the Tychonoff property) one example has become of paramount importance.

**17.12 Example.** *The Tychonoff plank.* Our basic building blocks are the ordinal spaces $\mathbf{\Omega}$, with which we are familiar, and $\Omega(\omega) = \mathbf{N} \cup \{\omega\}$, where $\omega$ is the first infinite ordinal. When $\Omega(\omega)$ is given its order topology, the points of $\mathbf{N}$ are isolated (open) and the point $\omega$ has for basic nhoods the sets $\{n, n + 1, \ldots\} \cup \{\omega\}$.

The product space $\mathbf{\Omega} \times \Omega(\omega)$ will be denoted $\mathbf{T}^*$ and the corner point $(\omega_1, \omega)$ in $\mathbf{T}^*$ will be denoted $t$. The *Tychonoff plank* is the subspace $\mathbf{T} = \mathbf{T}^* - \{t\}$ of $\mathbf{T}^*$. Since $\mathbf{T}^*$ is a compact Hausdorff space, $\mathbf{T}$ is a Tychonoff space.

But $\mathbf{T}$ is not normal. To develop this fact, some terminology will be useful. For each $n \in \mathbf{N}$, let $\Omega_n = \mathbf{\Omega}_0 \times \{n\}$, and for each $\alpha \in \mathbf{\Omega}_0$, let $\Omega_n(\alpha)$ be the tail $\{(\beta, n) \mid \beta \geq \alpha\}$ in $\Omega_n$. Also, we will call the set $A = \{(n, \omega_1) \mid n \in \mathbf{N}\}$ the *right edge* of $\mathbf{T}$, and the set $B = \{(\omega, \alpha) \mid \alpha \in \Omega_0\}$ the *top edge* of $\mathbf{T}$. Now $A$ and $B$ are closed sets in $\mathbf{T}$, since they are the intersections with $\mathbf{T}$ of closed sets in $\mathbf{T}^*$. Hence

APRIL

# 14

WEDNESDAY

Binker—what I call him—is a
secret of my own,
And Binker is the reason why
I never feel alone.
Playing in the nursery, sitting
on the stair,
Whatever I am busy at, Binker
will be there.

*Now We Are Six*

Darla —

Thanks for the loan.

— Bernie

if $\mathbf{T}$ were normal, there would be a continuous $f: \mathbf{T} \to \mathbf{I}$ with $f(A) = 0$ and $f(B) = 1$. But for each $n \in \mathbf{N}$, $f$ is constant on some tail $\Omega_n(\alpha_n)$ of $\Omega_n$ since $\Omega_n$ is just a copy of $\mathbf{\Omega}_0$ (see 17.2c). If we let $\alpha = \sup \{\alpha_1, \alpha_2, \ldots\}$, then $\alpha < \omega_1$ and $f$ takes some constant value on $\Omega_n(\alpha)$ for each $n$. But since $f(A) = 0$, this constant value must be 0 for each $n$. Thus $f(\alpha, n) = 0$ for each $n \in \mathbf{N}$, and hence $f(\alpha, \omega) = 0$. But $(\alpha, \omega) \in B$, contradicting the fact that $f(B) = 1$. Thus no continuous function separates the right edge of $\mathbf{T}$ from the top edge, so $T$ cannot be normal.

**17.13 Theorem.** *A continuous real-valued function on a countably compact space is bounded.*

*Proof.* If $f: X \to \mathbf{R}$ is continuous and $X$ is countably compact, then the open cover of $X$ by the sets $f^{-1}(-n, n)$ has a finite subcover. ∎

**17.14 Theorem.** *A one–one continuous map from a compact space $X$ onto a Hausdorff space $Y$ is a homeomorphism.*

*Proof.* If $f$ is such a map, then for each closed set $E \subset X$, $E$ is compact, so $f(E)$ is compact, and thus closed, in $Y$. Thus $f$ is a closed map, and hence a homeomorphism. ∎

Neither of the properties above is characteristic of compact spaces and each has been intensively investigated for noncompact spaces as well. The technique is a familiar one. By making the property the subject of a definition, its study becomes the study of a class of topological spaces (somewhat wider than the class of compact spaces). Exercises 17J, 17K, 17L and 17M are devoted to the development of this line of thought.

## Problems

17A. *Examples on compactness*

1. An infinite set $X$ with the cofinite topology is compact.

2. Which subsets of the Sorgenfrey line $\mathbf{E}$ are compact?

3. Which subsets of the slotted plane (4C) are compact?

4. Which subsets of the Moore plane $\mathbf{\Gamma}$ are compact?

5. The sequence space $\mathbf{m}$ (2H) is not compact [an uncountable subset $A$ of $\mathbf{m}$ exists any two of whose points are at distance 1].

17B. *Compact subsets*

1. A subset $E$ of $X$ is compact iff every cover of $E$ by open subsets of $X$ has a finite subcover. (But note that compactness is not a relative property; that is, if $E$ is compact, it is compact in whatever space it is embedded.)

2. The union of a finite collection of compact subsets of $X$ is compact.

3. The intersection of any collection of compact subsets of a Hausdorff space $X$ is compact; "Hausdorff" is necessary, even for intersections of two compact sets.

4. A compact subset of a non-Hausdorff space need not be closed (compare with 17.5).

5. In a Hausdorff space, disjoint compact sets can be separated by disjoint open sets. (This is an illustration of the general rule, "compact sets behave like points." The next two parts of this exercise are examples of the same principle.)

6. In a regular space, a compact set and a disjoint closed set can be separated by disjoint open sets.

7. If $A \times B$ is a compact subset of $X \times Y$ contained in an open set $W$ in $X \times Y$, then there exist open sets $U \subset X$ and $V \subset Y$ such that $A \times B \subset U \times V \subset W$.

### 17C. *Maximal compact spaces*

A compact space $X$ is *maximal compact* iff every strictly larger topology on $X$ is noncompact.

1. A compact space $X$ is maximal compact iff every compact subset is closed.

2. Every compact Hausdorff space is maximal compact and every maximal compact space is $T_1$ (so maximal compactness acts like a separation axiom for compact spaces).

### 17D. *z-filters in compact spaces*

A variant of the convergence characterization of compactness (17.4) is important in studying the interplay between compactness and the Tychonoff separation axiom. To give it, we must review the language of $z$-filters.

A nonempty collection $\mathscr{F}$ of nonempty zero sets in a topological space $X$ is a *z-filter* on $X$ iff

a) if $Z_1, Z_2 \in \mathscr{F}$, then $Z_1 \cap Z_2 \in \mathscr{F}$,
b) if $Z \in \mathscr{F}$ and $Z'$ is a zero set containing $Z$, then $Z' \in \mathscr{F}$.

Thus a $z$-filter is almost a filter, but the superset property has been altered to that only zero sets will belong. Convergence for $z$-filters is easily defined, once we recall that the zero-set nhoods of a point in a Tychonoff space form a nhood base (14C). We say a $z$-filter $\mathscr{F}$ in a Tychonoff space $X$ *converges* to a point $x$ in $X$, written $\mathscr{F} \to x$, iff each zero-set nhood of $x$ belongs to $\mathscr{F}$. We say $\mathscr{F}$ has $x$ as a *cluster point* iff $x \in F$ for each $F \in \mathscr{F}$ (since $\mathscr{F}$ consists of closed sets, we needn't take closures here). Finally, a *z-ultrafilter* is a $z$-filter which is contained in no strictly larger $z$-filter. Parts 1 and 3 below are repeats of parts of the Exercise 12E on $\mathscr{P}$-filters.

1. Every $z$-filter is contained in some $z$-ultrafilter.

2. For a Tychonoff space $X$, the following are equivalent:

   a) $X$ is compact,
   b) every $z$-filter on $X$ has a cluster point (i.e., has nonempty intersection),
   c) every $z$-ultrafilter on $X$ converges.

3. If $Z_1$ and $Z_2$ are zero sets, $\mathscr{F}$ is a $z$-ultrafilter and $Z_1 \cup Z_2 \in \mathscr{F}$, then one of $Z_1$ or $Z_2$ belongs to $\mathscr{F}$.

### 17E. *Compact ordered spaces*

Call an ordered space $X$ *lattice complete* iff each nonempty subset has a supremum and an infimum. Recall that $X$ is Dedekind complete iff every subset of $X$ having an upper bound has a least upper bound. Then the following are equivalent:

a) $X$ is compact,

b) $X$ is lattice complete,

c) $X$ is Dedekind complete and has a first and a last element.

### 17F. *Countably compact spaces*

1. A space is countably compact iff each sequence has a cluster point. (Hence, iff each sequence has a convergent *subnet*. This does *not* necessarily mean each sequence has a convergent *subsequence*, see 11B. Spaces in which each sequence has a convergent subsequence are studied in 17G.)

2. A $T_1$-space is countably compact iff every infinite subset has a cluster point.

3. The product of a compact space and a countably compact space is countably compact. (The result fails for two countably compact factors; see the notes.)

4. If $X_1, X_2, \ldots$ are all first countable, then $\prod X_n$ is countably compact iff each $X_n$ is countably compact.

5. Continuous images and closed subspaces of countably compact spaces are countably compact.

6. For metric spaces, compactness and countable compactness are equivalent.

7. Let $X$ be a countably compact space, $x \in X$. If $U_1, U_2, \ldots$ is a sequence of open sets in $X$ such that $U_{n+1} \subset U_n$ for all $n$ and $\bigcap_{n=1}^{\infty} U_n = \{X\}$, then $\{U_1, U_2, \ldots\}$ is a nhood base at $x$. (Compare with 16A.4.)

### 17G. *Sequentially compact spaces*

A space $X$ is *sequentially compact* iff every sequence in $X$ has a convergent subsequence. (Compare with countable compactness; see 17F.)

1. Not every compact space is sequentially compact. [Consider an uncountable product of copies of **I**.]

2. Every sequentially compact space is countably compact, but not every sequentially compact space is compact. Hence, together with part 1, sequential compactness is neither stronger nor weaker than compactness; just different. [Use $\mathbf{\Omega}_0$.]

3. A first-countable space is sequentially compact iff it is countably compact. (Thus, for metric spaces, sequential compactness is equivalent to compactness, by 17F.6.)

4. A second-countable $T_1$-space is sequentially compact iff it is compact.

5. The countable product of sequentially compact spaces is sequentially compact. (It is also true, but difficult to prove, that the product of $\leq \aleph_1$ sequentially compact spaces is countably compact. See the notes.)

6. Assuming the continuum hypothesis, the product of any uncountable family of $T_1$-spaces, each having more than one point, is never sequentially compact.

### 17H. *Realcompact spaces*

Every compact Hausdorff space is Tychonoff, and thus embeddable in some cube. This makes it clear that a space $X$ is a compact Hausdorff space iff it is embeddable as a closed subset of some product of copies of the unit interval **I** and leads to the following generalization of compactness: $X$ is *realcompact* iff it can be embedded as a closed subset of a product of copies of the real line **R**.

1. Every compact Hausdorff space is realcompact.
2. Every intersection of realcompact subsets of $X$ is realcompact.
3. Every product of realcompact spaces is realcompact.

### 17I. $\sigma$-*Compact spaces*

A space $X$ is $\sigma$-*compact* iff $X$ can be written as the union of countably many compact subsets. $X$ is said to be *hemicompact* (or *denumerable at infinity*) iff there is a sequence $K_1, K_2, \ldots$ of compact subsets of $X$ such that if $K$ is any compact subset of $X$, then $K \subset K_n$ for some $n$.

1. Every hemicompact space is $\sigma$-compact; the converse fails.
2. Every $\sigma$-compact space is Lindelöf.
3. The product of finitely many $\sigma$-compact spaces is $\sigma$-compact. This cannot be extended to infinitely many factors. [Consider $\mathbf{N}^{\aleph_0}$.]

### 17J. *Pseudocompact spaces*

A space $X$ is *pseudocompact* iff every continuous real-valued function on $X$ is bounded.

1. Every countably compact space is pseudocompact.
2. In a Tychonoff space $X$ the following are all equivalent:
   a) $X$ is pseudocompact,
   b) if $U_1 \supset U_2 \supset \cdots$ is a decreasing sequence of nonempty open sets in $X$, then $\bigcap \overline{U_n} \neq \emptyset$,
   c) every countable open cover of $X$ has a finite subcollection whose closures cover $X$. (Compare with 17K.2.)
3. A pseudocompact $T_4$-space is countably compact. [If $X$ is not countably compact, it has a denumerable closed discrete subset $D$. Use 15.8.]

### 17K. *H-closed spaces*

A Hausdorff space is *H-closed* (*absolutely closed*) iff it is closed in every Hausdorff space in which it can be embedded. This generalizes a property of compact Hausdorff spaces.

An *open filter* in a topological space is a collection of open sets satisfying the axioms for a filter, except that only *open* supersets of elements must belong. See Exercises 12E and 12G for elementary facts about open filters.

For the duration of this problem, all spaces are Hausdorff.

1. A space $X$ is $H$-closed iff every open filter has a cluster point. [If some open filter fails to have a cluster point, a point can be added to $X$ whose nhoods are the elements of the open filter (together with the point itself), and the result is a Hausdorff extension of $X$ in which $X$ is not closed. The reverse implication is also done by contradiction.]

2. A space is $H$-closed iff every open cover has a finite subcollection whose closures cover (i.e., a *finite dense subsystem*). [If an open filter does not have a cluster point, the

complements of closures of its elements form an open cover with no finite dense subsystem.] (Compare with 17J.2.)

3. An $H$-closed space is compact iff it is regular. [One way is trivial. For the reverse, let $\mathscr{U}$ be an open cover and use regularity to prove the existence of a cover $\mathscr{V}$ such that for each $V \in \mathscr{V}$, there is some $U \in \mathscr{U}$ containing $\overline{V}$. Then a finite dense subsystem of $\mathscr{V}$ induces a finite subcover from $\mathscr{U}$.]

4. Let $\mathbf{N}^*$ be the subspace $\{0\} \cup \{1/n \mid n \in \mathbf{N}\}$ of $\mathbf{R}$, and to the space $\mathbf{N} \times \mathbf{N}^*$ adjoin a point $q$ whose nhoods have the form $U_{n_0}(q) = \{(n, 1/m) \in \mathbf{N} \times \mathbf{N}^* \mid n \geq n_0\}$. Use part 2 above to prove that the resulting (Hausdorff) space $X$ is $H$-closed and show that $X$ is not compact.

## 17L. *More on H-closed spaces*

1. A regularly closed subset of an $H$-closed space is $H$-closed.

2. A descending chain $\mathscr{A}$ of nonempty $H$-closed subsets of an $H$-closed space $X$ has nonempty intersection. [Let $\mathscr{G}$ be the collection of all open sets $G$ in $X$ such that $\overline{G} \supset A$ for some $A \in \mathscr{A}$. Show that $\mathscr{G}$ has the finite intersection property and thus (17K.1) has a cluster point $p$. Then $p \in \bigcap \mathscr{A}$.]

3. An $H$-closed space is compact iff each closed subset is $H$-closed. [If $\mathscr{F}$ is a closed filter, well-order $\mathscr{F}$ and use this well-ordering to find a descending chain of closed sets with the same intersection. Apply part 2.]

4. A continuous Hausdorff image of an $H$-closed space is $H$-closed. [Use 17K.2.]

5. A nonempty product is $H$-closed iff each factor is $H$-closed.

## 17M. *Minimal Hausdorff spaces*

A Hausdorff space $X$ is *minimal Hausdorff* iff every one–one continuous map of $X$ to a Hausdorff space is a homeomorphism [i.e., iff there is no strictly weaker Hausdorff topology on $X$]. This, again, generalizes a property of compact, Hausdorff spaces.

1. A Hausdorff space $X$ is minimal Hausdorff iff every open filter with a unique cluster point converges (necessarily to that point). [The key is the statement in brackets after the definition of minimal Hausdorff space. Thus if a nonconvergent open filter with a unique cluster point exists, construct a strictly weaker Hausdorff topology on the space (by enlarging nhoods of the unique cluster point).]

2. Every minimal Hausdorff space is $H$-closed. [Construct a weaker Hausdorff topology for a nonabsolutely closed space.] Thus a minimal Hausdorff space is compact iff it is regular. Also, if every closed subset of a minimal Hausdorff space is minimal Hausdorff, the space is compact. [See 17K.3 and 17L.3.]

3. Every $H$-closed space $X$ has a unique weaker topology which is minimal Hausdorff. [Use the complements of the regularly closed sets in $X$ as a base for a new topology on $X$.]

4. A space is minimal Hausdorff iff it is semiregular and $H$-closed.

More is known. A product of minimal Hausdorff spaces is minimal Hausdorff. Every Hausdorff space can be embedded (as a closed, nowhere dense subspace) in a minimal Hausdorff space. See the notes.

## 17N. *Hausdorffness of closed images*

1. If $f$ is a closed map of $X$ onto $Y$ and $f^{-1}(y)$ is compact for each $y \in Y$, then $Y$ is Hausdorff (or regular) if $X$ is.

2. For a compact Hausdorff space $X$, if $f$ is a quotient map of $X$ onto $Y$, the following are equivalent:

a) $Y$ is Hausdorff,
b) $f$ is closed,
c) $\{(x_1, x_2) \in X \times X \mid f(x_1) = f(x_2)\}$ is closed in $X \times X$.

## 17O. *The Tychonoff theorem is equivalent to the axiom of choice*

1. How does the Tychonoff theorem rely for its proof on the axiom of choice?

2. The Tychonoff theorem implies the axiom of choice [allowable reference: any paper of Kelley written in 1950].

## 17P. *Onto maps of compact spaces*

The basic question we raise here is the following. Given a map $f$ of a compact space $X$ onto a compact space $Y$, when is it possible to throw away part of the domain in such a way that the restriction of $f$ to what remains is a homeomorphism?

Let $X$ and $Y$ be compact Hausdorff spaces and let $f$ be a continuous map of $X$ onto $Y$.

1. There is a compact subset $X_0$ of $X$ such that $f(X_0) = Y$, but $f$ maps no proper closed subset of $X_0$ onto $Y$. [Use Zorn's lemma.]

2. If $Y$ is extremally disconnected and no proper closed subset of $X$ is mapped onto $Y$, then $f$ is a homeomorphism. [It suffices to show $f$ is one–one. If $x_1 \neq x_2$, pick disjoint open $G_1, G_2$ such that $x_1 \in G_1, x_2 \in G_2$. Then $A - f(X - G_1)$ and $A - f(X - G_2)$ are disjoint and open in $Y$ and hence so are $\overline{A - f(X - G_1)}$ and $\overline{A - f(X - G_2)}$, by 15G.1. But

$$f(G_i) \subset \overline{A - f(X - G_i)},$$

for $i = 1, 2$, and hence $f(x_1) \neq f(x_2)$.]

## 17Q. *Projective spaces*

A compact space $X$ is called *projective* in the category of compact spaces and continuous maps provided whenever $f: X \to Z$ is continuous and $g: Y \to Z$ is continuous and onto, then there is a continuous map $h: X \to Y$ such that $f = g \circ h$.

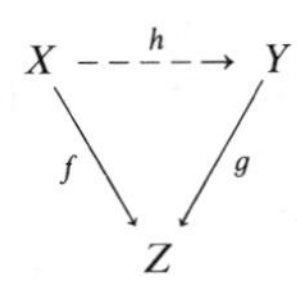


Recall that a space is extremally disconnected (15G) iff the closure of every open set is open. We will draw on parts of Problems 15G and 17P in the course of presenting the following material.

1. Every projective space is extremally disconnected. [Let $G$ be an open set in the projective space $X$. Let $Y$ be the disjoint union of $X - G$ and $\bar{G}$, and $g: Y \to X$ the obvious map (essentially the identity) while $f: X \to X$ is the identity. By projectivity of $X$, there is a map $h: X \to Y$ such that $f = g \circ h$. Conclude that $\bar{G}$ is open.]

2. Every extremally disconnected space is projective. [Let $X$ be extremally disconnected, and let $f: X \to Z$ and $g: Y \to Z$ be as in the introductory paragraph. In $X \times Y$, let $D = \{(x, y) \mid f(x) = g(y)\}$. Then $D$ is compact and the projection $\pi_1$ of $X \times Y$ onto $X$ carries $D$ onto $X$. Apply 17P to get a homeomorphism $\pi_1 \mid E$ of a closed subset $E$ of $D$ with $X$. Let $h = \pi_2 \circ (\pi_1 \mid E)^{-1}$, where $\pi_2$ is the projection of $X \times Y$ onto $Y$.]

### 17R. *Compact subsets of* **R**

There are uncountably many nonhomeomorphic compact subsets of **R**. [Use ordinals.]

### 17S. *The Alexander subbase theorem*

When describing compactness of $X$ in terms of open covers, it is evident that it suffices to restrict attention to a fixed base for $X$. That is, $X$ is compact iff there is a base $\mathscr{B}$ for the topology of $X$ such that any cover of $X$ by elements of $\mathscr{B}$ has a finite subcover. The corresponding assertion for subbases remains true, if we assume the axiom of choice, but is much less obvious. It is interesting, since it can be used to prove the Tychonoff theorem.

A family $\mathscr{B}$ of subsets of $X$ will be called *inadequate* iff it fails to cover $X$, and *finitely inadequate* iff no finite subfamily of $\mathscr{B}$ covers $X$.

1. Given any finitely inadequate family $\mathscr{B}$, there is a finitely inadequate family $\mathscr{B}^* \supset \mathscr{B}$ which is maximal in the order $\mathscr{B}_1 \leq \mathscr{B}_2$ iff $\mathscr{B}_1 \subset \mathscr{B}_2$ on the set of all finitely inadequate families.

2. A maximal finitely inadequate family $\mathscr{B}^*$ has the following property: if $C_1, \ldots, C_n$ are subsets of $X$ and $C_1 \cap \cdots \cap C_n$ belongs to $\mathscr{B}^*$, then $C_k$ belongs to $\mathscr{B}^*$ for some $k = 1, \ldots, n$. [The proof is by contradiction.]

3. The following are all equivalent, for a topological space $X$:
   a) there is a subbase $\mathscr{C}$ for $X$ such that each cover of $X$ by elements of $\mathscr{C}$ has a finite subcover,
   b) there is a subbase $\mathscr{C}$ for $X$ such that each finitely inadequate subfamily of $\mathscr{C}$ is inadequate,
   c) every finitely inadequate family of open subsets of $X$ is inadequate,
   d) $X$ is compact.

[The only hard part is (b) implies (c).]

4. Use part 3 to provide a proof of the Tychonoff theorem.

## 18 Locally compact spaces

Analysts who deal with abstract spaces often appreciate the presence of some form of compactness. Quite often, it is enough that the spaces in question be locally compact.

**18.1 Definition.** A space $X$ is *locally compact* iff each point in $X$ has a nhood base consisting of compact sets.

Recalling that a space is regular iff each point has a nhood base consisting of closed sets, we see immediately that every locally compact Hausdorff space is regular. (In the next section, we will see from a slightly better angle that every locally compact Hausdorff space is, in fact, completely regular.)

Definition 18.1 provides many compact nhoods of each point in a locally compact space, but for most spaces, we can stop as soon as one has been found, according to the next theorem.

***18.2 Theorem.*** *A Hausdorff space $X$ is locally compact iff each point in $X$ has a compact nhood.*

*Proof.* Suppose $x$ has a compact nhood $K$. Let $U$ be any nhood of $x$, and let $V = \text{Int}(K \cap U)$. Then $V$ is an open nhood of $x$. Now $\text{Cl}_X V$ is compact and Hausdorff, so $\text{Cl}_X V$ is regular. Then, since $V$ is a nhood of $x$ in $\text{Cl}_X V$, there is an open nhood $W$ of $x$ in $\text{Cl}_X V$ with $\text{Cl}_{\text{Cl}_X V} W \subset V$. Now $W$ is open in $V$ and hence in $X$, and $\text{Cl}_{\text{Cl}_X V} W$ is closed in $\text{Cl}_X V$ and hence compact; this makes it a compact nhood of $x$ in $X$ which is contained in $U$. Hence $x$ has a base of compact nhoods in $X$. The other implication is easy. ■

Theorem 18.2 provides us with the usual path to proving local compactness, or nonlocal compactness, of the familiar examples of topological spaces. For one thing, it implies that *every compact Hausdorff space is locally compact.* Here are some other examples.

**18.3 Examples.** a) The real line $\mathbf{R}$ is locally compact.

b) The space $\mathbf{Q}$ of rationals and the space $\mathbf{P}$ of irrationals are not locally compact.

c) *Manifolds.* A *topological n-manifold* is a Hausdorff space $X$ such that for each $x \in X$ there is a homeomorphism $\varphi_x$ carrying an open set $U$ in $X$ which contains $x$ onto an open subset of $\mathbf{R}^n$ (for $X$ to be a $C^\infty$*-manifold*, or a *differentiable manifold*, it must also be true that whenever (domain $\varphi_x$) $\cap$ (domain $\varphi_y$) $\neq \varnothing$, then $\varphi_y \circ \varphi_x^{-1}$ is a $C^\infty$-function from $\mathbf{R}^n$ to $\mathbf{R}^n$). Using Theorem 18.2, every topological $n$-manifold is locally compact. Other properties of manifolds are mentioned in 18H.

We turn now to the usual questions about subspaces, products and continuous maps of locally compact spaces, beginning with subspaces.

***18.4 Theorem.*** *In a locally compact Hausdorff space, the intersection of an open set with a closed set is locally compact. Conversely, a locally compact subset of a Hausdorff space is the intersection of an open set and a closed set.*

*Proof.* Let $X$ be locally compact and $T_2$. If $A$ is open in $X$ and $a \in A$, then $a$ has a compact nhood $K$ in $X$ contained in $A$, and $K$ is then a compact nhood of $a$ in $A$, so $A$ is locally compact. If $B$ is closed in $X$ and $b \in B$, then $b$ has a compact nhood

$K$ in $X$ and $K \cap B$ is a compact nhood of $b$ in $B$, so $B$ is locally compact. Hence, open subsets and closed subsets of $X$ are locally compact. But (easily) the intersection of two locally compact subsets of $X$ is locally compact so, in particular, the intersection of an open set with a closed set in $X$ is locally compact.

Conversely, suppose $Y$ is Hausdorff and $X$ is a locally compact subset of $Y$. It will suffice to show $X$ is open in $\mathrm{Cl}_Y X$ (Why?). Let $x \in X$ and find a nhood $U$ of $x$ in $X$ such that $\mathrm{Cl}_X U$ is compact, by local compactness. Say $U = X \cap V$ where $V$ is open in $Y$. Then

$$\mathrm{Cl}_Y (X \cap V) \cap X = \mathrm{Cl}_Y U \cap X = \mathrm{Cl}_X U$$

and the latter is compact. Thus $\mathrm{Cl}_Y (X \cap V) \cap X$ is closed in $Y$. But it contains $X \cap V$ and thus $\mathrm{Cl}_Y (X \cap V)$; i.e., $\mathrm{Cl}_Y (X \cap V) \cap X \supset \mathrm{Cl}_Y (X \cap V)$. But then $\mathrm{Cl}_Y (X \cap V) \subset X$, and hence $(\mathrm{Cl}_Y X) \cap V \subset X$. Thus $(\mathrm{Cl}_Y X) \cap V$ is a nhood of $x$ in $\mathrm{Cl}_Y X$ which is contained in $X$, so $X$ is open in $\mathrm{Cl}_Y X$. ∎

There is one consequence of the previous theorem which crops up often. *A dense subset of a compact Hausdorff space is locally compact iff it is open.*

Quotients of locally compact spaces need not be locally compact. In fact, the spaces which are quotients of locally compact spaces are studied for their intrinsic interest. They are called "$k$-spaces," or "compactly generated spaces," and are dealt with in some detail in Section 43.

Some quotient maps *do* preserve local compactness, as the next theorem shows.

***18.5 Theorem.*** *If $f$ is a continuous, open map of $X$ onto $Y$ and $X$ is locally compact, then so is $Y$.*

*Proof.* Suppose $y \in Y$ and $V$ is a nhood of $y$. Pick $x \in f^{-1}(y)$ and, by continuity and local compactness, find a compact nhood $K$ of $x$ such that $f(K) \subset V$. Now $x \in \mathrm{Int}_X K$, so $y \in f(\mathrm{Int}_X K) \subset f(K)$ and, since $f$ is open, $f(\mathrm{Int}_X K)$ is open. It follows that $f(K)$ is a compact nhood of $y$ contained in $V$. ∎

Local compactness behaves well only with respect to finite products, essentially, according to the next theorem.

***18.6 Theorem.*** *Suppose $X_\alpha$ is nonempty for each $\alpha \in A$. Then $\prod X_\alpha$ is locally compact iff*

a) *each $X_\alpha$ is locally compact,*

b) *all but finitely many $X_\alpha$ are compact.*

*Proof.* $\Rightarrow$: Projections are continuous and open, so part a) follows from 18.5. For b), let $x \in \prod X_\alpha$ and let $W$ be a compact nhood of $x$. Then $W$ contains a basic nhood of the form

$$\pi_{\alpha_1}^{-1}(U_{\alpha_1}) \cap \cdots \cap \pi_{\alpha_n}^{-1}(U_{\alpha_n}),$$

and it follows that, if $\alpha \neq \alpha_1, \ldots, \alpha_n$, then $\pi_\alpha(W) = X_\alpha$. Thus $X_\alpha$ is compact for all $\alpha$ except possibly $\alpha_1, \ldots, \alpha_n$.

$\Leftarrow$: Let $x \in \prod X_\alpha$, and let $U$ be a basic nhood of $x$; say

$$U = U_{\alpha_1} \times \cdots \times U_{\alpha_n} \times \prod \{X_\alpha \mid \alpha \neq \alpha_1, \ldots, \alpha_n\}$$

where we assume the set $\{\alpha_1, \ldots, \alpha_n\} = S$ is expanded to include all $\alpha$ for which $X_\alpha$ is not compact. It suffices to find a compact nhood contained in $U$. But, for each $\alpha_i$, $i = 1, \ldots, n$, there is a compact nhood $K_{\alpha_i}$ of $x_{\alpha_i}$ with $K_{\alpha_i} \subset U_{\alpha_i}$. Then, since $X_\alpha$ is compact for $\alpha \notin S$,

$$K = K_{\alpha_1} \times \cdots \times K_{\alpha_n} \times \prod \{X_\alpha \mid \alpha \notin S\}$$

is a compact nhood of $x$ and $K \subset U$. ∎

**18.7 Examples.** a) $\mathbf{R}^n$ is locally compact for each positive integer $n$, $\mathbf{R}^{\aleph_0}$ is not.

b) *Hilbert space* $\mathbf{H}$ is the collection of all real sequences $\mathbf{x} = (x_1, x_2, \ldots)$ such that $\sum x_k^2 < \infty$, with the metric

$$d(\mathbf{x}, \mathbf{y}) = \sqrt{\sum (x_k - y_k)^2}$$

The proof that $d$ is actually a metric requires the classical Schwarz inequality and is left to Exercise 18B. Let $\mathbf{0}$ denote the sequence $(0, 0, \ldots)$ in $\mathbf{H}$. We will show now that *the closed $\epsilon$-disk $B_\epsilon = \{\mathbf{x} \in \mathbf{H} \mid d(\mathbf{x}, \mathbf{0}) \leq \epsilon\}$ in $\mathbf{H}$ is not compact.* For $n = 1, 2, \ldots$, let $\mathbf{x}^n$ be the sequence in $\mathbf{H}$ whose $k$th coordinate is 0 if $k \neq n$ and whose $k$th coordinate is $\epsilon$ if $k = n$. Then $\mathbf{x}^1, \mathbf{x}^2, \ldots$ is a sequence in $B_\epsilon$, having no cluster point since, if $n \neq m$, $d(\mathbf{x}^n, \mathbf{x}^m) = \sqrt{2}\,\epsilon$. Thus $B_\epsilon$ is not compact. It follows that $\mathbf{H}$ *is not locally compact.* For if $K$ were a compact nhood of $\mathbf{0}$ in $\mathbf{H}$, then for sufficiently small $\epsilon$, $B_\epsilon$ would be a closed subset of $K$ and thus compact. Other properties of $\mathbf{H}$ are given in Exercise 18B.

## Problems

### 18A. *Examples on local compactness*

1. $\mathbf{Q}$ is not locally compact.
2. The Moore plane $\mathbf{\Gamma}$ is not locally compact.
3. The Sorgenfrey line $\mathbf{E}$ is not locally compact.
4. The slotted plane (4C) is not locally compact.
5. Let $A \subset X$ and let $\tau$ be the topology for $X$ consisting of ø together with all subsets of $X$ containing $A$. Is $(X, \tau)$ locally compact?
6. Discuss local compactness of the radial plane (3A.4).

### 18B. *Hilbert space*

Recall that $\mathbf{H}$ denotes all real sequences $\mathbf{x} = (x_1, x_2, \ldots)$ with $\sum x_n^2 < \infty$.

1. The distance function $d(\mathbf{x}, \mathbf{y}) = \sqrt{\sum (x_k - y_k)^2}$ is a metric for $\mathbf{H}$. [Use the *Schwarz inequality*: $(\sum x_k y_k)^2 \leq \sum x_k^2 \sum y_k^2$.]

2. If $(\mathbf{x}^n)_{n\in\mathrm{N}}$ is a sequence in $\mathbf{H}$, then $\mathbf{x}^n \to \mathbf{x}$ in $\mathbf{H}$ implies $x_i^n \to x_i$ in $\mathbf{R}$ for each $i = 1, 2, \ldots$ . The converse fails.

3. $\mathbf{H}$ is separable.

4. The topology on $\mathbf{H}$ differs from the topology it would inherit as a subspace of $R^{\aleph_0}$. [See part 2.]

5. $\mathbf{R}^n$ is isometric to the subspace of $\mathbf{H}$ consisting of all sequences $(x_1, x_2, \ldots)$ such that $x_k = 0$ for $k > n$.

6. $\mathbf{H}$ is isometric to a nowhere dense subset of itself.

Part 6 above shows that, for subsets of $\mathbf{H}$, the property of being open in $\mathbf{H}$ is not a topological (or even a metric) invariant. The corresponding result for $\mathbf{R}^n$ is true: if $U$ and $V$ are homeomorphic subsets of $\mathbf{R}^n$ and $U$ is open, then $V$ is open. This result, due originally to Brouwer and called *invariance of domain*, is most elegantly proved using the machinery of algebraic topology.

### 18C. *Quotients of locally compact spaces*

Compare with Theorem 18.5.

1. The closed continuous image of a locally compact space need not be locally compact. [Let $X$ be the plane, $A$ the $x$-axis in $X$, and $\mathscr{D}$ the decomposition of $X$ whose elements are $A$ and the sets $\{x\}$ for $x \in X - A$. The projection $P$ of $X$ onto $X/\mathscr{D}$ is closed because $\mathscr{D}$ is upper semicontinuous (see 9E).]

2. The closed continuous image of a locally compact space *is* locally compact provided the pre-image of each point is compact (so noncompactness of $A$ was needed in part 1).

3. The condition of 2 is not necessary. [Identify $[1, \infty)$ in $\mathbf{R}$.]

### 18D. *Subsets and subgroups of topological groups*

Let $G$ be a topological group. (13G.)

1. If $U$ and $V$ are open in $G$, so is $UV$. If $A$ and $B$ are closed in $G$, $AB$ need not be closed. [An example can be found in $\mathbf{R}$ with its usual topology and addition (caution: then $AB$ becomes $A + B$).] If one of $A$ or $B$ is compact, then $AB$ *is* closed.

2. If $F$ is compact and $U$ open in $G$, with $F \subset U$, then for some nhood $V$ of the identity in $G$, $FV \subset U$.

3. If $F$ is compact in $G$, then for each nhood $U$ of $e$, there is a nhood $V$ of $e$ such that $xVx^{-1} \subset U$, for each $x \in F$. (Compare with 13G.6(d).)

4. For $x, y \in G$ and $A, B \subset G$ we have

   a) $\bar{A} \cdot \bar{B} \subset \overline{AB}$

   b) $(\overline{A^{-1}}) = (\bar{A})^{-1}$

   c) $\overline{xAy} = x\bar{A}y$

   d) if $ab = ba$, for each $a \in A$, $b \in B$, then $ab = ba$ for each $a \in \bar{A}$, $b \in \bar{B}$.

5. If $H$ is a subgroup (Abelian subgroup, normal subgroup) of $G$, so is $\bar{H}$.

6. A subgroup is discrete iff it has an isolated point.

7. Every open subgroup is closed.

8. Every locally compact subgroup is closed. (This is difficult.)

## 18E. *Quotients and products of topological groups*

Let $G$ be a topological group.

1. The product $\prod G_\alpha$ of topological groups is a topological group when given the product topology and pointwise multiplication $(\pi_\alpha(x \cdot y) = \pi_\alpha(x)\pi_\alpha(y))$. The projection $\pi_\alpha$ is a continuous open homomorphism.

2. Let $H$ be a closed normal subgroup of $G$. Then $G/H$, the set of all left cosets $xH$ of $H$, is a topological group when given the quotient topology and factor group structure. The natural map $P: G \to G/H$, $P(x) = xH$, is continuous and open.

3. $G/H$ is discrete iff $H$ is open.

4. If $G$ is locally compact and $K \subset G/H$ is compact, a compact set $F \subset G$ exists with $P(F) = K$.

5. If $G$ is compact, so is $G/H$. Conversely, for locally compact $G$, if $H$ and $G/H$ are compact, so is $G$. [See 4.] A similar theorem holds for local compactness (i.e., if $H$ and $G/H$ are locally compact, so is $G$), but the proof (due to Gleason) is difficult. See the notes.

## 18F. *Character groups*

Let $G$ be a locally compact, Abelian group. A *character* on $G$ is a homomorphism $\chi: G \to T$ where $T$ is the *circle group* (the unit circle in $\mathbf{R}^2$ with the usual topology and complex multiplication).

1. The set $\hat{G}$ of all continuous characters on $G$ is a topological group when given pointwise multiplication and the topology for which the sets

$$P(F, \epsilon) = \{\chi \in \hat{G} \mid |\chi(x) - 1| < \epsilon, \text{ for all } x \in F\},$$

for compact $F \subset G$ and $\epsilon > 0$, form a base at the identity $\mathbf{1}$ ($\mathbf{1}(x) = 1$, for all $x \in G$). $\hat{G}$ is called the *character group* of $G$.

2. $\hat{G}$ is locally compact and Abelian.

3. If $G$ is compact, $\hat{G}$ is discrete. If $G$ is discrete, $\hat{G}$ is compact.

4. $\hat{\mathbf{R}} = \mathbf{R}$. $\hat{\mathbf{N}} = T$. $\hat{T} = \mathbf{N}$.

5. $\hat{\hat{G}} = G$. [Map $G \to \hat{\hat{G}}$ by $x \to \mathscr{E}_x$, where $\mathscr{E}_x$ is the character on $\hat{G}$ defined by $\mathscr{E}_x(\chi) = \chi(x)$. Assume the fact that if $a \neq e$ in $G$, then for some $\chi \in \hat{G}$, $\chi(a) \neq 1$ (this is very hard to prove!) and show $x \to \mathscr{E}_x$ is a topological isomorphism (a homeomorphism and an isomorphism) of $G$ onto $\hat{\hat{G}}$.] This is the *Pontryagin duality theorem.*

## 18G. *A regular space not completely regular*

Recall that $\mathbf{T}$ denotes the Tychonoff plank (17.12) $(\mathbf{\Omega} \times \mathbf{N}^*) - \{(\omega_1, \omega_0)\}$. Let $Z$ be the set of all integers, positive, negative and zero and form the product $\mathbf{T} \times Z$. Identify points in $\mathbf{T} \times Z$ as follows: if $n$ is odd, the right edges of $\mathbf{T} \times \{n\}$ and $\mathbf{T} \times \{n + 1\}$ (which are copies of $\mathbf{T}$) are identified point for point and if $n$ is even, the top edges of $\mathbf{T} \times \{n\}$ and $\mathbf{T} \times \{n + 1\}$

are identified point for point. The image $T_n$ of $\mathbf{T} \times \{n\}$ under the resulting quotient map is clearly a homeomorphic copy of $\mathbf{T}$.

Now add points $a$ and $b$ to the quotient space obtained, the basic nhoods of $a$ being of the form $U_n(a) = \{a\} \cup \bigcup_{m=n}^{\infty} T_m$ and the basic nhoods of $b$ being of the form

$$U_n(b) = \{b\} \cup \bigcup_{m=n}^{\infty} T_{-m},$$

as $n$ ranges over all integers. Let the space $\bigcup_{n=-\infty}^{\infty} T_n \cup \{a, b\}$ be denoted $K$.

1. $K$ is regular and $T_1$.

2. Let $f: K \to \mathbf{R}$ be continuous, let $n$ be an even integer, and let $p$ be a positive integer. If $f \geq 1/p$ at all but finitely many points on the right edge of $T_n$, then $f \geq 1/(p+1)$ at all but finitely many points on the right edge of $T_{n-2}$. [Otherwise, $f < 1/(p+1)$ at infinitely many points on the right edge of $T_{n-2}$, and hence on the right edge of $T_{n-1}$, and then (see 17.12) there is some $\beta_0 < \omega_1$ such that, in $T_{n-1}$, $f(\beta, \omega_0) \leq 1/(p+1)$ for all $\beta > \beta_0$. Since the top edge of $T_{n-1}$ coincides with the top edge of $T_n$, we would have $f(\beta, \omega_0) \leq 1/(p+1)$ for all $\beta > \beta_0$ in $T_n$. This is impossible, since $f \geq 1/p$ on most of the right edge of $T_n$ entails $f \geq 1/p$ on most of the top edge of $T_n$ (again using 19F(1)).] Similarly, if $f \leq -1/p$ at all but finitely many points on the right edge of $T_n$, then $f \leq -1/(p+1)$ at all but finitely many points on the right edge of $T_{n-2}$.

3. Any continuous real-valued function on $K$ has the same value at $a$ and $b$. [It is enough to show every such function has the same sign at $a$ and $b$ (why?). But if $f(a) > 0$, then $f \geq 1/p$ at all but finitely many points on the right edge of $T_n$ for some even $n$. Use part 2 to conclude $f(b) \geq 0$. Similarly, if $f(a) < 0$, then $f(b) \leq 0$.]

Thus $K$ is a regular $T_1$-space which is not completely regular (not even completely Hausdorff).

By modifying this example, E. Hewitt manufactured *a regular $T_1$-space on which every continuous real-valued function is constant*! See the notes.

### 18H. *Manifolds*

Topological $n$-manifolds were introduced in 18.3(c). Let $X$ be a compact $n$-manifold.

1. If $U$ is an open subset of $X$ which is homeomorphic to $\mathbf{R}^n$, the quotient of $X$ obtained by collapsing $X - U$ to a single point is homeomorphic to the $n$-sphere $\mathbf{S}^n$.

2. $X$ can be embedded in a finite product of spheres (and hence in some Euclidean space $\mathbf{R}^m$). [You need an evaluation map.]

## 19 Compactification

Since compact Hausdorff spaces behave nicely, it is of interest to study the process of "compactification" that is, the process of embedding a given space as a dense subset of some compact Hausdorff space.

**19.1 Definition.** A *compactification* of a space $X$ is an ordered pair $(K, h)$ where $K$ is a compact Hausdorff space and $h$ is an embedding of $X$ as a dense subset of $K$.

In many cases $h$ will be an inclusion map, so that $X \subset K$. In other cases, we can agree to write $X$ when we mean $h(X)$ (referring to our earlier remarks that homeomorphic spaces are, to a topologist, the same), so that we can again write $X \subset K$. Whenever one of these situations occurs we say simply that $K$ is a compactification of $X$, and think of $K$ as containing $X$ as a dense subspace.

Many examples of compactifications lie at hand. To mention a few, $[0, 1]$ is a compactification of $[0, 1)$, $\mathbf{S}^1$ is a compactification of $\mathbf{R}$ (under stereographic projection), the ordinal space $\mathbf{\Omega}$ is a compactification of $\mathbf{\Omega}_0$. These are all obtained by adding one point to the space $X$ to be compactified; this process can be generalized to arbitrary locally compact Hausdorff spaces.

**19.2 Definition.** Let $X$ be a locally compact, noncompact Hausdorff space, $p$ a point not in $X$ (for example, $p = X$). Let $X^* = X \cup \{p\}$, and let the basic nhoods of $p$ be the sets of the form $\{p\} \cup (X - L)$, where $L$ is a compact set in $X$. Nhoods of points in $X$ are unchanged in $X^*$. In Exercise 19A, you will verify that this is a valid assignment of nhoods in $X^*$. Clearly $X^*$ is compact (since the element of an open cover which contains $p$ will cover all but a compact subset of $X$) and $X$ is open and dense in $X^*$. Moreover, $X^*$ is Hausdorff (precisely because $X$ is locally compact and Hausdorff, see 19A). We call $X^*$ the *one-point compactification* (*Alexandroff compactification*) of $X$.

This embedding of a locally compact Hausdorff space in a compact Hausdorff space has the following consequence.

***19.3 Theorem.*** *Every locally compact Hausdorff space is a Tychonoff space.*

We have just used the fact that if a space has a compactification, it is a Tychonoff space. To establish the converse, that every Tychonoff space has a compactification, we recall the details of the embedding of any Tychonoff space in a cube. The procedure we will outline here is a modification of that used in the original embedding theorem (14.13) in that here we use the bounded continuous functions from $X$ to $\mathbf{R}$ while there we used the continuous functions from $X$ to $\mathbf{I}$. Since a bounded continuous function from $X$ to $\mathbf{R}$ can be regarded as a function from $X$ to some closed bounded interval, the difference is not great.

Let $C^*(X)$ denote the collection of all bounded continuous real-valued functions on $X$; the range of each $f \in C^*(X)$ can be taken as a closed bounded interval $I_f$ in $\mathbf{R}$. Since $X$ is Tychonoff, the collection $C^*(X)$ separates points from closed sets in $X$ and thus, by 8.16, the evaluation map $e\colon X \to \prod \{I_f \mid f \in C^*(X)\}$ defined by

$$[e(x)]_f = f(x)$$

is an embedding of $X$ in $\prod I_f$. Note that under the embedding $e$, the element $f$ of $C^*(X)$ is transformed into the restriction to $e(X)$ of the $f$th projection map $\pi_f$; that is, for $f\colon X \to I_f$, $f = \pi_f \circ e$. (Fig. 19.1.)

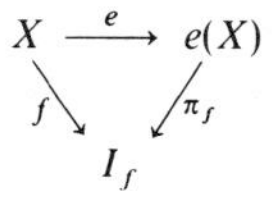

$$\begin{array}{ccc} X & \xrightarrow{e} & e(X) \\ & {\scriptstyle f}\searrow \quad \swarrow{\scriptstyle \pi_f} & \\ & I_f & \end{array}$$

**Figure 19.1**

**19.4 Definition.** The *Stone–Čech compactification* of $X$ is the closure $\beta X$ of $e(X)$ in the product $\prod I_f$. (More formally, $(\beta X, e)$ is the Stone–Čech compactification of $X$.)

The central useful fact about the Stone–Čech compactification is an extension property, given by the following theorem.

***19.5 Theorem.*** *If $K$ is a compact Hausdorff space and $f: X \to K$ is continuous, there is a continuous $F: \beta X \to K$ such that $F \circ e = f$.*

*Proof.* $K$ is a Tychonoff space and thus can be embedded by an evaluation map $e'$ in a cube $\prod \{I_g \mid g \in C^*(K)\}$. The situation is illustrated in Fig. 19.2.

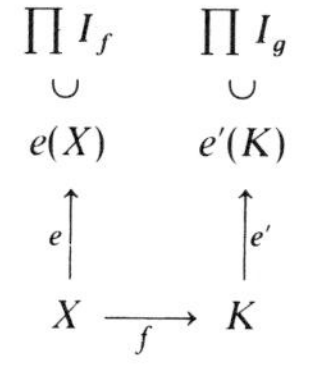

$$\begin{array}{ccc} \prod I_f & & \prod I_g \\ \cup & & \cup \\ e(X) & & e'(K) \\ \uparrow{\scriptstyle e} & & \uparrow{\scriptstyle e'} \\ X & \xrightarrow[f]{} & K \end{array}$$

**Figure 19.2**

We can define a map $H: \prod I_f \to \prod I_g$ as follows; for each $t \in \prod I_f$, $[H(t)]_g = t_{g \circ f}$. This map is continuous when followed by each projection $\pi_g$, in fact $(\pi_g \circ H)(t) = \pi_{g \circ f}(t)$, so $H$ is continuous. Now $H$ takes $e(X)$ into $e'(K)$, for an element of $e(X)$ has the form $e(x)$ for some $x \in X$ and

$$H[e(x)]_g = [e(x)]_{g \circ f} = g \circ f(x) = [e'(f(x))]_g$$

so that $H[e(x)] = e'(f(x))$. But $e(X)$ is dense in $\beta X$, so $H[e(X)]$ is dense in $H(\beta X)$ and thus, since $e'(K)$ is closed and contains $H[e(X)]$, $H(\beta X) \subset e'(K)$. Finally, define $F = e'^{-1} \circ (H \mid \beta X)$. Then $F: \beta X \to K$ is continuous and $F \circ e = f$ since, for $x \in X$,

$$F \circ e(x) = e'^{-1}[H(e(x))] = e'^{-1}[e'(f(x))] = f(x). \blacksquare$$

Very often it is possible to deal with $e(X)$ directly (as, for example, when dealing with preservation of a topological property in the passage from $X$ to $\beta X$). Then $X$ is often written for $e(X)$, so that $X \subset \beta X$, and the above theorem becomes: *every continuous function from $X$ to a compact space $K$ can be extended to $\beta X$.*

Theorem 19.5 actually characterizes the Stone–Čech compactification, up to what is called a topological equivalence. We need some preliminary terminology and lemmas.

**19.6 Definition.** If $(K_1, h_1)$ and $(K_2, h_2)$ are compactifications of $X$, we write $(K_1, h_1) \leq (K_2, h_2)$ iff there exists a continuous $F: K_2 \to K_1$ such that $F \circ h_2 = h_1$ (Fig. 19.3). When emphasis on $F$ is needed we write $F: (K_1, h_1) \leq (K_2, h_2)$. Note that $F$ is just an extension to $K_2$ of the canonical homeomorphism $h_1 \circ h_2^{-1}$ of $h_2(X)$ with $h_1(X)$.

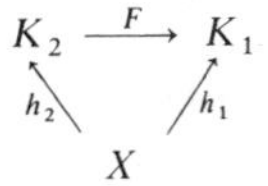

$$\begin{array}{ccc} K_2 & \xrightarrow{F} & K_1 \\ {\scriptstyle h_2}\nwarrow & & \nearrow{\scriptstyle h_1} \\ & X & \end{array}$$

**Figure 19.3**

(In case $h_1$ and $h_2$ are inclusion maps, this says $(K_1, h_1) \leq (K_2, h_2)$ iff there is a continuous $F: K_2 \to K_1$ such that $F \mid X$ is the identity.) If both

$$(K_1, h_1) \leq (K_2, h_2) \qquad \text{and} \qquad (K_2, h_2) \leq (K_1, h_1),$$

we say $(K_1, h_1)$ and $(K_2, h_2)$ are *topologically equivalent.* Topologically equivalent compactifications of $X$ are regarded by any topologist as the same (for example, any compactification of $X$ topologically equivalent to $(\beta X, e)$ is called *the* Stone–Čech compactification of $X$), because of the following result.

**19.7 Lemma.** *$(K_1, h_1)$ and $(K_2, h_2)$ are topologically equivalent compactifications of $X$ iff there is a homeomorphism $H$ of $K_2$ with $K_1$ such that $H \circ h_2 = h_1$.*

*Proof.* Exercise 19E. ■

**19.8 Lemma.** *If $F: (K_1, h_1) \leq (K_2, h_2)$ then*

a) *$F \mid h_2(X)$ is a homeomorphism of $h_2(X)$ with $h_1(X)$,*

b) *$F$ carries $K_2 - h_2(X)$ onto $K_1 - h_1(X)$.*

*Proof.* a) In fact $F \mid h_2(X) = h_1 \circ h_2^{-1}$.

b) From (a), $F$ is onto. Thus we can prove (b) by proving, more generally, that whenever $S$ is Hausdorff and $f: S \to T$ is a continuous map whose restriction to a dense subset $A$ of $S$ is a homeomorphism, then $f(S - A) \subset T - f(A)$. Suppose not. Then for some $x \in A$ and $y \in S - A$, $f(x) = f(y)$. Pick disjoint nhoods $U$ of $x$ and $V$ of $y$. Now $f(U \cap A)$ is a nhood of $f(x)$ in $f(A)$, since $f$ is a homeomorphism, so $f(U \cap A) = W \cap f(A)$ where $W$ is a nhood of $f(x)$ in $T$. But any nhood $V'$ of $y$ contained in $V$ contains points of $A$ not in $U$, so $f(V') \not\subset W$. Thus $f$ is not continuous at $y$, a contradiction. ■

The proofs of the following theorems are now easy exercises (19E).

***19.9 Theorem.*** *If $(K_1, h_1)$ and $(K_2, h_2)$ are compactifications of $X$ and $(K_2, h_2)$ has the extension property of Theorem 19.5, then $(K_1, h_1) \leq (K_2, h_2)$.*

***19.10 Corollary.*** *$(\beta X, e)$ is characterized up to topological equivalence by the extension property.*

Thus $\beta X$ is (up to topological equivalence) the only compactification of $X$ with the extension property and, by 19.9, it is the largest element in the collection of compactifications of $X$ partially ordered by $\leq$. Note that, if $X \subset \beta X$ and $X \subset K$, 19.9 provides a continuous $F: \beta X \to K$ such that $F \mid X$ is the identity while 19.8 says $F(\beta X - X) = K - X$. We will use this fact later.

More light can be shed on the nature of the Stone–Čech compactification using the following terminology.

**19.11 Definition.** A subset $A$ of a space $T$ is *$C^*$-embedded* in $T$ iff every bounded continuous real-valued function on $A$ can be extended to $T$.

Either regarded as a consequence of 19.5 or taken directly from the fact that the bounded real-valued continuous functions on $e(X)$ are just the restrictions to $e(X)$ of the projection maps, we see that $e(X)$ is $C^*$-embedded in $\beta X$. This property also characterizes $\beta X$.

***19.12 Theorem.*** *If $(K, h)$ is a compactification of $X$ such that $h(X)$ is $C^*$-embedded in $K$, then $(K, h)$ is the Stone–Čech compactification of $X$.*

*Proof.* It suffices to show the extension property 19.5 holds for $(K, h)$. Let $f: X \to L$ be a continuous map of $X$ into a compact Hausdorff space $L$. Let $e: L \to \prod_{g \in C^*(L)} I_g$ be the cube embedding of $L$ (Fig. 19.4).

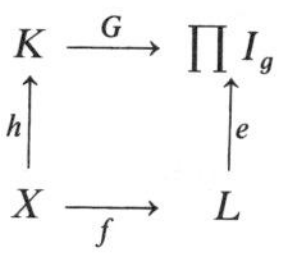

$$\begin{array}{ccc} K & \xrightarrow{G} & \prod I_g \\ \uparrow h & & \uparrow e \\ X & \xrightarrow[f]{} & L \end{array}$$

**Figure 19.4**

For each $g: L \to I_g$, the map $g \circ f \circ h^{-1}: h(X) \to I_g$ has a continuous extension $h_g: K \to I_g$. Define $G: K \to \prod I_g$ by

$$[G(t)]_g = h_g(t).$$

Then $G$ is continuous since for each projection $\pi_g$ of $\prod I_g$, $\pi_g \circ G(t) = h_g(t)$, so that $\pi_g \circ G$ is continuous. Moreover $G$ carries $h(X)$ into $e(L)$ since

$$G[h(x)]_g = h_g(h(x)) = (g \circ f \circ h^{-1})(h(x)) = g[f(x)] = e[f(x)]_g.$$

But $G[h(X)]$ is dense in $G(K)$ and $e(L)$ is compact, so $G(K) \subset e(L)$. Thus $F = e^{-1} \circ G$ carries $K$ into $L$ and, by the above computation, $F \circ h = f$. ■

Theorem 19.12 is most useful in proving that one familiar space is or is not the Stone–Čech compactification of another.

**19.13 Examples.** a) $\mathbf{I}$ is not the Stone–Čech compactification of (0, 1) since the bounded continuous real-valued function $\sin(1/x)$ on (0, 1) cannot be extended to $\mathbf{I}$.

b) From 17.2, every continuous real-valued function on the ordinal space $\mathbf{\Omega}_0$ can be extended to $\mathbf{\Omega}$, so $\beta\mathbf{\Omega}_0 = \mathbf{\Omega}$.

c) As an exercise (19F), you will show every continuous real-valued function on the Tychonoff plank $\mathbf{T}$ can be extended to $\mathbf{T}^*$ (see 17.12). Thus $\beta\mathbf{T} = \mathbf{T}^*$.

d) $|\beta\mathbf{N}| = 2^{\mathfrak{c}}$. From Theorem 16.4, the product $\mathbf{I}^{\mathfrak{c}}$ of $\mathfrak{c}$ copies of $\mathbf{I}$ has a countable dense set $A$. Any one–one map $f$ of $\mathbf{N}$ onto $A$ is continuous and hence has an extension $f^{\beta}: \beta\mathbf{N} \to \mathbf{I}^{\mathfrak{c}}$ (by 19.5). Since $f^{\beta}$ is onto a dense subset of $\mathbf{I}^{\mathfrak{c}}$, it is onto $\mathbf{I}^{\mathfrak{c}}$. Thus $|\beta\mathbf{N}| \geq |\mathbf{I}^{\mathfrak{c}}| = \mathfrak{c}^{\mathfrak{c}} = 2^{\mathfrak{c}}$. On the other hand, there are $\mathfrak{c}$ elements in $C^*(\mathbf{N})$ so $\beta\mathbf{N} \subset \mathbf{I}^{\mathfrak{c}}$ and thus $|\beta\mathbf{N}| \leq 2^{\mathfrak{c}}$.

## Problems

### 19A. *The one-point compactification: construction*

The procedure used to obtain the one-point compactification $X^*$ of a locally compact, noncompact Hausdorff space $X$ can be applied to any space $Y$. That is, $Y^* = Y \cup \{p\}$ with nhoods of $y \in Y$ unchanged in $Y^*$ while nhoods of $p$ have the form $\{p\} \cup (Y - L)$ where $L$ is a compact subset of $Y$. $Y^*$ is called the *Alexandroff extension* of $Y$.

1. This is a valid assignment of nhoods in $Y^*$.
2. $Y^*$ is compact and $Y$ is open in $Y^*$.
3. $Y$ is dense in $Y^*$ iff $Y$ is noncompact.
4. $Y^*$ is Hausdorff iff $Y$ is locally compact and Hausdorff.

### 19B. *The one-point compactification: examples*

1. The one-point compactification of $\mathbf{R}^n$ is homeomorphic to $\mathbf{S}^n$.
2. The one-point compactification of $\mathbf{N}$ is homeomorphic to the subspace $\{0\} \cup \{1/n \mid n = 1, 2, \ldots\}$ of $\mathbf{R}$.
3. The one-point compactification of the Tychonoff plank $\mathbf{T}$ is $\mathbf{T}^*$ (see 17.12).

### 19C. *Compactification in the plane*

The one-point compactification $X^*$ of $X$ has the property that $X^* - X$ is a discrete space.

Find a nonlocally compact subset of the plane which has a compactification $K$ such that $K - A$ is discrete.

### 19D. *Compactification of ordered spaces*

Every ordered space has an ordered compactification [use 17E].

### 19E. *Exercise on topological equivalence*

1. Compactifications $(K_1, h_1)$ and $(K_2, h_2)$ of $X$ are topologically equivalent iff there is a homeomorphism $H: K_2 \to K_1$ such that $H \circ h_2 = h_1$. [For necessity, if

$$F: (K_1, h_1) \le (K_2, h_2) \quad \text{and} \quad G: (K_2, h_2) \le (K_1, h_1),$$

then $F$ and $G$ are inverses and hence homeomorphisms.]

2. Prove 19.9 and 19.10.

### 19F. *The Tychonoff plank*

Show that the Tychonoff plank $\mathbf{T}$ is $C^*$-embedded in $\mathbf{T}^*$ [see 17.12].

### 19G. *$C^*$-embedding and $\beta X$*

Let $X$ and $T$ be Tychonoff spaces.

1. If $X$ is dense and $C^*$-embedded in $T$, the embedding $e: X \to \beta X$ can be extended to an embedding $E: T \to \beta X$.

2. If $X$ is $C^*$-embedded in $T$, then $\mathrm{Cl}_{\beta T}\, X = \beta X$ (up to topological equivalence).

### 19H. *Cardinality of $\beta X$*

1. $|\beta \mathbf{N}| \ge |\beta \mathbf{Q}|$. [Consider any one–one map of $\mathbf{N}$ onto $\mathbf{Q}$ and use 19.5.]
2. $|\beta \mathbf{Q}| \ge |\beta \mathbf{R}|$. [Consider the inclusion map of $\mathbf{Q}$ into $\mathbf{R}$ and use 19.5.]
3. $|\beta \mathbf{N}| = |\beta \mathbf{Q}| = \beta \mathbf{R}| = 2^{\mathbf{c}}$. [$\mathbf{N}$ is $C^*$-embedded in $\mathbf{R}$. See 19G.]

### 19I. $\beta(X \times X) \ne \beta X \times \beta X$

Exercise 15G on extremally disconnected spaces is a necessary prerequisite to this problem.

1. $\beta X$ is extremally disconnected iff $X$ is extremally disconnected. [15G.1c for sufficiency.] In particular, if $X$ is discrete, $\beta X$ is extremally disconnected.

2. If $D$ is any infinite discrete space, $\beta D \times \beta D$ is not homeomorphic to $\beta(D \times D)$. [Show $\beta D \times \beta D$ is not extremally disconnected by studying the closure of the open set

$$\{(x, x) \in \beta D \times \beta D \mid x \in D\}.]$$

The Stone–Čech compactification of a product has been intensively studied. The identity $\beta(\prod X_\alpha) = \prod \beta X_\alpha$ holds iff $\prod X_\alpha$ is pseudocompact; see the notes.

### 19J. *Filter description of $\beta X$*

In 17D we observed that a completely regular space is compact iff every $z$-ultrafilter converges (i.e., is fixed). A compact space containing a copy of $X$ can be obtained by "fixing" the free $z$-ultrafilters on $X$.

Let $BX$ be the space whose points are the $z$-ultrafilters in $X$. For each zero set $Z \subset X$, define $Z^* = \{\mathcal{F} \in BX \mid Z \in \mathcal{F}\}$.

1. The sets $Z^*$ can be used as a base for closed sets to obtain a topology on $BX$. [Use 5E.2 and 17D.3.]

2. For $x \in X$, let $h(x)$ be the unique $z$-ultrafilter converging to $x$. Then $h$ is an embedding of $X$ as a dense subset of $BX$. (Hereafter we identify $X$ with $h(X)$, so $X \subset BX$.)

3. For each zero set $Z$ in $X$, $\mathrm{Cl}_{BX}\, Z = Z^*$.

4. For zero sets $Z_1$ and $Z_2$ in $X$, $\mathrm{Cl}_{BX}\,(Z_1 \cap Z_2) = \mathrm{Cl}_{BX}\, Z_1 \cap \mathrm{Cl}_{BX}\, Z_2$.

5. $BX$ is a compact Hausdorff space. [It is enough to show each family of *basic* closed sets with the finite intersection property has nonempty intersection.]

6. Each continuous map $f$ of $X$ into a compact Hausdorff space $K$ can be extended to $BX$. (This should be compared with 19.5.) [If $p \in BX - X$ extend $f$ to $p$ as follows: $p$ is a unique $z$-ultrafilter in $X$ and $\mathscr{F} = \{Z \subset K \mid Z \text{ is a zero-set in } K \text{ and } f^{-1}(Z) \in p\}$. Show $\mathscr{F}$ is prime (12E) and thus has a unique cluster point $q$ (12E.6). Define $f(p) = q$. (This is essentially a use of 12F.)]

7. $BX = \beta X$.

## 19K. *Wallman compactification*

Let $X$ be a Hausdorff space, and let $\gamma X$ be the collection of all closed ultrafilters on $X$. For each closed set $D \subset X$, define $D^* \subset \gamma X$ to be the set

$$D^* = \{\mathscr{F} \in \gamma X \mid D \in \mathscr{F}\}.$$

Let $\mathscr{C} = \{D^* \mid D \text{ is a closed subset of } X\}$. It is somewhat surprising that the procedure used in 19J can be applied here, with results which are *not* always identical.

1. $\mathscr{C}$ is a base for the closed sets of a topology on $\gamma X$.

2. The mapping $h\colon X \to \gamma X$ which takes $x \in X$ to the (unique) ultrafilter in $\gamma X$ which converges to $x$ is an embedding of $X$ in $\gamma X$.

3. $X$ is dense in $\gamma X$ (more accurately, $h(x)$ is dense in $\gamma X$). More generally, if $D$ is closed in $X$, then $\mathrm{Cl}_{\gamma X}\, D = D^*$.

4. If $A$ and $B$ are closed subsets of $X$, $\mathrm{Cl}_{\gamma X}\,(A \cap B) = \mathrm{Cl}_{\gamma X}\, A \cap \mathrm{Cl}_{\gamma X}\, B$.

5. $\gamma X$ is compact. [Any collection of basic closed sets with the finite intersection property has nonempty intersection.]

6. Every continuous function on $X$ to a compact Hausdorff space $K$ can be extended to $\gamma X$. [Mimic the proof of 19J.6.]

7. $\gamma X$ is Hausdorff iff $X$ is normal. [Use part 4 for necessity.] Thus $\gamma X = \beta X$ iff $X$ is normal (by 6).

## 19L. *Wallman basis problem*

The procedure used to obtain $\beta X$ in 19J and $\gamma X$ in 19K can be generalized. Let $\mathscr{B}$ be any base for the closed sets of $X$ satisfying the following conditions:

a) for each closed set $F$ and $x \notin F$, there is some $A \in \mathscr{B}$ such that $x \in A$ and $A \cap F = \varnothing$,

b) finite unions and finite intersections of elements of $\mathscr{B}$ belong to $\mathscr{B}$,

c) if $A, B \in \mathscr{B}$ are disjoint, then for some $C$ and $D \in \mathscr{B}$, $A \subset X - C$, $B \subset X - D$ and $(X - C) \cap (X - D) = \varnothing$.

Then $\mathscr{B}$ is called a *Wallman base* for $X$; a space $X$ is *seminormal* iff it has a Wallman base.

Now let $\mathscr{A}$ be *any* base for the closed sets of $X$ and call $\mathscr{F} \subset \mathscr{A}$ an *$\mathscr{A}$-filter* iff $\mathscr{F}$ consists of nonempty sets and

i) $F_1 \cap F_2 \in \mathscr{F}$ whenever $F_1, F_2 \in \mathscr{F}$,

ii) $F \in \mathscr{F}$ whenever $F \supset G \in \mathscr{F}$ and $F$ belongs to $\mathscr{A}$.

An *$\mathscr{A}$-ultrafilter* is a maximal $\mathscr{A}$-filter. See 12E for basic results on $\mathscr{A}$-filters.

Let $\omega_{\mathscr{A}}(X)$ be the set of all $\mathscr{A}$-ultrafilters on $X$ and for each $A \in \mathscr{A}$, let

$$A^* = \{\mathscr{F} \in \omega_{\mathscr{A}}(X) \mid A \in \mathscr{F}\}.$$

Let $\tau_{\mathscr{A}} = \{A^* \mid A \in \mathscr{A}\}$.

1. $\tau_{\mathscr{A}}$ is a base for the closed sets of a topology on $\omega_{\mathscr{A}}(X)$. ($\omega_{\mathscr{A}}(X)$ with this topology is called the *Wallman space* of the Wallman base $\mathscr{A}$, whenever $\mathscr{A}$ is a Wallman base.)

2. The mapping $h: X \to \omega_{\mathscr{A}}(X)$ which takes $x \in X$ to the (unique) $\mathscr{A}$-ultrafilter in $\omega_{\mathscr{A}}(X)$ which converges to $x$ is an embedding of $X$ in $\omega_{\mathscr{A}}(X)$.

3. $X$ is dense in $\omega_{\mathscr{A}}(X)$. More generally, if $A \in \mathscr{A}$, then $\text{Cl}_{\omega_{\mathscr{A}}(X)} A = A^*$.

4. If $A, B \in \mathscr{A}$, then $\text{Cl}_{\omega_{\mathscr{A}}(X)} (A \cap B) = \text{Cl}_{\omega_{\mathscr{A}}(X)} A \cap \text{Cl}_{\omega_{\mathscr{A}}(X)} B$.

5. $\omega_{\mathscr{A}}(X)$ is Hausdorff iff $\mathscr{A}$ is a Wallman base for $X$.

A great deal of Exercises 19J and 19K is now subsumed in the following two propositions:

7. In a completely regular $T_1$-space the zero sets form a Wallman base. (Thus a $T_1$-space is seminormal iff it is completely regular.)

8. The collection of all closed subsets of $X$ form a Wallman base iff $X$ is normal.

For an arbitrary normal space $X$, it is still an open question whether *every* Hausdorff compactification of $X$ can be obtained from some Wallman base for $X$. A limitation on the search for a correct Wallman base in $X$ to produce a particular compactification $K$ is given in the next problem.

## 19M. *Wallman basis, continued*

Let $\mathscr{A}$ be a Wallman base for $X$. A bounded function $f: X \to \mathbf{R}$ is *$\mathscr{A}$-uniformly continuous* iff for each $\epsilon > 0$, a finite collection $A_1, \ldots, A_n \in \mathscr{A}$ exists such that $f(X - A_k)$ has diameter $<\epsilon$, for each $k = 1, \ldots, n$.

1. The continuous real-valued functions which extend from $X$ to $\omega_{\mathscr{A}}(X)$ are precisely the $\mathscr{A}$-uniformly continuous ones.

2. If $K$ is a compactification of $X$, the zero sets of those bounded continuous functions $f$ on $X$ to $\mathbf{R}$ which extend to $K$ form a Wallman base $\mathscr{A}$ for $X$.

3. The Wallman space resulting from the base in 2 need not be $K$ [Consider $K = \mathbf{R}^*$.]

## 19N. *H-closure*

An *H-closure* of a Hausdorff space $X$ is an $H$-closed space containing $X$ as a dense subset. The ultrafilter process introduced in 19J and 19K for describing compactifications of a topological space $X$ can be used here, with some modification, to describe the "best" $H$-closure of $X$. See 17K for the basic facts about $H$-closed spaces.

1. A Hausdorff space is $H$-closed iff every open ultrafilter converges.

Now let $\mathscr{C}$ be the collection of nonconvergent open ultrafilters on $X$ and for each $\mathscr{F} \in \mathscr{C}$, let the nhoods of $\mathscr{F}$ in the space $\alpha X = X \cup \mathscr{C}$ be the sets $\{\mathscr{F}\} \cup G$, where $G \in \mathscr{F}$. Nhoods of points in $X$ are unchanged.

2. $\alpha X$ is a topological space, containing $X$ as an open dense subset.

3. $\alpha X$ is $H$-closed.

4. $\alpha X$ is the largest $H$-closure of $X$; i.e., if $T$ is any $H$-closure of $X$, $T$ is the continuous image of $\alpha X$, under a map which is the identity on $X$.

19O. *Realcompactification*

1. Construct a realcompactification for any Tychonoff space $X$, that is, a realcompact space containing a dense subset homeomorphic to $X$, by following step for step the construction of $\beta X$ given in the text (but replacing $C^*(X)$ by $C(X)$). This is the *Hewitt realcompactification* (*Nachbin completion*) of $X$, denoted $\upsilon X$.

2. Show that every real-valued continuous function $f$ on $X$ can be extended to $\upsilon X$ (that is, that $X$ is $C$-embedded in $\upsilon X$). Conclude that $X \subset \upsilon X \subset \beta X$.

3. Describe the circumstances under which $\upsilon X = \beta X$.

That Greek letter, by the way, is *upsilon*, not *nu*, and $\upsilon X$ *is* the standard notation for the Hewitt realcompactification of $X$.

## 20 Paracompactness

Paracompact spaces were first introduced by Dieudonné in 1944 as a natural generalization of compact spaces still retaining enough structure to enjoy many of the properties of compact spaces, yet sufficiently general to include a much wider class of spaces. The notion of paracompactness gained stature with the proof, by A. H. Stone, that every metric space is paracompact and the subsequent use of this result in the solutions of the general metrization problem by Bing, Nagata and Smirnov. The central role played by paracompactness, or paracompact-like properties, in some of the current areas of intensive investigation in topology ensure it a permanent place alongside metrizability and compactness among the most important concepts in general topology.

To proceed, we need a great deal of terminology applying to coverings.

**20.1 Definition.** If $\mathscr{U}$ and $\mathscr{V}$ are covers of $X$, we say $\mathscr{U}$ *refines* $\mathscr{V}$, and write $\mathscr{U} < \mathscr{V}$, iff each $U \in \mathscr{U}$ is contained in some $V \in \mathscr{V}$. Then we say $\mathscr{U}$ is a *refinement* of $\mathscr{V}$.

If $\mathscr{U}$ is a cover of $X$ and $A \subset X$, the *star* of $A$ with respect to $\mathscr{U}$ is the set

$$\mathrm{St}\,(A, \mathscr{U}) = \bigcup \{U \in \mathscr{U} \mid A \cap U \neq \varnothing\}.$$

We say $\mathscr{U}$ *star-refines* $\mathscr{V}$, or $\mathscr{U}$ is a *star-refinement* of $\mathscr{V}$, written $\mathscr{U} \mathbin{*<} \mathscr{V}$, iff for each $U \in \mathscr{U}$, there is some $V \in \mathscr{V}$ such that $\mathrm{St}(U, \mathscr{U}) \subset V$. Finally, $\mathscr{U}$ is a *barycentric refinement* of $\mathscr{V}$, written $\mathscr{U} \Delta \mathscr{V}$, provided the sets $\mathrm{St}(x, \mathscr{U})$, for $x \in X$, refine $\mathscr{V}$. As an easy exercise, the reader should prove that a barycentric refinement of a barycentric refinement is a star-refinement; that is, if $\mathscr{U} \Delta \mathscr{V} \Delta \mathscr{W}$, then $\mathscr{U}^* < \mathscr{W}$. (See Exercise 20B.)

**20.2 Definition.** A collection $\mathscr{U}$ of subsets of $X$ is *locally finite* (or *nhood finite*) iff each $x \in X$ has a nhood meeting only finitely many $U \in \mathscr{U}$. We call $\mathscr{U}$ *point finite* iff each $x \in X$ belongs to only finitely many $U \in \mathscr{U}$. (We have already met point finite covers in Section 15 in connection with their shrinkability in normal spaces.) Apparently every locally finite collection is point finite. A notion related to local finiteness is that of a discrete collection of sets. A collection $\mathscr{U}$ of subsets of $X$ is *discrete* iff each $x \in X$ has a nhood meeting at most one element of $\mathscr{U}$. Clearly every discrete collection of sets is locally finite.

Finally, we point out that given any property of collections of sets in $X$, there is a corresponding "$\sigma$-property" which we illustrate with an example. A collection $\mathscr{V}$ of subsets of $X$ is *$\sigma$-locally finite* iff $\mathscr{V} = \bigcup_{n=1}^{\infty} \mathscr{V}_n$ where each $\mathscr{V}_n$ is a locally finite collection in $X$. The definition of a "$\sigma$-discrete" collection should now be clear. It is worth pointing out that if $\mathscr{V}$ is a $\sigma$-locally finite *cover* of $X$, the subcollections $\mathscr{V}_n$ which are locally finite and make up $\mathscr{V}$ will not usually be covers.

**20.3 Examples.** a) A point finite collection need not be locally finite. In fact, for any space $X$, $\{\{x\} \mid x \in X\}$ is a point finite cover, which is locally finite only under stringent conditions on $X$ (what are they?).

b) The cover of $\mathbf{R}$ by the sets $[n, n+1]$, as $n$ ranges through all integers, is point finite.

To illustrate the properties of locally finite collections, we prove some simple lemmas.

**20.4 Lemma.** *If $\{A_\lambda \mid \lambda \in \Lambda\}$ is a locally finite system of sets in $X$, then so is $\{\bar{A}_\lambda \mid \lambda \in \Lambda\}$.*

*Proof.* Pick $p \in X$ and find an open nhood $U$ of $p$ such that $U \cap A_\lambda = \varnothing$ except for finitely many $\lambda$. But then $U \cap \bar{A}_\lambda = \varnothing$ except for these same $\lambda$. This establishes the lemma. ∎

**20.5 Lemma.** *If $\{A_\lambda \mid \lambda \in \Lambda\}$ is a locally finite system of sets, then $\bigcup \bar{A}_\lambda = \overline{\bigcup A_\lambda}$. In particular, the union of a locally finite collection of closed sets is closed.*

*Proof.* Easily $\bigcup \bar{A}_\lambda \subset \overline{\bigcup A_\lambda}$. On the other hand, suppose $p \in \overline{\bigcup A_\lambda}$. Now some nhood of $p$ meets only finitely many of the sets $A_\lambda$, say $A_{\lambda_1}, \ldots, A_{\lambda_n}$. Since every nhood of $p$ meets $\bigcup A_\lambda$, every nhood of $p$ must then meet $A_{\lambda_1} \cup \cdots \cup A_{\lambda_n}$. Hence, $p \in \overline{A_{\lambda_1} \cup \cdots \cup A_{\lambda_n}} = \bar{A}_{\lambda_1} \cup \cdots \cup \bar{A}_{\lambda_n}$ so that, for some $k$, $p \in \bar{A}_{\lambda_k}$. Thus $\overline{\bigcup A_\lambda} \subset \bigcup \bar{A}_\lambda$, establishing the lemma. ∎

**20.6 Definition.** A Hausdorff space $X$ is *paracompact* iff each open cover of $X$ has an open locally finite refinement.

It should be pointed out that some writers do not require that a paracompact space be Hausdorff.

***20.7 Theorem.*** *If $X$ is a $T_3$-space, the following are equivalent:*

a) *$X$ is paracompact,*

b) *each open cover of $X$ has an open $\sigma$-locally finite refinement,*

c) *each open cover has a locally finite refinement (not necessarily open),*

d) *each open cover has a closed locally finite refinement.*

*Proof.* a) ⇒ b): A locally finite cover is $\sigma$-locally finite.

b) ⇒ c): Let $\mathscr{U}$ be an open cover of $X$. By (b), there is a refinement $\mathscr{V}$ of $\mathscr{U}$ such that $\mathscr{V} = \bigcup_{n=1}^{\infty} \mathscr{V}_n$, where each $\mathscr{V}_n$ is a locally finite collection of open sets, say $\mathscr{V}_n = \{V_{n\beta} \mid \beta \in B\}$. For each $n$, let $W_n = \bigcup_\beta V_{n\beta}$. Then $\{W_1, W_2, \ldots\}$ covers $X$. Define $A_n = W_n - \bigcup_{i<n} W_i$. Then $\{A_n \mid n \in \mathbf{N}\}$ is a locally finite refinement of $\{W_n \mid n \in \mathbf{N}\}$. Now consider $\{A_n \cap V_{n\beta} \mid n \in \mathbf{N}, \beta \in B\}$. This is a locally finite refinement of $\mathscr{V}$ and hence of $\mathscr{U}$.

c) ⇒ d): Let $\mathscr{U}$ be an open cover of $X$. For each $x \in X$, pick some $U_x$ in $\mathscr{U}$ such that $x \in U_x$, and, by regularity, find an open nhood $V_x$ of $x$ such that $\overline{V}_x \subset U_x$. Now $\{V_x \mid x \in X\}$ is an open cover of $X$ and so, by (c), has a locally finite refinement $\{A_\gamma \mid \gamma \in \Gamma\}$. Then $\{\overline{A_\gamma} \mid \gamma \in \Gamma\}$ is still locally finite, by Lemma 20.4, and for each $\gamma$, if $A_\gamma \subset V_x$, then $\overline{A}_\gamma \subset \overline{V}_x \subset U$ for some $U \in \mathscr{U}$. It follows that $\{\overline{A_\gamma} \mid \gamma \in \Gamma\}$ is a closed locally finite refinement of $\mathscr{U}$.

d) ⇒ a): Let $\mathscr{U}$ be an open cover of $X$, $\mathscr{V}$ a closed locally finite refinement. For each $x \in X$, let $W_x$ be a nhood of $x$ meeting only finitely many $V \in \mathscr{V}$. Now let $\mathscr{A}$ be a closed locally finite refinement of $\{W_x \mid x \in X\}$. For each $V \in \mathscr{V}$, let

$$V^* = X - \bigcup \{A \in \mathscr{A} \mid A \cap V = \varnothing\}.$$

Then $\{V^* \mid V \in \mathscr{V}\}$ is an open cover (the sets $V^*$ are open by Lemma 20.5) and is furthermore locally finite. For consider $x \in X$. There is a nhood $U$ of $x$ meeting only $A_1, \ldots, A_n$, say, from $\mathscr{A}$. But whenever $U \cap V^* \neq \varnothing$, we have $A_k \cap V^* \neq \varnothing$ for some $k = 1, \ldots, n$ which implies $A_k \cap V \neq \varnothing$. Since each $A_k$ meets only finitely many $V$, we must then have $U \cap V^* = \varnothing$ for all but finitely many of the $V^*$.

Now for each $V \in \mathscr{V}$, pick $U \in \mathscr{U}$ such that $V \subset U$, and form the set $U \cap V^*$. The collection of sets which results, as $V$ ranges through $\mathscr{V}$, serves as an open locally finite refinement of $\mathscr{U}$; the details are easily checked. ■

***20.8 Corollary.*** *Every Lindelöf $T_3$-space is paracompact.*

*Proof.* A countable subcover is a $\sigma$-locally finite refinement. ■

Regarded either as a consequence of 20.8 or taken directly from the definition (since a finite subcover is a locally finite refinement) we have the fundamental result that *a compact Hausdorff space is paracompact.* The following theorem establishes the importance of paracompact spaces as the smallest known class of spaces including both the compact and the metrizable spaces.

***20.9 Theorem.*** (A. H. Stone) *Every metric space is paracompact.*

*Proof.* Let $\mathscr{U}$ be an open cover of the metric space $(X, \rho)$. For each $n = 1, 2, \ldots$ and $U \in \mathscr{U}$, let $U_n = \{x \in U \mid \rho(x, X - U) \geq 1/2^n\}$. Then

$$\rho(U_n, X - U_{n+1}) \geq 1/2^n - 1/2^{n+1} = 1/2^{n+1}.$$

Let $\prec$ be a well-ordering of the elements of $\mathscr{U}$. For each $n = 1, 2, \ldots$ and $U \in \mathscr{U}$, let

$$U_n^* = U_n - \bigcup \{V_{n+1} : V \in \mathscr{U}, V \prec U\}.$$

For each $U, V \in \mathscr{U}$, and each $n = 1, 2, \ldots$, we have

$$U_n^* \subset X - V_{n+1}$$

or

$$V_n^* \subset X - U_{n+1}$$

(depending on which comes first in the well ordering). In either case,

$$\rho(U_n^*, V_n^*) \geq 1/2^{n+1}.$$

Hence, defining an open set $U_n^\sim$, for each $U \in \mathscr{U}$ and $n \in \mathbf{N}$, by

$$U_n^\sim = \{x \in X \mid \rho(x, U_n^*) < 1/2^{n+3}\},$$

we have $\rho(U_n^\sim, V_n^\sim) \geq 1/2^{n+2}$, so $\mathscr{V}_n = \{U_n^\sim \mid U \in \mathscr{U}\}$ is discrete for each $n$. Hence, $\mathscr{V} = \bigcup \mathscr{V}_n$ is $\sigma$-discrete, and thus $\sigma$-locally finite. Moreover, $\mathscr{V}$ refines $\mathscr{U}$ and covers $X$. (If $x \in X$, find the first $U \in \mathscr{U}$ to which $x$ belongs, and then $x \in U_n^\sim$ for some $n$.) ∎

Note (and we will use this fact) that the above proof can be used without change to conclude that any open cover of a pseudometrizable space has an open locally finite refinement.

The normal spaces also form a class of spaces including both the compact spaces and the metric spaces. The relationship between paracompactness and normality is given next.

***20.10 Theorem.*** *Every paracompact space is normal.*

*Proof.* We first establish regularity. Suppose $A$ is a closed set in a paracompact space $X$ and $x \notin A$. For each $y \in A$, find open $V_y$ containing $y$ such that $x \notin \overline{V}_y$. Then the sets $V_y$, $y \in A$, together with the set $X - A$, form an open cover of $X$. Let $\mathscr{W}$ be an open locally finite refinement and let $V = \bigcup \{W \in \mathscr{W} \mid W \cap A \neq \varnothing\}$. Then $V$ is open, contains $A$, and $\overline{V} = \bigcup \{\overline{W} \mid W \cap A \neq \varnothing\}$. But each such set

$W$ is contained in some $V_y$, and hence $\overline{W}$ is contained in $\overline{V}_y$ and thus does not contain $x$. Hence $x \notin \overline{V}$. Thus $x$ and $A$ are separated by open sets in $X$.

Now suppose $A$ and $B$ are disjoint closed sets in $X$. For each $y \in A$, by regularity, find open $V_y$ such that $y \in V_y$ and $\overline{V}_y \cap B = \varnothing$. Then proceeding exactly as above, we can produce an open set $V$ such that $A \subset V$ and $\overline{V} \cap B = \varnothing$. Thus $X$ is normal. ∎

Recall that $\mathbf{\Omega}_0$ denotes the set of ordinals less than the first uncountable ordinal $\omega_1$. The next theorem gives us the easiest example of a normal space which is not paracompact. Another example can be found in 20H.

**20.11 Example.** $\mathbf{\Omega}_0$ *is not paracompact.* Otherwise, the cover by sets

$$U_\beta = \{\gamma \in \mathbf{\Omega}_0 \mid \gamma < \beta\}, \qquad \beta \in \mathbf{\Omega}_0,$$

has a locally finite refinement $\{V_a \mid a \in A\}$. For each $\alpha \in \mathbf{\Omega}_0$, $\alpha \in V_a$ for some $a \in A$ and hence $(f(\alpha), \alpha] \subset V_a$ for some $f(\alpha) < \alpha$. We assert that some $\beta_0$ belongs to $(f(\alpha), \alpha]$ for a cofinal set of points $\alpha$. It is sufficient to prove this since then $\beta_0$ will necessarily belong to infinitely many $V_a$.

If no such $\beta_0$ exists, then for each $\beta_0 \in \mathbf{\Omega}_0$, the set $\{\beta \mid \text{for all } \alpha \geq \beta, f(\alpha) \geq \beta_0\}$ is nonempty. Hence, it has a least element $\alpha(\beta_0)$. Consider the sequence of points defined inductively by $\alpha_0 = \alpha(0)$ and $\alpha_n = \alpha(\alpha_{n-1})$ for $n \geq 1$. Note that for all $\alpha \geq \alpha_n$, $f(\alpha) \geq \alpha_{n-1}$. Now let $\alpha^* = \sup\{\alpha_n\}$. Then $\alpha^* \in \mathbf{\Omega}_0$, and since $\alpha^* \geq \alpha_n$ for each $n$, $f(\alpha^*) \geq \alpha_{n-1}$ for each $n$, from which it follows that $f(\alpha^*) \geq \alpha^*$. But this is impossible since for each $\alpha$ we chose $f(\alpha) < \alpha$.

By contradiction, then, some $\beta_0$ exists with the required property. ∎

Having established the position of paracompactness in the scheme of things, we proceed to investigate the usual questions involving subspaces, products and continuous images.

***20.12 Theorem.*** a) *An $F_\sigma$-subset of a paracompact space is paracompact (so closed subsets of paracompact spaces are paracompact).*

b) *The continuous closed image of a paracompact space is paracompact if it is Hausdorff.*

c) *The product of a paracompact space with a compact $T_2$-space is paracompact.*

*Proof.* a) Suppose $F = \bigcup_{n=1}^{\infty} F_n$ is an $F_\sigma$-subset of a paracompact space $X$, where each $F_n$ is closed in $X$. Let $\{U_\alpha \mid \alpha \in A\}$ be an open cover of $F$; say $U_\alpha = F \cap V_\alpha$, where $V_\alpha$ is open in $X$. For each $n$, $\{X - F_n\} \cup \{V_\alpha \mid \alpha \in A\}$ is an open cover of $X$ which has an open locally finite refinement $\omega_n$. Let

$$\mathscr{A}_n = \{W \cap F \mid W \in \omega_n\}.$$

Then $\mathscr{A}_n$ is a locally finite collection of open subsets of $F$ and $\bigcup_{n=1}^{\infty} \mathscr{A}_n$ clearly refines $\{U_\alpha \mid \alpha \in A\}$. Thus $\{U_\alpha \mid \alpha \in A\}$ has an open $\sigma$-locally finite refinement, so $X$ is paracompact by 20.7.

b) We will provide only a sketch of the proof of this result. The reader interested in the complete details is referred to the original proof as given by Michael (see the notes). Michael proves that a $T_1$-space is paracompact if every open cover $\mathscr{U}$ has a refinement $\mathscr{V}$ such that $\bigcup \mathscr{V}'$ is closed for each $\mathscr{V}' \subset \mathscr{V}$. ($\mathscr{V}$ is called a closure-preserving closed refinement of $\mathscr{U}$). The techniques used to prove this are similar to those you will see in the proof of 20.14 below. Note that a closed locally finite refinement of $\mathscr{U}$ would satisfy this requirement, so that for $T_3$-spaces the stated property is equivalent to paracompactness, by 20.7.

Now suppose $X$ is paracompact and $f$ is a closed continuous map of $X$ onto $Y$. Then $Y$ is a $T_1$-space, so it suffices to show every open cover $\mathscr{U}$ of $Y$ has a closure-preserving closed refinement. But $\{f^{-1}(U) \mid U \in \mathscr{U}\}$ is an open cover of $X$ and thus has a closed locally finite refinement $\mathscr{W}$. It is easily checked that, since $\bigcup \mathscr{W}'$ is closed for any $\mathscr{W}' \subset \mathscr{W}$, the cover $\mathscr{V} = \{f(W) \mid W \in \mathscr{W}\}$ is a closure-preserving closed refinement of $\mathscr{U}$. Thus $Y$ is paracompact.

c) Let $X$ be paracompact, $Y$ compact, and let $\mathscr{U}$ be an open cover of $X \times Y$. For fixed $x \in X$, a finite number of elements of $\mathscr{U}$, say $U^x_{\alpha_1}, \ldots, U^x_{\alpha_{nx}}$, cover $\{x\} \times Y$. Pick an open nhood $V_x$ of $x$ in $X$ such that $V_x \times Y \subset \bigcup_{i=1}^{n_x} U_{\alpha_i}$ (see 17.6c). The sets $V_x$, as $x$ ranges through $X$, form an open cover of $X$. Let $\mathscr{V}$ be an open locally finite refinement. For each $V \in \mathscr{V}$, $V \subset V_x$ for some $x$. Consider the sets $(V \times Y) \cap U^x_{\alpha_i}$, $i = 1, \ldots, n_x$, formed as $V$ ranges through $\mathscr{V}$. This is a refinement of $\mathscr{U}$ and an open cover $\mathscr{W}$ of $X \times Y$. Moreover, given $(x, y) \in X \times Y$, there is a nhood $U$ of $x$ which meets only finitely many $V \in \mathscr{V}$ and the nhood $U \times Y$ of $(x, y)$ can then meet only finitely many sets of $\mathscr{W}$. ∎

**20.13 Examples.** a) $\mathbf{\Omega}_0$ is a nonparacompact open subspace of the paracompact space $\mathbf{\Omega}$. (But if every open subspace of $X$ is paracompact every subspace is paracompact; see Exercise 20E.)

b) The Sorgenfrey line $\mathbf{E}$ is regular Lindelöf and thus paracompact, while $\mathbf{E} \times \mathbf{E}$ is not normal and thus not paracompact. So products of paracompact spaces need not be paracompact. See also Exercise 20F.

c) Every discrete space is paracompact and every topological space is the continuous, one–one image of a discrete space. Thus continuous images of paracompact spaces need not be paracompact. Another example is given in 13.9(b). Note there that $X$ is paracompact and $Y$ is the image of $X$ under an open continuous map, but $Y$ is not $T_2$.

We close this section with a final property of paracompact spaces which will prove useful later on, in the material on uniform spaces. The proof embodies the actual approach used by A. H. Stone to prove every metric space is paracompact.

***20.14 Theorem.*** *A $T_1$-space $X$ is paracompact iff each open covering of $X$ has an open barycentric refinement.*

*Proof.* Suppose $X$ is paracompact. $\mathcal{U}$ is any open cover of $X$. Let $\mathcal{V} = \{V_\alpha \mid \alpha \in A\}$ be an open locally finite refinement of $\mathcal{U}$, and by 15.10, since $X$ is normal, let

$$\mathcal{W} = \{W_\alpha \mid \alpha \in A\}$$

be a shrinking of $\mathcal{V}$. Now $\mathcal{W}$ must be locally finite also. Pick $x \in X$ and let $A_x = \bigcap \{V_\alpha \mid x \in \overline{W}_\alpha\}$. Since each such $V_\alpha$ contains $x$, this is really a finite intersection, so $A_x$ is open. Let $B_x = \bigcup \{\overline{W}_\alpha \mid x \notin \overline{W}_\alpha\}$. Since $\mathcal{W}$ is locally finite, $B_x$ is a closed set. Now set $C_x = A_x - B_x$. We assert $\mathcal{E} = \{C_x \mid x \in X\}$ is the required open barycentric refinement.

Fix $y \in X$ and pick $\alpha$ such that $y \in \overline{W}_\alpha$. We claim $\text{St}(y, \mathcal{E}) \subset V_\alpha$. Suppose $y \in C_x$ (i.e., $C_x$ is part of $\text{St}(y, \mathcal{E})$). But then since $y \in \overline{W}_\alpha$, we have $x \in \overline{W}_\alpha$ also (otherwise $\overline{W}_\alpha \subset B_x$, so $y \notin C_x$). But if $x \in \overline{W}_\alpha$, then $A_x \subset V_\alpha$ and hence $C_x \subset V_\alpha$. So $\text{St}(y, \mathcal{E}) \subset V_\alpha$.

Thus $\mathcal{E}$ is a barycentric refinement of $\mathcal{V}$ and hence of $\mathcal{U}$.

Suppose, conversely, that $X$ is $T_1$ and each open cover has an open barycentric refinement.

First we show $X$ is regular. Let $p \in X$ and let $A$ be a closed set in $X$ not containing $p$. Then $\{X - p, X - A\}$ is an open cover of $X$. Let $\mathcal{V}_1$ be an open barycentric refinement of $\{X - p, X - A\}$ and $\mathcal{V}_2$ an open barycentric refinement of $\mathcal{V}_1$. Then $\mathcal{V}_2$ star-refines $\{X - p, X - A\}$. We claim $\text{St}(p, \mathcal{V}_2)$ and $\text{St}(A, \mathcal{V}_2)$ are the required disjoint nhoods of $p$ and $A$. If not, for some $V, V' \in \mathcal{V}_2$, $V$ contains $p$ and $V'$ meets $A$ while $V \cap V' \neq \emptyset$. Then $\text{St}(V, \mathcal{V}_2)$ meets both $A$ and $p$, which is impossible.

Next we show every open cover $\mathcal{U} = \{U_\alpha \mid \alpha \in A\}$ has an open $\sigma$-locally finite refinement. Construct open covers $\mathcal{U}_1, \mathcal{U}_2, \ldots$ of $X$ such that $\mathcal{U}_1$ is a barycentric refinement of $\mathcal{U}$ and, for each $n = 1, 2, \ldots, \mathcal{U}_{n+1}$ is a barycentric refinement of $\mathcal{U}_n$. For each $\alpha \in A$, define $V_\alpha = \{x \in U_\alpha \mid \text{St}(x, \mathcal{U}_n) \subset U_\alpha$ for some $n\}$. Note that if $\text{St}(x, \mathcal{U}_n) \subset U_\alpha$ then because $\mathcal{U}_{n+1}$ is a barycentric refinement of $\mathcal{U}_n$, $\text{St}(x, \mathcal{U}_{n+1})$ consists of points of $V_\alpha$ [precisely, for each $y \in \text{St}(x, \mathcal{U}_{n+1})$, $\text{St}(y, \mathcal{U}_{n+1}) \subset U_\alpha$ so $y \in V_\alpha$]. Moreover, for each $x \in X$, $\text{St}(x, \mathcal{U}_1) \subset$ some $U_\alpha$ so $x \in$ some $V_\alpha$. Thus the sets $V_\alpha$ form an open refinement of $\mathcal{U}$ with the property that, if $x \in V_\alpha$ then $\text{St}(x, \mathcal{U}_m) \subset V_\alpha$ for some $m$. We will find a $\sigma$-locally finite refinement of the cover $\{V_\alpha \mid \alpha \in A\} = \mathcal{V}$.

Well-order $\mathcal{V}$ say as $V_1, V_2, \ldots, V_\alpha, \ldots$. For each fixed $n = 1, 2, \ldots$ define a sequence of closed sets $H_{n1}, H_{n2}, \ldots, H_{n\alpha}, \ldots$ as follows (see Fig. 20.1):

$$H_{n1} = X - \text{St}(X - V_1, \mathcal{U}_n)$$

and

$$H_{n\alpha} = X - \text{St}\left((X - V_\alpha) \cup \bigcup_{\beta<\alpha} H_{n\beta}, \mathcal{U}_n\right), \qquad \text{if } \alpha > 1.$$

Note (as a mildly intricate exercise) that $\text{St}(H_{n\alpha}, \mathcal{U}_n)$ is contained in $V_\alpha$ and does not meet $H_{n\beta}$ for any $\beta \neq \alpha$. Now, the sets $H_{n\alpha}$ for all $n = 1, 2, \ldots$ and $\alpha \in A$, cover $X$. For if $x \in X$, there is a first index $\alpha$ for which $x \in V_\alpha$. Then, from above,

St $(x, \mathscr{U}_m) \subset V_\alpha$ for some $m$. We claim $x \in H_{m\alpha}$. If not,

$$x \in \mathrm{St}\left((X - V_\alpha) \cup \bigcup_{\beta<\alpha} H_{m\beta}, \mathscr{U}_m\right),$$

and then St $(x, \mathscr{U}_m)$ meets $(X - V_\alpha) \cup \bigcup_{\beta<\alpha} H_{m\beta}$. Since St $(x, \mathscr{U}_m)$ is contained in $V_\alpha$, it must then meet some $H_{m\beta}$ for $\beta < \alpha$. But then $x \in$ St $(H_{m\beta}, \mathscr{U}_m) \subset V_\beta$, which is impossible since $\alpha$ was the first index for which $x \in V_\alpha$.

Finally, for each $n = 1, 2, \ldots$ expand the sequence $H_{n\alpha}$ of closed sets to a sequence of open sets by defining

$$G_{n\alpha} = \mathrm{St}\,(H_{n\alpha}, \mathscr{U}_{n+2}).$$

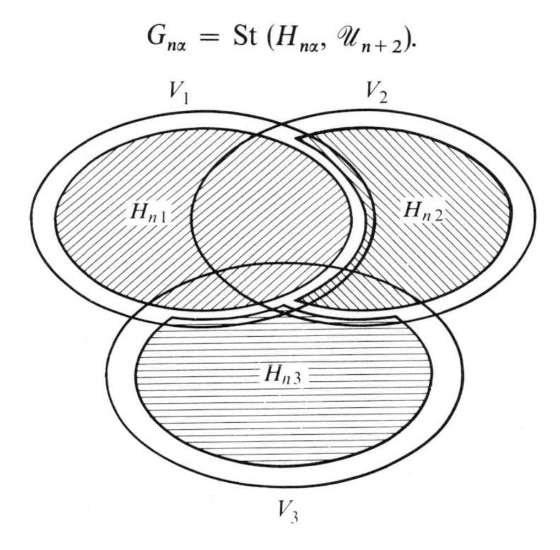


**Figure 20.1**

Then $G_{n\alpha} \subset V_\alpha$ for each $\alpha$ and $n$, and the $G_{n\alpha}$ for all $\alpha$ and $n$ form an open cover of $X$. It suffices, then, to show each subcollection $\{G_{n\alpha} \mid \alpha \in A\}$ is locally finite. In fact, it is discrete. Since $\mathscr{U}_{n+2}$ is a cover of $X$, it is sufficient to show no $U \in \mathscr{U}_{n+2}$ meets both $G_{n\alpha}$ and $G_{n\beta}$ for $\alpha \neq \beta$. Otherwise, there exist $V_1, V_2 \in \mathscr{U}_{n+2}$ such that $V_1$ meets both $H_{n\alpha}$ and $U$, and $V_2$ meets both $H_{n\beta}$ and $U$. But St $(U, \mathscr{U}_{n+2})$ then meets both $H_{n\alpha}$ and $H_{n\beta}$ and hence, since $\mathscr{U}_{n+2} \overset{*}{<} \mathscr{U}_n$, some $W \in \mathscr{U}_n$ meets both $H_{n\alpha}$ and $H_{n\beta}$. Then St $(H_{n\alpha}, \mathscr{U}_n)$ meets $H_{n\beta}$, which is impossible.

Thus $\{G_{n\alpha} \mid \alpha \in A\}$ is discrete, so $\{G_{n\alpha} \mid \alpha \in A, n = 1, 2, \ldots)$ is a $\sigma$-locally finite refinement of $\mathscr{V}$, and thus of $\mathscr{U}$. ■

***20.15 Corollary.*** *A $T_1$-space is paracompact iff every open cover has an open star-refinement.*

*Proof.* A barycentric refinement of a barycentric refinement is a star refinement. ■

## Problems

20A. *Examples on paracompactness*

1. The scattered line **S** (5C) is paracompact.
2. The Moore plane **Γ**, the slotted plane (4C) and the radial plane (3A) are not paracompact.

3. Discuss paracompactness of the sequence spaces **m**, **c** and $\mathbf{c}_0$ (2H).

## 20B. *Barycentric and star refinement*

1. A barycentric refinement of a barycentric refinement of a cover $\mathscr{U}$ is a star-refinement of $\mathscr{U}$.

2. If $\mathscr{U}_n$ is the cover of a metric space $X$ by $(1/3^n)$-spheres about each of its points, then $\mathscr{U}_{n+1} {}^* \!< \mathscr{U}_n$.

3. If $\mathscr{U}$ is an open cover of $X$, $\mathscr{V}$ is an open barycentric refinement of $\mathscr{U}$, and for each $U \in \mathscr{U}$ we define $F_U = X - \text{St}(X - U, \mathscr{V})$, then $\{F_U \mid U \in \mathscr{U}\}$ is a closed cover of $X$.

## 20C. *Partitions of unity*

A *partition of unity* on a space $X$ is a collection $\Phi$ of continuous functions from $X$ to $R^+$ (the nonnegative reals) such that, at each $x \in X$, $\varphi(x) \neq 0$ for only finitely many $\varphi \in \Phi$, and $\sum_{\varphi \in \Phi} \varphi(x) = 1$. $\Phi$ is called *locally finite* iff each $x \in X$ has a nhood on which all but finitely many $\varphi \in \Phi$ vanish. $\Phi$ is *subordinated* to a cover $\mathscr{U}$ of $X$ iff each $\varphi \in \Phi$ vanishes outside some $U \in \mathscr{U}$.

For a $T_1$-space $X$, the following are equivalent:

a) $X$ is paracompact,

b) Every open cover $\mathscr{U}$ of $X$ has a locally finite partition of unity subordinated to it,

c) Every open cover $\mathscr{U}$ of $X$ has a partition of unity $\Phi$ subordinated to it.

[For (a) $\Rightarrow$ (b), use normality to find a shrinking of a locally finite refinement of $\mathscr{U}$, then construct (and modify) Urysohn functions. For (c) $\Rightarrow$ (a), first show $X$ is completely regular. Then, let $\mathscr{V}_i$ be the collection of sets of the form $\{x \in X \mid \varphi(x) > 1/i\}$ for $\varphi \in \Phi$. Show $\mathscr{V}_i$ is locally finite and $\bigcup \mathscr{V}_i$ is a refinement of $\mathscr{U}$. Apply Theorem 20.7.]

## 20D. *Metacompact spaces*

A space is *metacompact* iff each open cover has an open point finite refinement.

1. If $\mathscr{U}$ is any point finite cover of $X$, then $\mathscr{U}$ has an *irreducible* subcover $\mathscr{V}$; i.e., no proper subcollection of $\mathscr{V}$ covers $X$.

2. A countably compact metacompact space is compact. [An irreducible open subcover of an open point finite cover of a countably compact space must be finite; use 17F.2.]

## 20E. *Subspaces of paracompact spaces*

Let $X$ be paracompact.

1. If every open subspace of $X$ is paracompact, then every subspace of $X$ is paracompact.

2. Every paracompact space with a dense Lindelöf subspace is Lindelöf. In particular, then, a separable paracompact space is Lindelöf. [Since a paracompact space is regular, to show it is Lindelöf, it is enough to show every open cover has a countable dense subsystem, by 16D.3.]

3. If $X$ is Lindelöf and $F$ is a closed subset of $\beta X$ which is not a $G_\delta$ and which is contained in $\beta X - X$, $\beta X - F$ is not paracompact [use 2].

## 20F. *Products of paracompact spaces*

The following result supplements the result (20.12) that the product of a paracompact space with a compact space is paracompact.

The product of a paracompact space with a metric space need not be paracompact. [In fact, if **P** denotes the space of irrationals and **S** is the scattered line, then **S** is paracompact (20A.1) and **P** is metric, but $\mathbf{S} \times \mathbf{P}$ is not even normal. For the sets $A = \{(x, y) \in \mathbf{S} \times \mathbf{P} \mid x$ is rational$\}$ and $B = \{(x, x) \in \mathbf{S} \times \mathbf{P} \mid x \in \mathbf{P}\}$ are closed and cannot be separated by disjoint open sets in $\mathbf{S} \times \mathbf{P}$.

## 20G. *Continuous images of paracompact spaces*

If $f$ is a *perfect* mapping of $X$ onto $Y$ (i.e., if $f$ is continuous, closed, and $f^{-1}(y)$ is compact for each $y \in Y$) then $X$ is paracompact iff $Y$ is paracompact.

## 20H. *A separable, normal nonparacompact space*

Recall that $\mathbf{\Omega}_0$ denotes the set of all ordinals $<\omega_1$, the first uncountable ordinal.

1. To each $\alpha \in \mathbf{\Omega}_0$, we can assign a function $f_\alpha: \mathbf{N} \to \mathbf{N}$ such that whenever $\alpha < \beta$, then eventually (i.e., for $n \geq N_{\alpha\beta}$) $f_\alpha(n) < f_\beta(n)$.

We will use the functions $f_\alpha$ to describe a topology on $X = (\mathbf{N} \times \mathbf{N}) \cup \mathbf{\Omega}_0$. For each $\alpha \in \mathbf{\Omega}_0$ and integer $n \in \mathbf{N}$, let $U_n(\alpha) = \{\alpha\} \cup \{(k, f_\alpha(k)) \mid k > n\}$; thus $U_n(\alpha)$ is $\{\alpha\}$ together with a portion of the graph of $f_\alpha$. Now we assign nhoods to points in $X$ as follows:

a) points of $\mathbf{N} \times \mathbf{N}$ are to be isolated,
b) if $\alpha$ is a nonlimit ordinal, nhoods of $\alpha$ will be the sets $U_n(\alpha)$, for $n = 1, 2, \ldots,$
c) if $\alpha$ is a limit ordinal, nhoods of $\alpha$ will be obtained by choosing $\beta < \alpha$, choosing an integer $n(\gamma)$ for each ordinal $\gamma$ with $\beta < \gamma \leq \alpha$ and letting $\bigcup_{\beta<\gamma\leq\alpha} U_{n(\gamma)}(\gamma)$ be a nhood of $\alpha$.

2. The above is a valid assignment of a nhood base to each point in $X$, making $X$ a Hausdorff, separable topological space.

3. $X$ is normal. [Of two disjoint closed sets $H$ and $K$ in $X$, one must be countable (consider their intersections with $\mathbf{\Omega}_0$). For this set, say $H$, find $\alpha_0 \in \mathbf{\Omega}_0$ such that no ordinal beyond $\alpha_0$ lies in $H$. For each $\alpha \in \mathbf{\Omega}_0$, pick an integer $n_\alpha$ as follows:

a) $n_\alpha > N_{\alpha\alpha_0}$ if $\alpha > \alpha_0$ (see 1 for the definition of $N_{\alpha\beta}$),
b) arrange the countably many ordinals $\leq\alpha_0$ in a sequence (beginning with $\alpha_0$), $\alpha_0, \alpha_1, \ldots$ and define $n_{\alpha_0}$ to be 1, $n_{\alpha_k}$ to be any integer larger than

$$\max(N_{\alpha_0\alpha_k}, N_{\alpha_1\alpha_k}, \ldots, N_{\alpha_{k-1}\alpha_k}).$$

Using the integers $n_\alpha$ thus defined, a nhood $U(\alpha)$ of $\alpha$ of the form $U_{n_\alpha}(\alpha)$ or $\bigcup_{\beta<\gamma\leq\alpha} U_{n_\alpha}(\gamma)$ can be contructed using the scheme in either (b) or (c) above, as is appropriate. If (c) is needed, $\beta$ is taken to be the largest ordinal $<\alpha$ which is not in $H$ (or $\beta = 1$ if $H$ contains all ordinals $<\alpha$).

Let $U = [H \cap (\mathbf{N} \times \mathbf{N})] \cup \bigcup_{\alpha\in H} U(\alpha)$. Then $U$ is an open set containing $H$ whose closure does not meet $K$.]

4. $X$ is not paracompact. [The cover of $X$ by the basic nhoods defined in (a), (b) and (c) can have no locally finite refinement.]

## 21 Products of normal spaces

In this section, all spaces are assumed to be Hausdorff (so that normality and the $T_4$-axiom are equivalent here). Sorgenfrey's example of a pair of normal spaces whose product is not normal is well known. We will discuss here subsequent work on the problem of suitably restricting spaces $X$ and $Y$ to make $X \times Y$ normal. Specifically, we will require that $X$ be normal and ask: *under what conditions on $Y$ will $X \times Y$ be normal*? The results are largely disappointing, although attempts to find positive theorems have led to a number of interesting insights and one pretty strange result. Our program will take us through three conditions on $Y$:

a) $Y$ metric,

b) $Y$ compact,

c) $Y$ compact metric.

The first condition is easily disposed of. In Exercise 20F, we provided an example, due to Michael, of a normal space $X$ and a metric space $Y$ such that the product $X \times Y$ was not normal. Alternatively, a study of Michael's paper would do no harm. He provides several examples of nonnormal products $X \times Y$ with conditions of varying strength on $X$ and $Y$. Among them: $X$ can be hereditarily paracompact and $Y$ can be separable metric.

The second condition on $Y$, compactness, is disposed of by a theorem of Tamano based on work of Corson. We will take the time now to present this theorem; it is interesting for other reasons also.

**21.1 Theorem.** *The following are equivalent, for a Tychonoff space $X$:*

a) *$X \times \beta X$ is normal,*

b) *for each compact $F \subset \beta X - X$, there is a locally finite open cover $\{U_\lambda \mid \lambda \in \Lambda\}$ of $X$ such that $(\mathrm{Cl}_{\beta X} U_\lambda) \cap F = \emptyset$, for each $\lambda \in \Lambda$,*

c) *$X$ is paracompact.*

*Proof.* a) $\Rightarrow$ b): Suppose $X \times \beta X$ is normal and let $F$ be a compact subset of $\beta X - X$. Then $\Delta_X = \{(x, x) \in X \times \beta X \mid x \in X\}$ and $X \times F$ are disjoint closed subsets of $X \times \beta X$, so there is a Urysohn function $f: X \times \beta X \to \mathbf{I}$ with $f(\Delta_X) = 0$ and $f(X \times F) = 1$. Let $f_x$ be the restriction of $f$ to $\{x\} \times \beta X$, for each $x \in X$, and define $d$ on $X \times X$ by

$$d(x, y) = \sup_{p \in \beta X} |f_x(p) - f_y(p)|.$$

Then $d$ is a pseudometric on $X$, which induces a topology $\tau$ on $X$ weaker than the original topology. Now the cover of $(X, \tau)$ by spheres $U_d(x, \frac{1}{2}) = U_x$ has locally finite refinement $\{U_\lambda \mid \lambda \in \Lambda\}$ by elements of $\tau$ (and each $U_\lambda$ is an open set in $X$

with its original topology). If $y \in U_x$, then $d(x, y) < \frac{1}{2}$, so

$$f_x(y) = |f_x(y) - f_y(y)| < \tfrac{1}{2}.$$

Hence $f_x(p) \leq \frac{1}{2}$ for each $p \in \mathrm{Cl}_{\beta X}\, U_x$. But $f_x(p) = f(x, p) = 1$ for each $p \in F$ so $(\mathrm{Cl}_{\beta X}\, U_x) \cap F = \varnothing$ for each $x \in X$. Hence $(\mathrm{Cl}_{\beta X}\, U_\lambda) \cap F = \varnothing$ for each $\lambda \in \Lambda$.

b) ⇒ c): Let $\{U_\alpha \mid \alpha \in A\}$ be any open covering of $X$. For each $\alpha$ fix an open set $U_\alpha^*$ in $\beta X$ such that $U_\alpha^* \cap X = U_\alpha$. Let $F_\alpha = \beta X - U_\alpha^*$ for each $\alpha$ and set $F = \bigcap F_\alpha$. Then $F$ is a compact subset of $\beta X - X$ so, by part b), there is a locally finite open cover $\{V_\lambda \mid \lambda \in \Lambda\}$ of $X$ such that $(\mathrm{Cl}_{\beta X}\, V_\lambda) \cap F = \varnothing$ for each $\lambda$. Then $\mathrm{Cl}_{\beta X}\, V_\lambda \subset \bigcup U_\alpha^*$ for each $\lambda$ and, since $\mathrm{Cl}_{\beta X}\, V_\lambda$ is compact, it is contained in the union of a finite subcollection $\{U_{\alpha_1}^*, \ldots, U_{\alpha_{n_\lambda}}^*\}$. It follows that $V_\lambda \subset \bigcup_{k=1}^{n_\lambda} U_{\alpha_k}$. If we now let $H_{\lambda,k} = V_\lambda \cap U_{\alpha_k}$ for each $\lambda \in \Lambda$ and $k = 1, \ldots, n_\lambda$, then $\{H_{\lambda,k}\}$ is a locally finite refinement of $\{U_\alpha \mid \alpha \in A\}$. Thus $X$ is paracompact.

c) ⇒ a): If $X$ is paracompact, then $X \times \beta X$ is paracompact (by 20.12) and thus normal. ■

As we have mentioned, the last theorem provides the answer to the question: is the product of a normal space and a compact space always normal? The answer, since there are nonparacompact normal spaces (e.g., $\mathbf{\Omega}_0$), is no.

The last theorem also provides what will probably be the conclusive result in a string of attempts to provide a global characterization of paracompactness. These attempts began with a conjecture by Kelley to the effect that the paracompact spaces were those which were completely uniformizable by the family of all nhoods of the diagonal. This conjecture was proved false by Corson,† who showed that paracompactness of $X$ was equivalent to the imposition of two global conditions:

1. the family of all nhoods of the diagonal is a uniformity for $X$, and
2. $X \times \beta X$ is normal.

Tamano's theorem eliminates any reference to uniformities for $X$, providing a completely topological characterization.

Returning to the main line of development in this section, we ask whether the product of a normal space $X$ with a compact metric space $Y$ is normal. To handle this case, the work of Dowker is significant; it requires a definition.

**21.2 Definition.** A space $X$ is *countably paracompact* iff every countable open covering has a locally finite refinement. A countably paracompact normal space is called a *binormal space.*

---

† The example referred to by Kelley in a footnote does not work.

**21.3 Theorem.** *Let $X$ be normal. The following are then equivalent:*

a) *$X$ is countably paracompact,*

b) *each countable open covering of $X$ has an open point-finite refinement,*

c) *each countable open covering $\{U_n \mid n = 1, 2, \ldots\}$ of $X$ is shrinkable; i.e., has an open refinement $\{V_n \mid n = 1, 2, \ldots\}$ with $\overline{V}_n \subset U_n$ for $n = 1, 2, \ldots$,*

d) *each sequence $F_1 \supset F_2 \supset \cdots$ of closed sets with empty intersection has an "expansion" to open sets $G_i \supset F_i$ with $\bigcap G_i = \varnothing$.*

*Proof.* a) $\Rightarrow$ b): A locally finite refinement is point finite.

b) $\Rightarrow$ c): Let $\{U_n \mid n = 1, 2, \ldots\}$ be a countable open cover of $X$, $\{V_\alpha \mid \alpha \in A\}$ a point finite refinement. For $n = 1, 2, \ldots$ let

$$V_n = \bigcup \{V_\alpha \mid V_\alpha \subset U_n, V_\alpha \not\subset U_j \text{ if } j < n\}.$$

Then $V_i \subset U_i$ for $i = 1, 2, \ldots$ and $\{V_n \mid n = 1, 2, \ldots\}$ is still point finite. But any point finite cover in a normal space is shrinkable (15.10).

c) $\Rightarrow$ d): If $\{F_n \mid n = 1, 2, \ldots\}$ is a decreasing sequence of closed sets with empty intersection, then $\{X - F_n \mid n = 1, 2, \ldots\}$ is an open cover of $X$. If $\{V_n \mid n = 1, 2, \ldots\}$ is a shrinking of this open cover, then $\{X - \overline{V}_n \mid n = 1, 2, \ldots\}$ will be an expansion of $\{F_n \mid n = 1, 2, \ldots\}$ with empty intersection.

d) $\Rightarrow$ a): Let $\{U_n \mid n = 1, 2, \ldots\}$ be an open covering of $X$ and, for each $n$, let $F_n = X - (U_1 \cup \cdots \cup U_n)$. Let $\{G_n \mid n = 1, 2, \ldots\}$ be an expansion of $\{F_n \mid n = 1, 2, \ldots\}$ with empty intersection (given by d)). Now pick $W_1, W_2, \ldots$ as follows:

$W_1$ is any open set with $X - G_1 \subset W_1$, $\overline{W}_1 \cap F_1 = \varnothing$,

$W_2$ is any open set with $\overline{W}_1 \cup (X - G_2) \subset W_2$, $\overline{W}_2 \cap F_2 = \varnothing$,

and so on. Then $\{W_n \mid n = 1, 2, \ldots\}$ is an open cover of $X$, since

$$\{X - G_n \mid n = 1, 2, \ldots\}$$

covers $X$, and moreover

i) $\overline{W}_n \subset W_{n+1}$,

ii) $X - G_n \subset W_n$,

iii) $W_n \subset \bigcup_{i=1}^n U_i$.

Now let $S_n = W_{n+1} - \overline{W}_{n-1}$ for $n \geq 2$ (and $S_1 = W_1$). Then since $\overline{W}_{n-1} \subset W_n$, $S_n \supset W_{n+1} - W_n$, for each $n$, so $\{S_n \mid n = 1, 2, \ldots\}$ is an open cover of $X$. Moreover, $S_i \cap S_j \neq \varnothing$ iff $|i - j| \leq 1$. Finally, consider the sets

$$\begin{array}{llll} S_1 \cap U_1, & S_1 \cap U_2 & & \\ S_2 \cap U_1, & S_2 \cap U_2, & S_2 \cap U_3 & \\ S_3 \cap U_1, & S_3 \cap U_2, & S_3 \cap U_3, & S_3 \cap U_4 \end{array}$$

and so on. These are all open, they cover $X$ (since the $S_n$ cover $X$ and the union across the $n$th row above is $S_n$), and they form a refinement of $\{U_n \mid n = 1, 2, \ldots\}$. Moreover, $S_i \cap U_j$ can meet at most the other sets on the same row and the rows one above and one below (in the scheme above). Thus if $x \in X$ and $S_n \cap U_m$ contains $x$, then $S_n \cap U_m$ is a nhood of $x$ meeting only finitely many sets of the form $S_i \cap U_j$. Thus $\{S_i \cap U_j \mid i \in \mathbf{N}, j = 1, \ldots, i + 1\}$ is a locally finite refinement of $\{U_n \mid n = 1, 2, \ldots\}$. ∎

With the last result, we are now ready for the fundamental result on products of normal spaces and compact metric spaces. One interesting aspect of the following theorem: it ties normality of such products to normality of the more special class of products $X \times \mathbf{I}$ where $X$ is normal; these products are of interest to those who do homotopy theory.

***21.4 Theorem.*** *The following are equivalent for any (Hausdorff) space $X$:*

a) $X \times \mathbf{I}$ *is normal,*

b) $X \times Y$ *is normal whenever $Y$ is compact metric,*

c) $X$ *is binormal.*

*Proof.* a) ⇒ c): Suppose $X \times \mathbf{I}$ is normal. Clearly $X$ will be normal. To show countable paracompactness, let $F_1 \supset F_2 \supset \cdots$ be a sequence of closed sets with $\bigcap F_n = \varnothing$. Let $W_n = X - F_n$. Let $A$ be the complement in $X \times \mathbf{I}$ of

$$[W_1 \times [0, 1)] \cup [W_2 \times [0, \tfrac{1}{2})] \cup \cdots$$

(Fig. 21.1) and let $B = X \times \{0\}$. Then $A$ and $B$ are disjoint closed sets in $X \times \mathbf{I}$. Let $U$ be an open set in $X \times \mathbf{I}$ containing $A$ such that $\overline{U} \cap B = \varnothing$. Let $G_n = \{x \in X \mid (x, 1/n) \in U\}$. Then $G_n$ is open, $G_n \supset F_n$ and $\bigcap G_n = \varnothing$. Thus $X$ is countably paracompact, by 21.3.

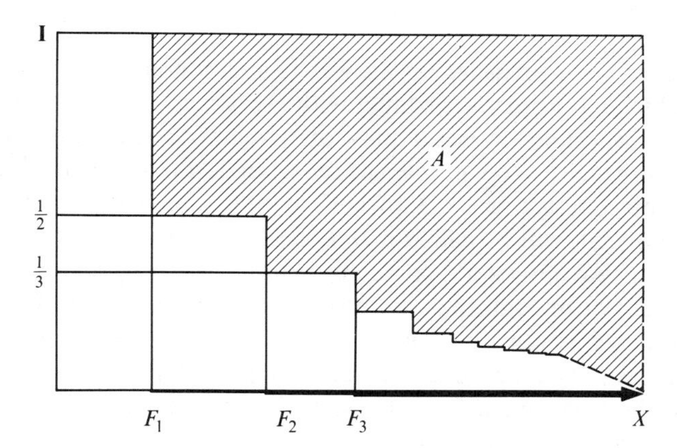


**Figure 21.1**

c) $\Rightarrow$ b): Let $A$ and $B$ be disjoint closed sets in $X \times Y$. Let $\{B_n \mid n = 1, 2, \ldots\}$ be a base for $Y$ and for each finite subset $\gamma$ of $\mathbf{N}$ let $H_\gamma = \bigcup_{n \in \gamma} B_n$. Let $A_x = \{y \in Y \mid (x, y) \in A\}$ and $B_x = \{y \in Y \mid (x, y) \in B\}$. For each $\gamma$, let

$$U_\gamma = \{x \in X \mid A_x \subset H_\gamma\} \cap \{x \in X \mid B_x \subset Y - H_\gamma\}.$$

Then each $U_\gamma$ is open. To show this, suppose $A_{x_0} \subset H_\gamma$. Then if $y \notin H_\gamma$, $(x_0, y) \notin A$. Since $A$ is closed, there is then a basic open nhood $N_y \times M_y$ of $(x_0, y)$ in $X \times Y$ which does not meet $A$. The sets $M_y$ thus obtained as $y$ ranges through $Y - H_\gamma$ form an open cover of $Y - H_\gamma$. Since $Y - H_\gamma$ is compact, we can find a finite subcover $\{M_{y_1}, \ldots, M_{y_n}\}$. Let $N = \bigcap_{i=1}^n N_{y_i}$. Then $N$ is a nhood of $x_0$, and $x \in N$ implies $A_x \subset H_\gamma$. Thus $\{x \in X \mid A_x \subset H_\gamma\}$ is an open set in $X$. Similarly, $\{x \in X \mid B_x \subset Y - H_\gamma\}$ is open in $X$. It follows that $U_\gamma$ is open in $X$.

Furthermore, the sets $U_\gamma$ cover $X$. For if $x \in X$, then for each $y \in A_x$ there is some $B_n$ such that $y \in B_n$ and $\bar{B}_n \cap B_x = \varnothing$. The sets $B_n$ thus obtained as $y$ ranges through $A_x$ form a cover of $A_x$, so a finite subcover can be extracted. Thus $A_x \subset H_\gamma$ and $\bar{H}_\gamma \cap B_x = \varnothing$ for some finite subset $\gamma$ of $\mathbf{N}$. Then $x \in U_\gamma$.

Now let $\mathscr{V}$ be any locally finite refinement of the cover formed by the sets $U_\gamma$ and for each $\gamma$ let $W_\gamma = \bigcup \{V \in \mathscr{V} \mid V \subset U_\gamma\}$. Then the sets $W_\gamma$ form a locally finite cover with the property that $W_\gamma \subset U_\gamma$ for each $\gamma$. Let $\{V_\gamma \mid \gamma$ is a finite subset of $\mathbf{N}\}$ be a shrinking of $\{W_\gamma \mid \gamma$ is a finite subset of $\mathbf{N}\}$; that is, $\bar{V}_\gamma \subset W_\gamma$ for each $\gamma$. Define $V$ to be the union of the sets $V_\gamma \times H_\gamma$, as $\gamma$ ranges through all finite subsets of $\mathbf{N}$. Then $V$ is open and $A \subset V$. Also

$$\bar{V} = \bigcup \overline{(V_\gamma \times H_\gamma)} = \bigcup (\bar{V}_\gamma \times \bar{H}_\gamma) \subset \bigcup (W_\gamma \times \bar{H}_\gamma) \subset \bigcup (U_\gamma \times \bar{H}_\gamma)$$

and this does not meet $B$. Thus $X \times Y$ is normal.

b) $\Rightarrow$ a): This is obvious. ■

The theorem above provides an answer to our fundamental question: that is, the spaces which have normal product with every compact metric space are the binormal spaces. But it also raises a question with an interesting history. Is every normal space binormal? We will refer to the assertion that this is so as *Dowker's conjecture*. A counterexample to Dowker's conjecture, that is, a normal space which is not binormal, will be called a *Dowker space*. In this terminology, M. E. Rudin has shown that Dowker's conjecture cannot be proved with the existing axioms of set theory (through the axiom of choice). In fact, from Dowker's conjecture she deduces a result (the *Souslin hypothesis*, that every compact ordered space with the countable chain condition is separable) which is known to be independent of these axioms (a recent result of Jech, Tennenbaum and Solovay). It is still unknown whether a Dowker space can be constructed using existing set-theoretic axioms through the choice axiom.

## Problems

### 21A. *Countable paracompactness*

1. Every perfectly normal space is countably paracompact.

2. A closed subset of a countably paracompact space is countably paracompact.

3. The product of a compact space and a countably paracompact space is countably paracompact. [Study the proof of 20.12(c).]

### 21B. *Semicontinuity in countably paracompact spaces*

1. Let $X$ be countably paracompact and normal. If $g$ is a real-valued lower semicontinuous function on $X$ and $h$ is a real-valued upper semicontinuous function on $X$ with $h(x) < g(x)$ for each $x \in X$, then there is a continuous real-valued function $f$ on $X$ with $h(x) < f(x) < g(x)$ for each $x \in X$. [For each rational $r$, let $G_r = \{x \mid h(x) < r < g(x)\}$ and let $\{U_r \mid r \in Q\}$ and $\{V_r \mid r \in Q\}$ be locally finite open coverings of $X$ such that $\bar{V}_r \subset U_r \subset G_r$. Define $f_r$ to be continuous from $X$ to $[-\infty, \infty]$ such that $f_r(x) = -\infty$ if $x \notin U_r$, $f_r(x) = r$ if $x \in \bar{V}_r$. Let $f(x) = \text{l.u.b.}\ f_r(x)$. Show $f$ has the required properties.]

2. If $X$ has the property expressed above, then $X$ is countably paracompact and normal. [Show $X$ is normal. Then let $(F_i)$ be a decreasing sequence of closed sets in $X$ with empty intersection. Set $g(x) = 1/(n + 1)$ for $x \in F_i - F_{i+1}$, $i = 0, 1, 2, \ldots$ (where $F_0 = X$) and set $h(x) = 0$ for all $x \in X$.]

### 21C. *Normality in infinite products*

Let $A$ be an uncountable set and, for each $\alpha \in A$, let $N_\alpha$ be a copy of the positive integers. Consider the space $T = \prod_{\alpha \in A} N_\alpha$. A typical basic nhood $U(t; a_1, \ldots, \alpha_n)$ of a point $t$ in $T$ consists of all points $t'$ for which $t'_\alpha = t_\alpha$ for $\alpha \in \{\alpha_1, \ldots, \alpha_n\}$.

1. For each positive integer $k$, let $A_k$ be the set of all points $t$ in $T$ such that each integer other than $k$ occurs at most once among the coordinates of $t$. Prove that the sets $A_k$ are closed and pairwise disjoint.

2. $T$ is not normal. [Suppose $A_1$ is contained in an open set $U$. Define a sequence $x_1, x_2, \ldots$ of points of $A_1$ as follows: let $x_1$ be the point all of whose coordinates are 1, and let $U(x_1; \alpha_1, \ldots, \alpha_n)$ be a nhood of $x_1$ contained in $U$. Let $x_2$ be the point all of whose coordinates are 1, except that the $\alpha_i$th coordinate of $x_2$ is $i$, for $i = 1, \ldots, n_1$ and let

$$U(x_2; \alpha_1, \ldots, \alpha_{n_1}, \alpha_{n_1+1}, \ldots, \alpha_{n_2})$$

be a nhood of $x_2$ contained in $U$. Continue, obtaining a sequence $x_1, x_2, \ldots$ of points of $A_1$ and a related sequence $\alpha_1, \alpha_2, \ldots$ of coordinate indices. Now let $x$ be the point of $A_2$ whose coordinates are 2 except that $x_{\alpha_i} = i$ for $i = 1, 2, \ldots$. Prove $A_1$ and $A_2$ cannot be separated by showing $x \in \text{Cl}_T\, U$.]

3. Every countably compact, $T_1$-space contains a closed copy of the integers (i.e., a countable, closed, relatively discrete set).

4. *If a product of nonempty $T_1$-spaces is normal, all but countably many of the factor spaces are countably compact.*

5. If $X$ is any product of metric spaces, the following are all equivalent:

   a) $X$ is normal,
   b) $X$ is paracompact,
   c) all but countably many of the factor spaces are compact.

Chapter 7

# Metrizable Spaces

## 22 Metric spaces and metrizable spaces

Our purpose in this section is twofold: we seek to establish the notational and conventional groundwork for the material to follow and to prove a few basic facts about metric and metrizable spaces. We will begin by investigating products and continuous images of metrizable spaces.

**22.1 Definition.** Two metrics $\rho_1$ and $\rho_2$ on the same set $M$ are said to be *equivalent* if they generate the same topology on $M$.

A topologist, then, is always willing to replace a given metric with an equivalent metric if it serves some purpose. One useful result in this direction is the following theorem.

**22.2 Theorem.** *Every metric $\rho$ on $M$ is equivalent to a bounded metric.*

*Proof.* In fact, there are two standard ways of replacing $\rho$ by a metric with a bound: define new functions $\rho_1$ and $\rho_2$ on $M \times M$ by

$$\rho_1(x, y) = \min\{1, \rho(x, y)\},$$
$$\rho_2(x, y) = \frac{\rho(x, y)}{1 + \rho(x, y)}.$$

The reader will verify (22F) that these are indeed metrics on $M$, giving the same topology as $\rho$ does. ∎

We are prepared to use 22.2 immediately.

**22.3 Theorem** *A nonempty product space $\prod_{\alpha \in A} M_\alpha$ is metrizable iff each $M_\alpha$ is metrizable and $M_\alpha$ is a single point for all but a countable set of indices.*

*Proof.* $\Rightarrow$: Each $M_\alpha$ is homeomorphic to a subspace of the product and hence metrizable. Moreover, the product is first countable, if metrizable, and thus can be at most a countable product (see 16A.2).

$\Leftarrow$: Let $M_1, M_2, \ldots$ be metrizable spaces. Using 22.2, let $\rho_i$ be a metric on $M_i$, bounded by 1, which gives the topology on $M_i$, for $i = 1, 2, \ldots$. Define $\rho$ on $\prod_{i=1}^{\infty} M_i$ as follows: for $x = (x_1, x_2, \ldots)$ and $y = (y_1, y_2, \ldots)$,

$$\rho(x, y) = \sum_{i=1}^{\infty} \frac{\rho_i(x_i, y_i)}{2^i}.$$

This is easily verified to be a metric. We will show that it gives the product topology on $\prod_{i=1}^{\infty} M_i$.

Let $x = (x_1, x_2, \ldots)$ be a point in $\prod_{i=1}^{\infty} M_i$. A basic nhood $U$ of $x$ in the Tychonoff product topology restricts only finitely many coordinates and thus can be written

$$U = U_{\rho_1}(x_1, \varepsilon_1) \times U_{\rho_2}(x_2, \varepsilon_2) \times \cdots \times U_{\rho_n}(x_n, \varepsilon_n) \times \prod \{M_k \mid k = n + 1, n + 2, \ldots\}.$$

Let $\varepsilon$ be chosen so that

$$\varepsilon = \min\left(\frac{\varepsilon_1}{2}, \frac{\varepsilon_2}{2^2}, \ldots, \frac{\varepsilon_n}{2^n}\right).$$

Now a routine calculation shows that if $\rho(x, y) < \varepsilon$, then $\rho_i(x_i, y_i) < \varepsilon_i$ for each $i = 1, \ldots, n$, so that apparently $U_\rho(x, \varepsilon) \subset U$. Thus the product topology on $\prod M_k$ is weaker than the topology induced by $\rho$. On the other hand, given $\varepsilon > 0$, we can choose $N$ large enough that $\sum_{i=N+1}^{\infty} 1/2^i < \varepsilon/2$. Then it is easily verified that

$$U_{\rho_1}\left(x_1, \frac{\varepsilon}{2N}\right) \times U_{\rho_2}\left(x_2, \frac{\varepsilon}{2N}\right) \times \cdots \times U_{\rho_n}\left(x_n, \frac{\varepsilon}{2N}\right) \times \prod \{M_k \mid k = N + 1, N + 2, \ldots\} \subset U_\rho(x, \varepsilon)$$

so that the topology induced by $\rho$ is weaker than the product topology. ■

**22.4 Example.** Among the spaces admitted to metrizability by the last theorem the most important are $\mathbf{R}^{\aleph_0}$, also called *Frechet space*, and its subspace $\mathbf{I}^{\aleph_0}$, the *Hilbert cube* (which we studied in 17.9). These two spaces, together with *Hilbert space* $\mathbf{H}$ (18.7), form the backbone of the theory of separable metric spaces. One easy result that is of particular interest: any one of these spaces can be homeomorphically embedded as a subspace of any other. This embedding property takes on additional significance once the Urysohn metrization theorem (23.1) is proved, since a part of that theorem asserts that every separable metric space is homeomorphic to a subset of $\mathbf{I}^{\aleph_0}$. Thus any one of $\mathbf{I}^{\aleph_0}$, $\mathbf{R}^{\aleph_0}$, or $\mathbf{H}$ can be used as a *universal space* for separable metric spaces. Still dealing with homeomorphisms between these three spaces, it is immediately clear that $\mathbf{I}^{\aleph_0}$ cannot be homeomorphic to either $\mathbf{R}^{\aleph_0}$ or $\mathbf{H}$, since it is compact and the others are not even locally compact. The question of whether $\mathbf{R}^{\aleph_0}$ is homeomorphic to $\mathbf{H}$ has been only recently settled (in a very general context) in the affirmative; in fact, R. D. Anderson has proved that *all separable infinite-dimensional Banach spaces are homeomorphic.* (See 24J for the definition of a Banach space.)

Turning to continuous maps of metric spaces, we limit ourselves to quoting results which will be proved later when better machinery is available.

Quotients of metrizable spaces need not be metrizable; they are studied in Exercise 23K. In Section 23, as a corollary to the Urysohn metrization theorem, we will prove that *the continuous image of a compact metric space is a compact metric space if it is Hausdorff.*

We close this section with one of the fundamental results about compact metric spaces. It is used both in dimension theory (an application we will not see) and in building the theory of uniform spaces (which we get to in Sections 35 through 41).

**22.5 Theorem** (Lebesgue covering lemma). *If $\{U_1, \ldots, U_n\}$ is a finite open cover of a compact metric space $X$, there is some $\delta > 0$ such that if $A$ is any subset of $X$ of diameter $< \delta$, then $A \subset U_i$ for some $i$.*

*Proof.* Suppose not. Then for each $n \in \mathbf{N}$, let $A_n$ be a set of diameter $< 1/n$ such that $A_n \not\subset U_i$ for any $i$. Pick $x_n \in A_n$ for each $n$, and let $x$ be a cluster point of the resulting sequence. Now $x \in U_i$, for some $i$, so for some $\delta > 0$, $U(x, \delta) \subset U_i$. Pick $n$ large enough that $1/n < \delta/2$, and find $m > n$ such that $x_m \in U(x, \delta/2)$. Now $x_m \in A_m$, so $A_m \cap U(x, \delta/2) \neq \varnothing$, while the diameter of $A_m$ is less than $\delta/2$. It follows that $A_m \subset U(x, \delta) \subset U_i$, a contradiction. ∎

Any number $\delta$ which works in the previous lemma is called a *Lebesgue number* for the cover $\{U_1, \ldots, U_n\}$.

## Problems

### 22A. *Results on metric spaces*

1. The collection of all metrics on a fixed set $M$ has cardinal number $2^{|M|}$.

2. Every 2-element metric space can be embedded isometrically in the real line $\mathbf{R}$. Every 3-element metric space can be embedded isometrically in $\mathbf{R}^2$. There are 4-element metric spaces which cannot be isometrically embedded in Hilbert space $\mathbf{H}$ and hence cannot be embedded in any $\mathbf{R}^n$ (since each $\mathbf{R}^n$ is isometric to a subspace of $\mathbf{H}$, by 18B.5).

### 22B. *Perfect normality*

1. Show that a compact Hausdorff space is metrizable iff the diagonal $\Delta$ in $X \times X$ is a zero set. [If $\Delta$ is a zero set, it is the zero set of a nonnegative function.]

2. Find a perfectly normal compact space $X$ which is not metrizable.

3. Conclude that the product of two perfectly normal compact spaces need not be perfectly normal.

### 22C. *Linear topological spaces*

A (real) *linear topological space* is a real linear space (vector space) $E$ with a Hausdorff topology such that:

*TL*-a) vector addition is continuous; that is, the map $a: E \times E \to E$ defined by $a(x, y) = x + y$ is continuous,

*TL*-b) scalar multiplication is continuous; that is, the map $s: \mathbf{R} \times E \to E$ defined by $s(\lambda, x) = \lambda x$ is continuous.

For $x$ and $y$ in $E$, denote by $L(x, y)$ the set of all points $z$ such that $z = \lambda_1 x + \lambda_2 y$ with $0 \le \lambda_i \le 1$ and $\lambda_1 + \lambda_2 = 1$. A subset $A$ of $E$ is *convex* iff whenever $x$ and $y$ belong to $A$, $L(x, y) \subset A$. A linear topological space is *locally convex* iff each point $p$ of $E$ has a base of convex nhoods.

1. Every normed linear space (2J) is a locally convex linear topological space.
2. If $A$ is convex and $x \in A^\circ$, $y \in \bar{A}$, then $L(x, y) - \{y\} \subset A$.
3. If $A$ and $B$ are convex, so are $A^\circ$, $\bar{A}$, $A + B$, $A \cap B$ and, for $\lambda \in \mathbf{R}$, $\lambda A$.
4. A convex open set $A$ in $L$ is regularly open (3D). So a locally convex linear topological space is semiregular (14E).
5. A linear topological space is a topological group (13G).

## 22D. *Metric-absolute retracts*

A space $Y$ is a *metric-absolute retract* iff whenever $A$ is a closed subset of a metric space $X$ and $f: A \to Y$ is continuous, then $f$ can be extended to all of $X$ (compare with 15D).

Let $(X, d)$ be a metric space, $A$ a closed subset of $X$. Let $L$ be a locally convex linear topological space (22C).

1. An open, locally finite cover $\mathcal{U}$ of $X - A$ can be found with the properties:
   a) if $a \in \operatorname{Fr}(A)$, each nhood of $a$ meets infinitely many sets from $\mathcal{U}$,
   b) if $a \in A$ and $W$ is any nhood of $a$, there is a nhood $W' \subset W$ of $a$ such that, for $U \in \mathcal{U}$, $U \cap W' \neq \varnothing \Rightarrow U \subset W$.

[To get $\mathcal{U}$, consider an open locally finite refinement of a set of disks in $X - A$ which get smaller as they get closer to $A$.] Such a cover $\mathcal{U}$ of $X - A$ will be called a *canonical cover* of $X - A$.

2. If $\mathcal{U}$ is a canonical cover of $X - A$, for each $U_0 \in \mathcal{U}$, define

$$\lambda_{U_0}(x) = \frac{d(x, X - U_0)}{\sum_{U \in \mathcal{U}} d(x, X - U)}.$$

Then $\lambda_{U_0}$ is continuous on $X - A$, and if $\alpha_U$ is a real constant for each $U \in \mathcal{U}$, then $\sum_{U \in \mathcal{U}_0} \alpha_U \cdot \lambda_U(x)$ is continuous on $X - A$.

3. Let $\mathcal{U}$ be a canonical cover of $X - A$. For each $U \in \mathcal{U}$, pick $x_U \in U$ and then find $a_U \in A$ such that $d(x_U, a_U) < 2 \cdot d(x_U, A)$. If $f: A \to L$ is any continuous function, define $F: X \to L$ by

$$F(x) = \sum_{U \in \mathcal{U}} \lambda_U(x) \cdot f(a_U), \qquad \text{for} \quad x \in X - A,$$

$$F(x) = f(x), \qquad \text{for} \quad x \in A.$$

Then $F$ is continuous. [Check continuity at points $a \in A$ as follows. Let $V$ be any convex nhood of $F(a)$ and, by continuity of $f$, find $\delta > 0$ such that $f$ maps the $\delta$-sphere about $a$ into $V$. Let $W$ be the $(\delta/3)$-sphere about $a$ in $X$. Apply (b) of part 1, to find $W' \subset W$ such that, for $U \in \mathcal{U}$, $U \cap W' \neq \varnothing \Rightarrow U \subset W$. Verify that $x_U \in W' \Rightarrow F(x_U) = f(a_U) \in V$. Then apply part 1 again, finding $W'' \subset W'$ such that $U \cap W'' \neq \emptyset \Rightarrow U \subset W'$. The claim is that $F(W'') \subset V$. (For $x \in W'' \cap (X - A)$, $F(x)$ will belong to the convex hull of a finite set of the $f(a_U)$, and hence to $V$. For $x \in W'' \cap A$, $F(x) \in V$ because $W'' \subset W$.)]

4. Thus *every locally convex linear topological space is a metric-absolute retract.*

### 22E. *Extending metrics*

Let $(X, d)$ be a metric space, $A$ a closed subset of $X$. For $a \in A$, define $r_a: A \to \mathbf{R}$ by $r_a(x) = d(a, x)$. Now fix a point $a \in A$ and define $\varphi$ on $A$ by $\varphi(x) = r_x - r_a$ (so $\varphi(x)$ is a function on $A$, for each $x \in A$). Provide $C^*(A)$ and $C^*(X)$ with their sup norms (2J.4).

1. $\varphi$ maps $A$ continuously into $C^*(A)$. Now by 22C.1 and 22D.4, there is a continuous extension $\Phi: X \to C^*(A)$ of $\varphi$. Let $L$ be the linear topological space $C^*(A) \times \mathbf{R} \times C^*(X)$ with norm defined by

$$\|(f, p, g)\| = \max(\|f\|, |p|, \|g\|).$$

Map $X$ onto $L$ as follows: for $x \in X$, let $\alpha_x(y) = d(x, y)$ and let

$$F(x) = (\Phi(x), d(x, A), d(x, A) \cdot \alpha_x).$$

Clearly $F: X \to L$ and $F$ is continuous.

2. $F$ is an isometry on $A$.

3. $F$ is a homeomorphism on $X$.

4. *If $X$ is any metrizable space, $A$ is a closed subset of $X$, and $\rho$ is a compatible metric on $A$, then $\rho$ can be extended to a compatible metric on $X$.*

### 22F. *Bounded metrics*

1. If $\rho$ is a metric on $X$, then both

$$\rho_1(x, y) = \min\{1, \rho(x, y)\} \qquad \text{and} \qquad \rho_2(x, y) = \rho(x, y)/[1 + \rho(x, y)]$$

are metrics equivalent to $\rho$ on $X$.

2. Every metric generating the topology of a compact metrizable space is bounded.

3. Conversely, if every metric generating the topology of a metrizable space $X$ is bounded, then $X$ is compact. [Otherwise, a sequence $(x_n)$ exists in $X$ with no cluster point. Define $\rho$ on $(x_n)$ by $\rho(x_n, x_m) = |n - m|$ and apply 22E.4.]

## 23 Metrization

A natural question follows the statement that metrics generate topologies, namely, "which topologies?" More precisely, can a condition be found which is equivalent to metrizability but which deals only with open sets? The search for such conditions was long and was not satisfactorily concluded until the early 1950's when Bing, Nagata and Smirnov independently provided similar characterizations. Our general metrization theorem (23.9) is given in the form proved by Nagata and Smirnov.

Before giving the main metrization theorem, we will provide some other useful results on metrization. The first is the classical theorem of Urysohn, characterizing the *separable* metric spaces.

**23.1 Urysohn's metrization theorem.** *The following are equivalent for a $T_1$-space $X$:*

a) *$X$ is regular and second countable,*

b) *$X$ is separable and metrizable,*

c) *$X$ can be embedded as a subspace of the Hilbert cube $\mathbf{I}^{\aleph_0}$.*

*Proof.* a) $\Rightarrow$ c): Let $\mathscr{B}$ be a countable base for $X$, and let $\mathscr{A} = \{(U, V) \mid U, V \in \mathscr{B}$ and $\overline{U} \subset V\}$. $\mathscr{A}$ is countable and, since $X$ is a regular Lindelöf space and thus normal, for each pair $(U, V)$ in $\mathscr{A}$, there is a function $f_{UV}: X \to \mathbf{I}$ such that $f(\overline{U}) = 0$, $f(X - V) = 1$. If $\mathscr{F} = \{f_{UV} \mid (U, V) \in \mathscr{A}\}$, then $\mathscr{F}$ is countable, and certainly $\mathscr{F}$ separates points from closed sets in $X$. It follows, by 8.16, that if $I_f$ is a copy of $\mathbf{I}$ for each $f \in \mathscr{F}$, the evaluation map $e: X \to \prod_{f \in \mathscr{F}} I_f$ defined by giving coordinates:

$$[e(x)]_f = f(x),$$

is an embedding. Since $\mathscr{F}$ is countable, $\prod_{f \in \mathscr{F}} I_f = \mathbf{I}^{\aleph_0}$, and we have established (c).

c) $\Rightarrow$ b): $\mathbf{I}^{\aleph_0}$ is separable and metric and thus so is every subspace of $\mathbf{I}^{\aleph_0}$.

b) $\Rightarrow$ a): This is obvious. ■

Apparently second countability is a strong axiom, differing from metrizability only by a separation axiom.

**23.2 Corollary.** *The continuous image of a compact metric space in a Hausdorff space is metrizable.*

*Proof.* Let $f$ be a continuous map of a compact space $X$ onto a Hausdorff space $Y$. Then $Y$ is compact and thus regular so, by Urysohn's theorem, it suffices to show $Y$ is second countable. Let $\mathscr{B}$ be a countable base for $X$ and let $\mathscr{C}$ be the collection of all finite unions of sets from $\mathscr{B}$. Then $\mathscr{D} = \{Y - f(X - C) \mid C \in \mathscr{C}\}$ is a countable collection of open sets in $Y$; we claim it is a base for $Y$. Let $U$ be open in $Y$ and suppose $p \in U$. Then $f^{-1}(p) \subset f^{-1}(U)$ and $f^{-1}(p)$ is compact. Now a simple argument shows that there are sets $B_1, \ldots, B_n$ in $\mathscr{B}$ such that $f^{-1}(p) \subset B_1 \cup \cdots \cup B_n \subset f^{-1}(U)$. Let $C = B_1 \cup \cdots \cup B_n$. Then $C \in \mathscr{C}$ and (easily) $p \in Y - f(X - C) \subset U$. Thus $\mathscr{D}$ is a base for $Y$. ■

The next metrization theorem will be useful later in our work with uniform spaces. We need some terminology.

**23.3 Definition.** A *normal sequence* in a space $X$ is a sequence $\mathscr{U}_1, \mathscr{U}_2, \ldots$ of open covers of $X$ such that $\mathscr{U}_{n+1}$ star-refines $\mathscr{U}_n$, for $n = 1, 2, \ldots$. It will be called a *compatible* normal sequence in $X$ iff $\{\text{St}(x, \mathcal{U}_n) \mid n = 1, 2, \ldots\}$ is a nhood base at $x$, for each $x \in X$. Any open cover of $X$ which is $\mathcal{U}_1$ in some normal sequence in $X$ will be called a *normal cover*. (Thus, every cover in a normal sequence is a normal cover.)

**23.4 Theorem.** *A topological space $X$ is pseudometrizable iff it has a compatible normal sequence. (Hence, a $T_0$-space is metrizable iff it has a compatible normal sequence.)*

*Proof.* If $X$ is pseudometrizable, its topology generated by the pseudometric $\rho$, define $\mathcal{U}_n = \{U_\rho(x, 1/3^n) \mid x \in X\}$. Then the sequence $\mathcal{U}_1, \mathcal{U}_2, \ldots$ is a compatible normal sequence in $X$. (Certainly the sets $\text{St}(x, \mathcal{U}_n)$ form a nhood base at $x$, for each $x$ in $X$. It is also pretty clear that $\text{St}\left(U_\rho(x, 1/3^n), \mathcal{U}_n\right) \subset U_\rho(x, 1/3^{n-1})$, so that $\ldots \mathcal{U}_3 \mathrel{*<} \mathcal{U}_2 \mathrel{*<} \mathcal{U}_1$.)

Conversely, suppose we have a compatible normal sequence $(\mathcal{U}_n)$ for $X$. Define $t$ on $X \times X$ as follows:

$$\begin{aligned}
t(x, y) &= 0 && \text{if } y \in \text{St}(x, \mathcal{U}_n), \text{ for all } n,\\
t(x, y) &= 1 && \text{if } y \notin \text{St}(x, \mathcal{U}_1),\\
t(x, y) &= \tfrac{1}{2} && \text{if } y \in \text{St}(x, \mathcal{U}_1),\ y \notin \text{St}(x, \mathcal{U}_2),\\
t(x, y) &= \frac{1}{2^n} && \text{if } y \in \text{St}(x, \mathcal{U}_n),\ y \notin \text{St}(x, \mathcal{U}_{n+1}).
\end{aligned}$$

Now for $x, y \in X$, let $\mathcal{S}(x, y)$ be all finite sequences $s = \{x_1, \ldots, x_n\}$ of points of $X$ such that $x_1 = x$, $x_n = y$ or $x_1 = y$, $x_n = x$. Define

$$\rho(x, y) = \inf\left\{\sum_{i=2}^{n} t(x_{i-1}, x_i) \cdot \;\middle|\; \{x_1, \ldots, x_n\} \in \mathcal{S}(x, y)\right\}.$$

The reader will have no trouble verifying that $\rho(x, y)$ is a pseudometric. It remains to show that $\rho$ is compatible with the topology on $X$.

Let $\mathcal{V}_n$ be the cover of $X$ by the spheres $U_\rho(x, 1/2^n)$. It will suffice to show that, for any $n$,

a) $\mathcal{U}_n < \mathcal{V}_{n-1}$,

b) $\mathcal{V}_n < \mathcal{U}_{n-1}$,

since it will then be clear that the topologies generated by the two sequences are the same. (Compare the nhood bases at any point.)

a) Suppose $U \in \mathcal{U}_n$. Pick $x \in U$. If $y \in U$, then $y \in \text{St}(x, \mathcal{U}_n)$ so $t(x, y) \le 1/2^n$ and hence $\rho(x, y) \le 1/2^n < 1/2^{n-1}$. Thus $y \in U_\rho(x, 1/2^{n-1})$, so $U \subset U_\rho(x, 1/2^{n-1})$. Thus $\mathcal{U}_n < \mathcal{V}_{n-1}$.

b) To show $\mathcal{V}_n < \mathcal{U}_{n-1}$, it is enough to prove that whenever $\rho(x, y) < 1/2^n$, then $x$ and $y$ lie together in some element of $\mathscr{U}_n$, since then

$$U_\rho(x, 1/2^n) \subset \text{St}(x, \mathscr{U}_n) \subset U$$

for some $U \in \mathscr{U}_{n-1}$.

Hence, suppose $\rho(x, y) < 1/2^n$. Then

$$\inf_{s \in \mathscr{S}(x,y)} \sum_{i=2}^{k} t(x_{i-1}, x_i) < 1/2^n$$

and consequently, for some sequence $\{x_1, \ldots, x_k\}$ from $\mathscr{S}(x, y)$,

$$\sum_{i=1}^{k} t(x_{i-1}, x_i) < \frac{1}{2^n}.$$

We proceed now by induction on the length $k$ of this sequence. If $k = 2$, then $t(x, y) < 1/2^n$ so that $y \in \text{St}(x, \mathscr{U}_m)$, $y \notin \text{St}(x, \mathscr{U}_{m+1})$ for some $m > n$. Hence, in particular, $y \in \text{St}(x, \mathscr{U}_{n+1})$, from which it follows that $x, y \in U$ for some $U \in \mathscr{U}_{n+1}$ in fact, so that certainly $x, y$ lie together in some $U' \in \mathscr{U}_n$. (Recall, then, that if $t(x, y) < 1/2^n$, we have $x$ and $y$ together in some element of $\mathscr{U}_{n+1}$; we will use this again.)

Suppose the result is true for sequences of length $<k$, and suppose $\sum_{i=2}^{k} t(x_{i-1}, x_i) < 1/2^n$. Let $j$ be the last number, $2 \le j \le k$, such that

$$\sum_{i=2}^{j} t(x_{i-1}, x_i) < \frac{1}{2^{n+1}}.$$

Then

$$\sum_{i=2}^{j+1} t(x_{i-1}, x_i) \ge \frac{1}{2^{n+1}},$$

so that

$$\sum_{i=j+2}^{k} t(x_{i-1}, x_i) < \frac{1}{2^{n+1}}.$$

Now by the inductive hypothesis $x_1, x_j$ lie in some $U_1 \in \mathscr{U}_{n+1}$ while the argument above shows, since $t(x_j, x_{j+1}) < 1/2^n$, that $x_j, x_{j+1}$ lie in some $U_2 \in \mathscr{U}_{n+1}$, and finally, using the inductive hypothesis again, $x_{j+1}, x_k$ lie in some $U_3 \in \mathscr{U}_{n+1}$. Then $x_1$ and $x_k$ lie in $\text{St}(U_2, \mathscr{U}_{n+1}) \subset U$ for some $U \in \mathscr{U}_n$. This establishes our claim, by induction. ■

The above construction should be studied carefully. It and the theorem it proves are fundamental building blocks in the theory of uniform spaces, which we will develop in Sections 35 through 41.

We exhibit a use of the above theorem by proving the following elegant neighborhood characterization of metrizable spaces, a slight alteration of a result of Nagata.

**23.5 Theorem.** *A $T_0$-space $X$ is metrizable iff each $x \in X$ possesses a countable* *nhood base* $\{U_{xn} \mid n \in \mathbf{N}\}$ *with the following properties:*

a) $y \in U_{xn} \Rightarrow U_{yn} \subset U_{xn-1}$

b) $y \notin U_{xn-1} \Rightarrow U_{yn} \cap U_{xn} = \varnothing$.

*Proof.* $\Rightarrow$: This part is easy, since the properties (a) and (b) are obviously satisfied if $U_{xn}$ is the disk of radius $1/2^n$ about $x$.

$\Leftarrow$: Let $\mathscr{U}_n = \{U_{xn} \mid x \in X\}$. We claim $\operatorname{St}(U_{xn}, \mathscr{U}_n) \subset U_{xn-2}$, for any $n > 2$. Suppose $U_{zn} \cap U_{xn} \neq \varnothing$. Then, by property (b), $z \in U_{xn-1}$. Hence, by property (a), $U_{zn} \subset U_{xn-2}$, and thus $\operatorname{St}(U_{xn}, \mathscr{U}_n) \subset U_{xn-2}$ as asserted. It now follows that $\mathscr{U}_n$ star-refines $\mathscr{U}_{n-2}$ for any $n > 2$, so that $\mathscr{U}_1, \mathscr{U}_3, \ldots$ is a normal sequence. It also follows that $\operatorname{St}(x, \mathscr{U}_n) \subset U_{xn-2}$ for any $n > 2$, so that $\mathscr{U}_1, \mathscr{U}_3, \ldots$ is compatible with $X$. Thus, by 23.4, $X$ is metrizable. ∎

We introduce now an idea which is obviously related to the notion of a compatible normal sequence; it will subsequently be used in Theorem 23.7.

**23.6 Definition.** A *development* for a space $X$ is a sequence $\mathscr{U}_1, \mathscr{U}_2, \ldots$ of open covers of $X$ such that $\mathscr{U}_n$ refines $\mathscr{U}_{n-1}$, and, at each $x \in X$, $\{\operatorname{St}(x, \mathscr{U}_n) \mid n = 1, 2, \ldots\}$ is a nhood base. A space having a development is called *developable.* A *Moore space* is a regular, Hausdorff space having a development.

The requirement that $\mathscr{U}_n$ refine $\mathscr{U}_{n-1}$ is not crucial. An otherwise satisfactory sequence without this property can easily be made (the reader should do it!) to give rise to a development.

The *normal Moore space conjecture* states that every normal Moore space is metrizable. Whether or not this is true is an unsolved question; it may, in fact, be unsolvable. See the notes. A related theorem on metrizability of developable spaces can easily be given here. It is the first recorded metrization theorem, due to Alexandroff and Urysohn in 1923.

**23.7 Theorem.** *A $T_0$-space $X$ is metrizable iff it has a development $\mathscr{U}_1, \mathscr{U}_2, \ldots$ with the additional property that whenever $U, V \in \mathscr{U}_n$ and $U \cap V \neq \varnothing$, then $U \cup V \subset W$ for some $W \in \mathscr{U}_{n-1}$.*

*Proof.* Necessity is easy. If $X$ is metrizable, take for $\mathscr{U}_n$ the collection of $1/4^n$ spheres in $X$.

To prove sufficiency, we employ the nhood metrization theorem, 23.5. Let $\mathscr{U}_1, \mathscr{U}_2, \ldots$ be a development for $X$ with the required property. Then easily, for each $n > 1$, we find that if $U \in \mathscr{U}_n$ and $x \in U$, then $\operatorname{St}(U, \mathscr{U}_n) \subset \operatorname{St}(x, \mathscr{U}_{n-1})$. Now for $n = 1, 2, \ldots$ and $x \in X$, define $U_{xn} = \operatorname{St}(x, \mathscr{U}_n)$. Then we need only verify properties a) and b) of Theorem 23.5.

a) If $y \in U_{xn}$, then for some $V \in \mathscr{U}_n$, $x \in V$ and $y \in V$. But then

$$U_{yn} = \operatorname{St}(y, \mathscr{U}_n) \subset \operatorname{St}(V, \mathscr{U}_n) \subset \operatorname{St}(x, \mathscr{U}_{n-1}) = U_{xn-1},$$

using the comment above for the next-to-the-last step.

b) If $U_{xn} \cap U_{yn} \neq \emptyset$, then for some $U, V \in \mathcal{U}_n$, $U \cap V \neq \emptyset$. But then $U \cup V \subset W$ for some $W \in \mathcal{U}_{n-1}$, and hence $y \in \text{St}(x, \mathcal{U}_{n-1}) = U_{xn-1}$. Thus, if $y \notin U_{xn-1}$, then $U_{yn} \cap U_{xn} = \emptyset$. ■

Each of the metrization theorems so far given possesses unique advantages. The Urysohn theorem is an indispensable part of the theory of separable metric spaces; 23.4 (which is a variant of the "uniform metrization theorem") will play a key role in building a theory of uniform spaces in Chapter 8; the nhood metrization theorem, in addition to having a unique visual appeal, is clearly well suited to dealing with spaces whose primary description is a nhood description; the Alexandroff–Urysohn theorem is historically important and takes on additional significance in investigation of questions involving metrization of Moore spaces.

The last three named theorems are general, in the sense that they apply, unlike the Urysohn theorem, to any topological space. The next theorem, which also possesses this advantage, is usually called *the* "general metrization theorem," however, because it alone provides the Urysohn theorem as an easy corollary. It was discovered and proved in the early 1950's by Nagata, Smirnov and in a slightly different form, Bing. Our treatment is essentially Smirnov's. Note the key role played by A. H. Stone's theorem that every metric space is paracompact.

The vehicle for proving the general metrization theorem is, as with the Urysohn theorem, an embedding. This time the "universal space" is a generalization of Hilbert space.

**23.8 Definition.** Let $\tau$ be an infinite cardinal number. The *generalized Hilbert space of weight* $\tau$, $\mathbf{H}^\tau$, is described as follows (compare with 18.7b):

Let $A$ be an index set of cardinal $\tau$. Then $\mathbf{H}^\tau$ consists of all functions $x: A \to \mathbf{R}$ such that

a) $x_a \neq 0$ for at most countably many $a \in A$,

b) $\sum_{a \in A} x_a^2$ converges,

where we are writing $x_a$ instead of $x(a)$. Note that the sum in (b) makes sense, since it is really a countable sum. The distance function in $\mathbf{H}^\tau$ is defined, just as it was in Hilbert space $\mathbf{H}$, by

$$d(x, y) = \sqrt{\sum (x_a - y_a)^2}.$$

Recall that a collection $\mathcal{U}$ of sets in $X$ is $\sigma$-*locally finite* provided $\mathcal{U} = \bigcup_{n=1}^\infty \mathcal{U}_n$ where each $\mathcal{U}_n$ is a locally finite collection.

**23.9 Theorem.** *A topological space is metrizable iff it is $T_3$ and has a $\sigma$-locally finite base.*

*Proof.* Necessity follows from the fact that every metric space is paracompact.

Thus, let $\mathscr{U}_n$ be the cover of $X$ by $1/2^n$ spheres, and let $\mathscr{V}_n$ be a locally finite refinement of $\mathscr{U}_n$. Then $\bigcup \mathscr{V}_n$ is a $\sigma$-locally finite base for $X$. Since every metric space is $T_3$, necessity is proved.

We now prove sufficiency. Let $X$ be a space with a $\sigma$-locally finite base $\mathscr{B} = \bigcup \mathscr{B}_n$. It is apparent that $X$ is paracompact, since every open cover has a $\sigma$-locally finite refinement consisting of basis elements, and hence $X$ is normal.

Next, we show $X$ is perfectly normal. Let $G$ be open in $X$. By regularity, for each $x \in G$, there is a basis element $B_x$ such that $\bar{B}_x \subset G$. Let

$$B_n = \bigcup \{\bar{B}_x \mid B_x \in \mathscr{B}_n\}.$$

Then $B_n$ is the union of a locally finite collection of closed sets and hence closed, and $G = \bigcup_{n=1}^{\infty} B_n$. Thus every open set in $X$ is an $F_\sigma$, so $X$ is perfectly normal. (See 15C.1.)

Now each basis element $B_{n\alpha}$ has the property that for some continuous $f_{n\alpha}: X \to \mathbf{I}$, $B_{n\alpha} = \{x \in X \mid f_{n\alpha}(x) \neq 0\}$, by perfect normality. Let $\tau$ be the cardinal number of the base $\mathscr{B}$, and let $\mathbf{H}^\tau$ be the generalized Hilbert space of weight $\tau$; we can use the pairs $n$, $\alpha$ as the index set $A$ in the definition of $\mathbf{H}^\tau$. Define $F: X \to \mathbf{H}^\tau$ by giving coordinate functions $F_{n\alpha}(x) = [F(x)]_{n\alpha}$ as follows:

$$F_{n\alpha}(x) = \frac{1}{(\sqrt{2})^n} \frac{f_{n\alpha}(x)}{\sqrt{1 + \sum_\beta f_{n\beta}^2(x)}}.$$

The denominator here makes sense because for any $x$ in $X$, $x \in B_{n\alpha}$ for only finitely many $B_{n\alpha} \in \mathscr{B}_n$, so that $f_{n\alpha}(x) \neq 0$ for only finitely many $\alpha$, if $n$ is fixed. This also shows that $F_{n\alpha}(x) \neq 0$ for only countably many pairs $n$, $\alpha$. Since

$$\sum_\alpha F_{n\alpha}^2(x) < \frac{1}{2^n},$$

we find that

$$\sum_{n,\alpha} F_{n\alpha}^2(x) < \sum_n \frac{1}{2^n} = 1$$

so that $F(x)$ is indeed an element of $\mathbf{H}^\tau$. We claim $F$ is a homeomorphism of $X$ with a subset of $\mathbf{H}^\tau$.

First, if $x \neq y$ in $X$, then for some $B_{n\alpha} \in \mathscr{B}$, $x \in B_{n\alpha}$ and $y \notin B_{n\alpha}$. Then $f_{n\alpha}(x) \neq 0$ and $f_{n\alpha}(y) = 0$, from which it follows that $F_{n\alpha}(x) \neq F_{n\alpha}(y)$, and hence $F(x) \neq F(y)$. Thus, $F$ is one–one.

Continuity is harder. First, note that each $F_{n\alpha}$ is continuous as a map of $X$ into $\mathbf{R}$. Now let $x_0 \in X$ and $\epsilon > 0$ be given. Choose $N$ so large that

$$\sum_{n=N+1}^{\infty} \frac{1}{2^n} < \frac{\epsilon^2}{4}.$$

Now let $U$ be a nhood of $x_0$ meeting only finitely many $B_{n\alpha}$ for $n \leq N$; say, $U$

meets $B_{n_1\alpha_1}, \ldots, B_{n_k\alpha_k}$. Let $V \subset U$ be a nhood of $x_0$ such that for $x \in V$,

$$|F_{n_i\alpha_i}(x) - F_{n_i\alpha_i}(x_0)| < \frac{\epsilon}{\sqrt{2k}}$$

for $i = 1, \ldots, k$. Now for $x \in V$ and any pair $n$, $\alpha$ other than $n_i$, $\alpha_i$ for $i = 1, \ldots, k$, we have $F_{n\alpha}(x) = F_{n\alpha}(x_0) = 0$. Hence, for $x \in V$,

$$\sum_{n \leq N} \sum_{\alpha} |F_{n\alpha}(x) - F_{n\alpha}(x_0)|^2 = \sum_{i=1}^{k} |F_{n_i\alpha_i}(x) - F_{n_i\alpha_i}(x_0)|^2 < \frac{\epsilon^2}{2}.$$

But we also have

$$\sum_{n > N} \sum_{\alpha} |F_{n\alpha}(x) - F_{n\alpha}(x_0)|^2 \leq \sum_{n > N} \sum_{\alpha} (F_{n\alpha}^2(x) + F_{n\alpha}^2(x_0))$$

$$< \sum_{n > N} \left( \frac{1}{2^n} + \frac{1}{2^n} \right) = 2 \sum_{n > N} \frac{1}{2^n} < \frac{\epsilon^2}{2}$$

by choice of $N$. It now follows that, for $x \in V$,

$$\sum_{n,\alpha} |F_{n\alpha}(x) - F_{n\alpha}(x_0)|^2 < \epsilon^2.$$

Hence, for $x \in V$, $d(F(x), F(x_0)) < \epsilon$, proving continuity of $F$.

Finally, we show $F$ is closed. If $A$ is closed in $X$, we assert $F(A) = \overline{F(A)}$. Suppose $F(x) \notin F(A)$; i.e., $x \notin A$. Then for some $n\alpha$, $x \in B_{n\alpha}$ and $B_{n\alpha} \cap A = \emptyset$. Hence $f_{n\alpha}(x) \neq 0$ and $f_{n\alpha}(A) = 0$. It follows that $F_{n\alpha}(x) \neq 0$ and $F_{n\alpha}(A) = 0$ and then, obviously, $d(F(x), F(A)) > 0$ so that $F(x) \notin \overline{F(A)}$. Thus $\overline{F(A)} \subset F(A)$, so $F(A)$ is closed. ∎

The proof of the general metrization theorem just given is that of Smirnov. Nagata's proof of the same theorem is accomplished by converting a $\sigma$-locally finite base for $X$ to a countable collection of locally finite covers (using perfect normality), proving that a locally finite cover of a normal space is a normal cover, and then applying the uniform metrization theorem (23.4).

## Problems

23A. *Examples on metrizability*

1. The looped line (4D) is metrizable.

2. The scattered line **S** (5C) is not metrizable.

3. The disjoint union of metrizable spaces is metrizable.

4. Let $A$ be any infinite set and for each $\alpha \in A$, let $I_\alpha$ be a copy of **I**. Let $Z$ be the disjoint union of the spaces $I_\alpha$ and let $X$ be the quotient of $Z$ obtained by identifying all the left-hand endpoints. Let $a$ denote the common left-hand endpoint of the spaces $I_\alpha$ in $X$. Does

the following metric:

$$\rho(x, y) = |x - a| + |a - y| \quad \text{if} \quad x \in I_\alpha, y \in I_\beta, \alpha \neq \beta,$$
$$\rho(x, y) = |x - y| \quad \text{if} \quad x, y \in I_\alpha,$$

generate the topology of $X$? The set $X$ with the metric $\rho$ is the *hedgehog space* (of *spininess* $|A|$), and $\rho$ is the *hedgehog metric.* Metrizability of quotient spaces in general is discussed in Exercise 23K.

5. Find a countable Hausdorff space which is not metrizable.

### 23B. *Exercise on normal sequences and covers*

1. Let $\mathscr{G}$ be an open cover of $X$. If there is a normal open cover $\{U_\lambda \mid \lambda \in \Lambda\}$ of $X$ such that, for each $\lambda$, $\{G \cap U_\lambda \mid G \in \mathscr{G}\}$ is a normal cover of $U_\lambda$, then $\mathscr{G}$ is a normal cover of $X$.

2. If $\mathscr{G}$ is a normal cover of $X$, then $\{G \times Y \mid G \in \mathscr{G}\}$ is a normal cover of $X \times Y$.

3. Every locally finite open cover of a $T_4$-space is a normal cover.

4. If a normal sequence $\mathscr{U}_1, \mathscr{U}_2, \ldots$ is compatible with $X$, then $\bigcup \mathscr{U}_n$ is a base for $X$. The converse fails.

### 23C. *Metrizability of* $X^*$

The following are all equivalent, for a locally compact metric space $X$:

a) $X$ is separable.

b) $X = \bigcup_{n=1}^{\infty} K_n$, where $K_n$ is compact and $K_n \subset \operatorname{Int} K_{n+1}$.

c) $X^*$ is metrizable.

(Recall that $X^*$ is the one-point compactification of $X$.)

### 23D. *Metrizability of* $\beta X$

1. If $p \in X$ has a countable base of nhoods in $X$, it has a countable base of nhoods in $\beta X$.

2. No point in $\beta X - X$ can be a $G_\delta$ in $\beta X$. [Otherwise $\{p\}$ is a zero set in $\beta X$. Use the resulting function $f$ to construct disjoint zero sets in $X$ whose closures in $\beta X$ are not disjoint. Then refer to 19J.4.]

### 23E. *Urysohn's theorem*

1. Prove that $X$ is $T_3$ and second countable iff $X$ is a separable metric space by appealing to the general metrization theorem (23.9). (Compare with 23.1.)

2. Give an example of a second countable Hausdorff space which is not metrizable (thus showing regularity is needed in 23.1).

3. Show that a regular Lindelöf space need not be metrizable (so that second countability cannot be weakened in 23.1). Recall that a regular separable space need not even be normal (16G) so that improvement of 23.1 in this direction is not possible either.

### 23F. *Semimetrization*

A *semimetric* on a set $X$ is a function $d: X \times X \to \mathbf{R}$ satisfying the requirements: for all $x$ and $y$ in $X$,

a) $d(x, y) = 0$ iff $x = y$, and

b) $d(x, y) = d(y, x)$.

One can define open sets in a semimetric space just as if $d$ were a metric, and the result is a topology on $X$. The question then arises: which topological spaces are *semimetrizable*?

1. $X$ is semimetrizable iff at each $x \in X$, a countable nhood base $\{U_{xn} \mid n = 1, 2, \ldots\}$ can be found such that $y \in U_{xn} \Leftrightarrow x \in U_{yn}$.

2. Not every first countable space is semimetrizable.

## 23G. *Piecewise metrizability*

1. If a Tychonoff space $X$ is the union of a locally finite collection of closed, metrizable subspaces, then $X$ is metrizable.

2. If a $T_4$-space $X$ is the union of any locally finite collection of metrizable subspaces, then $X$ is metrizable. [Use 15.10.]

3. A locally metrizable, Hausdorff space is metrizable iff it is paracompact. (Thus, every paracompact $n$-manifold is metrizable.)

4. A space can be the union of two metrizable subsets without being metrizable. [Let $X$ be the one-point compactification of an uncountable discrete space.] For further results, see the notes.

5. If $X$ is $T_2$ and the union of two compact metrizable subsets, then $X$ is metrizable. (This is the *addition theorem* for compacta. It is also true for countable unions; see the notes.)

## 23H. *The nhood metrization theorem*

Provide examples to show that neither one of the conditions of 23.5 is by itself sufficient to ensure metrizability.

## 23I. *The general metrization theorem*

1. Exhibit a specific $\sigma$-locally finite base for $\mathbf{R}$.

2. Show that regularity is needed in the general metrization theorem; that is, that $T_3$ cannot be replaced by $T_2$ in 23.9. [See 23E.2.]

## 23J. *Frink's metrization theorem*

Use the uniform metrization theorem (23.4) to prove the following metrization theorem, due to A. H. Frink.

A $T_1$-space $X$ is metrizable iff there is a nhood base $\{U_{xn} \mid n \in N\}$ at each $x \in X$ such that:

a) $U_{x1} \supset U_{x2} \supset \cdots$,

b) for each $n \in \mathbf{N}$, there is some $m > n$ such that $U_{xm} \cap U_{ym} \neq \varnothing \Rightarrow U_{xm} \subset U_{yn}$.

## 23K. *Metrizability of quotient spaces*

Let $f$ be a closed continuous map of a metric space $M$ onto a space $Y$.

1. If $p \in Y$ has a countable nhood base, then $f^{-1}(p)$ has compact frontier. [Let $\{V_n \mid n = 1, 2, \ldots\}$ be a countable nhood base at $p$. If $\mathrm{Fr}\,(f^{-1}(p))$ is not compact, let $(x_n)$ be a sequence in $\mathrm{Fr}\,(f^{-1}(p))$ with no cluster point. For each $n$, find $y_n \in f^{-1}(V_n) - \mathrm{Fr}\,(f^{-1}(p))$ within $1/n$ of $x_n$. Then $E = \{y_n \mid n = 1, 2, \ldots\}$ is closed and hence $f(E)$ is closed in $Y$. But $y \in \overline{f(E)} - f(E)$.]

2. Suppose for each $p \in Y$, $f^{-1}(p)$ has compact frontier. Let $F_p = [f^{-1}(p)]$ and define sets $U_{pn}$ as follows:

$$W_{pn} = \{x \in X \mid d(x, \text{Fr } F_p) < 1/n\}$$
$$V_{pn} = W_{pn} \cup \text{Int } F_p$$
$$U_{pn} = f(V_{pn}) = Y - f(X - W_{pn}).$$

Verify that $\{U_{pn} \mid n = 1, 2, \ldots\}$ is a nhood base at $p \in Y$ satisfying the conditions of 23J.

3. The following are equivalent:

a) $Y$ is metrizable,
b) $Y$ is first countable,
c) For each $p \in Y$, $f^{-1}(p)$ has compact frontier.

### 23L. *Metrizability of continuous images*

According to 23.2, the Hausdorff continuous image of a compact metric space is metrizable. This result cannot be improved by weakening the conditions on the space, according to part 2 below.

1. Every closed continuous image of a metric space $X$ in a Hausdorff space is metrizable iff the set of accumulation points of $X$ is compact. [Use 23K.3.]

2. Every continuous image of a metric space $X$ in a Hausdorff space is metrizable iff $X$ is compact. [Use part 1.]

## 24 Complete metric spaces

Compact spaces enjoy nice properties, but compactness is itself a strong property, tailored to overcome the weak structure available in a topological space. When a metric is present, it is possible to gain many of the advantages of compactness with a weaker property, tailored to the metric structure. As with compactness, it provides for the existence of certain limits and, as with compactness, this makes it interesting to "existential" analysts.

**24.1 Definition.** A sequence $(x_n)$ in a metric space $(M, \rho)$ is *Cauchy* (or, where confusion is possible, $\rho$-*Cauchy*) iff for each $\epsilon > 0$, there is some positive integer $N$ such that $\rho(x_n, x_m) < \epsilon$ whenever $m, n \geq N$.

It is apparent that every convergent sequence in $(M, \rho)$ is Cauchy. For if $\epsilon > 0$ is given, as soon as the terms of the sequence pass the point beyond which they are within $\epsilon/2$ of their limit, they will all be within $\epsilon$ of each other, by the triangle inequality.

A Cauchy sequence need not converge, however. For example, the sequence $(1/n)$ is Cauchy in the open interval $(0, 1)$ with its usual metric, but fails to converge (in that space). In some metric spaces, every Cauchy sequence converges. This is true of $\mathbf{R}$ with its usual metric, for example, by the classical Cauchy criterion for convergence.

**24.2 Definition.** A metric space $(M, \rho)$ is *complete* iff every Cauchy sequence in $M$ converges. We also say $\rho$ is a *complete metric* for $M$. A topological space $X$

is *completely metrizable* iff there is a complete metric for $X$ which generates its topology. Thus $X$ is completely metrizable iff it is homeomorphic to some complete metric space.

Completeness is a property of metric spaces, complete metrizability is a property of topological spaces. For example, $(0, 1)$ with its usual metric is not a complete metric space, but it is completely metrizable since it is homeomorphic to the complete space $\mathbf{R}$. Some metrizable spaces are not completely metrizable; one example is the space $\mathbf{Q}$ of rationals, as we will see in 25A.4.

In showing $(0, 1)$ was not complete, we produced a nonconvergent sequence which was Cauchy because it did converge in a larger space. The next theorem shows that all examples of noncomplete spaces have the same property, by providing the fundamental result that every metric space has a *completion*; that is, a complete space containing it as a dense subspace. We require a definition.

**24.3 Definition.** Metric spaces $(M, \rho)$ and $(N, \sigma)$ are *isometric* iff there is a one–one function $f$ of $M$ onto $N$ such that $\sigma(f(x), f(y)) = \rho(x, y)$, for all $x$ and $y$ in $M$. The mapping $f$ is called an *isometry*.

> ***24.4 Theorem.*** *Every metric space $M$ can be isometrically embedded as a dense subset of a complete space. The resulting completion is unique up to an isometry which leaves $M$ pointwise fixed.*

*Proof.* The details of the following proof should be familiar. The process used is entirely analogous to the construction of the real line as a set of equivalence classes of Cauchy sequences of rational numbers.

Let $(M, \rho)$ be a metric space, $\mathscr{M}$ the set of all Cauchy sequences in $M$. Note that if $(x_n), (y_n) \in \mathscr{M}$, then $(\rho(x_n, y_n))$ forms a Cauchy sequence in $\mathbf{R}$ and hence converges. Thus we can define

$$d((x_n), (y_n)) = \lim_{n \to \infty} \rho(x_n, y_n).$$

Moreover, $d$ turns out to be a pseudometric. Let $(\mathscr{M}^*, d^*)$ be the associated metric space (see 2C.2). For reference, $\mathscr{M}^*$ has for points the equivalence classes $[(x_n)]$ consisting of all $(y_n)$ such that $\lim_{n\to\infty} \rho(x_n, y_n) = 0$, and $d^*$ is defined on $\mathscr{M}^*$ by

$$d^*([(x_n)], [(y_n)]) = \lim_{n \to \infty} \rho(x_n, y_n).$$

Now the map $g(x) = [(x, x, \ldots)]$ is an isometry of $M$ onto a dense subspace of $\mathscr{M}^*$. Moreover, $\mathscr{M}^*$ is complete (an easy diagonal process shows every Cauchy sequence converges), so $\mathscr{M}^*$ is the desired completion of $M$.

Uniqueness of $\mathscr{M}^*$ is easy. If $\mathscr{M}'$ is any complete space containing $M$ as a dense subspace, then each point $x$ in $\mathscr{M}'$ is reached by a sequence $(x_n)$ in $M$. Define $f: \mathscr{M}' \to \mathscr{M}^*$ by $f(x) = [(x_n)]$, where $(x_n)$ is a sequence in $M$ (necessarily Cauchy!) converging to $x$. Then $f$ is well defined and preserves distances, and if $x \in M$, $f(x) = [(x, x, \ldots)]$, so $f$ leaves $M$ pointwise fixed. ■

Perusing the proof of 24.4, we obtain the following corollary.

**24.5 Corollary.** *Every pseudometric space has a pseudometric completion; that is, can be isometrically embedded as a dense subset of a complete pseudometric space.*

**24.6 Examples.** a) The completion of $(0, 1)$ is $[0, 1]$.

b) The completion of the space of rationals $\mathbf{Q}$ is the real line.

c) Let $X$ be any topological space, $C_{00}(X)$ the space consisting of all bounded continuous real-valued functions $f: X \to \mathbf{R}$ which are 0 except on some compact subset of $X$; i.e., $C_{00}(X)$ is all real-valued continuous functions with compact support. Define

$$\rho(f, g) = \sup_{x \in X} |f(x) - g(x)|.$$

Then $C_{00}$ with this distance function is a metric space, but is not complete. Its completion is the set of functions $C_0(X)$ which are *small off compact sets*; i.e., $C_0(X) = \{f: X \to \mathbf{R} \mid f$ continuous and for each $\epsilon > 0$ there is a compact $K_f \subset X$ such that $|f(x)| < \epsilon$ for all $x \notin K_f\}$.

Next on our program is the development of Lavrentieff's theorem (24.9), one of the more important embedding theorems useful in dealing with complete metric spaces.

**24.7 Definition.** Suppose $f: A \to M$, where $M$ is a metric space and $A$ is a subset of $X$. We define osc $(f, U)$, the *oscillation of $f$ on $U$*, for any $U \subset X$, as follows:

$$\operatorname{osc}(f, U) = \sup \{\rho(f(x), f(y)) \mid x, y \in U \cap A\},$$

and we accept the convention that osc $(f, U) = \infty$ if $U \cap A$ is empty.

If $X$ is a topological space and $x \in \bar{A}$, we define the *oscillation of $f$ at $x$* to be

$$\operatorname{osc}(f, x) = \inf \{\operatorname{osc}(f, U) \mid U \text{ nhood of } x\}.$$

**24.8 Lemma.** *Let $X$ be a metric space, $Y$ a complete metric space and $A \subset X$. If $f: A \to Y$ is continuous, then $f$ can be extended to a continuous function $f^*: A^* \to Y$, where $A^*$ is a $G_\delta$-set in $X$ and $A \subset A^* \subset \bar{A}$.*

*Proof.* Let $A^* = \{x \in \bar{A} \mid \operatorname{osc}(f, x) = 0\}$. For $x \in A^*$, let $(x_n)$ be any sequence in $A$ converging to $x$. Given $\epsilon > 0$, since osc $(f, x) = 0$, there is some nhood $U$ of $x$ such that osc $(f, U) < \epsilon$. Since $x_n \to x$, there is some $N$ such that $m, n \geq N \Rightarrow x_n, x_m \in U \Rightarrow \rho(f(x_n), f(x_m)) < \epsilon$. Thus, $(f(x_n))$ is a Cauchy sequence in $Y$ and since $Y$ is complete, $f(x_n) \to y$ for some $y$. Now define $f^*(x) = y$. The reader should check that this definition of $f^*: A^* \to Y$ is independent of the choice of the sequence $(x_n)$ converging to $x$, and that $f^*$ as defined is continuous.

It remains, then, to show that $A^*$ is a $G_\delta$-set in $X$. But if we let

$$A_n = \{x \in \bar{A} \mid \operatorname{osc}(f, x) < 1/n\},$$

then $A_n$ is open in $\bar{A}$. For if $y \in A_n$, then there is some open nhood $U$ of $y$ such that $\operatorname{osc}(f, U) < 1/n$, and it is clear that $U \cap \bar{A} \subset A_n$. Since $A^* = \bigcap_{n=1}^{\infty} A_n$, $A^*$ is a $G_\delta$-set in $\bar{A}$ and thus in $X$. ■

**24.9 Theorem.** (Lavrentieff) *If $X$ and $Y$ are complete metric spaces and $h$ is a homeomorphism of $A \subset X$ onto $B \subset Y$, then $h$ can be extended to a homeomorphism $h^*$ of $A^*$ onto $B^*$ where $A^*$ and $B^*$ are $G_\delta$-sets in $X$ and $Y$, respectively, and $A \subset A^* \subset \bar{A}$, $B \subset B^* \subset \bar{B}$.*

*Proof.* Since $h: A \to Y$ is continuous, it can be extended to a continuous map $h^*: A_1 \to Y$, where $A_1 \subset \bar{A}$ and $A_1$ is a $G_\delta$-set in $X$.

Since $h^{-1}: B \to X$ is continuous, it can be extended to a continuous map $g^*: B_1 \to X$, where $B_1 \subset \bar{B}$, and $B_1$ is a $G_\delta$-set in $Y$.

Let $A^* = \{x \in A_1 \mid h^*(x) \in B_1\}$. This is the inverse image of a $G_\delta$-set, and thus a $G_\delta$-set, in $A_1$ and hence in $X$. We claim $h^* \mid A^*$ is a homeomorphism of $A^*$ onto the $G_\delta$-set $B^* = \{x \in B_1 \mid g^*(x) \in A_1\}$ in $Y$. We can prove this by showing

a) $h^*(A^*) = B^*$.

b) $(h^*)^{-1} = g^*$.

If $x \in A^*$, then $h^*(x) \in B_1$ and $g^*(h^*(x)) = x \in A_1$, so $h^*(x) \in B^*$. Thus $h^*(A^*) \subset B^*$. If $y \in B^*$, then $g^*(y) \in A_1$ and $h^*[g^*(y)] = y$, so $y \in h^*(A^*)$. Thus $h^*(A^*) = B^*$. Moreover, since $h^*[g^*(y)] = y$, for each $y \in B^*$, and $g^*[h^*(x)] = x$, for each $x \in A^*$, $h^*$ and $g^*$ are inverses. Thus $h^*$ is a homeomorphism of $A^*$ onto $B^*$. ■

Making good use of Lavrentieff's theorem, we turn to the question of manufacturing new complete spaces from old. The product theorem is not difficult, but the full force of Lavrentieff's theorem will be needed to obtain a pleasant subspace theorem. We can easily prove a weak subspace theorem now.

**24.10 Theorem.** *A closed subset $A$ of a complete metric space $(M, \rho)$ is complete.*

*Proof.* If $(a_n)$ is a $\rho$-Cauchy sequence in $A$, it is also Cauchy in $M$ and hence converges, say to $a$. But $A$ is closed, so $a \in A$. Thus every Cauchy sequence in $A$ converges (to a point in $A$). ■

**24.11 Theorem.** *Suppose $X_n$ is a nonempty metric space for $n = 1, 2, \ldots$. Then $\prod X_n$ is completely metrizable iff each $X_n$ is completely metrizable.*

*Proof.* $\Rightarrow$: Pick $a_i \in X_i$, $i = 1, 2, \ldots$. Then $X_n$ is isometric to the closed subset

$$X_n^* = \{(x_1, x_2, \ldots) \in \prod X_n \mid x_i = a_i \text{ except for } i = n\}$$

of $\prod X_n$, and it follows that $X_n$ is completely metrizable.

$\Leftarrow$: Suppose $\rho_n$ is the complete metric for $X_n$, $n = 1, 2, \ldots$. The bounded metric

$$\rho_n^* = \min(\rho_n, 1)$$

already introduced in 22.2 as equivalent to $\rho_n$ is easily verified to be complete (24A.3). Define $\rho$ on $\prod X_n$ by

$$\rho(x, y) = \sum_{n=1}^{\infty} \frac{\rho_n^*(x_n, y_n)}{2^n}.$$

We know that this gives a compatible metric on $\prod X_n$, so only completeness remains to be checked. Suppose $x^1, x^2, \ldots$ is a $\rho$-Cauchy sequence in $\prod X_n$. Then for each $i$, $x_i^1, x_i^2, \ldots$ is a $\rho_i^*$-Cauchy sequence in $X_i$, and hence converges, say to $y_i$. We assert $x^1, x^2, \ldots$ converges to $y = (y_1, y_2, \ldots)$. Let $\epsilon > 0$ be given. Choose $N$ so large that $\sum_{n=N+1}^{\infty} (1/2^n) < \epsilon/2$. Then pick $N_\epsilon$ so large that when $n > N_\epsilon$,

$$\rho_i^*(x_i^n, y_i) < \frac{\epsilon \cdot 2^i}{2N}$$

for $i = 1, \ldots, N$. Then for $n > N_\epsilon$ we find

$$\begin{aligned} \rho(x^n, y) &= \sum_{i=1}^{\infty} \frac{\rho_i^*(x_i^n, y_i)}{2^i} \\ &\leq \sum_{i=1}^{N} \frac{\rho_i^*(x_i^n, y_i)}{2^i} + \frac{\epsilon}{2} \\ &< \sum_{i=1}^{N} \frac{\epsilon \cdot 2^i}{2N \cdot 2^i} + \frac{\epsilon}{2} = \frac{\epsilon}{2} + \frac{\epsilon}{2} = \epsilon \end{aligned}$$

so that $(x^n)$ converges to $y$, as claimed. ■

We are now ready for the subspace theorem. The first part is due to Alexandroff, the second to Mazurkiewicz. Both are classical results from the 1920's.

***24.12 Theorem.*** *A $G_\delta$-set in a complete space is completely metrizable. Conversely, if a subset $A$ of a metric space $M$ is completely metrizable, it is a $G_\delta$-set.*

*Proof.* First, suppose $G$ is open in the complete space $(M, \rho)$. Define $f(x) = 1/[\rho(x, M - G)]$ for each $x \in G$. Then $f$ is continuous on $G$ (24E). Now define

$$\rho^*(x, y) = \rho(x, y) + |f(x) - f(y)|$$

for $x, y \in G$. Then $\rho^*$ is a metric on $G$, and if a sequence $(x_n)$ in $G$ is $\rho^*$-Cauchy, then it is $\rho$-Cauchy. Also, for any $\epsilon > 0$ there is some integer $N$ such that

$$n \geq N \Rightarrow |f(x_N) - f(x_n)| < \epsilon \Rightarrow \left| \frac{1}{\rho(x_n, M - G)} - \frac{1}{\rho(x_N, M - G)} \right| < \epsilon.$$

An easy computation with this last inequality shows that $\rho(x_n, M - G)$ must be bounded away from 0; thus, for some $\delta > 0$,

$$(x_n) \subset M_\delta = \{x \in M \mid \rho(x, M - G) \geq \delta\}.$$

But $M_\delta$ is closed in $M$ and thus complete, and $(x_n)$ is $\rho$-Cauchy, so $(x_n)$ converges in $M_\delta$ and hence in $G$. Thus, every $\rho^*$-Cauchy sequence converges and we have established that $G$ is completely metrizable, provided $\rho^*$ gives the same topology as does $\rho$ on $G$. This is left as an exercise (24E).

Now suppose $H$ is a $G_\delta$-set in $M$, say $H = \bigcap_{n=1}^{\infty} H_n$, where each $H_n$ is open. From the above, $H_n$ is completely metrizable, for each $n$, and hence $\prod H_n$ is completely metrizable. But the set

$$\Delta = \{(x_1, x_2, \ldots) \in \prod H_n \mid x_1 = x_2 = \cdots\}$$

is closed in $\prod H_n$ and thus completely metrizable, and by Exercise 24I, the map

$$f(x) = (x, x, \ldots)$$

is a homeomorphism of $H$ with $\Delta$. Thus $H$ is completely metrizable.

Conversely, suppose $A$ is a completely metrizable subspace of a metric space $M$, and let $\hat{M}$ denote the completion of $M$. The inclusion $i\colon A \to \hat{M}$ is a homeomorphism, and thus, by Lavrentieff's theorem, has an extension to a homeomorphism of $G_\delta$-sets. But $i$ can have only itself for an extension so evidently $i(A) = A$ must itself be a $G_\delta$ in $\hat{M}$. Since the intersection with $M$ of a $G_\delta$ in $\hat{M}$ is a $G_\delta$, $A$ is a $G_\delta$ in $M$. ∎

Thus, the completely metrizable spaces are precisely those metric spaces which are $G_\delta$-sets in whatever metric space they are embedded ("absolute $G_\delta$-sets"). Next we see that they retain this property, to a certain extent, in nonmetric embeddings.

**24.13 Theorem.** *For a metric space $X$ the following are all equivalent:*

a) *$X$ is completely metrizable,*

b) *$X$ is a $G_\delta$ in its completion $\hat{X}$,*

c) *$X$ is a $G_\delta$ in every metric embedding,*

d) *$X$ is a $G_\delta$ in $\beta X$,*

e) *$X$ is a $G_\delta$ whenever densely embedded in a Tychonoff space.*

*Proof.* The equivalence of (a), (b), and (c) has already been established in Theorem 24.12.

c) $\Rightarrow$ d): Let $\rho$ be a bounded metric on $X$ compatible with the topology. For each $x \in X$, let $\varphi_x\colon X \to \mathbf{R}$ be the function

$$\varphi_x(y) = \rho(x, y).$$

This is a bounded continuous real-valued function on $X$ so by the mapping property for $\beta X$, there is an extension $\bar{\varphi}_x$ of $\varphi_x$ to all of $\beta X$. Define $\rho^*$ on $\beta X$ by

$\rho^*(a, b) = \inf_{x \in X} \{|\bar{\varphi}_x(a) - \bar{\varphi}_x(b)|\}$. Then $\rho^*$ is a pseudometric on $\beta X$, for

i) $\rho^*(a, a) = 0$
ii) $\rho^*(a, b) = \rho^*(b, a)$
iii) if $a, b, c \in \beta X$, then

$$\begin{aligned}\rho^*(a, c) &= \inf_{x \in X} |\bar{\varphi}_x(a) - \bar{\varphi}_x(c)| \\ &\leq \inf_{y \in X} |\bar{\varphi}_y(a) - \bar{\varphi}_y(b)| + \inf_{z \in X} |\bar{\varphi}_z(b) - \bar{\varphi}_z(c)| \\ &= \rho^*(a, b) + \rho^*(b, c).\end{aligned}$$

Moreover, the restriction of $\rho^*$ to $X$ is $\rho$, and $\rho^*$ is a "continuous pseudometric" on $\beta X$; that is, the topology it induces is weaker than the usual topology on $\beta X$.

Now perform the usual metric identification on $(\beta X, \rho^*)$. The result is a metric space $K$ which contains $X$ ($X \subset \beta X$ is not affected by the identification since $\rho^*$ is already a metric there). Let $h: \beta X \to K$ be the identification map. Now $X$ is a $G_\delta$ in $K$, and hence $h^{-1}(X)$ is a $G_\delta$ in $(\beta X, \rho^*)$ and hence in $\beta X$. But $h^{-1}(X) = X$.

d) $\Rightarrow$ e): Suppose $X = \bigcap G_n$, each $G_n$ open in $\beta X$, and let $f: X \to Y$ be an embedding of $X$ as a dense subset of a Tychonoff space $Y$. Then $f$ has an extension $f^\beta: \beta X \to \beta Y$. Consider the sets $\beta Y - f^\beta(\beta X - G_n) = H_n$, for $n = 1, 2, \ldots$. They are open, since each $f^\beta(\beta X - G_n)$ is compact and, moreover, for each $n$, $X \subset (f^\beta)^{-1}(H_n) \subset G_n$ so that

$$X = \bigcap (f^\beta)^{-1}(H_n) = (f^\beta)^{-1}(\bigcap H_n).$$

Thus $f^\beta(X) = f^\beta[(f^\beta)^{-1}(\bigcap H_n)] = \bigcap H_n$, so $X$ is a $G_\delta$ in $\beta Y$ and hence in $Y$. Note that we did make use of the fact that $X$ was dense in $Y$ in our tacit use of the assumption that $\beta Y$ was a compactification of $X$, so that $f^\beta$ was onto.

e) $\Rightarrow$ c): If $X$ is embedded in a metric space $M$, then by (e) $X$ is a $G_\delta$ in $\bar{X}$ and $\bar{X}$ is, of course, a $G_\delta$ in $M$. It follows that $X$ is a $G_\delta$ in $M$. ∎

We conclude this section with an important fixed point theorem for complete spaces.

**24.14 Definition.** If $f: X \to X$, a *fixed point* of $f$ is a point $x \in X$ such that $f(x) = x$.

**24.15 Definition.** A map $f: X \to X$, where $(X, d)$ is a metric space, is *d-contractive* provided $d(fx, fy) \leq \alpha \cdot d(x, y)$ for some $\alpha < 1$ and all pairs $(x, y)$ in $X \times X$.

***24.16 Theorem.*** (Banach) *If $X$ is complete in the metric $d$ and $f: X \to X$ is d-contractive, then $f$ is continuous and has precisely one fixed point.*

*Proof.* That $f$ is continuous is clear (any distance decreasing map is continuous). If $x, y$ are both fixed points of $f$ in $X$, then $f(x) = x$, $f(y) = y$ so

$$d(x, y) = d(fx, fy) \leq \alpha \cdot d(x, y),$$

but since $\alpha < 1$, this can be so only if $d(x, y) = 0$; i.e., if $x = y$. Hence $f$ has at most one fixed point.

Choose $x \in X$. Consider the sequence $x_1, x_2, \ldots$ defined as follows: $x_1 = x$, $x_2 = f(x_1), \ldots, x_n = f(x_{n-1})$. Then $x_1, x_2, \ldots$ is a Cauchy sequence and hence converges, say to $x_0$. We claim $x_0$ is the required fixed point. In fact, since $x_n \to x_0$ and $f$ is continuous, we have $f(x_n) \to f(x_0)$. But the sequence $f(x_1), f(x_2), \ldots$ is just $x_2, x_3, \ldots$ so that $x_n \to f(x_0)$. It follows that $f(x_0) = x_0$, as claimed. ■

Fixed-point theorems, such as the one just given, are useful in proving certain existence theorems in differential and integral equations. One example is given in 24L.

### Problems

#### 24A. *Examples on completeness and completion*

1. Hilbert space **H** (18.7) is complete.
2. The completion of $C_{00}(X)$ is $C_0(X)$ (see 24.6c).
3. If $\rho$ is a complete metric on $X$, so is the metric $\rho^*$ defined by

$$\rho^*(x, y) = \min \{1, \rho(x, y)\}.$$

4. If $I$ is any closed interval in **R**, the space $C^*(I)$ of bounded continuous functions on $I$, with the sup metric $\rho(f_1, f_2) = \sup_{x \in I} |f_1(x) - f_2(x)|$, is complete.
5. The space **P** of irrationals is completely metrizable.

#### 24B. *Totally bounded metric spaces*

A metric space $M$ is *totally bounded* iff for each $\epsilon > 0$, a finite number of $\epsilon$-disks will cover $M$.

1. Every totally bounded metric space is bounded. The converse fails.
2. A metric space is separable iff it is homeomorphic to a totally bounded metric space.
3. A metric space is totally bounded iff each sequence has a Cauchy subsequence.
4. A metric space is compact iff it is complete and totally bounded.

The results of this exercise, particularly 3 and 4, have generalizations to uniform spaces. See Section 39.

#### 24C. *Equivalent conditions for completeness*

In a metric space $(X, d)$, define the *diameter* of $A \subset X$ to be $\delta(A) = \sup \{d(x, y) \mid x, y \in A\}$.

1. The following are equivalent:
   a) $X$ is complete.
   b) each decreasing sequence $C_1 \supset C_2 \supset \cdots$ of closed sets with $\delta(C_n) \to 0$ has nonempty intersection.
   c) each infinite totally bounded (24B) subset has an accumulation point.

2. The condition that $\delta(C_n) \to 0$ in b) above is necessary.

3. A metrizable space is compact iff it is complete in every compatible metric. [Use 22E.4 for sufficiency.]

## 24D. *Completion*

Check the details in the proof of 24.4. Specifically:

1. $d$, as defined on $\mathscr{M}$, is a pseudometric.

2. The map $g(x) = [(x, x, \ldots)]$ is an isometry of $M$ with a subspace of $\mathscr{M}^*$.

3. If $\mathscr{M}'$ is a complete space containing $M$ as a dense subspace, for each $x \in \mathscr{M}'$, let $(x_n) \subset M$ be a sequence converging to $x$ and define $f(x) = [(x_n)]$. Verify that $f$ is an isometry of $\mathscr{M}'$ with $\mathscr{M}^*$, such that $f(z) = z$ for each $z \in M$ (i.e., $f(z) = g(z)$, see part 2).

## 24E. *Equivalent metrics on open subsets*

Let $G$ be an open subset of a metric space $(M, \rho)$. Define $f(x) = 1/[\rho(x, M - G)]$, for $x \in G$. Then for $x$ and $y$ in $G$, define $\rho^*(x, y) = \rho(x, y) + |f(x) - f(y)|$.

1. $f$ is continuous.

2. $\rho^*$ is a metric on $G$.

3. $\rho^*$ is equivalent to $\rho$ on $G$.

## 24F. *Topologically complete spaces*

Certain of the assertions in 24.13 do not require metrizability of $X$. In particular: a completely regular space $X$ is a $G_\delta$ in $\beta X$ iff $X$ is a $G_\delta$ in every completely regular space in which it is densely embedded.

A space which is a $G_\delta$ in its Stone–Čech compactification is called *topologically complete.*

## 24G. *Pseudometric completion*

Given a pseudometric space $(X, p)$, we can form the metric identification of the completion of $X$, i.e., $(\hat{X})^*$, or the completion of the metric identification of $X$, i.e., $\widehat{X^*}$.

1. The metric identification of a complete pseudometric space is a complete metric space.

2. $(\hat{X})^*$ is isometric to $\widehat{X^*}$.

## 24H. *Extending maps*

Give an example of a subset $A$ of a metric space $X$ and a continuous map $f$ of $A$ into a complete space $Y$ which cannot be extended to all of $\bar{A}$ (compare with 24.8).

## 24I. *Embedding an intersection in a product*

If $X$ is a topological space and $X_n \subset X$ for each $n = 1, 2, \ldots$ then $\bigcap X_n$ is homeomorphic to $\{(x_1, x_2, \ldots) \in \prod X_n \mid x_1 = x_n, n = 2, 3, \ldots\}$.

## 24J. *Banach spaces*

A normed linear space (2J) is called a *Banach space* iff its norm metric is complete.

A sequence $x_1, x_2, \ldots$ of points in a normed linear space is *summable* iff the associated sequence $x_1, x_1 + x_2, \ldots$ of partial sums converges (in the norm metric) and *absolutely summable* if $\sum \|x_n\| < \infty$.

1. A normed linear space is a Banach space iff every absolutely summable sequence is summable.

2. If $Y$ is any Banach space, the space $L(X, Y)$ of all bounded linear operators (7L) from a normed linear space $X$ to $Y$ is a Banach space. In particular, the dual space $X^*$ of any normed linear space $X$ is a Banach space.

3. $\mathbf{R}^n$, with any of the norms given in 2J.6, is a Banach space. [(a) gives the usual metric, which we already know is complete.]

4. The space $l_2$ of all real sequences $(x_n)$ such that $\sum |x_n|^2 < \infty$, with the norm $\|(x_n)\| = [\sum |x_n|^2]^{1/2}$ is a Banach space. Compare with 18.7(b).

5. The space $s$ of all sequences of real numbers, with the norm

$$\|(x_n)\| = \sum \frac{1}{2^n} \cdot \frac{|x_n|}{1 + |x_n|}$$

is a Banach space.

6. For any topological space $X$, the space $C^*(X)$ of bounded real-valued functions on $X$, with the sup norm $\|f\| = \sup |f(x)|$, is a Banach space.

The *dimension* of a Banach space is the least cardinal of a base for the underlying vector space.

7. If a Banach space is $\aleph_0$-dimensional, it is separable. The spaces described in (3), (4) and (5) above are separable, while (6) need not be.

## 24K. *The irrationals as a product*

The space $\mathbf{P}$ of irrational numbers (with the relative topology in $\mathbf{R}$) is homeomorphic to the product of denumerably many copies of $\mathbf{N}$. [Enumerate the rationals in $\mathbf{R}$ as $r_1, r_2, \ldots$ . Now partition $\mathbf{P}$ into countably many intervals $I_1, I_2, \ldots$ each having rational endpoints and length $\leq \frac{1}{2}$. Also, so determine $I_1, I_2, \ldots$ that one of the endpoints of one of the intervals is $r_1$. Next partition each $I_n$ into countably many intervals $I_{n1}, I_{n2}, \ldots$ each having rational endpoints and length $\leq \frac{1}{4}$. Also, we may so determine these intervals that $r_2$ is an endpoint of some interval of the form $I_{n_1 n_2}$, while $r_1$ is not. Continue, at the $k$th stage using intervals of length $\leq (1/2^k)$ with rational endpoints and requiring that $r_n$ be an endpoint of some interval $I_{n_1 n_2 \ldots n_k}$ while none of $r_1, \ldots, r_{n-1}$ are. For an irrational number $p$, consider the sequence $I_{n_1}, I_{n_1 n_2}, \ldots$ of intervals containing $p$. Using 24C to prove that it is onto, show that the map $f(p) = (n_1, n_2, \ldots)$ is a homeomorphism of $\mathbf{P}$ onto the product $\mathbf{N}^{\aleph_0}$.]

## 24L. *Picard's theorem*

Let $f(x, y)$ be a continuous real-valued function defined on an open set $A$ in the plane containing $(x_0, y_0)$ and suppose $f$ satisfies a Lipschitz condition with respect to $y$:

$$|f(x, y_1) - f(x, y_2)| \leq M\,|y_1 - y_2|.$$

We assert that the integral equation

$$\varphi(x) = y_0 + \int_{x_0}^{x} f(t, \varphi(t))\,dt, \tag{1}$$

which is equivalent to the differential equation

$$\frac{dy}{dx} = f(x, y)$$

$$y(x_0) = y_0,$$

has a unique solution defined on some closed interval $[x_0 - K, x_0 + K]$.

Let $B$ be an open set such that $(x_0, y_0) \in B \subset A$ and such that $|f(x, y) - f(x_0, y_0)| < L$, for some constant $L$, on $B$. Let $K$ be a positive constant $< 1/M$ such that

$$\{(x, y) \mid |x - x_0| \leq K, |y - y_0| \leq KL\} \subset B.$$

For each $\varphi \in C^*[x_0 - K, x_0 + K]$, define $A\varphi$ by

$$A\varphi(x) = y_0 + \int_{x_0}^{x} f(t, \varphi(t))\, dt.$$

1. $A$ maps a closed subspace of $C^*[x_0 - K, x_0 + K]$ into itself.
2. $A$ is a contraction mapping, if $C^*[x_0 - K, x_0 + K]$ is endowed with the sup metric: $\rho(\varphi_1, \varphi_2) = \sup\{|\varphi_1(x) - \varphi_2(x)| \mid x \in [x_0 - K, x_0 + K]\}$.
3. Conclude that the integral equation (1) has a unique solution defined on $[x_0 - K, x_0 + K]$.

### 24M. *Lavrentieff's theorem*

Show that Lavrentieff's theorem (24.9) is equivalent to the following complement to 22E.4: if $A$ is a subset of a metrizable space $X$ and $\rho$ is a compatible metric on $A$, then $\rho$ can be extended to a compatible metric on a $G_\delta$-set in $X$ which contains $A$.

## 25 The Baire theorem

The applications of topology to analysis are usually manifested in the form of an "existence theorem" of some sort and the major share of the work in this direction is borne, directly or indirectly, by two theorems: the Tychonoff theorem and the Baire category theorem. We turn now to the development necessary to introduce the latter.

**25.1 Definition.** $X$ is a *Baire space* iff the intersection of each countable family of dense open sets in $X$ is dense. A set $A \subset X$ is *nowhere dense* in $X$ iff $\text{Int}_X \text{Cl}_X A = \varnothing$. A set $A \subset X$ is *first category* in $X$ iff $A = \bigcup_{n=1}^{\infty} A_n$, where each $A_n$ is nowhere dense in $X$. All other subsets of $X$ are called *second category* in $X$. You can visualize first category sets as being "thin", second category sets as being "thick".

Every Baire space is second category in itself. In fact:

**25.2 Theorem.** *$X$ is second category in itself iff the intersection of every countable family of dense open sets in $X$ is nonempty.*

*Proof.* $\Rightarrow$: Let $G_1, G_2, \ldots$ be dense open sets. Then $X - G_1, X - G_2, \ldots$ are nowhere dense closed sets, so $\bigcup (X - G_i)$ is first category. Hence $X - \bigcup (X - G_i) = \bigcap G_i \neq \emptyset$.

$\Leftarrow$: If $X = \bigcup A_n$, each $A_n$ closed and nowhere dense, then

$$X - \bigcup A_n = \bigcap (X - A_n)$$

is an intersection of open dense sets and hence $\neq \emptyset$, a contradiction. Thus $X \neq \bigcup A_n$ for any sequence of closed nowhere dense sets $A_n$. ■

**25.3 Theorem.** (Baire) *A $G_\delta$-set in a compact Hausdorff space is a Baire space.*

*Proof.* We begin by proving that a compact $T_2$-space is Baire. Let $J_1, J_2, \ldots$ be dense open sets in the compact space $K$, and let $U$ be any open set in $K$. Now $U \cap J_1 \neq \emptyset$ so there is a nonempty open set $V_1$ with $\overline{V}_1 \subset U \cap J_1$ (using regularity). Similarly, a nonempty open set $V_n$ can be found, $n = 2, 3, \ldots$ such that $\overline{V}_n \subset V_{n-1} \cap J_n$. Now $\overline{V}_1, \overline{V}_2, \ldots$ is a decreasing sequence of compact sets, so $\bigcap \overline{V}_n \neq \emptyset$. But $\bigcap \overline{V}_n \subset U \cap (\bigcap J_n)$. Thus every open set $U$ meets $\bigcap J_n$, establishing that $\bigcap J_n$ is dense in $K$. Hence, $K$ is a Baire space.

Now suppose $X = \bigcap H_n$ where each $H_n$ is open in a compact Hausdorff space $K$. We can assume $X$ is dense in $K$ (otherwise replace $K$ by $\mathrm{Cl}_K X$). Now if $G_1, G_2, \ldots$ is a sequence of dense open sets in $X$, then for each $i$, $G_i = J_i \cap X$, where $J_i$ is dense and open in $K$. But now $J_1, H_1, J_2, H_2, \ldots$ is a sequence of dense open sets in $K$, and hence

$$\bigcap (J_i \cap H_i) = (\bigcap J_i) \cap (\bigcap H_i) = (\bigcap J_i) \cap X = \bigcap G_i$$

is dense in $K$ and therefore in $X$. Hence $X$ is a Baire space. ■

**25.4 Corollary.** a) *Every locally compact Hausdorff space is Baire.*

b) *Every completely metrizable space is Baire.*

*Proof.* A locally compact space $X$ is open in $\beta X$ (18.4) and a completely metrizable space $X$ is a $G_\delta$ in $\beta X$ (24.13). ■

The corollary above, rather than 25.3, is often referred to as the Baire theorem since it deals with the spaces of most interest to analysts. Its importance is well documented. Two of the most powerful theorems in functional analysis, the open mapping principle and the uniform boundedness principle (25D) are direct consequences of application of the Baire theorem. The example we give next is typical of an existence theorem based on the Baire theorem; we show that some element of a space must have a given property by showing that the space is second category while the elements which do not have the property form a set of first category.

**25.5 Theorem.** *There is a continuous real-valued function $f$ on* **I** *having a derivative at no point.*

*Proof.* We will show that

a) $C(\mathbf{I})$ = all real continuous functions on $[0, 1]$ is complete in the uniform metric $d$, and

b) the set $\mathscr{E}$ of functions in $C(\mathbf{I})$ which have a derivative somewhere is first category in $C(\mathbf{I})$.

It will follow that $C(\mathbf{I}) - \mathscr{E}$ is nonempty; in fact, it must then be second category.

a) Let $f_1, f_2, \ldots$ be a Cauchy sequence of functions from $C(\mathbf{I})$ in the uniform metric. Then, for each $x \in \mathbf{I}$, $f_1(x), f_2(x), \ldots$ is a Cauchy sequence of real numbers and hence converges, say to $f(x)$. The resulting function $f$ defined on $\mathbf{I}$ is easily verified to be the uniform limit of the continuous functions $f_1, f_2, \ldots$ and thus continuous. Since every Cauchy sequence thus converges, $C(\mathbf{I})$ is complete.

b) Define $\mathscr{E}_n$ for $n = 1, 2, \ldots$ by

$$\mathscr{E}_n = \left\{ f \in C(\mathbf{I}) \,\middle|\, \begin{array}{l} \text{for some } x \in [0, 1 - 1/n], \\ \text{whenever } h \in (0, 1/n], \left| \dfrac{f(x+h) - f(x)}{h} \right| \leq n \end{array} \right\}.$$

If a function $f \in C(\mathbf{I})$ has a derivative at some point, then for some $n$ large enough, $f \in \mathscr{E}_n$; hence $\mathscr{E} \subset \bigcup_{n=1}^{\infty} \mathscr{E}_n$. Thus we can establish (b) by showing each $\mathscr{E}_n$ is closed and has no interior.

1. $\mathscr{E}_n$ has no interior. Given $f \in \mathscr{E}_n$ and $\epsilon > 0$, we will find a continuous function $g$ such that $d(f, g) < \epsilon$ and $g \notin \mathscr{E}_n$; that is, for all $x \in [0, 1 - 1/n]$, there is some $h \in (0, 1/n]$ with

$$\left| \frac{g(x + h) - g(x)}{h} \right| > n.$$

We sketch the construction of $g$. Find a polynomial function $P(x)$ on $[0, 1]$ such that $d(f, P) < \epsilon/2$. Let $M$ be the maximum slope of $P(x)$ in $[0, 1]$, and let $Q(x)$ be a continuous function consisting of straight-line segments of slope $\pm$ $(M + n + 1)$ constrained so that $|Q(x)| < \epsilon/2$.

Define $g(x) = P(x) + Q(x)$. Then $d(f, g) \leq d(f, P) + d(P, g) < \epsilon/2 + \epsilon/2 = \epsilon$ and

$$\begin{aligned} \left| \frac{g(x + h) - g(x)}{h} \right| &= \left| \frac{P(x + h) + Q(x + h) - P(x) - Q(x)}{h} \right| \\ &\geq \left| \frac{Q(x + h) - Q(x)}{h} \right| - \left| \frac{P(x + h) - P(x)}{h} \right|. \end{aligned}$$

But for $x \in [0, 1 - 1/n]$, an $h \in (0, 1/n]$ can be found for which the right-hand side is $\geq (M + n + 1) - M = n + 1$. Thus $g \notin \mathscr{E}_n$.

2. $\mathscr{E}_n$ is closed. The (evaluation) map $e: C(\mathbf{I}) \times \mathbf{I} \to \mathbf{R}$ defined by $e(f, x) = f(x)$ is continuous. It follows easily that, if $h_0$ is a fixed element of $(0, 1/n]$, the map $E_{h0}: C(\mathbf{I}) \times [0, 1 - 1/n] \to \mathbf{R}$ defined by

$$E_{h_0}(f, x) = \left| \frac{f(x + h_0) - f(x)}{h_0} \right|$$

is continuous. Thus $E_{h_0}^{-1}[0, n]$ is closed in $C(\mathbf{I}) \times [0, 1 - 1/n]$. Let

$$D_{h_0} = \{f \in C(\mathbf{I}) \mid (f, x) \in E_{h_0}^{-1}[0, n),$$

for some $x \in [0, 1 - 1/n]\}$.

Then $D_{h_0}$ is closed in $C(\mathbf{I})$. For if $f_m \in D_{h_0}$ for $m = 1, 2, \ldots$ and $f_m \to f$, then the sequence $(x_m)$ in $[0, 1 - 1/n]$ such that $(f_m, x_m) \in E_{h_0}^{-1}[0, n]$ has a cluster point $x$; easily, $(f, x) \in E_{h_0}^{-1}[0, n]$, so that $f \in D_{h_0}$. Moreover,

$$D_{h_0} = \left\{ f \in C(\mathbf{I}) \mid \text{for some } x \in [0, 1 - 1/n], \left| \frac{f(x + h_0) - f(x)}{h_0} \right| \leq n \right\}$$

so that $\mathscr{E}_n = \bigcap \{D_{h_0} \mid h_0 \in (0, 1/n]\}$, establishing that $\mathscr{E}_n$ is closed. ■

## Problems

### 25A *Exercise on category*

1. The union of finitely many nowhere dense subsets of $X$ is nowhere dense.
2. The frontier of any open subset of $X$ is nowhere dense.
3. Every open subset of a Baire space is a Baire space. The result fails for second category spaces.
4. The space $\mathbf{Q}$ of rationals is not completely metrizable.
5. The space $\mathbf{P}$ of irrationals is a Baire space.

### 25B. *Category in $\sigma$-compact spaces*

A topological space $X$ is *$\sigma$-compact* iff $X$ is a countable union of compact subsets. For $\sigma$-compact spaces, there is a partial converse to the Baire theorem. To state it succinctly, we will define $X$ to be locally compact at one of its points $x$ iff $x$ has a compact nhood in $X$. Note that the set of points at which $X$ is locally compact is always open.

*A $\sigma$-compact space is second category* (*Baire*) *iff the set of points at which $X$ is locally compact is nonempty* (*dense*) *in $X$.*

## 25C. *Continuous functions on Baire spaces*

Let $X$ be a Baire space and $f: X \to \mathbf{R}$ a real-valued continuous function on $X$. Then every nonempty open subset of $X$ contains a nonempty open set on which $f$ is bounded. (If you have done Exercise 7K on semicontinuous functions, you can prove similar results for (1) lower semicontinuous functions and upper bounds, and (2) upper semicontinuous functions and lower bounds, which together imply the result for continuous functions.)

## 25D. *Category in Banach spaces*

The Baire category theorem plays an integral role in the proof of 1 below, and thus indirectly in the proofs of three important theorems in analysis: the open mapping theorem, the closed graph theorem and the uniform boundedness principle.

The definitions and elementary facts about Banach spaces needed here are found in Problems 2J, 7L and 24J.

1. Let $X$ and $Y$ be Banach spaces and $\Gamma$ a bounded linear operator from $X$ onto $Y$. For some $\epsilon > 0$, the image under $\Gamma$ of $\{x \in X \mid ||x|| < 1\}$ covers $\{y \in Y \mid ||y|| < \epsilon\}$. [Let $B_n = \{x \in X \mid ||x|| < 1/2^n\}$ for $n = 1, 2, \ldots$. Use the Baire category theorem to conclude some $n \cdot \Gamma(B_1)$, and hence $\Gamma(B_1)$, is *not* nowhere dense in $Y$. Then for some $y \in Y$ and $\delta > 0$, $\{z \in Y \mid ||z - y|| < \delta\} \subset \overline{\Gamma(B_1)}$ and hence, $\{z \in Y \mid ||z|| < \delta\} \subset \Gamma(B_0)$. Conclude, using completeness of $X$, that $\{z \in Y \mid ||z|| < \delta/2\} \subset \Gamma(B_0)$.]

2. *Open mapping theorem.* If $X$ and $Y$ are Banach spaces and $\Gamma$ is a bounded linear operator of $X$ onto $Y$, then $\Gamma$ is open. [Use part 1.] Hence, if $\Gamma$ is a one–one bounded operator of $X$ onto $Y$, it is a homeomorphism.

3. If $X$ is a vector space with norms $||\cdot||_1$ and $||\cdot||_2$, each of which makes $X$ a Banach space, and if a constant $C$ exists such that $||x_1|| \leq C\,||x_2||$, for all $x \in X$, then $||\cdot||_1$ and $||\cdot||_2$ are *equivalent*; that is, they generate the same topology on $X$.

4. Conclude that the norms $||\cdot||$, $||\cdot||_1$ and $||\cdot||$ given in 2J.6 for $\mathbf{R}^n$ are all equivalent. [In 24J.3 you showed each of these is complete.]

5. *Uniform boundedness principle* (version 1). Let $\mathscr{F}$ be any family of continuous, real-valued functions on a complete metric space $X$ such that for each $x \in X$, there is some constant $M_x$ such that $|f(x)| \leq M_x$ for all $f \in \mathscr{F}$. Then there is some constant $M$ and a nonempty open set $U$ in $X$ such that $|f(x)| \leq M$ for each $x \in U$ and each $f \in \mathscr{F}$. [Let

$$E_n = \{x \in X \mid |f(x)| \leq n \text{ for each } f \in \mathscr{F}\}.$$

Show $E_n$ is closed and apply the Baire category theorem to conclude some $E_n$ contains a nonempty open set $U$.]

6. *Uniform boundedness principle* (version 2). Let $\mathscr{F}$ be any family of bounded linear operators from a Banach space $X$ into a normed linear space $Y$ such that at each $x \in X$, there is a constant $M_x$ such that $||\Gamma(x)|| \leq M_x$ for each $\Gamma \in \mathscr{F}$. Then for some constant $M$, $||\Gamma|| \leq M$ for all $\Gamma \in \mathscr{F}$. [Use part 4. Note that what you want to show is that $||\Gamma(x)|| \leq M$ for each $x$ with $||x|| \leq 1$.]

## 25E. *Hilbert space*

A linear space $X$ becomes an *inner product space* when to every pair $x$, $y$ of elements of $X$ a real number (or, for a complex inner product space, a complex number) $\langle x, y\rangle$ is assigned, subject to the following rules:

(IP1) $\langle x, x \rangle \geq 0$; $\langle x, x \rangle = 0$ iff $x = 0$,

(IP2) $\langle x, y \rangle = \langle y, x \rangle$ (or, in the complex case $\langle x, y \rangle = \overline{\langle y, x \rangle}$),

(IP3) $\langle \alpha x + \beta y, z \rangle = \alpha \langle x, z \rangle + \beta \langle y, z \rangle$.

The number $\langle x, y \rangle$ is called the *inner product* of $x$ and $y$.

1. Every inner product space is a normed linear space, (2J), when the norm is defined by $||x|| = \langle x, x \rangle^{1/2}$. When the resulting normed linear space is a Banach space, we call the inner product space a *Hilbert space.*

2. *Cauchy–Schwarz inequality.* In any inner product space, $\langle x, y \rangle \leq ||x|| \cdot ||y||$. [Set $\lambda = ||x||/||y||$ and work with the inequality $0 \leq ||x - \lambda y||^2$.]

Elements $x$ and $y$ in an inner product space $X$ are *orthogonal* iff $\langle x, y \rangle = 0$. A subset $A$ of $X$ is an *orthonormal system* iff any two elements of $X$ are orthogonal and $||x|| = 1$ for each $x \in A$. An orthonormal system which is maximal (with respect to inclusion) is called *complete.*

3. An orthonormal system $A$ is complete iff whenever $\langle x, a \rangle = 0$ for each $a \in A$, then $x = 0$. Every inner product space has a complete orthonormal system.

4. If $A$ is a complete orthonormal system in a Hilbert space $H$, and $x \in H$, then $x$ has a unique representation of the form

$$x = \sum_{n=1}^{\infty} \alpha_n x_n$$

for some sequence $x_1, x_2, \ldots$ of elements of $A$. [Show that if $x_1, x_2, \ldots$ is any sequence from $A$, then $\sum \langle x, x_n \rangle^2 \leq ||x||^2$. Use this to conclude that only countably many of the inner products $\langle x, z \rangle$, for $z \in A$, are nonzero. Let $x_1, x_2, \ldots$ be the resulting sequence of elements of $A$ and set $\alpha_n = \langle x, x_n \rangle$.]

## 25F. *An application of the Baire theorem*

1. Suppose that for each irrational $p$, an equilateral triangle $A_p$ (with interior) is constructed with a vertex at $(p, 0)$ and the opposite side parallel to and above the $x$-axis. Use the Baire category theorem and 25A.5 to show that $\bigcup A_p$ contains a rectangle of the form $\{(x, y) \in \mathbf{R}^2 \mid a \leq x \leq b, 0 < y < 1/n\}$ for some $a < b$ and some positive integer $n$. [It is enough to show that, for some $n$, $\{p \in \mathbf{P} \mid A_p \text{ has height } \geq 1/n\}$ is dense in some interval $[a, b]$ with $a < b$.]

2. Let $D = \{(x, 0) \mid x \text{ is rational}\}$ and $E = \{(x, 0) \mid x \text{ is irrational}\}$. Then $D$ and $E$ are disjoint closed sets in the Moore plane $\boldsymbol{\Gamma}$. Apply part 1 to show that $D$ and $E$ cannot be contained in disjoint open sets in $\boldsymbol{\Gamma}$.

Chapter 8

# Connectedness

## 26 Connected spaces

The topological study of connectedness is heavily geometric (or visual). Thus connectedness-like properties play an important role in most topological characterization theorems, as well as in the study of obstructions to the extension of functions. The use of connectedness in characterization theorems is exemplified in later sections of this chapter; its use in obstruction theory is appropriate subject matter for a book on algebraic topology.

**26.1 Definition.** A space $X$ is *disconnected* iff there are disjoint nonempty open sets $H$ and $K$ in $X$ such that $X = H \cup K$. We then say that $X$ is disconnected by $H$ and $K$. When no such disconnection exists, $X$ is *connected.*

Note that we can replace "open" in this definition by "closed". It is apparent, then, that $X$ is connected iff there are no open–closed subsets of $X$ other than $\emptyset$ and $X$ or, equivalently, iff $\emptyset$ and $X$ are the only subsets of $X$ with empty frontier.

**26.2 Examples.** a) The Sorgenfrey line $\mathbf{E}$ is disconnected.

b) Any discrete space of more than one point is disconnected. In fact, any $T_1$-space having an isolated (open) point is disconnected. In particular, the ordinal spaces $\mathbf{\Omega}_0$ and $\mathbf{\Omega}$ are disconnected.

c) $\mathbf{I}$ is connected. For if $\mathbf{I}$ is disconnected by $H$ and $K$, with $1 \in H$, then $H$ contains some nhood of 1, so $c = \sup K$ cannot be 1. Now $c$ belongs to either $H$ or $K$ and hence some nhood of $c$ is contained in $H$ or $K$. But any nhood of $c$ contains points of $H$ (to the right of $c$) and points of $K$ (to the left of $c$), a contradiction.

d) *The long line.* The ordinal space $\mathbf{\Omega}$, as we have mentioned, is not connected. A connected space can be obtained from $\mathbf{\Omega}$ by inserting between each pair of consecutive ordinals a copy of (0, 1) and giving the resulting ordered set the order topology. This space is called the *long line*, $\mathbf{W}$. $\mathbf{W}$ is connected, since a disconnection of $\mathbf{W}$ would either disconnect a copy of [0, 1] or isolate a limit ordinal, neither of which is acceptable. $\mathbf{W}$ is also compact (this can be proved in the same way we proved $\mathbf{\Omega}$ is compact, or else use the criterion for compactness of ordered spaces given in 17E).

We turn now to the usual questions, involving continuous maps, subspaces and products of connected spaces.

**26.3 Theorem.** *The continuous image of a connected space is connected.*

*Proof.* Suppose $X$ is connected and $f$ is a continuous map of $X$ onto $Y$. If $Y$ were disconnected by $H$ and $K$, then $X$ would be disconnected by $f^{-1}(H)$ and $f^{-1}(K)$, so $Y$ must be connected. ■

Subspaces of connected spaces are not usually connected; examples abound in **I**. In fact, the only subspace theorem available dealing with connectedness is just a useful way of rephrasing the definition so that it can be applied to a subspace without passing to the relative topology. Note that connectedness of $X$ is not a part of 26.5.

**26.4 Definition.** Sets $H$ and $K$ in $X$ are *mutually separated* in $X$ iff

$$H \cap \overline{K} = \overline{H} \cap K = \varnothing.$$

**26.5 Theorem.** *A subspace $E$ of $X$ is connected iff there are no nonempty, mutually separated sets $H$ and $K$ in $X$ with $E = H \cup K$.*

*Proof.* If $E$ is disconnected by $H$ and $K$, then $H$ and $K$ are mutually separated in any $X$ containing $E$, since

$$\begin{aligned} H \cap \mathrm{Cl}_X K &= (H \cap E) \cap \mathrm{Cl}_X K \\ &= H \cap (E \cap \mathrm{Cl}_X K) \\ &= H \cap \mathrm{Cl}_E K = \varnothing \end{aligned}$$

and similarly for $(\mathrm{Cl}_X H) \cap K$.

Conversely, if $H$ and $K$ are mutually separated in $X$ and $E = H \cup K$, then

$$\begin{aligned} \mathrm{Cl}_E H &= E \cap \mathrm{Cl}_X H = (H \cup K) \cap \mathrm{Cl}_X H \\ &= (H \cap \mathrm{Cl}_X H) \cup (K \cap \mathrm{Cl}_X H) \\ &= H \end{aligned}$$

and hence $H$ is closed in $E$. Similarly $K$ is closed in $E$. ■

**26.6 Corollary.** *If $H$ and $K$ are mutually separated in $X$ and $E$ is a connected subset of $H \cup K$, then $E \subset H$ or $E \subset K$.*

*Proof.* If $H$ and $K$ are mutually separated in $X$, so are $E \cap H$ and $E \cap K$. ■

The last theorem and its corollary provide us with some neat ways of proving a given space $X$ is connected.

**26.7 Theorem.** a) *If $X = \bigcup X_\alpha$, where each $X_\alpha$ is connected and $\bigcap X_\alpha \neq \varnothing$, then $X$ is connected.*

b) *If each pair of $x$, $y$ of points of $X$ lies in some connected subset $E_{xy}$ of $X$, then $X$ is connected.*

c) *If $X = \bigcup_{n=1}^{\infty} X_n$ where each $X_n$ is connected and $X_{n-1} \cap X_n \neq \varnothing$ for each $n \geq 2$, then $X$ is connected.*

*Proof.* a) Suppose $X = H \cup K$ where $H$ and $K$ are mutually separated in $X$. Then, since $X_\alpha$ is a connected subset of $H \cup K$ for each $\alpha$, we have $X_\alpha \subset H$ or $X_\alpha \subset K$. Since the $X_\alpha$ are not disjoint, while $H$ and $K$ are, we must have $X_\alpha \subset H$ for all $\alpha$ or $X_\alpha \subset K$ for all $\alpha$; say the former. Then $X \subset H$, so $K = \varnothing$. Thus $X$ can never be the union of two nonempty mutually separated sets in $X$, so $X$ is connected.

b) Fix $a \in X$. Then $X = \bigcup_{x \in X} E_{ax}$ and the latter union satisfies the conditions of part (a).

c) $X_1$ is connected, and if $X_1 \cup \cdots \cup X_{n-1}$ is connected, so is $X_1 \cup \cdots \cup X_n$ by part (a). Thus $A_n = X_1 \cup \cdots \cup X_n$ is connected, for $n = 1, 2, \ldots$. Since $\bigcap A_n = X_1$ is nonempty, $\bigcup A_n = X$ is connected by part (a). ■

**26.8 Theorem.** *If $E$ is a connected subset of $X$ and $E \subset A \subset \bar{E}$, then $A$ is connected.*

*Proof.* It is enough to show $\bar{E}$ is connected (since if $E \subset A \subset \bar{E}$, then $A = \mathrm{Cl}_A E$ and we can replace $X$ by $A$). Suppose $\bar{E} = H \cup K$ where $H$ and $K$ are disjoint, nonempty open sets in $\bar{E}$. Then $E = (H \cap E) \cup (K \cap E)$, and the latter are disjoint, nonempty open sets in $E$. Thus if $\bar{E}$ is disconnected, so is $E$. ■

The two theorems just proved give nice ways of leap-frogging from connectedness of some familiar spaces (we already know $\mathbf{I}$ is connected) to connectedness of others.

**26.9 Examples.** a) $\mathbf{R}$ is connected. For $\mathbf{R} = \bigcup_{n=1}^{\infty} [-n, n]$ and each set $[-n, n]$ is homeomorphic to $\mathbf{I}$ and hence connected, while their intersection is nonempty, so connectedness of $\mathbf{R}$ is a simple application of 26.7(a).

b) $\mathbf{R}^n$ is connected. We can use the same theorem. $\mathbf{R}^n$ is the union of the family of all straight lines through its origin; each such line is homeomorphic to $\mathbf{R}$ and thus connected, so $\mathbf{R}^n$ is connected.

We turn now to the problem of deciding connectedness for product spaces. The last theorem will be useful here.

**26.10 Theorem.** *A nonempty product space is connected iff each factor space is connected.*

*Proof.* If the product is connected and no factor space is empty, then the projections are continuous and onto and hence each factor space is connected.

Conversely, suppose each factor space $X_\alpha$, $\alpha \in A$, is connected. Pick $a \in \prod X_\alpha$ and denote by $E$ the set of all points in the product which lie together with $a$ in some connected subset of the product. Then $E$ is connected, so it suffices by the previous theorem to show $E$ is dense in the product.

Let $U = \pi_{\alpha_1}^{-1}(U_{\alpha_1}) \cap \cdots \cap \pi_{\alpha_n}^{-1}(U_{\alpha_n})$ be a basic open set in the product. Pick $b_{\alpha_i} \in U_{\alpha_i}$ for $i = 1, \ldots, n$ and define sets $E_1, \ldots, E_n$ as follows:

$$E_1 = \{c \in \prod X_\alpha \mid c_{\alpha_1} \text{ arbitrary, } c_\alpha = a_\alpha \text{ otherwise}\},$$
$$E_2 = \{c \in \prod X_\alpha \mid c_{\alpha_1} = b_{\alpha_1}, c_{\alpha_2} \text{ arbitrary, } c_\alpha = a_\alpha \text{ otherwise}\},$$
$$\vdots$$
$$E_n = \{c \in \prod X_\alpha \mid c_{\alpha_i} = b_{\alpha_i} \text{ for } i = 1, \ldots, n-1, c_{\alpha_n} \text{ arbitrary,}$$
$$c_\alpha = a_\alpha \text{ otherwise}\}.$$

Then $E_k$ is homeomorphic to $X_{\alpha_k}$ and thus connected. Moreover, $E_k \cap E_{k+1} \neq \emptyset$ for $k = 1, \ldots, n-1$ so $\bigcup_{k=1}^n E_k = F$ is connected. But $a \in F$ and $F$ meets $U$. Thus every basic open set $U$ contains points of $E$. ■

The importance of connectedness for us lies almost wholly with its use in characterization theorems. In particular, it is not usually possible to deduce the presence of other topological properties in a space from the fact that the space is connected, or vice versa. In fact, if one needs connectedness of $X$, and $X$ is not itself connected, we can usually just look at the individual "components" (maximal connected pieces) of $X$, as described now.

**26.11 Definition.** If $x \in X$, the largest connected subset $C_x$ of $X$ containing $x$ is called the *component* of $x$. It exists, being just the union of all connected subsets of $X$ containing $x$.

If $x \neq y$ in $X$, then either $C_x = C_y$ or $C_x \cap C_y = \emptyset$; otherwise $C_x \cup C_y$ would be a connected set containing $x$ and $y$ and larger than $C_x$ or $C_y$, which is impossible. Thus the components of points in $X$ form a partition of $X$ into maximal connected subsets. This justifies referring to them as components of $X$.

***26.12 Theorem.*** *The components of $X$ are closed sets.*

*Proof.* If $C$ is the component of $x$ in $X$, then $\bar{C}$ is a connected set containing $x$ and thus $\bar{C} \subset C$, showing that $C$ is closed. ■

**26.13 Examples.** a) In the space $\mathbf{Q}$ of rational numbers, the component of each point $q$ is $\{q\}$. We would say, somewhat imprecisely, "the components in $\mathbf{Q}$ are the points." This example shows, incidentally, that components need not be open.

b) Recall (17.9c) the construction of the Cantor set $\mathbf{C}$: we define

$$C_1 = \mathbf{I} - (\tfrac{1}{3}, \tfrac{2}{3})$$
$$C_2 = C_1 - [(\tfrac{1}{9}, \tfrac{2}{9}) \cup (\tfrac{7}{9}, \tfrac{8}{9})]$$

and so on, with $C_n$ being obtained by removing the open middle thirds of the $2^{n-1}$ closed intervals which comprise $C_{n-1}$. Then $\mathbf{C} = \bigcap_{n=1}^{\infty} C_n$.

It is easy to see that the components of $\mathbf{C}$ are the points, for if $x \in \mathbf{C}$, then among the intervals removed from $\mathbf{I}$ in the process of constructing $\mathbf{C}$ there are intervals arbitrarily close to $x$ on either side, and each such interval induces a disconnection of $\mathbf{C}$.

We give now an important theorem, asserting that connectedness of a space implies "chain-connectedness" with respect to any open cover. This result will be useful later in theorems asserting existence of "paths" between points of certain connected spaces.

**26.14 Definition.** A *simple chain* connecting two points $a$ and $b$ of a space $X$ is a sequence $U_1, \ldots, U_n$ of open sets of $X$ such that $a \in U_1$ only, $b \in U_n$ only, and $U_i \cap U_j \neq \emptyset$ iff $|i - j| \leq 1$.

> **26.15 Theorem.** *If $X$ is connected and $\mathcal{U}$ is any open cover of $X$, then any two points $a$ and $b$ of $X$ can be connected by a simple chain consisting of elements of $\mathcal{U}$.*

*Proof.* Let $Z$ be the set of all points of $X$ which are connected to $a$ by a simple chain of elements of $\mathcal{U}$. Then $Z$ is obviously an open set and, since $a \in Z$, $Z$ is nonempty. We can prove the theorem by showing $Z$ is closed.

Let $z \in \bar{Z}$. Then $z \in U$ for some $U \in \mathcal{U}$ and, since $U$ is open, $U \cap Z$ contains some point $b$. Now $a$ is connected to $b$ by a simple chain $U_1, \ldots, U_n$ of elements of $\mathcal{U}$. If $z \in U_k$ for some $k$, then the smallest such $k$ produces a simple chain $U_1, \ldots, U_k$ from $a$ to $z$. If $z \notin U_k$ for any $k$, pick the smallest $l$ such that $U_l \cap U \neq \emptyset$ (e.g., $n$ is such an $l$). Then $U_1, \ldots, U_l, U$ is a simple chain from $a$ to $z$. Either way, $z \in Z$. ∎

## Problems

### 26A. *Examples on connectedness*

1. The Sorgenfrey line $\mathbf{E}$ is not connected.
2. The slotted plane (4C) and the radial plane (3A.4) are connected. [See 6A.]
3. Any infinite set with the cofinite topology is connected.
4. No countable subset of $\mathbf{R}$ is connected.

### 26B. *Quasicomponents*

Define $\sim$ in any space $X$ by $x \sim y$ iff $x$ and $y$ lie together in some connected subset of $X$. Define $\approx$ in $X$ by $x \approx y$ iff there is no decomposition $X = U \cup V$ into disjoint open sets, one containing $x$, the other containing $y$.

1. $\sim$ is an equivalence relation on $X$. The equivalence class $[x]$ of $x$ is just the component $C_x$ of $x$ in $X$.

2. $\approx$ is an equivalence relation on $X$. We call the equivalence class of $x$ the *quasicomponent* of $x$ in $X$. The quasicomponent of $x$ in $X$ is the intersection of all open–closed subsets of $X$ which contain $x$.

3. The component of $x$ is contained in the quasicomponent of $x$.

4. In the space $X$ in Fig. 26.1, the quasicomponent of the point $x$ shown is strictly larger than the component of $x$.

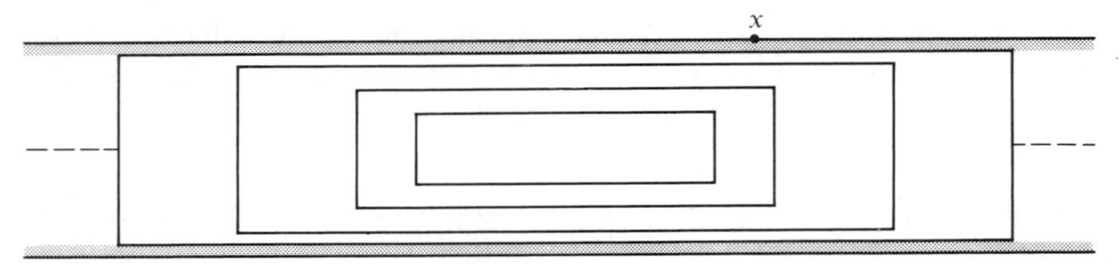


**Figure 26.1**

## 26C. *Cardinality of connected spaces*

1. A connected, Tychonoff space having more than one point has $\geq \mathfrak{c}$ points.

2. A connected, separable, metric space has either one point or $\mathfrak{c}$ points.

3. Let $X$ be the set of all points in the closed upper half plane both of whose coordinates are rational. Describe a topology for $X$ as follows: for each point $p$ (rational or irrational) on the $x$-axis, let $V_{p\epsilon}$ denote the set of all rational points in the interval $(p - \epsilon, p + \epsilon)$ on the $x$-axis. Now for $p \in X$, if $p$ lies on the $x$-axis, the nhoods of $p$ will be the sets $V_{p\epsilon}$, $\epsilon > 0$; if $p$ lies above the $x$-axis, let $p_1$ and $p_2$ be the uniquely determined points on the $x$-axis such that $p$, $p_1$ and $p_2$ are the vertices of an equilateral triangle (note that $p_1$ and $p_2$ will have irrational first coordinate, since the slopes of the lines joining them to $p$ are irrational). The nhoods of $p$ will be the sets $\{p\} \cup V_{p_1\epsilon} \cup V_{p_2\epsilon}$ for $\epsilon > 0$. Then $X$ is a countable, connected Hausdorff space. [To prove connectedness, show any nonempty open–closed subset $H$ of $X$ must be all of $X$.]

4. The space $X$ described in part 3 is not regular.

## 26D. *Subspaces*

Among the criteria for a subspace $E$ of a space $X$ to be connected, the following was absent: $E \subset X$ is disconnected iff there are disjoint open subsets $H$ and $K$ in $X$, each meeting $E$, such that $E \subset H \cup K$. Find a counterexample. (Thus 26.5 represents the best we can do along the lines of expressing connectedness of $E$ in terms of the topology on $X$.)

## 26E. *Nonhomeomorphism*

Some use of connectedness lies at the heart of most proofs that two spaces are not homeomorphic. Use connectedness to show that $X$ is not homeomorphic to $Y$ when:

1. $X = \mathbf{R}$, $Y = \mathbf{R}^n$ for $n > 1$, (compare with 28C);
2. $X = [0, \infty)$, $Y = \mathbf{R}$;
3. $X = \mathbf{I}$, $Y = \mathbf{S}^1$;
4. $X = \mathbf{S}^1$, $Y = \mathbf{S}^n$ for $n > 1$.

Note that in none of the above cases can we distinguish between $X$ and $Y$ using any of the forms of compactness available to us.

26F. *The Cantor set*

Every closed subset $A$ of $\mathbf{C}$ is a retract (7J) of $\mathbf{C}$.

26G. *Connectedness in ordered spaces*

1. An ordered space $X$ (6D) is connected iff it is Dedekind complete and whenever $x < y$ in $X$, then $x < z < y$ for some $z$ in $X$.

2. Every ordered space can be embedded in a connected ordered space. [First, embed in a Dedekind complete ordered space. Then whenever $x < y$ in this space, and no $z$ exists with $x < z < y$, put a copy of $(0, 1)$ between $x$ and $y$.]

3. Let $\mathbf{I}$ and $\{0, 1\}$ have their usual orders, and let $X = \mathbf{I} \times \{0, 1\}$ have the lexicographic order. Then $X$ is Dedekind complete. What space results from applying the process in part 2 to $X$?

26H. *Uses of connectedness*

1. Any continuous $f: \mathbf{I} \to \mathbf{I}$ has a fixed point (i.e., a point $x$ such that $f(x) = x$).

2. If $P(x)$ is a polynomial of odd degree, then the equation $P(x) = 0$ has at least one real root.

## 27 Pathwise and local connectedness

The definition of connectedness is negative in nature; it provides for the non-existence of a certain kind of splitting of the space. A more positive approach to the same sort of problem is provided by pathwise (or arcwise) connectedness, in which it is required that it be possible to reach any point in the space from any other point along a connected path. This approach is especially useful in studying connectivity properties from an algebraic point of view, e.g., via homotopy theory.

**27.1 Definition.** A space $X$ is *pathwise connected* iff for any two points $x$ and $y$ in $X$, there is a continuous function $f: \mathbf{I} \to X$ such that $f(0) = x$, $f(1) = y$. Such a function $f$ (as well as its range $f(\mathbf{I})$, when confusion is not possible) is called a *path* from $x$ to $y$.

We call $X$ *arcwise connected* iff for any two points $x$ and $y$ in $X$, there is a homeomorphism $f: \mathbf{I} \to X$ such that $f(0) = x$, $f(1) = y$. The function $f$ (as well as its range) is called an *arc* from $x$ to $y$.

We will observe in 31.6 that every Hausdorff path from $x$ to $y$ contains an arc from $x$ to $y$. Thus a $T_2$-space *is pathwise connected iff it is arcwise connected*!

**27.2 Theorem.** *Every pathwise connected space is connected.*

*Proof.* If $H$ and $K$ disconnect the pathwise connected space $X$, let $f: \mathbf{I} \to X$ be any path between points $x \in H$ and $y \in K$. Then $f^{-1}(H)$ and $f^{-1}(K)$ disconnect $\mathbf{I}$, which is impossible. ■

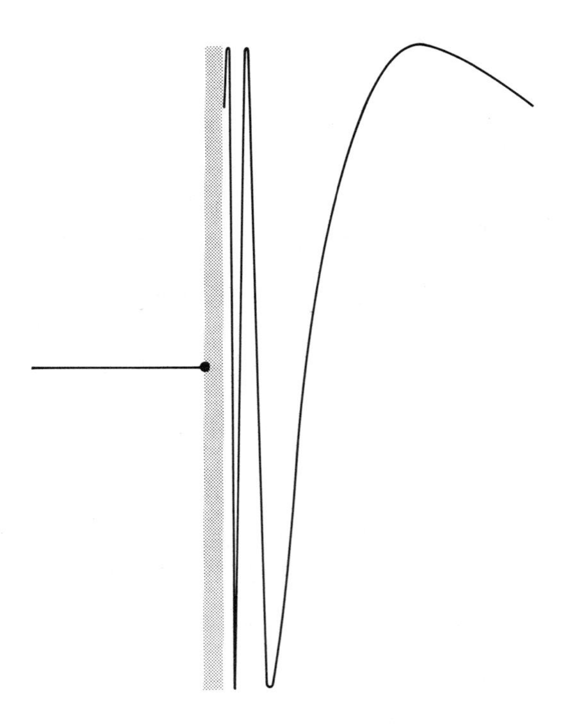

Figure 27.1

**27.3 Examples.** a) The *topologist's sine curve* (Fig. 27.1)

$$\mathbf{V} = \{(x, 0) \mid x \leq 0\} \cup \left\{\left(x, \sin\frac{1}{x}\right) \,\middle|\, x > 0\right\}$$

is a connected space, but no path can be found from (0, 0) to any point $(x, \sin(1/x))$ with $x > 0$. Verification is left to 27A.

b) Closed line segments are arcs, so $\mathbf{R}^n$ is pathwise connected.

c) If $E$ is any countable subset of $\mathbf{R}^2$, then the space $\mathbf{R}^2 - E$ is pathwise connected. In fact, if $a$ and $b$ are points in $\mathbf{R}^2 - E$, then $\mathbf{R}^2 - E$ contains uncountably many straight lines through each point and two of these will intersect, giving an arc from $a$ to $b$.

Paths can be "added," in the following sense. If $a, b, c \in X$, and $f_1 : \mathbf{I} \to X$ is a path from $a$ to $b$, while $f_2 : \mathbf{I} \to X$ is a path from $b$ to $c$, then the function $f : \mathbf{I} \to X$ defined by

$$f(t) = \begin{cases} f_1(2t) & \text{if } 0 \leq t \leq \frac{1}{2} \\ f_2(2t - 1) & \text{if } \frac{1}{2} \leq t \leq 1, \end{cases}$$

is a path from $a$ to $c$, obtained by "putting the paths $f_1$ and $f_2$ end-to-end". (For example, $f$ is continuous because it is continuous on each of the closed sets $[0, \frac{1}{2}]$ and $[\frac{1}{2}, 1]$.)

This path addition provides a way to associate with each pathwise connected space $X$ a group $\pi_1(X)$ in such a way that homeomorphic spaces have isomorphic groups. The branch of algebraic topology which is concerned with the relationships between $X$ and $\pi_1(X)$ is homotopy theory (a piece of which is developed in Sections 32 through 34). Other branches of algebraic topology study connectivity properties of a topological space $X$ by associating algebraic structures with $X$ in other ways. In particular, the ordinary covering notion of connectedness is studied using Čech homology theory, while singular homology theory (and homotopy theory) are suited to the study of pathwise connectedness.

For the time being, we will use the addition of paths defined above only to provide a partial converse to Theorem 27.2. We require a definition.

**27.4 Definition.** A space $X$ is *locally pathwise connected* iff each point has a nhood base consisting of pathwise connected sets. (We should point out here that a subset $A$ of $X$ is pathwise connected iff any two points in $A$ can be joined by a path *lying in A*.)

**27.5 Theorem.** *A connected, locally pathwise connected space X is pathwise connected.*

*Proof.* Let $a \in X$ and let $H$ be the set of all points of $X$ which can be joined to $a$ by a path. Now $H$ is nonempty since $a \in H$, so if $H$ is open–closed it must be all of $X$.

But $H$ is open. For if $b \in H$, let $U$ be any pathwise connected nhood of $b$. Then any point $z \in U$ can be joined to $b$ by a path and hence can be joined to $a$ by adding the path from $b$ to $a$.

Also, $H$ is closed. For if $b \in \overline{H}$, let $U$ be any pathwise connected nhood of $b$. Then $U \cap H \neq \varnothing$; say $z \in U \cap H$. Now $b$ can be joined to $z$ by a path and $z$ can be joined to $a$ by a path so, by addition of paths again, $b \in H$. ■

**27.6 Corollary.** *An open connected subset of* $\mathbf{R}^n$ *is pathwise connected.*

We turn now to the study of locally connected spaces. Unlike most other localized properties, there is no generally discernible relationship between connectedness and local connectedness.

**27.7 Definition.** A space $X$ is *locally connected* iff each $x \in X$ has a nhood base of open connected sets.

**27.8 Examples.** a) The space $[0, 1) \cup (1, 2]$ is locally connected but not connected.

b) Consider the space $X$ consisting of the vertical lines $x = 0$ and $x = 1$ in the plane, together with the horizontal line segments $\{(x, 1/n) \mid 0 \leq x \leq 1\}$ for $n = \pm 1, \pm 2, \ldots$ and the unit interval $\mathbf{I}$ on the $x$-axis (Fig. 27.2). This space is typical of connected spaces which are not locally connected. $X$ is, in fact,

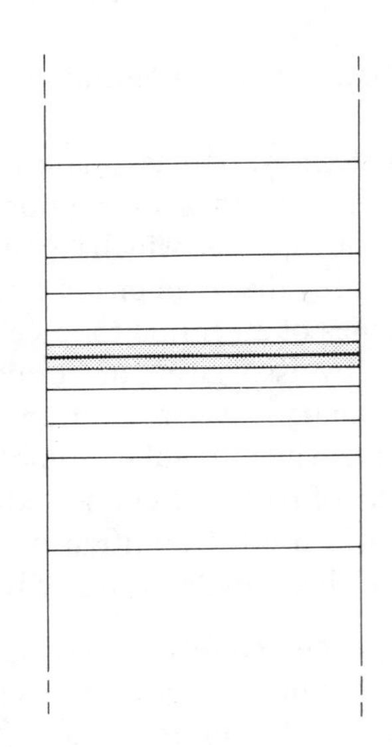

Figure 27.2

arcwise connected, but no point in **I** other than the endpoints will have a base of connected nhoods.

**27.9 Theorem.** *X is locally connected iff each component of each open set is open.*

*Proof.* Suppose $X$ is locally connected and $x \in C$, where $C$ is a component of the open set $U$ in $X$. There is, by local connectedness, an open connected set $V$ with $x \in V \subset U$. Now we must have $V \subset C$, so $C$ is open.

Conversely, suppose each component of each open set in $X$ is open. If $U$ is any open nhood of $x$ in $X$, then the component of $U$ containing $x$ is an open connected nhood of $x$ contained in $U$. Thus $X$ is locally connected. ■

**27.10 Corollary.** *The components of a locally connected space are open–closed.*

**27.11 Corollary.** *A compact locally connected space has a finite number of components.*

**27.12 Theorem.** *Every quotient of a locally connected space is locally connected.*

*Proof.* Let $f$ be a quotient map of $X$ onto $Y$. Suppose $U$ is an open set in $Y$, $C$ a component of $U$. For $x \in f^{-1}(C)$, let $C_x$ be the component of $x$ in the open set $f^{-1}(U)$. Now $f(C_x)$ is connected and contains $f(x) \in C$, so $f(C_x) \subset C$. Thus $x \in C_x \subset f^{-1}(C)$. Since $C_x$ is open, $f^{-1}(C)$ is open and thus, since $Y$ has the quotient topology, $C$ is open in $Y$. ■

The theorem above is one of the nicest dealing with preservation of a local property by continuous maps. For example, it follows that both continuous open images and continuous closed images of locally connected spaces are locally connected.

**27.13 Theorem.** *A nonempty product space is locally connected iff*

a) *each factor is locally connected,*

b) *all but finitely many factors are connected.*

*Proof.* The proof is obtained by substituting "connected" for "compact" in the proof of 18.6, the corresponding theorem for local compactness. See 27F. ■

**27.14 Definition.** $X$ is *connected im kleinen* at $x$ iff each open nhood $U$ of $x$ contains an open nhood $V$ of $x$ such that any pair of points in $V$ lie in some connected subset of $U$.

Certainly every locally connected space is connected *im kleinen.* At first it is easy to believe the converse, but the following example shows that the two notions are different. The theorem after that shows that they are not much different.

**27.15 Example.** At the point $x$, the space shown in Fig. 27.3 is connected *im kleinen*, but has no base of open connected nhoods.

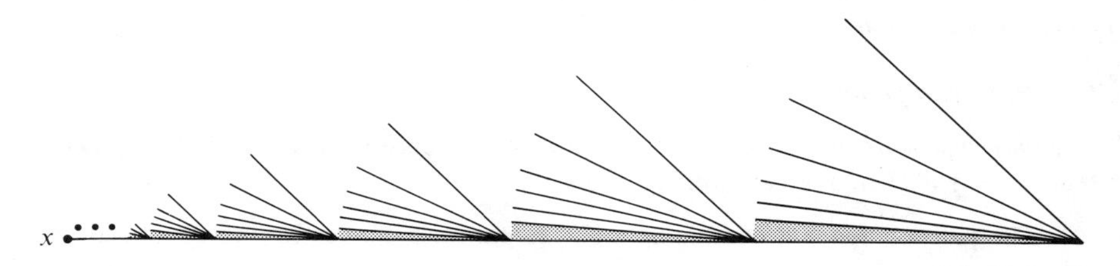


**Figure 27.3**

**27.16 Theorem.** *If $X$ is connected im kleinen at each point, then $X$ is locally connected.*

*Proof.* Let $U$ be an open set in $X$, $C$ a component of $U$. If $x \in C$, then there is an open set $V_x$ containing $x$ and lying in $U$ such that each two points in $V_x$ lie in a connected subset of $U$. It follows that $V_x \subset C$. Thus $C$ is open and $X$ is locally connected. ■

## Problems

### 27A. *The topologist's sine curve*

Let $\mathbf{V} = \{(x, 0) \mid x \leq 0\} \cup \{(x, \sin(1/x)) \mid x > 0\}$ with the relative topology in $\mathbf{R}^2$ and let $T$ be the subspace $\{(x, \sin(1/x)) \mid x > 0\}$ of $\mathbf{V}$.

1. $\mathbf{V}$ is connected. [Use 26.7 and 26.8.]

2. **V** is not pathwise connected. [If $f$ is a path from $(0, 0)$ to $(x, \sin(1/x))$, then $f(\mathbf{I})$ is compact and connected.]

3. $T$ is pathwise connected, but the closure of $T$ in **V** is not. (Compare with 26.8.)

## 27B. *Combinations of pathwise connected spaces*

1. The continuous image of a pathwise connected space is pathwise connected.

2. A nonempty product of finitely many spaces is pathwise connected iff each factor space is pathwise connected.

## 27C. *Pathwise connectification*

Let $X$ be any space and define a topology on $Y = X \times \mathbf{I}$ as follows: basic nhoods of points $(x, \alpha)$ for $\alpha \neq 0$ will be the sets of the form $\{(x, \beta) \mid \alpha - \epsilon < \beta < \alpha + \epsilon\}$ for $\epsilon > 0$ (that is, usual linear nhoods of $(x, \alpha)$ in the appropriate copy of **I**), and basic nhoods of $(x, 0)$ will have the form $(U \times \{0\}) \cup \bigcup_{z \in U} J_{z,\epsilon_z}$ where $U$ is a nhood of $x$ in $X$ and for each $z \in U$, $\epsilon_z > 0$ and $J_{z,\epsilon_z} = \{(z, \alpha) \mid 0 \leq \alpha < \epsilon_z\}$. Let $X^*$ be the quotient of $Y$ obtained by identifying all the points $(x, 1)$, $x \in X$.

1. $X$ is embedded in $X^*$ as the closed nowhere dense set $\{(x, 0) \mid x \in X\}$.

2. $X^*$ is pathwise connected.

3. If $f\colon X \to Z$ is continuous, where $Z$ is pathwise connected, then $f$ can be extended to a continuous function $F\colon X^* \to Z$.

## 27D. *Path components*

The *path components* of a space $X$ are the equivalence classes in $X$ under the equivalence relation $x \sim y$ iff there is a path joining $x$ to $y$.

1. The path component containing $x \in X$ is pathwise connected and contained in the component of $x$.

2. $X$ is locally pathwise connected iff each path component of each open set is open.

3. A path component of $X$ need not be closed. But if $X$ is locally pathwise connected, the path components of $X$ are both open and closed.

## 27E. *Examples on local connectedness*

1. The Sorgenfrey line **E** is not locally connected.

2. The topologist's sine curve **V** is not locally connected.

3. The space of Example 27.15 is not locally connected.

## 27F. *Combinations of locally connected spaces*

1. The continuous image of a locally connected space need not be locally connected.

2. A nonempty product space is iff locally connected

   a) each factor space is locally connected,

   b) all but finitely many factor spaces are connected.

## 27G. *Property S*

A topological space $X$ has *property S* iff every open cover of $X$ can be refined by a cover consisting of a finite number of connected sets. The property was introduced by Sierpinski in 1920.

1. If $X$ has property $S$, then $X$ is connected *im kleinen* at each point, and thus locally connected.

2. A compact, Hausdorff space is locally connected iff it has property $S$.

3. Not every locally connected Hausdorff space has property $S$.

4. The continuous Hausdorff image of a compact locally connected space is compact and locally connected. Is property $S$ preserved by *all* continuous maps?

Property $S$ assumes special importance in deciding questions about local connectivity of certain subsets of $\mathbf{R}^2$. In particular (see Whyburn: Analytic Topology, p. 112), if $A$ is a connected open subset of $\mathbf{R}^2$ such that $\text{Fr}(A)$ is a continuum, then $A$ has property $S$ iff $\text{Fr}(A)$ is locally connected. As a corollary, $\text{Fr}(A)$ locally connected $\Rightarrow A$ locally connected.

## 28 Continua

Compactness and connectedness are powerful, but dissimilar, properties. When they are combined to generate the notion of a continuum, the result is an extensive collection of interesting theorems (not all of which we will be able to give here).

**28.1 Definition.** A *continuum* is a compact, connected Hausdorff space. Among the continua we find many familiar spaces. Thus the unit interval $\mathbf{I}$, the circle $\mathbf{S}^1$, the torus $\mathbf{S}^1 \times \mathbf{S}^1$ (and, in fact, any product of continua) are all continua. Our main goal is to find topological criteria which will enable us to characterize the unit interval and the unit circle as continua.

**28.2 Theorem.** *Let $\{K_\alpha \mid \alpha \in A\}$ be a collection of continua in $X$ directed by inclusion. Then $\bigcap K_\alpha$ is a continuum.*

*Proof.* The intersection is a closed subset of each $K_\alpha$ and thus is compact. Suppose disjoint closed sets $H$ and $K$ can be found with $\bigcap K_\alpha = H \cup K$, and $x \in H$, $y \in K$. For any fixed $\alpha_0$, $X$ can be replaced by $K_{\alpha_0}$, and each $K_\alpha$ by $K_{\alpha_0} \cap K_\alpha$, without affecting the intersection, so we may assume $X$ is compact and Hausdorff. Then $H$ and $K$ are closed in $X$ and can be separated by open sets $U$ and $V$ in $X$. For each $K_\alpha$, $K_\alpha \not\subset U \cup V$ since otherwise $U \cap K_\alpha$ and $V \cap K_\alpha$ would disconnect $K_\alpha$. Thus we can pick $x_\alpha \in K_\alpha - (U \cup V)$. The result is a net $(x_\alpha)$ which has a cluster point $z$ in $X$, by compactness. Now if $W$ is any nhood of $z$ and $K_\alpha$ is given, then for some $K_\beta \subset K_\alpha$, $x_\beta \in W$. Thus $W \cap K_\alpha \neq \varnothing$ for each nhood $W$ of $z$, so $z \in \overline{K}_\alpha = K_\alpha$, for each $\alpha$. Then $z \in \bigcap K_\alpha \subset U \cup V$. But $U \cup V$ is then a nhood of $z$ inside which $(x_\alpha)$ never gets, by choice of the $x_\alpha$. We have a contradiction.

Thus $\bigcap K_\alpha$ must be connected. ∎

**28.3 Definition.** A continuum $K$ in $X$ is *irreducible* about a subset $A$ of $X$ provided

$A \subset K$ and no proper subcontinuum of $K$ contains $A$. If $A = \{a, b\}$, we say $K$ is *irreducible between a and b*.

**28.4 Theorem.** *If K is any continuum, any subset A of K lies in a subcontinuum irreducible about A.*

*Proof.* The set $\mathscr{K}$ of all subcontinua of $K$ containing $A$ is partially ordered by inclusion; i.e., $K_1 \leq K_2$ iff $K_2 \subset K_1$. By 28.2, each chain in this partially ordered set has an upper bound (the intersection of its elements) and hence, by Zorn's lemma, $\mathscr{K}$ has a maximal element $K'$. Clearly $K'$ is a subcontinuum of $K$ irreducible about $A$. ■

In particular, $K$ will contain subcontinua irreducible between any two of its points. In the plane, for example, any arc joining $a$ and $b$ is a continuum which is irreducible between $a$ and $b$ (and so, in general, a continuum irreducible about a set $A$ will not be unique).

**28.5 Definition.** Let $X$ be a connected $T_1$-space. A *cut point* of $X$ is a point $p \in X$ such that $X - \{p\}$ is not connected. If $p$ is not a cut point of $X$, we call it a *noncut point* of $X$. A *cutting* of $X$ is a set $\{p, U, V\}$ where $p$ is a cut point of $X$ and $U$ and $V$ disconnect $X - \{p\}$ (i.e., where $U$ and $V$ are disjoint nonempty open subsets of $X$ whose union is $X - \{p\}$).

The property of being a cut point (in fact, of being a cutting), is preserved under homeomorphism; but continuous maps can destroy cut points. Consider the map $f(x) = (\cos x, \sin x)$ of $[0, 2\pi]$ onto the unit circle in $\mathbf{R}^2$.

Cut points are critical in the characterizations of the interval and circle as continua having certain additional properties. One property relating to cut points is shared by all continua, however; they all have at least two noncut points. This follows easily from the second of the following lemmas.

**28.6 Lemma.** *If K is a continuum and $\{p, U, V\}$ is a cutting of K, then $U \cup \{p\}$ and $V \cup \{p\}$ are connected (and thus are continua).*

*Proof.* It suffices to prove the lemma for $U \cup \{p\}$. But the map $f$ defined on $K$ by

$$f(x) = \begin{cases} x & \text{if } x \in U \cup \{p\} \\ p & \text{if } x \in V \end{cases}$$

carries $K$ onto $U \cup \{p\}$, and $f$ is continuous on each of the closed sets $U \cup \{p\}$ and $V \cup \{p\}$, so $f$ is continuous. Thus $U \cup \{p\}$ is the continuous image of a connected space and therefore connected. (Since $U \cup \{p\} = K - V$, $U \cup \{p\}$ is closed in $K$ and thus compact. The part of the theorem in parentheses follows.) ■

**28.7 Lemma.** *If K is a continuum and $\{p, U, V\}$ is a cutting of K, then each of U and V contains a noncut point of K.*

*Proof.* Suppose each point $x$ in $U$ is a cut point, inducing a cutting $\{x, U_x, V_x\}$

of $K$. If both $U_x$ and $V_x$ meet $V \cup \{p\}$, they disconnect $V \cup \{p\}$ which is impossible by the previous lemma. So one, say $U_x$, is contained in $U$. Now $U_x \cup \{x\}$ is a continuum for each $x \in U$, by the previous lemma. Since $\{U_x \cup \{x\} \mid x \in U\}$ is directed by inclusion, $\bigcap_{x\in U} [U_x \cup \{x\}]$ is a nonempty continuum contained in $U$, by 28.2.

Pick $q \in \bigcap_{x\in U} [U_x \cup \{x\}]$. Then $U_q \subset U$ (as above), and if $r \in U_q$, then $U_r$ does not contain $q$ (otherwise $U_r$ and $V_r$ both meet $V_q \cup \{q\}$ and disconnect it). Then $U_r \cup \{r\}$ does not contain $q$. But this contradicts the fact that

$$q \in \bigcap_{x\in U} (U_x \cup \{x\}). \blacksquare$$

**28.8 Theorem.** *Every continuum $K$ of more than one point has at least two noncut points.*

*Proof.* If $p$ is a cut point of $K$, then a cutting $\{p, U, V\}$ of $K$ exists, and each of $U$ and $V$ contains a noncut point of $K$, by the previous lemma. On the other hand, if no cut point of $K$ exists, certainly there are two noncut points. ■

As we will see shortly, the property of continua expressed by Theorem 28.8 is the key to the characterization of the unit interval; it is the only metric continuum blessed with *exactly* two noncut points. For this, we need a series of results, the first of which says that you cannot get a new continuum from an old one by excision without excising some noncut points.

**28.9 Theorem.** *A continuum $K$ is irreducible about the set of its noncut points.*

*Proof.* Let $N$ be the set of noncut points of $K$ and suppose a proper subcontinuum $L$ of $K$ contains $N$. If $x \in K - L$, then a cutting $\{x, U, V\}$ of $K$ exists, and $L$ must lie in one or the other of $U$ and $V$, say $L \subset U$. Then $V \cup \{x\}$, being a continuum itself, has two noncut points and thus has a noncut point $y \neq x$. Then $[V \cup \{x\}] - \{y\}$ is connected, and $U \cup \{x\}$ is connected and these sets meet, so their union is connected. But their union is $K - \{y\}$, while $y$ lies in $V$, hence not in $U$, hence not in $L$; this is a contradiction since $L$ contains all the noncut points of $K$. ■

An order relation can be introduced on certain subspaces of a continuum. It is the last tool we need to reach our characterization theorems.

**28.10 Definition.** A cut point $p$ in a connected space $X$ *separates $a$ from $b$* iff a cutting $\{p, U, V\}$ exists with $a \in U$, $b \in V$. The set consisting of $a$, $b$ and all points $p$ which separate $a$ from $b$ is denoted $E(a, b)$. The *separation order* on $E(a, b)$ is defined by: $p_1 \leq p_2$ iff $p_1 = p_2$ or $p_1$ separates $a$ from $p_2$. This is easily seen to be a partial order on $E(a, b)$.

The basis for our proofs of the continuum characterization theorems (28.13, 28.14) will be the fact that the set $E(a, b)$ is linearly ordered by the separation order.

**28.11 Theorem.** *The separation order on $E(a, b)$ is a linear order.*

*Proof.* For each $p \in E(a, b)$, let $\{p, U_p, V_p\}$ be a cutting of $X$ such that $a \in U_p$ and $b \in V_p$.

If $r$ and $s$ are distinct points of $E(a, c) - \{a, b\}$, then either $s \in U_r$ or $s \in V_r$. If the latter, then $r$ separates $a$ from $s$, so $r < s$. Hence, suppose $s \in U_r$. Now $V_r \cup \{r\}$ is connected (28.6) and contained in the union of $U_s$ and $V_s$, so it must be contained in one of these. Since $b \in V_r \cup \{r\}$, we must then have $V_r \cup \{r\} \subset V_s$. Now $r \in V_s$ so that $s$ separates $a$ from $r$; i.e., $s < r$.

This completes the proof that $\leq$ is a total order on $E(a, b)$. ■

It is natural to ask, at this point, whether any connection exists between the order topology on $E(a, b)$ and its subspace topology relative to $X$.

**28.12 Theorem.** a) *If $E(a, b)$ has more than two points, its order topology is weaker than its subspace topology.*

b) *If $K$ is a continuum with exactly two noncut points $a$ and $b$, then $E(a, b) = K$, and the topology on $K$ is the order topology.*

*Proof.* a) It suffices to note that, for $p \in E(a, b)$, the sets $U_p \cap E(a, b)$ and $V_p \cap E(a, b)$ (in the notation of the previous proof) are open in $E(a, b)$ and

$$U_p \cap E(a, b) = \{q \in E(a, b) \mid q < p\}$$
$$V_p \cap E(a, b) = \{q \in E(a, b) \mid q < p\}.$$

b) If $p \in K$ and $p$ is not one of $a$ or $b$, then given any cutting $\{p, U, V\}$ of $K$, by Lemma 28.7, $U$ and $V$ each contain one of $a$ and $b$. Thus $p \in E(a, b)$, so $E(a. b) = K$.

From (a), the order topology is weaker than the given topology on $K$. Suppose, conversely, that $U$ is open in $K$ and $p \in U$. First assuming that $p$ is not one of $a$ or $b$, we will show that $U$ contains some interval $(r, s) = \{q \in K \mid r < q < s\}$ containing $p$. If not, then whenever $p \in (r, s)$, the closed interval $[r, s] = \{q \in K \mid r \leq q \leq s\}$ meets $K - U$. But the sets $[r, s] \cap (K - U)$ then form a family of closed subsets of $K$ with the finite intersection property (each $[r, s]$ is closed in $K$ by part a)).

Thus their intersection (in the compact space $K$) is nonempty. But $p \in U$ and

$$\bigcap \{[r, s] \mid p \in (r, s)\} = \{p\}$$

which leads to a contradiction. If $p$ is one of $a$ or $b$, the argument is similar. ■

Now every continuum with exactly two noncut points is a totally ordered set with the order topology induced by its separation order. Using the order, we are ready to characterize the metric continua with two noncut points as homeomorphs of the unit interval.

**28.13 Theorem.** *If $K$ is a metric continuum with exactly two noncut points, then $K$ is homeomorphic to the unit interval* $\mathbf{I}$.

*Proof.* Let $D$ be a countable dense subset of $K$ not containing the noncut points $a$ and $b$. Note that:

a) $D$ has no smallest or largest element,

b) given $p$ and $q$ in $D$ with $p < q$, there is an element $r$ of $D$ with $p < r < q$.

In Exercise 28B we show that every countable totally ordered set $D$ with these properties is order isomorphic, and thus homeomorphic, to the dyadic rationals $P$ in the interval $(0, 1)$. Let $f$ be an order isomorphism of $D$ onto $P$.

But each point $p$ of $K$ other than $a$ or $b$ is a cut point, dividing $K$ into sets $A_p$ and $B_p$ with $A_p < B_p$ (i.e., $x < y$ whenever $x \in A_p$ and $y \in B_p$). It follows that $f(A_p \cap D)$ and $f(B_p \cap D)$ form a Dedekind cut of the dyadic rationals, and thus uniquely determine an element $F(p)$ of $(0, 1)$. Defining $F(a) = 0$ and $F(b) = 1$, we have completed the job of extending $f$ to what is obviously an order isomorphism, and thus a homeomorphism, of $K$ onto $\mathbf{I}$. ■

With the notation and methods we have available now, the characterization of the circle comes fairly easily.

***28.14 Theorem.*** *If $K$ is a metric continuum such that for any two points $a$ and $b$, $K - \{a, b\}$ is not connected, then $K$ is homeomorphic to the unit circle.*

*Proof.* First we show $K$ has no cut points. For if $\{p, U_0, V_0\}$ is a cutting, then since $U_0 \cup \{p\}$ and $V_0 \cup \{p\}$ are continua, each has noncut points; say $y$ is a noncut point of $U_0 \cup \{p\}$ and $z$ is a noncut point of $V_0 \cup \{p\}$. But now the connected sets $(U_0 \cup \{p\}) - \{y\}$ and $(V_0 \cup \{p\}) - \{z\}$ intersect, and their union, $K - \{y, z\}$, is thus connected, contrary to the hypotheses of the theorem. Hence $K$ has no cut points.

Now let $a$ and $b$ be distinct points of $K$. Then $K - \{a, b\} = U \cup V$ where $U$ and $V$ are disjoint nonempty open subsets of $K$. We set $U^* = U \cup \{a, b\}$, $V^* = V \cup \{a, b\}$ and assert that $U^*$ and $V^*$ are arcs, each having $a$ and $b$ for endpoints and that $U^* \cap V^* = \{a, b\}$. This will obviously establish $K = U^* \cup V^*$ as a homeomorphic image of a circle.

First, $U^*$ and $V^*$ are connected. For suppose $U^* = S \cup T$ where $S$ and $T$ are disjoint, nonempty and open in $U^*$. If $S$ contains both $a$ and $b$, then $T$ is open in $U$ and hence in $K$. This is impossible, since $T$ is already closed in $K$ (being closed in the closed set $U^*$). Thus we can suppose $a \in S$, $b \in T$. But now using the same argument, $S - \{a\}$ is open and closed in the connected set $K - \{a\}$, which is impossible. Thus $U^*$ and $V^*$ are connected.

Second, $a$ and $b$ are both noncut points of $U^*$ (and similarly $V^*$). For if $S$ and $T$ disconnect $U^* - \{a\}$, and if $b \in S$ say, then (by arguments similar to those above) $T$ is both open and closed in $K - \{a\}$, which is impossible.

Finally, to show each of $U^*$ and $V^*$ has precisely two noncut points (namely, $a$ and $b$), we proceed in two stages: (1) Suppose each has a third; say $p$ is a noncut point of $U^*$ and $q$ is a noncut point of $V^*$, each different from $a$ or $b$. Then the sets $U^* - \{p\}$ and $V^* - \{q\}$ are connected, intersect, and their union is $K - \{p, q\}$, a nonconnected set. With this contradiction, we have dispensed with case 1. (2) Suppose one, say $U^*$, has a third noncut point $p$. Then if $q$ is any point in $V$, we have a cutting $\{q, A, B\}$ of $V^*$, where $A$ and $B$ are connected and, say, $a \in A$, $b \in B$. (Easily $a$ and $b$ cannot both belong to one.) Now $U^* - \{p\}$, $A$ and $B$ form a chain of connected sets whose union is $K - \{x, y\}$, a contradiction.

Thus each of $U^*$ and $V^*$ is a metric continuum with precisely two noncut points, $a$ and $b$, and $U^* \cap V^* = \{a, b\}$. It follows that $K = U^* \cup V^*$ is homeomorphic to the unit circle. ■

## Problems

### 28A. *Indecomposable continua*

A continuum $K$ is *decomposable* iff it is the union of two proper subcontinua; otherwise $K$ is *indecomposable.* For $p \in K$, consider the set $C_p$ of all points $x$ of $K$ such that a proper subcontinuum of $K$ contains both $p$ and $x$ (i.e., such that $K$ is not irreducible between $p$ and $x$). We call $C_p$ the *composant* of $p$ (or, the composant of $K$ containing $p$).

1. Describe the composants of the unit interval.

2. Every decomposable continuum is a composant for some one of its points.

3. A continuum $K$ is decomposable iff $K$ contains a proper subcontinuum $L$ with $\text{Int}_K L \neq \varnothing$.

4. Let $a, b, c$ be three points in $\mathbf{R}^2$. Construct simple chains $\mathscr{C}_1, \mathscr{C}_2, \ldots$ of connected open sets such that the sets in $\mathscr{C}_n$ have diameter less than $1/n$ and have closures contained in sets of $C_{n-1}$, with the following properties: $\mathscr{C}_1$ is a simple chain from $a$ to $c$ through $b$, $\mathscr{C}_2$ is a simple chain from $a$ to $b$ through $c$, $\mathscr{C}_3$ is a simple chain from $b$ to $c$ through $a$. Then repeat the process (Fig. 28.1). Let $C_n = \bigcup \{C \mid C \in \mathscr{C}_n\}$, and let $C = \bigcap C_n$. Then $C$ is an indecomposable continuum.

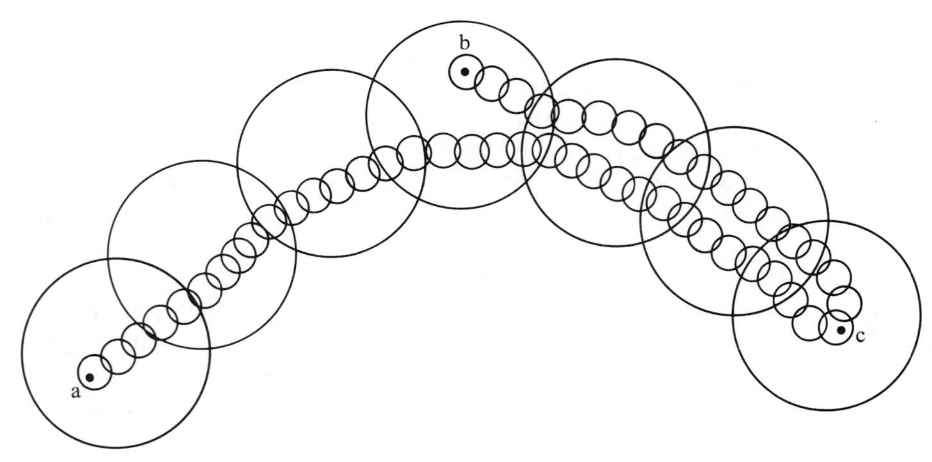


**Figure 28.1**

### 28B. *Order isomorphism*

Let $X$ and $Y$ be ordered spaces. A map $f$ of $X$ onto $Y$ is an *order isomorphism* iff $f$ is one–one and $x < y \Leftrightarrow f(x) < f(y)$.

1. Every order isomorphism is a homeomorphism relative to the order topologies on $X$ and $Y$.

2. Let $P$ denote the set of dyadic rationals in $(0, 1)$; i.e., $P$ consists of all numbers of the form $k/2^n$ for $n = 1, 2, \ldots$ and $k = 1, \ldots, 2^n - 1$. Then

   a) $P$ has no largest or smallest element,

   b) if $p, q \in P$ with $p < q$, then for some $r \in P$, $p < r < q$.

3. Any countable linearly ordered set $D$ with the properties (a) and (b) given in 2 is order isomorphic to $P$. (Thus $P$ is order isomorphic, and homeomorphic, to the set $\mathbf{Q}$ of all rationals in $\mathbf{R}$.)

### 28C. **R** *as a product*

The real line $\mathbf{R}$ can easily be written as a product space $X \times Y$, by taking $X$ to be a one-point space. Is $\mathbf{R}$ homeomorphic to any product $X \times Y$ with $X$ and $Y$ each having more than one point?

### 28D. *Continua of convergence*

Let $A_1, A_2, \ldots$ be a sequence of subsets of a space $X$. We define

$$\lim\sup A_n = \{x \in X \mid \text{each nhood of } x \text{ meets infinitely many } A_n\}$$
$$\lim\inf A_n = \{x \in X \mid \text{each nhood of } x \text{ meets all but finitely many } A_n\}$$

so that always $\lim\inf A_n \subset \lim\sup A_n$. When $\lim\inf A_n = \lim\sup A_n$, we denote their common value by $\lim A_n$.

1. $\lim\inf A_n$ and $\lim\sup A_n$ are closed sets.

2. If $X$ is compact and each $A_i$ is connected, and $\lim A_i$ exists, then $\lim A_i$ is connected.

3. If $X$ is a metric continuum which is not locally connected at one of its points $p$, there is a nhood $U$ of $p$ such that a sequence $K_1, K_2, \ldots$ of distinct components of $\overline{U}$ converges to a continuum $K$ containing $p$ and disjoint from the $K_i$. (Briefly, non-local connectedness of a metric continuum implies the existence of a "continuum of convergence," a result which is supported by reference to examples of non-locally connected spaces.)

### 28E. *Structure of continua*

1. Let $K$ be a continuum contained in $X$ and let $U$ be an open set in $X$ which meets both $K$ and $X - K$. Then every component of $\overline{U} \cap K$ meets Fr $(U)$.

2. No continuum can be written as the union of countably many disjoint closed sets. [Suppose $L = K_1 \cup K_2 \cup \cdots$. Let $G_1$ be an open set containing $K_2$ such that $\overline{G}_1 \cap K_1 = \emptyset$, and let $L_1$ be a component of $\overline{G}$ meeting $K_2$. Then $L_1 \cap K_1 = \emptyset$, but $L_1$ meets some $K_{n2}$ with $n_2 > 2$. Let $G_2$ be an open set containing $K_{n_2}$ such that $\overline{G}_2 \cap K_2 = \emptyset$ and let $L_2$ be a component of $L_1 \cap \overline{G}_2$ meeting $L_{n_2}$. Continue. Show that $L_1 \supset L_2 \supset \cdots$ but $\bigcap L_n = \emptyset$, obtaining a contradiction.]

## 29 Totally disconnected spaces

A connected space has one component. At the opposite extreme we have an important class of spaces, typified by the Cantor set.

**29.1 Definition.** A space $X$ is *totally disconnected* iff the components in $X$ are the points. Equivalently then, $X$ is totally disconnected iff the only nonempty connected subsets of $X$ are the one-point sets.

The Cantor set, the space $\mathbf{Q}$ of rationals, the space $\mathbf{P}$ of irrationals and any discrete space are all totally disconnected. We give an outline now of a famous example, due to Knaster and Kuratowski, of a connected space $\mathbf{K}$ and a point $p$ in $\mathbf{K}$ such that $\mathbf{K} - \{p\}$ is totally disconnected!

**29.2 Example.** Recall that the Cantor set $\mathbf{C}$ is obtained by deleting a countable collection of open intervals from $\mathbf{I}$. Let $Q$ be the set of endpoints of these intervals (so $Q \subset \mathbf{C}$) and set $P = \mathbf{C} - Q$. Let $p \in \mathbf{R}^2$ be the point $(\frac{1}{2}, \frac{1}{2})$ and for each $x \in \mathbf{C}$, denote by $L_x$ the straight-line segment joining $p$ and $x$. Define

$$L_x^* = \{(x_1, x_2) \in L_x \mid x_2 \text{ rational}\}, \qquad \text{if } x \in \mathbf{Q},$$
$$L_x^* = \{(x_1, x_2) \in L_x \mid x_2 \text{ irrational}\}, \qquad \text{if } x \in \mathbf{P}.$$

Then the subspace $\mathbf{K} = \bigcup_{x \in C} L_x^*$ of $\mathbf{R}^2$ is connected, while $\mathbf{K} - \{p\}$ is totally disconnected. See Exercise 29B.

**29.3 Theorem.** a) *Every product of totally disconnected spaces is totally disconnected.*

b) *Every subspace of a totally disconnected space is totally disconnected.*

*Proof.* a) Suppose $C$ is a nonempty connected subset of a product $\prod X_\alpha$ of totally disconnected spaces. Then, for each $\alpha$, $\pi_\alpha(C)$ is connected and hence must be a one-point set. It follows that $C$ is a one-point set.

b) is even easier. ∎

Continuous images of totally disconnected spaces need not be totally disconnected. In fact, one of the amazing results in topology is given in Section 30: every compact metric space is a continuous image of the Cantor set.

We now introduce a concept obviously related to total disconnectedness, but slightly stronger in the general case as examples and theorems will show.

**29.4 Definition.** A space $X$ is *0-dimensional* iff each point of $X$ has a nhood base consisting of open–closed sets. Equivalently, $X$ is 0-dimensional iff for each point $x$ in $X$ and closed set $A$ not containing $x$, there is an open–closed set containing $x$ and not meeting $A$.

The reformulation of the definition makes the following theorem clear.

**29.5 Theorem.** *Every 0-dimensional $T_1$-space is totally disconnected.*

To formulate a partial converse to this theorem we need a lemma.

**29.6 Lemma.** *A compact $T_2$-space $X$ is totally disconnected iff whenever $x \neq y$ in $X$, there is an open–closed set in $X$ containing $x$ and not $y$.*

*Proof.* This is left to Exercise 29D. ∎

**29.7 Theorem.** *A locally compact $T_2$-space is 0-dimensional iff it is totally disconnected.*

*Proof.* It suffices to prove a locally compact, totally disconnected $T_2$-space $X$ is 0-dimensional. Let $A$ be a closed set in $X$, $x \notin A$. Let $U$ be an open nhood of $x$ with compact closure disjoint from $A$. For each $p \in \text{Fr}(U)$, let $V_p$ be an open–closed subset of $\overline{U}$ containing $x$ but not $p$. The sets $X - V_p$ form an open cover of $\text{Fr}(U)$ so a finite subcover exists, say corresponding to the points $p_1, \ldots, p_n$. Let $V = V_{p_1} \cap \cdots \cap V_{p_n}$. Then $V$ is an open–closed set in $\overline{U}$ containing $y$ and disjoint from $\text{Fr}(U)$. But then $V \subset U$ and hence is an open–closed set in $X$ containing $x$ and not meeting $A$. Thus $X$ is 0-dimensional. ∎

**29.8 Examples.** a) The set $\mathbf{Q}$ of rationals is 0-dimensional.

b) The Cantor set $\mathbf{C}$ is 0-dimensional.

c) If $\mathbf{K}$ is the example of Knaster and Kuratowski, then $\mathbf{K} - \{p\}$ is a totally disconnected metric space which is not 0-dimensional. See Exercise 29B. Thus Theorem 29.7 cannot be much improved.

An infinite product of nontrivial discrete spaces is never discrete. According to Theorem 29.3, such products (and their subspaces) are, however, totally disconnected. We close this section with the important theorem providing a converse to this for an important class of totally disconnected spaces: that is, every totally disconnected compact metric space is homeomorphic to a subset of a countable product of discrete spaces.

The development requires the following notion.

**29.9 Definition.** Let $X_0, X_1, \ldots$ be topological spaces and, for each $n = 1, 2, \ldots$, let $f_n$ be a continuous map of $X_n$ into $X_{n-1}$. The sequence

$$X_0 \xleftarrow{f_1} X_1 \xleftarrow{f_2} X_2 \leftarrow \cdots,$$

which we abbreviate $\langle X_n, f_n \rangle$, is called an *inverse limit sequence.* The *inverse limit space* of this sequence is the following subset of $\prod X_n$:

$$X_\infty = \{(x_0, x_1, \ldots) \mid f_n(x_n) = x_{n-1} \text{ for each } n\}.$$

**29.10 Example.** Suppose $X_0 \supset X_1 \supset \cdots$ and $f_n: X_n \to X_{n-1}$ is the injection mapping. Then $X_\infty$ is homeomorphic to $\bigcap_{n=1}^\infty X_n$. The map is a natural one since, as a set,

$$\begin{aligned} X_\infty &= \{(x_0, x_1, \ldots) \mid x_n = x_{n-1} \text{ for each } n\} \\ &= \{(x_0, x_0, \ldots) \mid x_0 \in \textstyle\bigcap X_n\}. \end{aligned}$$

That the map $f(x) = (x, x, \ldots)$ is actually a homeomorphism of $\bigcap X_n$ onto $X_\infty$ is left as an easy exercise.

This example is the model for inverse limit sequences. Thus it is clear, e.g., that many inverse limit spaces will be empty. This does not suit our purposes and hence provokes the following theorem, generalizing the result that a decreasing intersection of nonempty compact Hausdorff spaces is nonempty.

**29.11 Theorem.** *If $\langle X_n, f_n \rangle$ is an inverse limit sequence of nonempty compact Hausdorff spaces, then the inverse limit space $X_\infty$ is a nonempty compact Hausdorff space.*

*Proof.* $X_\infty$ is obviously Hausdorff, since it is a subspace of $\prod X_n$. Moreover, if we let $Y_n = \{(x_0, x_1, \ldots) \in \prod X_n \mid f(x_n) = x_{n-1}\}$, then $X = \bigcap_{n=1}^{\infty} Y_n$ and each $Y_1 \cap \cdots \cap Y_n$ is nonempty, so it suffices to show $Y_1 \cap \cdots \cap Y_n$ is compact, for which it is enough to show each $Y_n$ is closed in the compact space $\prod X_n$.

If $z = (z_0, z_1, \ldots)$ is not in $Y_n$, then $f(z_n) \neq z_{n-1}$ so there are disjoint nhoods $U$ of $f_n(z_n)$ and $V$ of $z_{n-1}$ in $X_{n-1}$. Let $W$ be a nhood of $z_n$ in $X_n$ such that $f_n(W) \subset U$. Then $W \times V \times \prod \{X_k \mid k \neq n-1, n\}$ is a nhood of $(z_0, z_1, \ldots)$ not meeting $Y_n$. Thus $Y_n$ is closed in $\prod X_n$, as desired. ∎

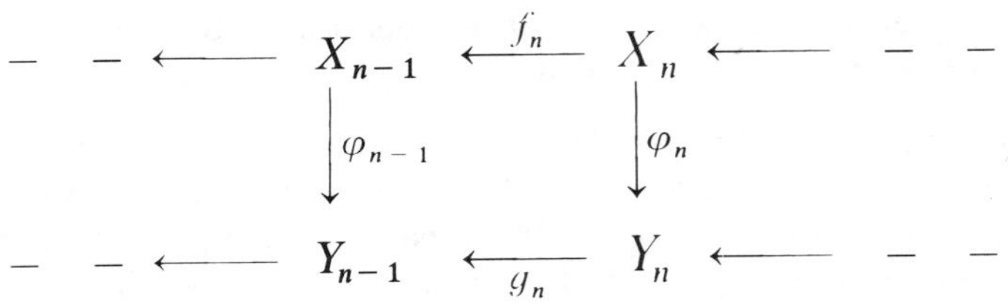

$$
\begin{array}{ccccccc}
- \; - \longleftarrow & X_{n-1} & \xleftarrow{\;f_n\;} & X_n & \longleftarrow - \; - \\
 & \downarrow \varphi_{n-1} & & \downarrow \varphi_n & \\
- \; - \longleftarrow & Y_{n-1} & \xleftarrow[\;g_n\;]{} & Y_n & \longleftarrow - \; -
\end{array}
$$

**Figure 29.1**

**29.12 Definition.** Let $\langle X_n, f_n \rangle$ and $\langle Y_n, g_n \rangle$ be inverse limit sequences. A *mapping* $\Phi$ of $\langle X_n, f_n \rangle$ to $\langle Y_n, g_n \rangle$ is a sequence $(\varphi_n)$ of mappings $\varphi_n : X_n \to Y_n$ such that $\varphi_{n-1} \circ f_n = g_n \circ \varphi_n$, for $n = 1, 2, \ldots$ (Fig. 29.1). We call $\Phi$ *continuous* iff each $\varphi_n$ is continuous, *onto* iff each $\varphi_n$ is onto, and so on. The *induced mapping* $\varphi : X_\infty \to Y_\infty$ is defined by

$$\varphi(x_0, x_1, \ldots) = (\varphi_0(x_0), \varphi_1(x_1), \ldots).$$

We know $(\varphi_0(x_0), \varphi_1(x_1), \ldots)$ belongs to $Y_\infty$ if $(x_0, x_1, \ldots)$ belongs to $X_\infty$, by virtue of the requirement $\varphi_{n-1} \circ f_n = g_n \circ \varphi_n$.

**29.13 Theorem.** a) *If $\Phi$ is continuous, so is the induced mapping $\varphi$.*

b) *If $\Phi$ is onto, so is the induced mapping $\varphi$, provided the $X_n$ and $Y_n$ are all compact Hausdorff spaces.*

*Proof.* a) Suppose $\Phi$ is continuous. Then denoting the $n$th projection in $\prod X_n$

by $\pi_n$ and the $n$th projection in $\prod Y_n$ by $\pi'_n$,

$$\pi'_n \circ \varphi(x_0, x_1, \ldots) = \varphi_n(x_n) = \varphi_n \circ \pi_n(x_0, x_1, \ldots).$$

Thus $\varphi$ is continuous when followed by each projection $\pi'_n$ and hence $\varphi$ is continuous.

b) Let $(y_0, y_1, \ldots) \in Y_\infty$ and for each $n$, let $A_n = \varphi_n^{-1}(y_n)$. Then $A_n$ is a nonempty compact subset of $X_n$. If $h_n = f_n \mid A_n$, the sequence

$$A_0 \xleftarrow{h_1} A_1 \xleftarrow{h_2} A_2 \leftarrow - - -$$

is an inverse limit sequence of nonempty compact spaces (stop to check that $h_n$ takes $A_n$ into $A_{n-1}$) and hence has nonempty limit space $A_\infty$. But if $(x_0, x_1, \ldots) \in A_\infty$, then $\varphi(x_0, x_1, \ldots) = (y_0, y_1, \ldots)$. Thus $\varphi$ is onto. ■

**29.14 Definition.** A partition of a set $X$ is a collection of disjoint sets in $X$ which cover $X$. If $\mathscr{U}_1, \mathscr{U}_2, \ldots$ is a sequence of partitions of $X$ such that $\mathscr{U}_{n+1}$ refines $\mathscr{U}_n$ for each $n \geq 0$, then the *derived sequence* obtained from $\mathscr{U}_1, \mathscr{U}_2, \ldots$ is the inverse limit sequence

$$Y_0 \xleftarrow{f_1} Y_1 \xleftarrow{f_2} \cdots$$

where $Y_n$ is the discrete space having the sets of $\mathscr{U}_n$ as elements and $f_n$ takes each set in $\mathscr{U}_n$ to the unique set in $\mathscr{U}_{n-1}$ which contains it.

**29.15 *Theorem.*** *Let $X$ be a totally disconnected compact metric space. Then*

a) *For each $n = 0, 1, 2, \ldots$ there is a finite open cover $\mathscr{U}_n$ of $X$ by disjoint open sets of diameter $< 1/2^n$ such that $\mathscr{U}_{n+1} < \mathscr{U}_n$ for each $n \geq 0$.*

b) *If $Y_0 \xleftarrow{f_1} Y_1 \leftarrow \cdots$ is the derived sequence of any such sequence $\mathscr{U}_0, \mathscr{U}_1, \ldots$ of covers, then $X$ is homeomorphic to the resulting inverse limit space $Y_\infty$.*

*Proof.* a) Since $X$ is compact and totally disconnected, it is 0-dimensional, so a cover $\mathscr{U}$ of $X$ by open–closed sets of diameter $< 1$ certainly exists. By compactness, $\mathscr{U}$ can be taken finite, $\mathscr{U} = \{U_1, \ldots, U_n\}$. Define

$$U'_1 = U_1, \quad U'_2 = U_2 - U_1, \quad \ldots, \quad U'_n = U_n - (U_1 \cup \cdots \cup U_{n-1}).$$

Then $\mathscr{U}_0 = \{U'_1, \ldots, U'_n\}$ is a finite cover by disjoint open sets of diameter $< 1$.

Having obtained $\mathscr{U}_0, \ldots, \mathscr{U}_{n-1}$, we can refine $\mathscr{U}_{n-1}$ by a finite cover of open–closed sets $\mathscr{U} = \{U_1, \ldots, U_k\}$ of diameter $< 1/2^n$ and then

$$\mathscr{U}_n = \left\{U_1, U_2 - U_1, \ldots, U_k - \bigcup_{j<k} U_j\right\}$$

is the desired $n$th cover.

b) Since the spaces $Y_n$ are nonempty, compact, Hausdorff spaces, $Y_\infty$ is compact and nonempty by 29.11. For each $n$, define $\varphi_n \colon X \to Y_n$ by letting $\varphi_n(x)$ be the set in $\mathscr{U}_n$ (i.e., element of $Y_n$) containing $x$. Then $(\varphi_n)$ is a mapping of the sequence $X \xleftarrow{i} X \xleftarrow{i} X \leftarrow \cdots$ (where $i$ is the identity) to $Y_0 \xleftarrow{f_1} Y_1 \xleftarrow{f_2} Y_2 \leftarrow \cdots$, because if $x \in X$, then

$$\begin{aligned} \varphi_{n-1} \circ i(x) &= \varphi_{n-1}(x) = \text{element of } \mathscr{U}_{n-1} \text{ containing } x, \\ f_n \circ \varphi_n(x) &= f_n \text{ (element of } \mathscr{U}_n \text{ containing } x) \\ &= \text{element of } \mathscr{U}_{n-1} \text{ containing } x, \end{aligned}$$

so that the desired commutativity relation holds. Moreover, each $\varphi_n$ is continuous and onto and hence so is the induced map $\varphi: X \to Y_\infty$ (it is obvious that the inverse limit space of $X \xleftarrow{i} X \xleftarrow{i} \cdots$ is $X$). Since $X$ is compact, and $Y_\infty$ is Hausdorff, $\varphi$ is also a closed map. Hence we need only show $\varphi$ is one–one.

But if $x \neq y$ in $X$, then say $\rho(x, y) = \epsilon$. Pick $n$ large enough that $1/2^n < \epsilon/2$. Then since each element of $\mathscr{U}_n$ has diameter $< 1/2^n$, $x$ and $y$ cannot belong to the same element of $\mathscr{U}_n$; i.e., $\varphi_n(x) \neq \varphi_n(y)$. Thus, easily, $\varphi(x) \neq \varphi(y)$. ∎

## Problems

### 29A. *Examples on totally disconnected and 0-dimensional spaces*

1. The Sorgenfrey line **E** is 0-dimensional.
2. The set **P** of irrationals is 0-dimensional.
3. $\beta$**N** and $\beta$**Q** are totally disconnected [allowable reference: Gillman and Jerison].

### 29B. *The example of Knaster and Kuratowski*

Recall the construction of the space **K** (29.2). **K** consists of the "rational points" on the lines joining endpoints of **C** to $p = (\frac{1}{2}, \frac{1}{2})$ and the "irrational points" on the lines joining other points of **C** to $p$.

1. **K** is connected. [If $U$ is an open–closed subset of **K** containing $p$, $U$ has open–closed intersection with each line $L_x^*$. Deduce that $U =$ **K**.]

2. **K** $- \{p\}$ is totally disconnected.

3. **K** $- \{p\}$ is not 0-dimensional. [The open set $\{(x, y) \mid y < \frac{1}{4}\}$ of **K** $- \{p\}$ cannot contain any open–closed set (otherwise, this set would be a proper open–closed subset of **K**).]

### 29C. *Inverse limit spectra*

Inverse limit sequences and their limit spaces have a natural generalization, obtained by replacing the integers as index set with any directed set. Specifically, let $A$ be any directed set and suppose $X_\alpha$ is a topological space for each $\alpha \in A$. For each $\alpha$ and $\beta$ with $\alpha < \beta$, let $f_{\beta\alpha}: X_\beta \to X_\alpha$ be a continuous map. The collection of spaces $X_\alpha$ and maps $f_{\beta\alpha}$ will be called an *inverse limit spectrum*, denoted $\langle X_\alpha; f_{\beta\alpha}\rangle$, provided the following condition is satisfied: if $\alpha < \beta < \gamma$, then $f_{\gamma\alpha} = f_{\beta\alpha} \circ f_{\gamma\beta}$.

The *inverse limit space* of an inverse limit spectrum $\langle X_\alpha; f_{\beta\alpha}\rangle$ is the set

$$X_\infty = \{x \in \prod X_\alpha \mid \text{whenever } \alpha < \beta, x_\alpha = f_{\beta\alpha}(x_\beta)\}.$$

1. If each $X_\alpha$ is $T_2$, then $X_\infty$ is closed in $\prod X_\alpha$.

2. If each $X_\alpha$ is a (nonempty) compact Hausdorff space, then $X_\infty$ is a (nonempty) compact Hausdorff space.

3. The projection $\pi_\alpha$ restricted to $X_\infty$ still maps $X_\infty$ *onto* $X_\alpha$, and the sets $\pi_\alpha^{-1}(U)$ for $\alpha \in A$ and $U$ open in $X_\alpha$ form a base (rather than just a subbase!) for $X_\infty$.

4. Suppose $\langle X_\alpha; f_{\beta\alpha}\rangle$ and $\langle Y_\alpha; g_{\beta\alpha}\rangle$ are two inverse limit spectra with the same index set $A$, and for each $\alpha \in A$, let $h_\alpha: X_\alpha \to Y_\alpha$ be continuous. If the $h_\alpha$ satisfy the appropriate composition condition, then a unique map $h_\infty: X_\infty \to Y_\infty$ is induced such that the diagram in Fig. 29.2 commutes (i.e., such that $h_\alpha \circ \pi_\alpha = \pi_\alpha \circ h_\infty$) for each $\alpha \in A$.

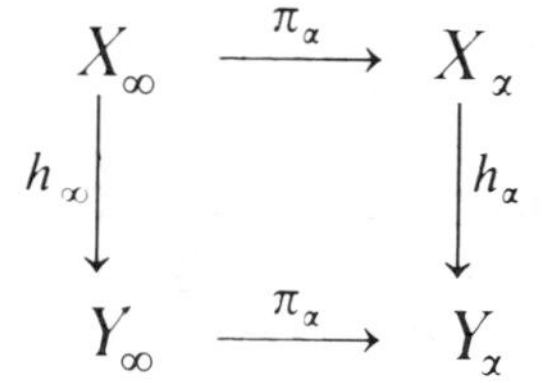

$$\begin{array}{ccc} X_\infty & \xrightarrow{\pi_\alpha} & X_\alpha \\ {\scriptstyle h_\infty}\downarrow & & \downarrow{\scriptstyle h_\alpha} \\ Y_\infty & \xrightarrow{\pi_\alpha} & Y_\alpha \end{array}$$

**Figure 29.2**

5. If each $h_\alpha$ is a homeomorphism of $X_\alpha$ with $Y_\alpha$, then $h_\infty$ is a homeomorphism of $X_\infty$ onto $Y_\infty$.

Inverse limit spectra and their limit spaces are important in the extension of homology and cohomology theory from simplicial objects to the more general Čech theory, applicable to a wide class of spaces. See the book by Spanier on algebraic topology.

### 29D. *Totally disconnected compact Hausdorff spaces*

In a compact Hausdorff space, the quasicomponents (26B) are the components. Conclude that a compact Hausdorff space is totally disconnected iff distinct points can be separated by an open–closed set containing one and not the other.

### 29E. *Connectedness in topological groups*

Let $G$ be a topological group.

1. The component $C$ of the identity in $G$ is a closed normal subgroup.

2. $G/C$ is totally disconnected (so if $G$ is locally compact, $G/C$ is 0-dimensional).

3. An open–closed compact nhood $U$ of $e$ in $G$ contains an open–closed subgroup $H$. [Use 18D.2 to find a (symmetric) nhood $V$ of $e$ such that $UV \subset U$. It follows that $V^n \subset U$ for any $n$. Then $\bigcup_{n=1}^\infty V^n$ is an open (hence closed by 18D.7) subgroup contained in $U$.]

4. If $G$ is compact, an open–closed compact nhood $U$ of $e$ in $G$ contains an open–closed *normal* subgroup $N$. [Let $H$ be the subgroup given by 3 and let $N = \bigcap_{x \in G} xHx^{-1}$.]

5. If $G$ is locally compact and totally disconnected, the open–closed subgroups of $G$ form a base at $e$. [The open–closed nhoods of $e$ are a base. See part 3.]

6. If $G$ is locally compact, $C$ is the intersection of all open–closed subgroups. [$G/C$ is locally compact and totally disconnected.]

7. In a locally compact group, the following are equivalent:

a) $G$ is connected,
b) $G$ has no proper open–closed subgroups,
c) $G = \bigcup_{n=1}^\infty V^n$ for any open nhood $V$ of $e$.

29F. *Cantor spaces*

As a corollary to Theorem 29.15, every totally disconnected compact metric space can be embedded in a product of countably many finite discrete spaces. The corollary can be strengthened. Show that every 0-dimensional $T_1$-space (hence, every locally compact totally disconnected $T_1$-space) which has a base $\mathscr{B}$ of cardinal $\aleph$ can be embedded in the product of $\aleph$ copies of the discrete space with two points. (Recall that a product of two-point discrete spaces is called a *Cantor space*.) [The base $\mathscr{B}$ can be taken to consist of open–closed sets, by an extension of 16B.2. For each $B \in \mathscr{B}$, consider the characteristic function of $B$. Apply 8.16.]

## 30 The Cantor set

The Cantor set **C** is a totally disconnected compact metric space. By adding one more property to this list, we can completely characterize **C**. Our goal in this section is the proof of this useful fact, and one of its startling corollaries: every compact metric space is a continuous image of **C**.

**30.1 Definition.** A set $A$ in a space $X$ is *perfect* in $X$ iff $A$ is closed and dense in itself; i.e., each point of $A$ is an accumulation point of $A$.

The whole space $X$, then, is perfect iff it is dense in itself. In particular, the Cantor set **C** is perfect.

***30.2 Lemma.*** *If $U$ is any nonempty open set in a compact totally disconnected, perfect $T_2$-space and $n$ is any positive integer, then $U = U_1 \cup \cdots \cup U_n$ for some choice of nonempty disjoint open sets $U_1, \ldots, U_n$.*

*Proof.* It suffices to check the case $n = 2$, since all others will follow by induction. But if $U$ is any nonempty open set in $X$, then $U$ cannot be a single point since $X$ is perfect. Now if $p$ and $q$ are different points of $U$, then there is an open–closed set $V$ in $X$ which contains $p$ but not $q$, by 29.6. Setting $U_1 = U \cap V$ and $U_2 = U - V$ gives the desired separation of $U$. ■

Now, given two totally disconnected, perfect, compact metric spaces, we can approximate them by inverse limit sequences of discrete spaces by using Theorem 29.15 and we can keep the discrete spaces in the two sequences the same size at each stage, using 30.2. The result is the following theorem.

***30.3 Theorem.*** *Any two totally disconnected, perfect compact metric spaces are homeomorphic.*

*Proof.* Let $X$, $Y$ be such spaces. Let $(\mathscr{U}_n)$, $(\mathscr{V}_n)$ be sequences of finite covers of $X$ and $Y$, respectively, by disjoint open sets, the sets of the $n$th covers having diameter $< 1/2^n$. The existence of these is guaranteed by the Theorem 29.15. By using Lemma 30.2 in order to split sets where necessary, we may assume $\mathscr{U}_n$ and $\mathscr{V}_n$ have the same number of elements for each $n$.

Now if $\mathscr{U}_1 = \{U_{11}, \ldots, U_{1n}\}$ and $\mathscr{V}_1 = \{V_{11}, \ldots, V_{1n}\}$, then each $U_{ij}$ is a union of elements of $\mathscr{U}_2$, and each $V_{ij}$ is a union of elements of $\mathscr{V}_2$. Again, using

Lemma 30.2 we can assume $U_{1j}$ and $V_{1j}$ are the union of the same number of elements of $\mathscr{U}_2, \mathscr{V}_2$, respectively, in such a way that $U_{2k} \subset U_{1j}$ iff $V_{2k} \subset V_{ij}$. Continue in this fashion, matching the covers of $\mathscr{U}_n$ and $\mathscr{V}_n$ for all $n$.

Now let $X_0 \xleftarrow{f_1} X_1 \leftarrow \cdots$ and $Y_0 \xleftarrow{g_1} Y_1 \leftarrow \cdots$ be the derived sequences of $(\mathscr{U}_n)$ and $(\mathscr{V}_n)$, respectively. Define $\varphi_n : X_n \to Y_n$ by $\varphi_n(U_{nj}) = V_{nj}$. Then $\varphi_n$ is a homeomorphism from $X_n$ to $Y_n$, and it is easily verified that $\varphi : X_\infty \to Y_\infty$ is also then a homeomorphism. But $X_\infty$ is homeomorphic to $X$, and $Y_\infty$ is homeomorphic to $Y$. ■

**30.4 Corollary.** *The Cantor set is the only totally disconnected, perfect compact metric space (up to homeomorphism).*

The previous result provides us with some interesting and easily proved results (some of which we already know). Recall that $2^{\aleph_0}$ denotes the product of $\aleph_0$ copies of the two-point discrete space.

**30.5 Corollary.** *The Cantor set* $\mathbf{C}$ *is homeomorphic to* $2^{\aleph_0}$.

**30.6 Corollary.** *The Cantor set* $\mathbf{C}$ *is homeomorphic to* $\mathbf{C}^{\aleph_0}$.

The next result is a much deeper (and more startling) application of the characterization theorem.

**30.7 Theorem.** *Every compact metric space* $X$ *is a continuous image of the Cantor set.*

*Proof.* Let $\mathscr{U}_1, \mathscr{U}_2, \ldots$ be a sequence of finite covers of $X$ by the closures of open sets, the sets of $\mathscr{U}_n$ being of diameter $< 1/2^n$, such that $\mathscr{U}_n < \mathscr{U}_{n-1}$ for $n = 2, 3, \ldots$ . Say $\mathscr{U}_n = \{U_{n1}, \ldots, U_{nk_n}\}$. For each $U_{1i} \in \mathscr{U}_1$, define $V_{1i} = \{(u, i) \mid u \in U_{1i}\}$ so that $V_1 = V_{11} \cup \cdots \cup V_{1k_1}$ is the disjoint union of the $U_{1i}$. Now each $U_{2j} \in \mathscr{U}_2$ is contained in some $U_{1k} \in \mathscr{U}_1$. Define $V_{2ij} = \{(u, i, j) \mid u \in U_{2j}\}$ whenever $U_{2j} \subset U_{1i}$, and let $V_2 = \bigcup_{j=1}^{k_2} \bigcup_{U_{1i} \supset U_{2j}} V_{2ij}$. Then $V_2$, it is worth pointing out, is somewhat more than the disjoint union of the $U_{2j}$. Each $U_{2j}$ occurs in the disjoint union once for each $U_{1i}$ such that $U_{2j} \subset U_{1i}$. Now define $f_2 : V_2 \to V_1$ by $f_2((u, i, j)) = (u, i)$. Then $f_2$ is continuous on each piece $V_{2ij}$ and thus continuous on $V_2$. Also, there is a map $\varphi_1 : V_1 \to X$ defined by $\varphi_1(u, i) = u$ and a map $\varphi_2 : V_2 \to X$ defined by $\varphi_2(u, i, j) = u$.

Continue the process. The result is a pair of inverse sequences and a mapping between them (Fig. 30.1), where $i$ is the identity map on $X$. The reader should

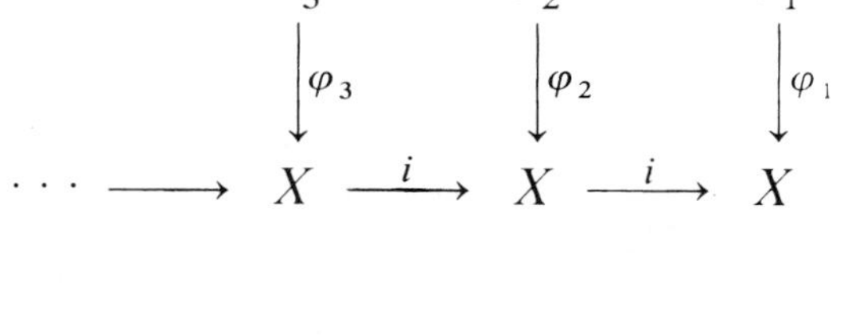

$$
\begin{array}{ccccccc}
\cdots \longrightarrow & V_3 & \xrightarrow{f_3} & V_2 & \xrightarrow{f_2} & V_1 \\
& \downarrow{\scriptstyle \varphi_3} & & \downarrow{\scriptstyle \varphi_2} & & \downarrow{\scriptstyle \varphi_1} \\
\cdots \longrightarrow & X & \xrightarrow{i} & X & \xrightarrow{i} & X
\end{array}
$$

**Figure 30.1**

check that $(\varphi_n)$ satisfies the composition condition necessary to be a map of inverse limit sequences. The result is a map $\varphi: V_\infty \to X$ of the inverse limit spaces, which is continuous and onto because $X$ and each $V_n$ is a compact Hausdorff space and each $\varphi_n$ is continuous and onto.

It is worth pointing out, at this stage, that each $V_n$ is a compact metric space, being a disjoint union of a finite number of compact metric spaces. Let $d_n$ be the metric on $V_n$ induced by the metrics on the $U_{nj}$. We also need the obvious fact that if $(x_1, x_2, \ldots) \in V_\infty$, then we must have $\varphi(x_1) = \varphi_2(x_2) = \cdots$, and, if $z_x$ denotes this common value, then for any $(y_1, y_2, \ldots) \in V_\infty$, $d_n(x_n, y_n) \geq d(z_x, z_y)$.

We would like to show $V_\infty$ is the Cantor set. It is compact because each $V_n$ is compact, and metric because it is a subset of the metric space $\prod_{n=1}^\infty V_n$. If $x = (x_0, x_1, \ldots)$ and $y = (y_0, y_1, \ldots)$ are distinct points of $V_\infty$, then for some $n$, $x_n \neq y_n$. Now $x_n$ and $y_n$ must correspond to distinct points of $X$ (under $\varphi_n: V_n \to X$), say to $z_x$ and $z_y$. Now if $d_m$ is the metric on $V_m$, then clearly $d_m(x_m, y_m) \geq d(z_x, z_y)$ for all $m \geq n$. Since the diameters of the sets $V_{m1}, \ldots, V_{mk_n}$ which compose $V_m$ approach 0 as $m \to \infty$, it follows that beyond some point $N$, $x_m$ and $y_m$ belong to different sets of $V_m$; say, $x_m \in V_{m1}$, $y_m \notin V_{m1}$. But $V_{m1}$ is open–closed in $V_m$, and hence $\{(z_0, z_1, \ldots) \in V_\infty \mid z_m \in V_{m1}\}$ is an open–closed nhood of $x$ in $V_\infty$ not containing $y$. Thus $V_\infty$ is totally disconnected.

But $V_\infty$ need not, in general, be perfect. However, if **C** is the Cantor set, $V_\infty \times \mathbf{C}$ is a perfect, totally disconnected compact metric space which has $V_\infty$, and hence $X$, for a continuous image. ■

## Problems

### 30A. *Properties of* C

1. **C** is nowhere dense in **I**.

2. **C** is homogeneous (i.e., given $x$ and $y$ in **C**, a homeomorphism of **C** onto itself can be found which carries $x$ to $y$). Thus, the property of being an endpoint in **C** is not topological, but merely reflects a peculiarity of the embedding of **C** in **I**.

3. Any totally disconnected, compact metric space is homeomorphic to a subset of **C**. [See 29.15(b).]

### 30B. *Perfect sets*

1. Every perfect set in a complete metric space contains a compact perfect set.

2. A compact Hausdorff space which is countable is not perfect. "compact Hausdorff" can be replaced by "complete metric."

3. If $A \subset X$ has no isolated points, then $\bar{A}$ is perfect in $X$.

### 30C. *Open subsets of* C

1. Every open subset of **C** can be written as the union of a finite or infinite sequence of disjoint open–closed subsets of **C**.

2. Every open subset of **C** is homeomorphic either to **C** or to **C** $-$ $\{0\}$. [If $G$ is a finite union as in (1), then $G$ is a 0-dimensional, perfect, compact metric space and is thus homeomorphic to **C**. If $G = \bigcup G_n$ as in (1), then we can write $\mathbf{C} - \{0\}$ as $\bigcup C_n$ where the $C_n$ are disjoint nonempty open–closed subsets of **C** and for each $n$, $C_n$ and $G_n$ are homeomorphic, so $\mathbf{C} - \{0\}$ and $G$ are homeomorphic (see 7H).]

### 30D. $\beta\mathbf{N}$ *versus* C

1. Every compact metric space is the continuous image of $\beta\mathbf{N}$.
2. Is $\beta\mathbf{N}$ a continuous image of **C**?
3. Why is **C** preferable to $\beta\mathbf{N}$ as a universal mapping space for compact metric spaces?

### 30E. *Scattered sets; the perfect kernel*

A topological space $X$ is *scattered* iff it contains no nonempty dense-in-itself subset.

1. Every discrete space is scattered. Exhibit a nondiscrete scattered space. [There are infinite compact subsets of **R** which are scattered.]

2. Every topological space $X$ can be written as the union of two disjoint sets, one perfect, the other scattered. (The perfect set in this union is called the *perfect kernel* of $X$.)

### 30F. *Homeomorphism and product spaces*

Let $X$ be a compact space such that $X \times X$ is homeomorphic to $X$. Must $X^{\aleph_0}$ be homeomorphic to $X$? [Add an isolated point to **C**.] Note that noncompact spaces $X$ with this property are easy to find. These results complement the observations made in Exercise 8J.

### 30G. *Convex sets in* $\mathbf{R}^n$

A subset $E$ of $\mathbf{R}^n$ is *convex* iff whenever $x$ and $y$ belong to $E$, then $E$ contains the closed line segment joining $x$ to $y$. Show that every closed bounded convex set in $\mathbf{R}^n$ is the continuous image of **I**. [There is a continuous map $f$ of **C** onto $E$, by 30.7. How can $f$ be extended to all of **I**?]

## 31 Peano spaces

Here we give a topological characterization of those spaces which are continuous images of the unit interval **I**.

**31.1 Definition.** A *Peano space* is a compact, connected, locally connected metric space.

The next three results are directed specifically at the proof of the Hahn–Mazurkiewicz theorem, which characterizes the continuous images of **I** as precisely the Peano spaces.

***31.2 Theorem.*** *Every Peano space is arcwise connected.*

*Proof.* Suppose $a$ and $b$ are points in a Peano space $X$. Using Theorem 26.15, there is a simple chain $U_{11}, \ldots, U_{1n}$ of open connected sets of diameter $<1$ from

$a$ to $b$. About each point $p$ of $U_{1i}$ there is an open connected set $V$ of diameter $<\frac{1}{2}$ whose closure is contained in $U_{1i}$, and if $p \in U_{1i+1}$ we can arrange that $\bar{V} \subset U_{1i+1}$ also. Do this for each $i = 1, \ldots, n$. We wish now to obtain a simple chain of such sets $V$ from $a$ to $b$.

Pick $x_i \in U_{1i} \cap U_{1i+1}$ for $i = 1, \ldots, n-1$ and for each $i = 0, \ldots, n-1$ (with $a = x_0$ and $b = x_n$) find a simple chain of the sets $V$ from $x_i$ to $x_{i+1}$ in $U_{1i+1}$. We cannot simply join these together to get a simple chain from $a$ to $b$, because of doubling back (Fig. 31.1), but we can obtain the desired simple chain as follows: take all elements of the first chain (from $a$ to $x_1$) up to and including the first one $U$ meeting some element $V$ of the second chain (from $x_1$ to $x_2$), then omit all elements of the first chain after $U$ and all elements of the second chain before $V$. Repeat this at all other intersections.

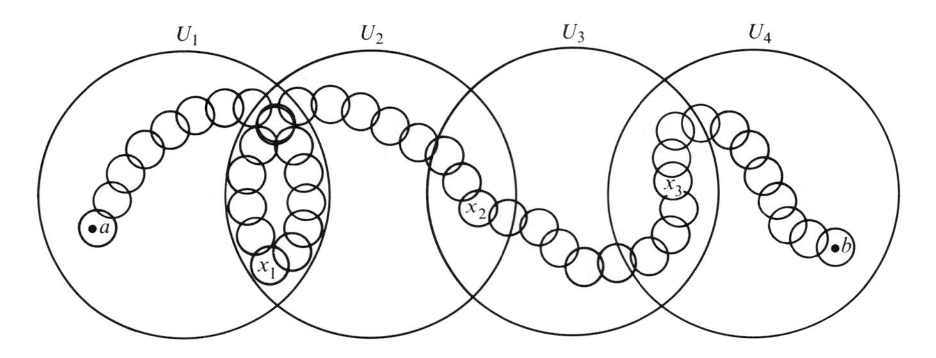


**Figure 31.1**

The result, then, is a chain $U_{21}, \ldots, U_{2n_2}$ of open connected sets of diameter $<\frac{1}{2}$ such that for each $i$, $\bar{U}_{2i} \subset U_{1j}$ for some $j$. Now continue this process, obtaining a simple chain of open connected sets of diameter $<1/2^n$ whose closures lie in elements of the previous chain for each $n > 1$.

For each $n$, let $C_n$ be the union of the closures of the elements of the $n$th chain. Then $C = \bigcap C_n$ is a compact, connected metric space containing $a$ and $b$. We have finished if we show that no points other than $a$ and $b$ are noncut points, since then $C$ is an arc by 28.13.

Let $x \in C - \{a, b\}$. For given $n$, at most one or two links of the $n$th chain contain $x$. Let $A_n$ be the union of all the links preceding these, $B_n$ the union of all the links following these. Then

$$A = \bigcup_{n=1}^{\infty} (A_n \cap C) \qquad \text{and} \qquad B = \bigcup_{n=1}^{\infty} (B_n \cap C)$$

form a separation of $C - \{x\}$ into disjoint, nonempty open sets. Thus $x$ is a cut point of $C$. ■

The proof just given can easily be modified to show that *every open connected subset of a Peano space is arcwise connected* (see 31C.1). We will use this fact a little later, in the proof of Lemma 31.4.

***31.3 Lemma.*** *A compact locally connected metric space is "uniformly locally connected"; that is, for any $\epsilon > 0$, there is some $\delta > 0$ such that whenever $\rho(x, y) < \delta$, then $x$ and $y$ both lie in some connected subset of $X$ of diameter $<\epsilon$.*

*Proof.* Given $\epsilon > 0$, cover $X$ by open connected nhoods of diameter $<\epsilon$. Reduce this to a finite subcover $\{V_{x_1}, \ldots, V_{x_n}\}$ and let $\delta$ be a Lebesgue number (22.5) for this cover. Then if $p(x, y) < \delta$, both $x$ and $y$ belong to some $V_{x_i}$. ■

***31.4 Theorem.*** *A Peano space $X$ is uniformly locally arcwise connected; i.e., for each $\epsilon > 0$, there is a $\delta > 0$ such that whenever $p(x, y) < \delta$, then $x$ and $y$ are joined by an arc of diameter $<\epsilon$.*

*Proof.* First, $X$ is uniformly locally connected, by 31.3. Thus if $\epsilon > 0$ is given, there is a $\delta > 0$ such that if $p(x, y) < \delta$, then $x$ and $y$ lie together in a connected set $B$ of diameter $<\epsilon/2$. Each $x \in B$ has an open connected nhood $U_x$ of diameter $<\epsilon/4$. Then $U = \bigcup_{x \in B} U_x$ is an open connected subset of $X$, and hence, (see Exercise 31C.1), $U$ is arcwise connected. Thus, if $p(x, y) < \delta$, then $x$ and $y$ lie in an arcwise connected subset $U$ of diameter $<\epsilon$. ■

We are now ready to prove the Hahn–Mazurkiewicz theorem, classifying the continuous images of the unit interval as the Peano spaces. Proving that continuous images of **I** have the properties of a Peano space is no trouble; all the necessary theorems are already at hand. But to prove the converse is significantly more difficult. The basic idea is that given any Peano space $X$, there is a continuous map of the Cantor set onto $X$ by 30.7 and, using the small arcs in $X$ provided by the previous theorem, we can extend this map to the whole unit interval. The details, of course, are painful.

***31.5 Theorem.*** (Hahn and Mazurkiewicz) *A Hausdorff space $X$ is a continuous image of the unit interval* **I** *iff it is a Peano space.*

*Proof.* Let $f$ be a continuous map of **I** onto $X$. By 23.2, $X$ must be compact and metric. Moreover, $X$ is the continuous image of a connected space and a quotient (in fact, a closed, continuous image) of a locally connected space, so $X$ has these properties itself. Thus, $X$ is a Peano space.

Now suppose $X$ is any Peano space. Recall **C** is the Cantor set in **I**, with $I_1, I_2, \ldots$ being the intervals in **I** − **C** ordered by size, and for intervals of the same size, from left to right. Let $f$ be a continuous map of **C** onto $X$. Our problem is to extend $f$ continuously over each $I_n = (p_n, q_n)$. Now $f(p_n)$ and $f(q_n)$ are already defined. If $f(p_n) = f(q_n)$, define $f^*(p) = f(p_n)$ for each $p \in I_n$. Now for each $n = 1, 2, \ldots$ find $\delta_n > 0$ such that $\rho(x, y) < \delta_n$ in $X \Rightarrow x$ and $y$ are joined by an arc of diameter $<1/2^n$. But, for each $n$, find $\eta_n > 0$ such that $|p - q| < \eta_n$ in $C \Rightarrow \rho(f(p), f(q)) < \delta_n$.

Only finitely many intervals $I_1, \ldots, I_{n_1}$ have length $\geq \eta_1$ and for each such $I_j$ extend $f$ to $(p_j, q_j)$ by letting its values run over any arc from $f(p_j)$ to $f(q_j)$ in $X$. Then intervals $I_{n_1+1}, \ldots, I_{n_2}$ will have length $\eta_2 \leq |I_j| < \eta_1$. For these intervals $I_j$ we have $|p_j - q_j| < \eta_1$ so that $\rho(f(p_j), f(q_j)) < \delta_1$. Extend $f$ to $(p_j, q_j)$ by letting its values run over any arc of diameter $<1$ between $f(p_j)$ and $f(q_j)$. In general, for the intervals $I_j$ such that $\eta_{k+1} \leq |I_j| < \eta_k$, we can let the values of $f$ on $(p_j, q_j)$ run over an arc of diameter $<1/k$ between $f(p_j)$ and $f(q_j)$.

The result is a function from **I** onto **C** whose continuity can be easily checked, once you see what is going on. ■

***31.6 Corollary.*** *A $T_2$-space is pathwise connected iff it is arcwise connected.*

*Proof.* By the previous theorem and Theorem 31.2, every path is arcwise connected. ■

## Problems

### 31A. *Peano spaces*

1. If $X$ is a Peano space of more than one point and $Y$ is any Peano space, there is a continuous map of $X$ onto $Y$.

2. If $a$ and $b$ are distinct points in $X$ above, and $c$ and $d$ are distinct points in $Y$, the map $f$ can be so constructed that $f(a) = c$ and $f(b) = d$.

### 31B. *Uniform local connectedness*

1. Every uniformly locally connected space is locally connected. [By 27.16, it is enough to show such a space is connected *im kleinem* at each point.]

2. The converse fails. [Consider the graph of sin $(1/x)$ for $x > 0$.]

### 31C. *Subsets of Peano spaces*

1. An open connected subset of a Peano space is arcwise connected.

2. Is a compact, connected subset of a Peano space always a Peano space?

### 31D. *Mapping the Cantor set*

Show that the extension $F$ of the map $f$ of the Cantor set onto $X$ given in the proof of 32.5 is continuous.

## 32 The homotopy relation

In the next three sections, we will provide a brief introduction to homotopy theory, one of the branches of algebraic topology. Our limited aim is the development of sufficient machinery to prove the Brouwer fixed-point theorem (34.6).

In this first section, we will build the framework of basic definitions and theorems which will enable us to introduce the appropriate algebraic techniques in Section 33. The Brouwer theorem will follow easily after we have applied these techniques to study the unit circle in Section 34.

**32.1 Definition.** Let $f$ and $g$ be continuous functions from $X$ to $Y$. We say $f$ is *homotopic* to $g$, written $f \simeq g$, iff there is a continuous function $H: X \times \mathbf{I} \to Y$ such that $H(x, 0) = f(x)$ and $H(x, 1) = g(x)$ for all $x \in X$. The map $H$ is called a *homotopy* between $f$ and $g$. For clarity, we will sometimes write $H: f \simeq g$ when $H$ is a homotopy between $f$ and $g$.

Setting $f_t(x) = H(x, t)$ for $x \in X$ and $t \in \mathbf{I}$, the homotopy $H$ is seen to represent a family $\{f_t \mid t \in \mathbf{I}\}$ of maps from $X$ to $Y$, varying continuously with $t$, such that $f_0 = f$ and $f_1 = g$. Thus $H$ gives a continuous deformation of the map $f$ into the map $g$.

**32.2 Examples.** a) In $\mathbf{R}^n$, define $f(x) = x$ for all $x$ and $g(x) = 0$ for all $x$. Then $f \simeq g$, the homotopy being given by

$$H(x, t) = (1 - t)x.$$

b) Let $X$ be any space, $Y$ a convex subset of $\mathbf{R}^n$. Then any two maps $f, g: X \to Y$ are homotopic, the homotopy being given by

$$H(x, t) = t \cdot g(x) + (1 - t) \cdot f(x)$$

Note the importance of the range in questions of homotopy. For example, if $\mathbf{D}$ is the disk $\{(x_1, y_2) \mid x_1^2 + y_2^2 \leq 1\}$ in the plane, we can conclude from example b) above that any two maps from $\mathbf{S}^1$ to $\mathbf{D}$ are homotopic; for instance, the map $f(x) = x$ and the constant map $g(x) = (1, 0)$. But regarded as maps from $\mathbf{S}^1$ to $\mathbf{S}^1$, $f$ and $g$ are no longer homotopic (34.4).

> ***32.3 Theorem.*** $\simeq$ *is an equivalence relation in the set $C(X, Y)$ of all continuous maps from $X$ to $Y$.*

*Proof.* If $f \in C(X, Y)$, then $H: f \simeq f$, where $H$ is defined by $H(x, t) = f(x)$ for all $x \in X$ and $t \in I$.

If $f, g \in C(X, Y)$ and $H: f \simeq g$, then $H': g \simeq f$ where $H'(x, t) = H(x, 1 - t)$ for all $x \in X$ and $t \in I$.

If $f, g, h \in C(X, Y)$ and $H_1: f \simeq g$ while $H_2: g \simeq h$, then $H: f \simeq g$, where

$$H(x, t) = \begin{cases} H_1(x, 2t) & 0 \leq t \leq \frac{1}{2} \\ H_2(x, 2t - 1) & \frac{1}{2} \leq t \leq 1. \end{cases}$$

$H$ is continuous on $X \times \mathbf{I}$ since it is continuous on each of the closed subsets $X \times [0, \frac{1}{2}]$ and $X \times [\frac{1}{2}, 1]$. ■

**32.4 Definition.** The equivalence classes in $C(X, Y)$ under the relation $\simeq$ are called the *homotopy classes* in $C(X, Y)$.

**32.5 Theorem.** *Composites of homotopic maps are homotopic.*

*Proof.* Suppose $f_1$ and $g_1$ are homotopic maps from $X$ to $Y$ and $f_2$ and $g_2$ are homotopic maps from $Y$ to $Z$; say $H_1: f_1 \simeq g_1$ and $H_2: f_2 \simeq g_2$.

Then $f_2 \circ H_1: f_2 \circ f_1 \simeq f_2 \circ g_1$. By transitivity of the homotopy relation, it remains to construct a homotopy between $f_2 \circ g_1$ and $g_2 \circ g_1$. Define

$$H: X \times \mathbf{I} \to Z$$

by $H(x, t) = H_2(g_1(x), t)$. Then $H$ is a composite of continuous functions and hence continuous, and $H: f_2 \circ g_1 \simeq g_2 \circ g_1$. ∎

**32.6 Definition.** A space $X$ is *contractible* iff the identity map $i: X \to X$ is homotopic to some constant map $c(x) = x_0$, from $X$ to a point $x_0 \in X$.

It follows from Example 32.2(b) that any convex subset of a Euclidean space is contractible.

**32.7 Theorem.** *$X$ is contractible iff for any space $T$, any two continuous maps $f, g: T \to X$ are homotopic.*

*Proof.* Sufficiency is obtained by setting $T = X$ and letting $f$ and $g$ be, respectively, the identity and a constant map.

For necessity, suppose $X$ is contractible; say $i \simeq c$, where $c$ is a constant map from $X$ to itself. Let $f, g: T \to X$ be any two continuous maps. By the previous theorem, $f = i \circ f \simeq c \circ f$ and $g = i \circ g \simeq c \circ g$. But $c \circ f = c \circ g$, so apparently $f \simeq g$. ∎

**32.8 Definition.** Two spaces $X$ and $Y$ are said to be *homotopically equivalent* iff there are continuous functions $f: X \to Y$ and $g: Y \to X$ such that $f \circ g \simeq i_Y$ and $g \circ f \simeq i_X$. The maps $f$ and $g$ are called *homotopy equivalences* and $g$ is called a *homotopy inverse* of $f$ (and vice versa).

Homotopy equivalence is an equivalence relation on any set of topological spaces, and homeomorphic spaces are always homotopically equivalent. The converse to the last statement fails, as the following theorem shows.

**32.9 Theorem.** *$X$ is contractible iff it is homotopically equivalent to a one-point space.*

*Proof.* Suppose $X$ is contractible, say the identity $i: X \to X$ is homotopic to the constant function $c(x) = x_0$. Let $Y = \{x_0\}$, and let $j: Y \to X$ be the inclusion map. Then $c \circ j$ is the identity on $Y$ and $j \circ c = c$ is homotopic to the identity on $X$. Thus $j$ is a homotopy equivalence from $Y$ to $X$.

Conversely, suppose $f: X \to Y$ is a homotopy equivalence between $X$ and a one-point space $Y$, and let $g: Y \to X$ be a homotopy inverse. Then $g \circ f$ is a constant map from $X$ to $X$ which is homotopic to the identity on $X$, so $X$ is contractible. ∎

**32.10 Definition.** A subset $A$ of $X$ is a *retract* of $X$ iff there is a continuous map

$r: X \to A$, called a *retraction*, such that $r(a) = a$ for each $a \in A$. We call $A$ a *deformation retract* of $X$ iff there is a retraction $r: X \to A$ which is homotopic (as a map into $X$) to the identity function $i$ on $X$. If $H: r \simeq i$, $H$ is called a *deformation retraction.*

**32.11 Example.** A retract need not be a deformation retract. In fact, the one-point subsets of any space are retracts, but no one-point subspace of $\mathbf{S}^1$ is a deformation retract (34.4).

***32.12 Theorem.*** *If $A$ is a deformation retract of $X$, then $A$ is homotopically equivalent to $X$.*

*Proof.* Let $j: A \to X$ be inclusion and $r: X \to A$ be the retraction. Then $j \circ r$ is homotopic to the identity on $X$ and $r \circ j$ is the identity on $A$, so $r$ is a homotopy equivalence. ■

We conclude this section by introducing a generalization of the homotopy relation which will be useful in the next section.

**32.13 Definition.** A *topological pair* is an ordered pair $(X, A)$ where $A$ is a topological space and $A \subset X$. A *mapping* $f: (X, A) \to (Y, B)$ of topological pairs is a mapping $f: X \to Y$ such that $f(A) \subset B$; it is *continuous* if it is continuous in the usual sense from $X$ to $Y$.

Two continuous mappings $f, g: (X, A) \to (Y, B)$ are *homotopic* iff there is a continuous function $H: X \times \mathbf{I} \to Y$ such that $H(x, 0) = f(x)$ and $H(x, 1) = g(x)$ for all $x \in X$ and such that $H(a, t) = f(a) = g(a)$ for all $a \in A$. Thus for $f$ and $g$ to be homotopic mappings of the pair $(X, A)$ it is necessary that $f \mid A = g \mid A$. If $f$ and $g$ are homotopic mappings of $(X, A)$ we say "$f$ is homotopic to $g$ relative to $A$" and write $f \simeq g[A]$.

Two pairs $(X, A)$ and $(Y, B)$ are *homotopically equivalent* iff there are pair mappings $f: (X, A) \to (Y, B)$ and $g: (Y, B) \to (X, A)$ such that $f \circ g \simeq i_Y[B]$ and $g \circ f \simeq i_X[A]$. Apparently $f \mid A$ must in this case be a homeomorphism of $A$ onto $B$ and $g \mid B$ must be its inverse.

Clearly, if $f \simeq g[A]$, then $f$ and $g$ are homotopic as mappings of $X$ to $Y$. The converse may fail, even when $f$ and $g$ agree on $A$, as the following example shows.

**32.14 Example.** Let $X$ be the subspace of $\mathbf{R}^2$ consisting of the segment $\{x \mid 0 \le x \le 1\}$ of the $x$-axis, the segment $\{y \mid 0 \le y \le 1\}$ of the $y$-axis and each of the line segments $\{(1/n, y) \mid 0 \le y \le 1\}$ for $n = 1, 2, \ldots$. Let $A$ be the one-point subspace $\{(0, 1)\}$. $X$ is easily seen to be contractible so, by 32.7, the identity $i$ on $X$ and the constant map $g(x) = (0, 1)$ are homotopic. Moreover, these two maps agree on $A$. But no homotopy $H$ between $i$ and $g$ can have $A$ pointwise fixed, as required for relative homotopy.

Some of the other relationships between homotopy and relative homotopy will be explored in the exercises. In particular, homotopy relative to $A$ is an equivalence relation in the set of all maps from $X$ to $Y$ (see 32B).

We should introduce a note of caution here. The literature contains references to several notions of relative homotopy, no two of which are exactly alike. It would be wise to check definitions whenever such a notion is encountered.

### Problems

#### 32A. *Contractible spaces*

1. Every contractible space is pathwise connected.
2. Every retract in a contractible space is contractible.

#### 32B. *Relative homotopy*

Let $(X, A)$ and $(Y, B)$ be topological pairs.

1. The relation $f \simeq g[A]$ is an equivalence relation on the set of all mappings $f: (X, A) \to (Y, B)$.

2. If $(X, A)$ and $(Y, B)$ are homotopically equivalent as pairs, then $X$ and $Y$ are homotopically equivalent spaces. Is the converse true? That is, if $f$ is a homotopy equivalence from $X$ to $Y$ such that $f \mid A$ is a homeomorphism of $A$ onto $B$, is $f$ a homotopy equivalence between $(X, A)$ and $(Y, B)$?

#### 32C. *Homotopy in subspaces and products*

1. If $f_0, f_1 : X \to Y$ are homotopic and $A \subset X$, then $f_0 \mid A$ and $f_1 \mid A$ are homotopic.

2. Maps $f_0$ and $f_1$ of $X$ into a product space are homotopic iff they are homotopic when followed by each projection.

#### 32D. *Weak deformation retracts*

If $A \subset X$, a map $r: X \to A$ is a *weak retraction* of $X$ onto $A$ iff $r \circ j$ is homotopic to the identity on $A$, where $j$ is the inclusion map of $A$ in $X$. Then $A$ is a *weak retract* of $X$. A subset $B$ of $X$ is *deformable into A in X* iff there is a continuous map $D: B \times \mathbf{I} \to X$, called a *deformation*, such that $D(b, 0) = b$ for all $b \in B$ and $D(B \times 1) \subset A$. Finally, $A$ is a *weak deformation retract* of $X$ iff there is a map $D: X \times \mathbf{I} \to X$ such that $D(x, 0) = x$, for all $x \in X$, and $r(x) = D(x, 1)$ is a weak retraction of $X$ onto $A$.

1. Note that $r$ is a weak retraction of $X$ onto $A$ iff it is a left homotopy inverse to the inclusion map $j$. Show $X$ is deformable into $A$ (in $X$) iff $j$ has a right homotopy inverse.

2. Every retract is a weak retract. The converse fails. (But see 32F.)

3. The following are equivalent, for $A \subset X$:

   a) $A$ is homotopically equivalent to $X$,
   b) $A$ is a weak deformation retract of $X$,
   c) $A$ is a weak retract of $X$ and $X$ is deformable into $A$.

(Compare with 32.12.)

32E. *Deformation and retraction*

1. Any compact, convex subset of $\mathbf{R}^n$ is a deformation retract of $\mathbf{R}^n$.
2. If $A$ is a retract of $X$, then $A \times Y$ is a retract of $X \times Y$.
3. Not every weak deformation retract of $X$ is a deformation retract.
4. $A$ is a deformation retract of $X$ iff $A$ is a retract of $X$ and $X$ is deformable into $A$ (32D).

32F. *The homotopy extension property*

Let $X$ be a topological space, $A$ a subspace of $X$. We say the pair $(X, A)$ has the *homotopy extension property* with respect to a space $Y$ iff each continuous $F'$ defined on $(X \times 0) \cup (A \times \mathbf{I})$ to $Y$ has an extension to a continuous $F: X \times \mathbf{I} \to Y$.

1. If $(X, A)$ has the homotopy extension property with respect to $Y$, $A$ is a weak retract of $X$ iff $A$ is a retract of $X$.
2. $(X, A)$ has the homotopy extension property with respect to every space $Y$ iff $(X \times 0) \cup (A \times \mathbf{I})$ is a retract in $X \times \mathbf{I}$.

32G. *Null-homotopic maps*

A map $f: X \to Y$ is *null homotopic* iff $f$ is homotopic to some constant map of $X$ into $Y$. Recall that $\Lambda X$ denotes the cone over $X$ (9.12(f)).

1. Two null-homotopic maps $f$ and $g$ of $X$ to $Y$ need not be homotopic to one another.
2. A map $f: X \to Y$ is null homotopic iff $f$ can be extended to a continuous map $F: \Lambda X \to Y$.

## 33 The fundamental group

We are now in a position to use the "addition" of paths defined in Section 27 to associate with any topological space a group (actually, several groups). The basic idea, of course, is to regard the paths in $X$ as elements of the group, with path addition as the group operation. The first obstacle we encounter is that it is not possible to add *any* two paths in $X$; the first must end at the point where the second begins. This is taken care of (in Definition 33.1) by restricting attention to the paths which begin and end at some fixed point of $X$. The second obstacle is that, even for this restricted family of paths, the requirements of a group operation are not satisfied by path addition. This we overcome, with the help of the material of the previous section, by considering homotopy equivalence classes of paths, rather than individual paths.

The result will be a group assigned to each fixed $x_0 \in X$ which, intuitively, measures the number of two-dimensional holes in the path component of $x_0$.

**33.1 Definition.** Let $X$ be a topological space, $x_0$ a fixed point in $X$. A continuous function $f: \mathbf{I} \to X$ will be called a *loop based at* $x_0$ iff $f(0) = f(1) = x_0$. Two loops $f$ and $g$ based at $x_0$ will be called *loop homotopic* (or, where no confusion can result, simply *homotopic*) iff $f \simeq g[\{0, 1\}]$. Thus a loop homotopy between two loops based at $x_0$ must be a relative homotopy which at any stage carries the endpoints of $\mathbf{I}$ into $x_0$. We will signify this relation of homotopy between two loops by $f \simeq_{x_0} g$.

The relation $\simeq_{x_0}$ between loops based at $x_0$ is an equivalence relation and hence partitions the set $\Omega(X, x_0)$ of loops based at $x_0$ into equivalence classes. The equivalence class containing $f$ will be denoted $[f]$, and the set of all such equivalence classes of loops based at $x_0$ will be denoted $\Pi_1(X, x_0)$.

We can "add" loops just as we "added" paths in Section 27. If $f_1$ and $f_2$ are loops based at $x_0$, we define a new loop $f_1 * f_2$ as follows:

$$(f_1 * f_2)(t) = \begin{cases} f_1(2t) & \text{if } 0 \le t \le \frac{1}{2} \\ f_2(2t - 1) & \text{if } \frac{1}{2} \le t \le 1 \end{cases}$$

Then we can elevate the operation $*$ to the set $\Pi_1(X, x_0)$ of equivalence classes of loops by defining

$$[f_1] * [f_2] = [f_1 * f_2].$$

It is left as Exercise 33A for the reader to show that $*$ is then well defined in $\Pi_1(X, x_0)$. That is, if $f_1 \simeq_{x_0} g_1$ and $f_2 \simeq_{x_0} g_2$, then $f_1 * f_2 \simeq_{x_0} g_1 * g_2$.

Thus $*$ is a binary operation on $\Pi_1(X, x_0)$.

***33.2 Theorem.*** *$\Pi_1(X, x_0)$, with the operation $*$, is a group.*

*Proof.* We check associativity first. For this, it suffices to show that

$$(f * g) * h \simeq_{x_0} f * (g * h)$$

for loops $f$, $g$ and $h$ based at $x_0$. A pictorial approach will make the idea behind the necessary homotopy easy. In terms of its action on $\mathbf{I}$, $(f * g) * h$ is accomplished by completing the action of $f$ in the interval $[0, \frac{1}{4}]$, the action of $g$ in the interval $[\frac{1}{4}, \frac{1}{2}]$ and the action of $h$ in the interval $[\frac{1}{2}, 1]$. This is represented on the top line of Fig. 33.1. The bottom line represents $f * (g * h)$. A homotopy between $f * (g * h)$ and $(f * g) * h$ can then be constructed by allowing the action of $f$, $g$ and $h$ to be divided at time $t$ as shown. The details are left to the reader.

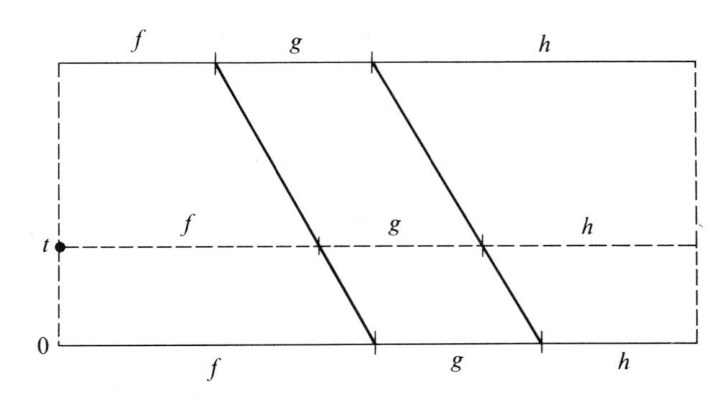


**Figure 33.1**

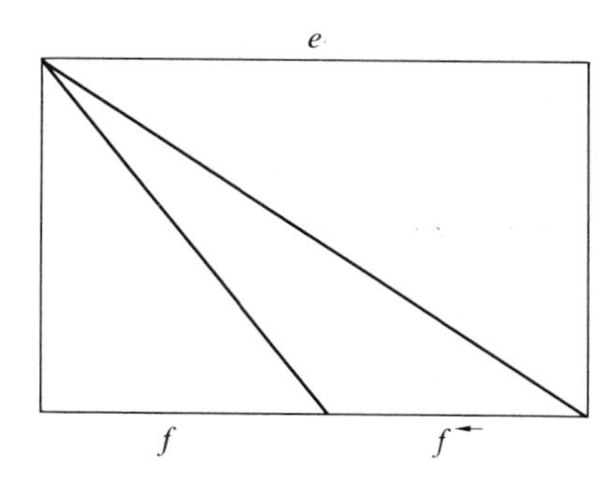


Figure 33.2

Now let $e$ denote the constant map $e(t) = x_0$ for all $t \in \mathbf{I}$. We claim $[e]$ serves as an identity in $\Pi_1(X, x_0)$. It suffices to show $f * e \simeq_{x_0} f$ and $e * f \simeq_{x_0} f$ for all $f \in \Omega(Y, y_0)$. To exhibit a homotopy for the first, define for each $t \in \mathbf{I}$,

$$H(x, t) = \begin{cases} f\left(\dfrac{2x}{2 - t}\right) & 0 \le x \le \dfrac{2 - t}{2} \\ x_0 & \dfrac{2 - t}{2} \le x \le 1. \end{cases}$$

(A picture like the one in Fig. 33.1 will help in understanding where this came from.) $H$ is continuous on $\mathbf{I} \times \mathbf{I}$ since it is continuous on each of the closed sets $\{(x, t) \mid x \le (2 - t)/2\}$ and $\{(x, t) \mid x \ge (2 - t)/2\}$ and it is easily checked that $H(x, 0) = f(x)$ and $H(x, 1) = (f * e)(x)$ for all $x \in \mathbf{I}$. The relation $e * f \simeq_{x_0} f$ is done in similar fashion.

Finally we must show existence of inverses. For each loop $f$ at $x_0$, define $f^{\leftarrow}$ to be the loop

$$f^{\leftarrow}(x) = f(1 - x), \qquad 0 \le x \le 1,$$

and let

$$[f]^{\leftarrow} = [f^{\leftarrow}].$$

The reader can check that this is well defined. To show $[f]^{\leftarrow}$ is an inverse for $f$, it suffices to check that $f * f^{\leftarrow} \simeq_{x_0} e$ and $f^{\leftarrow} * f \simeq_{x_0} e$. First (Fig. 33.2), let

$$H(x, t) = \begin{cases} f(x) & 0 \le x \le \dfrac{1 - t}{2} \\ f^{\leftarrow}(x + t) & \dfrac{1 - t}{2} \le x \le 1 - t \\ x_0 & 1 - t \le x \le 1 \end{cases}$$

The function $H$ is continuous on each of three closed sets which cover the square and thus continuous, and clearly $H(x, 0) = (f * f^{\leftarrow})(x)$ and $H(x, 1) = e(x)$ for all $x \in I$. The homotopy showing $f^{\leftarrow} * f \simeq_{x_0} e$ is similarly constructed. ■

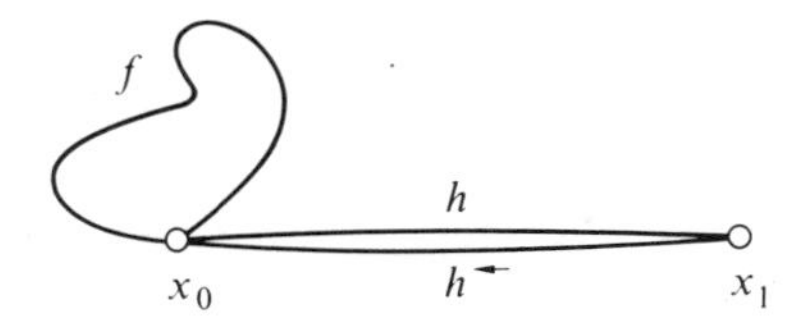


**Figure 33.3**

The dependence of $\Pi_1(X, x_0)$, called the *fundamental group of X based at* $x_0$, on the base point $x_0$ is not illusory in the general case (33B), but for an important special class of spaces it can be ignored.

***33.3 Theorem.*** *If X is an arcwise connected space, then for any pair of points $x_0$ and $x_1$ in X, $\Pi_1(X, x_0)$ and $\Pi_1(X, x_1)$ are isomorphic.*

*Proof.* Let $h: \mathbf{I} \to X$ be an arc from $x_0$ to $x_1$, $h^{\leftarrow}$ the arc $h$ traversed in the opposite direction. For each loop $f$ based at $x_0$, define $a(f)$ to be the following loop based at $x_1$ (Fig. 33.3):

$$a(f) = h^{\leftarrow} * f * h.$$

This induces a mapping $A[f] = [h^{\leftarrow} * f * h]$ of $\Pi_1(X, x_0)$ to $\Pi_1(X, x_1)$. We will show this is the desired isomorphism.

First, $A$ is single valued. That is, if $f \simeq_{x_0} g$, then $h^{\leftarrow} * f * h \simeq_{x_1} h^{\leftarrow} * g * h$. For if $H: f \simeq_{x_0} g$, then the function $G$ defined by

$$G(x, t) = \begin{cases} h^{\leftarrow}(3x) & 0 \leq x \leq \frac{1}{3} \\ H(3x - 1, t) & \frac{1}{3} \leq x \leq \frac{2}{3} \\ h(3x - 2) & \frac{2}{3} \leq x \leq 1 \end{cases}$$

is a homotopy between $h^{\leftarrow} * f * h$ and $h^{\leftarrow} * g * h$.

Second, $A$ is a homomorphism; that is $A([f] * [g]) = A[f] * A[g]$. But

$$\begin{aligned} A[f] * A[g] &= [h^{\leftarrow} * f * h] * [h^{\leftarrow} * g * h] \\ &= [h^{\leftarrow} * f * h * h^{\leftarrow} * g * h] = [h^{\leftarrow} * f * g * h] \\ &= A([f * g]) = A([f] * [g]). \end{aligned}$$

Finally, it is necessary to show $A$ is one–one and onto. This is left as Exercise 33B. ■

Thus for an arcwise connected space $X$, we can speak of *the fundamental group* $\Pi_1(X)$ *of* $X$. This will cause us no difficulty here, but would be an annoying oversimplification in a deeper study of the fundamental group. In point of fact, $\Pi_1(X)$ is a set of groups indexed by the points of $X$, any two of which are isomorphic under any one of a set of isomorphisms indexed by the paths between the two points in question. See Exercise 33B.

Later on, we will compute some simple homotopy groups (with the help of some by no means simple tools). Now we turn to the question of their homotopy (and thus topological) invariance.

**33.4 Definition.** A pair $(X, x_0)$ where $X$ is a topological space and $x_0 \in X$ will be called a *pointed space (space with base point)*. A mapping $f: (X, x_0) \to (Y, y_0)$ of pointed spaces is a continuous function from $X$ to $Y$ such that $f(x_0) = y_0$.

We have associated with each pointed space $(X, x_0)$ an algebraic object $\Pi_1(X, x_0)$. The power of the homotopy method in topology is largely traceable to the fact that mappings of pointed spaces induce homomorphisms of the associated algebraic structures.

**33.5 Theorem.** *Every continuous mapping $f: (X, x_0) \to (Y, y_0)$ induces a homomorphism $f^{\#}: \Pi_1(X, x_0) \to \Pi_1(Y, y_0)$.*

*Proof.* For each loop $g$ at $x_0$ in $X$, let $f'(g)$ be the loop at $y_0$ in $Y$ defined by $f'(g)(t) = f[g(t)]$. This defines a mapping $f'$ from $\Omega(X, x_0)$ to $\Omega(Y, y_0)$ which in turn induces a mapping $f^{\#}: \Pi_1(X, x_0) \to \Pi_1(Y, y_0)$ as follows:

$$f^{\#}([g]) = [f'(g)].$$

To see that $f^{\#}$ is well defined, note that if $H$ is a homotopy between $g_1$ and $g_2$ in $\Omega(X, x_0)$, then $f \circ H$ is a homotopy between $f'(g_1)$ and $f'(g_2)$ in $\Omega(Y, y_0)$.

It remains to show that $f^{\#}$ is a homomorphism, for which it suffices to establish the necessary algebraic property for $f'$. But

$$\begin{aligned} f'(g * h) &= \begin{cases} f[g(2x)] = f'(g)(2x) & 0 \le x \le \frac{1}{2} \\ f[h(2x - 1)] = f'(h)(2x - 1) & \frac{1}{2} \le x \le 1 \end{cases} \\ &= f'(g) * f'(h). \blacksquare \end{aligned}$$

**33.6 Theorem.** a) *If $f$ is the identity on $X$, $f^{\#}$ is the identity on $\Pi_1(X, x_0)$.*

b) *If $f$ and $g$ are continuous mappings from $(X, x_0)$ to $(Y, y_0)$ such that $f \simeq g[x_0]$, then $f^{\#} = g^{\#}$.*

c) *If $f: (X, x_0) \to (Y, y_0)$ and $g: (Y, y_0) \to (Z, z_0)$, then $(g \circ f)^{\#} = g^{\#} \circ f^{\#}$.*

d) *If $r: (X, x_0) \to (A, x_0)$ is a retraction and $i: (A, x_0) \to (X, x_0)$ is the inclusion map, then $i^{\#}$ is a monomorphism and $r^{\#}$ is an epimorphism.*

*Proof.* a) Obvious.

b) It suffices to show that if $h$ is a loop based at $x_0$ in $X$, then $f^*(h) \simeq_{y_0} g^*(h)$. But if $f$ and $g$ are homotopic relative to $x_0$, then $f \circ h$ and $g \circ h$ are homotopic; that is, $f^*(h)$ and $g^*(h)$ are homotopic.

c) If $h$ is any loop based at $x_0$ in $X$, then for $t \in I$,

$$\begin{aligned} [(g \circ f)^{\#}(h)](t) &= g \circ f(h(t)) = g[f(h(t))] \\ &= g^{\#}[f(h(t))] \\ &= g^{\#} \circ f^{\#}(h(t)). \end{aligned}$$

d) $r \circ i$ is the identity map on $(A, x_0)$, so $r^{\#} \circ i^{\#} = (r \circ i)^{\#}$ is the identity on $\Pi_1(A, x_0)$. Both results follow. $\blacksquare$

***33.7 Theorem.*** *If* $(X, x_0)$ *and* $(Y, y_0)$ *are homotopically equivalent, then* $\Pi_1(X, x_0)$ *and* $\Pi_1(Y, y_0)$ *are isomorphic.*

*Proof.* There are mappings $f: (X, x_0) \to (Y, y_0)$ and $g: (Y, y_0) \to (X, x_0)$ such that $f \circ g$ is homotopic to the identity on $Y$ and $g \circ f$ is homotopic to the identity on $X$. Then, from 33.6, $g^\# \circ f^\# = (g \circ f)^\#$ is the identity on $\Pi_1(Y, y_0)$ and $f^\# \circ g^\# = (f \circ g)^\#$ is the identity on $\Pi_1(X, x_0)$. Since $f^\#$ and $g^\#$ are homomorphisms, they are thus isomorphisms and the theorem is proved. ■

Note that the above theorem (and 33.6b) require relative homotopies as stated; this is a severe and unnecessary restriction, although the weak result obtained is sufficient for our purposes. One stronger result is stated in 33C. Even stronger results can be obtained; see the book by Massey, p. 82.

## Problems

### 33A. *The operation* $*$

1. Show that $[f_1] * [f_2] = [f_1 * f_2]$ is a well-defined operation in $\Pi_1(X, x_0)$. (Refer to 33.1.)

### 33B. $\Pi_1(X)$ *for arcwise connected spaces*

1. Construct a space $X$ with points $x_0$ and $y_0$ such that $\Pi_1(X, x_0)$ and $\Pi_1(X, y_0)$ are not isomorphic.

Now let $X$ be an arcwise connected space and, for each $x \in X$, let $G_x = \Pi_1(X, x)$. For each path $h$ from $x$ to $y$ in $X$, let $\alpha_h$ be the isomorphism $\alpha_h([f]) = [h^{\leftarrow} * f * h]$ of $G_x$ with $G_y$. If $h$ is a path from $x$ to $y$ and $k$ is a path from $y$ to $z$, let $h * k$ be the path from $x$ to $z$ defined by

$$(h * k)(t) = \begin{cases} h(2t), & \text{if } 0 \le t \le \frac{1}{2} \\ k(2t - 1), & \text{if } \frac{1}{2} \le t \le 1. \end{cases}$$

(This just extends the definition of $*$ to paths which are not loops.)

2. If $f$ is a loop at $x$, then $\alpha_f$ is an inner automorphism of $G_x$.

3. If $h$ is a path from $x$ to $y$ and $k$ is a path from $y$ to $z$, then $\alpha_{h*k} = \alpha_k \circ \alpha_h$.

This is intended to develop the categorical point of view of $\Pi_1(X)$ as an object in the category of groups and conjugacy classes of homomorphisms. A better understanding of this point of view can be gained by reading the relevant portions of Spanier's book, *Algebraic Topology*.

### 33C. *Homotopy equivalence*

Show that if $X$ and $Y$ are arcwise connected and homotopically equivalent, then $\Pi_1(X)$ and $\Pi_1(Y)$ are isomorphic. (This is difficult.)

### 33D. *The higher homotopy groups*

Let $X$ be a topological space. Let $\partial\mathbf{I}^n$ denote the boundary of the $n$-cube $\mathbf{I}^n$; that is, $\partial\mathbf{I}^n = \{(x_1, \ldots, x_n) \in \mathbf{I}^n \mid \text{some } x_i \text{ is 0 or 1}\}$. An *n-dimensional hyperloop based at* $y_0$ in $Y$

is a continuous function $f: \mathbf{I}^n \to Y$ such that $f(\partial \mathbf{I}^n) = \{y_0\}$. Define $f * g$ for hyperloops $f$ and $g$ by

$$\begin{aligned}(f * g)(x_1, \ldots, x_n) &= f(2x_1, x_2, \ldots, x_n) && \text{if } 0 \le x_1 \le \tfrac{1}{2}\\ &= g(2x_1 - 1, x_2, \ldots, x_n) && \text{if } \tfrac{1}{2} \le x_1 \le 1.\end{aligned}$$

Let $\Omega_n(Y, y_0)$ denote the set of $n$-dimensional hyperloops based at $y_0$ in $Y$ and let $\Pi_n(Y, y_0)$ denote the set of equivalence classes in $\Omega_n(Y, y_0)$ under the relation of homotopy relative to $\partial \mathbf{I}^n$. The equivalence class of $f$ will be denoted $[f]$.

1. $[f] * [g] = [f * g]$ is a well-defined operation in $\Pi_n(Y, y_0)$, making $\Pi_n(Y, y_0)$ a group, called the $n$th *homotopy group* of $(Y, y_0)$.

2. If $f: (X, x_0) \to (Y, y_0)$ is continuous, the induced map $f_*([h]) = [f \circ h]$ is a homomorphism of $\Pi_n(X, x_0)$ into $\Pi_n(Y, y_0)$.

3. a) If $i: (X, x_0) \to (Y, y_0)$ is the identity, then $i_*: \Pi_n(X, x_0) \to \Pi_n(Y, y_0)$ is the identity.

   b) If $f: (X, x_0) \to (Y, y_0)$ and $g: (Y, y_0) \to (Z, z_0)$, then $(g \circ f)_* = g_* \circ f_*$.

   c) If $f, g: (X, x_0) \to (Y, y_0)$ are homotopic relative to $x_0$, then $f_* = g_*$.

4. If $(X, x_0)$ and $(Y, y_0)$ are homotopically equivalent, then $\Pi_n(X, x_0)$ and $\Pi_n(Y, y_0)$ are isomorphic.

For more on the higher homotopy groups, see Exercise 43K.

## 34 $\Pi_1(\mathbf{S}^1)$

Let $f$ be a loop based at $(1, 0)$ in $\mathbf{S}^1$. We will, with some difficulty, assign a number $D(f)$ to $f$ which, intuitively, measures the number of times $f$ winds positively (counterclockwise) around $\mathbf{S}^1$ and which is an invariant of homotopy type. This will enable us to compute $\Pi_1(\mathbf{S}^1)$.

**34.1 Definition.** Let the loop $f$ be fixed; we will assume $f$ is nonconstant. A *proper partition* of $\mathbf{I}$ relative to $f$ is a partition $0 = a_0 < a_1 < \cdots < a_n = 1$ of $\mathbf{I}$ such that if $x \in [a_i, a_{i+1}]$, then $|f(x) - f(a_i)| < 1$ and such that $a_i \ne a_j \Rightarrow f(a_i) \ne f(a_j)$, except that $f(a_0) = f(a_n)$.

Uniform continuity of $f$ on $\mathbf{I}$ insures that proper partitions can be found. (Although we have neither defined nor studied uniform continuity as yet, we need here only the fact that a continuous function defined on a closed bounded interval with range in some metric space is uniformly continuous. Any course in real analysis should include a proof of this fact for real-valued functions and almost any proof for real-valued functions carries over without change to functions which take values in an arbitrary metric space. Alternatively, see Theorem 36.20.)

Given a proper partition $P = \{a_0, \ldots, a_n\}$ of $\mathbf{I}$ relative to $f$, the $P$-approximation to $f$ is the function $f_p$ which in each subinterval $[a_i, a_{i+1}]$ traverses from $f(a_i)$ to $f(a_{i+1})$ the shorter of the two subarcs of $\mathbf{S}^1$ determined by $f(a_i)$ and $f(a_{i+1})$. Each of the subarcs $A_i$ thus traversed is assigned a number $n(A_i)$: $+1$ if the arc is traversed in the positive (counterclockwise) direction and $-1$ if the arc is traversed in the negative direction.

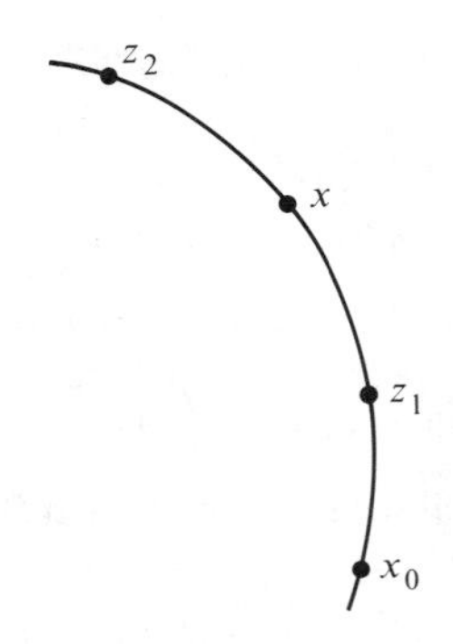


**Figure 34.1**

The degree of $f$ is then defined as follows: pick a point $x \neq f(a_i)$ for any $i$ and define

$$D(f) = \sum \{n(A_i) \mid x \in A_i\}.$$

We will show $D(f)$ is independent of the choice of the proper partition $P$ and the point $x$. For this purpose, we will denote the number just defined, which apparently depends on $P$ and $x$ as well as $f$, by $D(f, P, x)$.

**34.2 Theorem.** *For a proper partition $P$ of $\mathbf{I}$, $D(f, P, x)$ is independent of the choice of $x \neq f(a_i)$, and hence can be denoted $D(f, P)$.*

*Proof.* Let $x_0$ be any point in $\mathbf{S}^1$ not an $f(a_i)$ for some $i$ and let $z_1$ be the first point after $x_0$ (in the counterclockwise ordering of $\mathbf{S}^1$) which is an $f(a_i)$ for some $i$. Then for any point $x$ with $x_0 < x < z_1$, $x_0$ and $x$ must be in the same subarcs of $\mathbf{S}^1$ relative to the partition $P$, so $D(f, P, x_0) = D(f, P, x)$. Next let $z_2$ be the next point after $z_1$ which is an $f(a_i)$ for some $i$ and suppose $z_1 < x < z_2$ (Fig. 34.1). Suppose $z_1 = f(a_i)$. Consider $f(a_{i-1})$ and $f(a_{i+1})$. There are two cases:

*Case 1.* $f(a_{i-1})$ and $f(a_{i+1})$ are on opposite sides of $z_1$. Then the arcs $[f(a_{i-1}), f(a_i)]$ and $[f(a_i), f(a_{i+1})]$ lie in the same direction and one contains $x_0$, the other $x$; since all other subarcs of $\mathbf{S}^1$ relative to the partition $P$ contain both of $x_0$ and $x$ or neither, clearly $D(f, P, x) = D(f, P, x_0)$.

*Case 2.* $f(a_{i-1})$ and $f(a_{i+1})$ lie on the same side of $z_1$. Then the arcs $[f(a_{i-1}), f(a_i)]$ and $[f(a_i), f(a_{i+1})]$ lie in opposite directions and both contain one of the points $x_0$ or $x$; again, all other subarcs of $\mathbf{S}^1$ relative to the partition $P$ contain both of $x_0$ and $x$ or neither, so again $D(f, P, x) = D(f, P, x_0)$.

We can continue in this manner past all of the points $f(a_i)$ on $\mathbf{S}^1$. Hence $D(f, P, x)$ is independent of the choice of $x \neq f(a_i)$. ■

**34.3 Theorem.** *If $P_1$ and $P_2$ are proper partitions of $\mathbf{I}$, then $D(f, P_1) = D(f, P_2)$.*

*Proof.* A moment's reflection (enforced in 34C) should make it clear that it suffices to prove this when $P_2$ is obtained by adding one point to the partition $P_1$. Thus let $P_1 = \{a_0, a_1, \ldots, a_n\}$ and $P_2 = \{a_0, \ldots, a_i, b, a_{i+1}, \ldots, a_n\}$. But there must be a point $z \neq f(a_i)$ in $\mathbf{S}^1$not lying on either of the arcs $[f(a_i), f(b)]$ or $[f(b), f(a_{i+1})]$. Then clearly $D(f, P_1, z) = D(f, P_2, z)$ and hence

$$D(f, P_1) = D(f, P_2). \blacksquare$$

Thus we are justified in suppressing the role of the particular proper partition $P$ used and simply referring to the degree $D(f)$ of the loop $f: \mathbf{I} \to \mathbf{S}^1$. The proof of our next result establishes the importance of the notion of degree in questions of homotopy involving $\mathbf{S}^1$.

***34.4 Theorem.*** *$\Pi_1(\mathbf{S}^1)$ is infinite cyclic.*

*Proof.* Suppose first that $f_0$ and $f_1$ are homotopic loops based at $(1, 0)$ in $\mathbf{S}^1$, say $H: f_0 \simeq f_1$. We claim $D(f_0) = D(f_1)$. For each $t$, $0 \leq t \leq 1$, let $f_t$ be the loop defined by $f_t(x) = H(x, t)$ for $x \in \mathbf{I}$. Note that for $t = 0$ and 1, respectively, this gives $f_0$ and $f_1$ as it should. By compactness of $\mathbf{I}$, it is sufficient to show that for each $t \in \mathbf{I}$, there is some $\epsilon > 0$ such that $|s - t| < \epsilon$ implies $D(f_s) = D(f_t)$. Let $P = \{a_0, \ldots, a_n\}$ be any proper partition of $\mathbf{I}$ for the loop $f_t$. By uniform continuity of $H$, there is an $\epsilon_1 > 0$ such that $|s - t| < \epsilon_1$ implies $P$ is a proper partition for $f_s$. Now let $y \in \mathbf{S}^1$ be any point such that $y \neq f_t(a_i)$ for any $i$. Pick $\delta > 0$ small enough that both

a) $|z - y| < \delta$ implies $z \neq f_t(a_i)$ for any $i$, and

b) if $|z_1 - f_t(a_i)| < \delta$ and $|z_2 - f_t(a_{i+1})| < \delta$, then the shortest arc from $z_1$ to $z_2$ has the same orientation as the shortest arc from $f_t(a_i)$ to $f_t(a_{i+1})$, for $i = 0, 1, \ldots, n - 1$.

Now use uniform continuity of $H$ to pick $\epsilon < \epsilon_1$ such that $|s - t| < \epsilon$ implies $|f_s(a_i) - f_t(a_i)| < \delta$, for $i = 0, \ldots, n$. Then clearly $D(f_t, P, y) = D(f_s, P, y)$ whenever $|s - t| < \epsilon$, and hence $D(f_t) = D(f_s)$. Thus, by compactness of $\mathbf{I}$, $D(f_0) = D(f_1)$.

Now, regarding $\mathbf{S}^1$ as the unit circle in the complex plane, define the loop $p: \mathbf{I} \to \mathbf{S}^1$ by $p(t) = e^{i2\pi t}$. Then, for any integer $k$, $p^k = p * p * \cdots * p$ ($k$ times) is the loop $p^k(t) = e^{i2\pi kt}$. Note that the degree of $p^k$ is $k$, so that $p^k$ is not homotopic to $p^l$ for $k \neq l$. Thus the map $k \to [p^k]$ embeds the positive integers as a group in $\Pi_1(\mathbf{S}^1)$. To show this embedding is an isomorphism, we need only show every loop $f$ in $\mathbf{S}^1$ is homotopic to $p^k$ for some integer $k$.

To this end, let $f: \mathbf{I} \to \mathbf{S}^1$ be a loop based at $(1, 0)$ in $\mathbf{S}^1$. Define open sets $A_1$ and $A_2$ in $\mathbf{S}^1$ by:

$$A_1 = \{(x, y) \in \mathbf{S}^1 \mid y > -\tfrac{1}{10}\}$$
$$A_2 = \{(x, y) \in \mathbf{S}^1 \mid y < \tfrac{1}{10}\}.$$

Then $\{f^{-1}(A_1), f^{-1}(A_2)\}$ is an open cover of $\mathbf{I}$. Let $\delta$ be a Lebesgue number for this cover and let $\{a_0, \ldots, a_n\}$ be any partition of $\mathbf{I}$ into subintervals of length $<\delta$. Then each $f[a_{i-1}, a_i]$, $i = 1, \ldots, n$, is contained in either $A_1$ or $A_2$. By dropping partition points where necessary, we may assume $f[a_{i-1}, a_i]$ and $f[a_i, a_{i+1}]$ are never contained in the same set $A_j$, $j = 1$ or 2. Now each $f(a_i)$ is contained in a short arc of $A_1 \cap A_2$ containing either $(1, 0)$ or $(-1, 0)$. Let $\alpha_i$ be the map $f$ restricted to $[a_{i-1}, a_i]$, and for each $i$, let $\eta_i$ be the arc from $f(a_i)$ to the point $(1, 0)$ or $(-1, 0)$, whichever is closer. Now consider the arcs $\eta_1\alpha_1, \eta_2\alpha_2\overleftarrow{\eta_1}, \eta_3\alpha_3\overleftarrow{\eta_2}, \ldots, \alpha_n\overleftarrow{\eta_{n-1}}$. If $\beta_1$ is the positive arc from $(1, 0)$ to $(-1, 0)$ and $\beta_2$ is the positive arc from $(-1, 0)$ to $(1, 0)$, then each of the arcs

$$\eta_1\alpha_1, \eta_2\alpha_2\overleftarrow{\eta_1}, \eta_3\alpha_3\overleftarrow{\eta_2}, \ldots, \alpha_n\overleftarrow{\eta_{n-1}}$$

is homotopic to a constant map or one of $\beta_1$, $\beta_2$, $\overleftarrow{\beta_1}$, or $\overleftarrow{\beta_2}$. Hence, we have

$$f = \alpha_n\alpha_{n-1} \cdots \alpha_1 = (\alpha_n\overleftarrow{\eta_{n-1}})(\eta_{n-1}\alpha_{n-1}\overleftarrow{\eta_{n-2}} \cdots (\eta_2\alpha_2\overleftarrow{\eta_1})(\eta_1\alpha_1)$$

which, after cancellation, must reduce to

$$f \simeq_{x_0} \beta_2\beta_1\beta_2\beta_1\beta_2\beta_1 \cdots \beta_2\beta_1$$

or

$$f \simeq_{x_0} \overleftarrow{\beta_1}\,\overleftarrow{\beta_2}\,\overleftarrow{\beta_1}\,\overleftarrow{\beta_2} \cdots \overleftarrow{\beta_1}\,\overleftarrow{\beta_2}$$

and hence, for some $k$, $f \simeq_{x_0} p^k$. ■

***34.5 Theorem.*** *If* $\mathbf{D} = \{(x, y) \in \mathbf{R}^2 \mid x^2 + y^2 \leq 1\}$, *then* $\mathbf{S}^1$ *is not a retract of D.*

*Proof.* A retraction $r: \mathbf{D} \to \mathbf{S}^1$ would, by 33.6(d), induce an epimorphism $r^\#: \Pi_1(\mathbf{D}) \to \Pi_1(\mathbf{S}^1)$. But $\mathbf{D}$ is contractible, so $\Pi_1(\mathbf{D})$ is trivial, while $\Pi_1(\mathbf{S}^1)$ is infinite cyclic. ■

***34.6 Theorem.*** (*Brouwer Fixed-point Theorem.*) *Every continuous map*

$$f: \mathbf{D} \to \mathbf{D}$$

*has a fixed point.*

*Proof.* If $f(x) \neq x$ for each $x \in \mathbf{D}$, define $r(x)$ for each $x \in \mathbf{D}$ to be the point where the line from $f(x)$ through $x$ intersects $\mathbf{S}^1$. Then $r: \mathbf{D} \to \mathbf{S}^1$ would be a retraction, contradicting 34.5. ■

The higher-dimensional analogs to 34.6 are also true; we are not in a position to prove them.

## Problems

34A. *Application of the Brouwer fixed-point theorem*

1. Let $f_1$ and $f_2$ be continuous real-valued functions defined on

$$D = \{(x, y) \in \mathbf{R}^2 \mid x^2 + y^2 \leq 1\}.$$

Use the Brouwer fixed-point theorem to show the system

$$f_1(x, y) = 0$$
$$f_2(x, y) = 0$$

of equations has a solution under certain conditions on $f_1$ and $f_2$.

2. State and prove a similar theorem for functions of $n$ variables, using a higher-dimensional version of the Brouwer fixed-point theorem.

### 34B. *Examples of homotopy groups*

1. Let $X$ be the "punctured plane" $\mathbf{R}^2 - (0, 0)$. Show that $\Pi_1(X)$ is infinite cyclic.

2. Show that, if $X$ and $Y$ are arcwise connected spaces, then $\Pi_1(X \times Y)$ is isomorphic to $\Pi_1(X) \times \Pi_1(Y)$ (direct product).

3. Let $T$ be the torus (9.12c). Then $\Pi_1(T) = Z \times Z$, where $Z$ is the group of integers.

4. Let $M$ be the Moebius Band (9.12d). Then $\Pi_1(M) = Z$.

### 34C. *Proper partitions*

Show that if $D(f, P_1) = D(f, P_2)$ whenever $P_2$ is a proper partition obtained by adding one point to the proper partition $P_1$, then $D(f, P_1) = D(f, P_2)$ for any two proper partitions $P_1$ and $P_2$.

### 34D. *Retracts and the fixed-point property*

A space $X$ has the *fixed-point property* iff every continuous map $f: X \to X$ has a fixed point.

1. Every retract of a space with the fixed-point property has the fixed-point property.

2. Let $X_n$ be the space obtained by identifying the left-hand endpoints of $n$ disjoint copies of the unit interval $\mathbf{I}$. Show that $X_n$ has the fixed-point property.

### 34E. *The fundamental theorem of algebra*

Let $P(z) = a_0 z^n + \cdots + a_{n-1} z + a_n$ be a complex polynomial, with $a_0 \neq 0$. Let $Q(z) = z^n$. For each real number $r > 0$, define $S_r = \{z \mid |z| = r\}$ in $\mathbf{R}^2$.

1. For sufficiently large $r$, $P \mid S_r$ and $Q \mid S_r$ are homotopic maps of $S_r$ into $\mathbf{R}^2 - \{0\}$.

2. The polynomial $P(z)$ has a root. [For any $r > 0$, $Q \mid S_r$ is not a nullhomotopic map of $S_r$ into $R^2 - \{0\}$; hence, for sufficiently large $r$, neither is $P \mid S_r$.]

Chapter 9

# Uniform Spaces

## 35 Diagonal uniformities

Uniform spaces are the carriers for the notions of uniform convergence, uniform continuity and the like. These notions are easily defined in metric spaces (e.g., $f: M \to N$ is uniformly continuous iff for each $\epsilon > 0$, a $\delta > 0$ exists such that whenever $\rho(x, y) < \delta$, then $\sigma(f(x), f(y)) < \epsilon$), the important quality of metric spaces for this purpose being that distance is a notion which can be applied uniformly to pairs of points without regard to their location. This quality is not possessed by topological spaces, where the nhoods of a point (and hence the notion of "topological distance") depend on the location of the point, so uniform spaces will apparently need somewhat more structure than a topology provides, although we may be able to get away with less than a metric.

To introduce the first of two approaches we will take to uniform structure (a third will be developed in the exercises), we need some notation.

**35.1 Definition.** If $X$ is any set, we denote by $\Delta$ the diagonal $\{(x, x) \mid x \in X\}$ in $X \times X$. Where confusion might occur, we specify which set $X$ we are referring to by writing $\Delta(X)$.

If $U$ and $V$ are sets in $X \times X$, then $U \circ V$ is the set $\{(x, y) \mid \text{for some } z, (x, z) \in V \text{ and } (z, y) \in U\}$. Notice that $U$ and $V$ are just relations on $X$ and $\circ$ is a natural extension of the notion of composition of functions.

Our first definition of a uniform structure on $X$ has its roots in the observation that $x$ and $y$ are close together, in a metric space, iff the point $(x, y)$ is close to the diagonal in $X \times X$.

**35.2 Definition.** A *diagonal uniformity* on a set $X$ is a collection $\mathscr{D}(X)$, or just $\mathscr{D}$, of subsets of $X \times X$, called *surroundings*, which satisfy:

a) $D \in \mathscr{D} \Rightarrow \Delta \subset D$,

b) $D_1, D_2 \in \mathscr{D} \Rightarrow D_1 \cap D_2 \in \mathscr{D}$,

c) $D \in \mathscr{D} \Rightarrow E \circ E \subset D$ for some $E \in \mathscr{D}$,

d) $D \in \mathscr{D} \Rightarrow E^{-1} \subset D$ for some $E \in \mathscr{D}$,

e) $D \in \mathscr{D}, D \subset E \Rightarrow E \in \mathscr{D}$.

When $X$ has such a structure, we call $X$ a *uniform space*. The uniformity $\mathscr{D}$ is called *separating* (and $X$ is said to be *separated*) iff $\bigcap \{D \mid D \in \mathscr{D}\} = \Delta$.

A *base* for the uniformity $\mathscr{D}$ (also called a *base for the surroundings* on $X$) is any subcollection $\mathscr{E}$ of $\mathscr{D}$ from which $\mathscr{D}$ can be recovered by applying condition (e). Thus $\mathscr{E}$ is a base for $\mathscr{D}$ iff $\mathscr{E} \subset \mathscr{D}$ and each $D \in \mathscr{D}$ contains some $E \in \mathscr{E}$. Apparently, a collection $\mathscr{E}$ of subsets of $X \times X$ is a base for *some* uniformity iff its sets satisfy (a), (c), (d) and the following modified form of (b):

b′) $D_1, D_2 \in \mathscr{E} \Rightarrow D_3 \subset D_1 \cap D_2$ for some $D_3 \in \mathscr{E}$.

That is, all supersets of elements of $\mathscr{E}$ will then satisfy (a)–(e).

A *subbase* for $\mathscr{D}$ is a subcollection $\mathscr{E}$ of $\mathscr{D}$ such that all finite intersections of elements of $\mathscr{E}$ form a base for $\mathscr{D}$.

**35.3 Examples.** a) The *usual uniformity* on $\mathbf{R}$ is the uniformity having for a base the collection of sets $D_\epsilon$, $\epsilon > 0$, where

$$D_\epsilon = \{(x, y) \mid |x - y| < \epsilon\}.$$

b) More generally, any metric $\rho$ on a set $M$ generates a *metric uniformity* $\mathscr{D}_\rho$ on $M$, namely the uniformity having for a base the sets $D^\rho_\epsilon$, $\epsilon > 0$, where

$$D^\rho_\epsilon = \{(x, y) \in M \times M \mid \rho(x, y) < \epsilon\}.$$

The uniformities which can be generated in this way from metrics are called *metrizable*. They are characterized in Section 38.

This is an appropriate place to point out that, if $\rho$ is any metric on $X$, the uniformities generated by $\rho$ and $2\rho$ coincide, so that *different metrics may give rise to the same uniformity*. Thus a uniformity represents truly less structure on a set than a metric. (See also Exercise 35G.)

c) Given any set $X$, the collection $\mathscr{D}$ of all subsets of $X \times X$ which contain $\Delta$ is a uniformity on $X$, called the *discrete uniformity*. It has for a base the collection consisting of the single set $\Delta$.

d) Given any set $X$, the collection $\mathscr{D}$ consisting of the single set $X \times X$ is a uniformity on $X$, called the *trivial uniformity*.

e) For each $a \in \mathbf{R}$, let $D_a$ be the following subset of $\mathbf{R} \times \mathbf{R}$:

$$D_a = \Delta \cup \{(x, y) \mid x > a, y > a\}.$$

Then the sets $D_a$, $a \in \mathbf{R}$, form a base for a uniformity on $\mathbf{R}$.

**35.4 Remarks.** a) If $D \in \mathscr{D}$, then $D^{-1} \in \mathscr{D}$, for any uniformity $\mathscr{D}$ on $X$.

b) The requirements (c) and (d) in the definition of a uniformity are together equivalent to the single requirement: $D \in \mathscr{D} \Rightarrow E \circ E^{-1} \subset D$ for some $E \in \mathscr{D}$.

First suppose (c) and (d) hold. Then given $D \in \mathscr{D}$, find $E_1 \in \mathscr{D}$ such that $E_1 \circ E_1 \subset D$ and $E_2 \in \mathscr{D}$ such that $E_2^{-1} \subset E_1$. Let $E = E_1 \cap E_2$. Then $E \circ E^{-1} \subset D$. Thus the condition above holds.

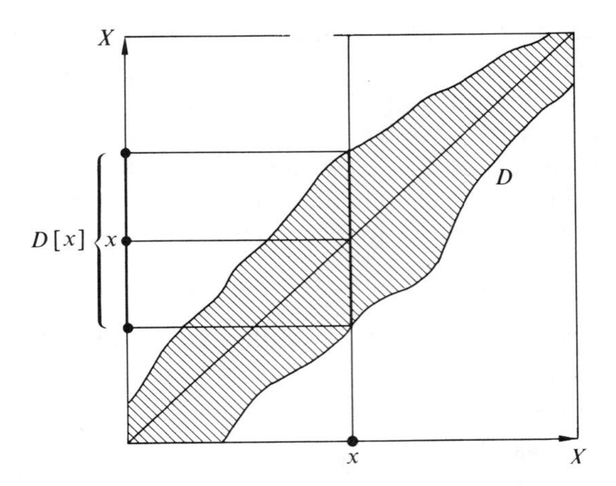


**Figure 35.1**

On the other hand, if the condition above holds, then given $D \in \mathscr{D}$, find $E \in \mathscr{D}$ such that $E \circ E^{-1} \subset D$. Then $E^{-1} \subset D$ easily, and if $F = E \cap E^{-1}$, then $F \in \mathscr{D}$ and $F \circ F \subset D$. Thus (c) and (d) hold.

c) The symmetric sets $D$ in $\mathscr{D}$ (i.e., those for which $D = D^{-1}$) form a base for $\mathscr{D}$.

In fact, if $E \in \mathscr{D}$, then by remark (a) above, $E^{-1} \in \mathscr{D}$, and then $D = E \cap E^{-1}$ is a symmetric element of $\mathscr{D}$ contained in $E$.

A diagonal uniformity represents strictly more structure on a set $X$ than a topology since, as we now proceed to show, every such uniformity generates a topology in a natural way, while different uniformities may produce the same topology.

**35.5 Definition.** For $x \in X$ and $D \in \mathscr{D}$, we define (see Fig. 35.1)

$$D[x] = \{y \in X \mid (x, y) \in D\}.$$

This is extended to subsets $A$ of $X$ as follows:

$$\begin{aligned} D[A] &= \bigcup_{x \in A} D[x] \\ &= \{y \in X \mid (x, y) \in D \text{ for some } x \in A\}. \end{aligned}$$

***35.6 Theorem.*** a) *For each $x \in X$, the collection $\mathscr{U}_x = \{D[x] \mid D \in \mathscr{D}\}$ forms a nhood base at $x$, making $X$ a topological space. The same topology is produced if any base $\mathscr{E}$ is used in place of $\mathscr{D}$.*

b) *The topology is Hausdorff iff $\mathscr{D}$ is separating.*

*Proof.* a) First note that $x \in D[x]$ for each $x$. Second,

$$D_1[x] \cap D_2[x] = (D_1 \cap D_2)[x],$$

so the intersection of nhoods is a nhood. Finally, if $D[x] \in \mathscr{U}_x$, find $E \in \mathscr{D}$ such that $E \circ E \subset D$. Then for any $y \in E[x]$, $E[y] \subset D[x]$, so this property of nhoods is satisfied.

The proof that only a base $\mathscr{E}$ for $\mathscr{D}$ need be used is left to the reader. See Exercise 35D.2.

b) Suppose $\mathscr{D}$ is separating. If $x \neq y$ in $X$, then for some $D \in \mathscr{D}$, $(x, y) \notin D$. Find a symmetric $E \in \mathscr{D}$ such that $E \circ E \subset D$. Then if $z \in E[x] \cap E[y]$, we have $(x, z) \in E$ and $(y, z) \in E$ so that $(z, y) \in E$, and hence $(x, y) \in E \circ E \subset D$. Since this is prohibited, apparently $E[x]$ and $E[y]$ are disjoint nhoods of $x$ and $y$.

Conversely, if the topology is Hausdorff, then if $(x, y) \notin \Delta$, $x \neq y$, so that $E[x] \cap D[y] = \varnothing$ for some $D, E \in \mathscr{D}$, and then $D \cap E$ is an element of $\mathscr{D}$ not containing $(x, y)$. ∎

**35.7 Definition.** The topology thus associated with a diagonal uniformity $\mathscr{D}$ will be called the *uniform topology* $\tau_{\mathscr{D}}$ *generated by* $\mathscr{D}$. Whenever the topology on a topological space $X$ can be obtained in this way from a uniformity, $X$ is called a *uniformizable* topological space.

**35.8 Examples.** a) The topology generated by the usual uniformity on **R** is the usual topology.

b) More generally, the metric uniformity on a metric space $M$ generates for its uniform topology the metric topology on $M$. This follows from the fact that, for $x \in M$,

$$D_\epsilon^\rho[x] = \{y \mid (x, y) \in D_\epsilon^\rho\} = \{y \mid \rho(x, y) < \epsilon\} = U_\epsilon(x)$$

It is reasonable to ask whether the converse is true; that is, if a uniformity has a metrizable topology, is the uniformity itself metrizable? Astonishingly, the answer is no! You will see, in Example 38.5, a nonmetrizable uniformity whose topology is the discrete topology.

c) The discrete uniformity on a set $X$ generates the discrete topology.

d) The trivial uniformity on a set $X$ generates the trivial topology.

e) Consider the uniformity for **R** a base for which consists of the sets

$$D_a = \Delta \cup \{(x, y) \mid x > a, y > a\}$$

for $a \in \mathbf{R}$. For any $x \in \mathbf{R}$, $D_a[x] = \{x\}$ whenever $a \geq x$ and consequently this uniformity generates the discrete topology on **R**.

This example, together with c), serves to establish that *different uniformities may give rise to the same topology.* Thus the correspondence between uniformities on $X$ and topologies on $X$ is many-to-one, so that a uniformity on $X$ represents truly more structure on $X$ than a topology. These comments are amplified by the results in Exercise 41F.

***35.9 Theorem.*** *The open, symmetric elements of* $\mathscr{D}$ *form a base for* $\mathscr{D}$.

*Proof.* An open symmetric set can be obtained by intersecting an open set with its inverse, so it suffices to show that the open sets form a base, for which purpose it is enough to verify that $D \in \mathscr{D} \Rightarrow D^\circ \in \mathscr{D}$. Pick a symmetric $E$ such that $E \circ E \circ E \subset D$. We have finished if we show $E \subset D^\circ$. But if $(x, y) \in E$, then $E[x] \times E[y] \subset D$, for if $(w, z) \in E[x] \times E[y]$, then $(x, w) \in E$, $(y, z) \in E$ and hence, since $(x, y) \in E$, $(w, z) \in E \circ E \circ E \subset D$. Thus each $(x, y) \in E$ has a nhood contained in $D$, so $E \subset D^\circ$. ∎

Now any uniform space we consider is automatically a topological space so we have there a notion of continuous function. We introduced uniform structures to provide a notion of uniform continuity, and we define this now.

**35.10 Definition.** Let $X$ and $Y$ be sets provided with diagonal uniformities $\mathscr{D}$ and $\mathscr{E}$. A function $f: X \to Y$ is *uniformly continuous* iff for each $E \in \mathscr{E}$, there is some $D \in \mathscr{D}$ such that $(x, y) \in D \Rightarrow (f(x), f(y)) \in E$. If $f$ is one–one, onto and both $f$ and $f^{-1}$ are uniformly continuous, we call $f$ a *uniform isomorphism* (*uniform equivalence*) and say $X$ and $Y$ are *uniformly isomorphic* (*uniformly equivalent*).

For the purpose of checking uniform continuity, it is clearly sufficient to restrict attention to bases for the uniformities $\mathscr{D}$ and $\mathscr{E}$.

***35.11 Theorem.*** *Every uniformly continuous function is continuous.*

*Proof.* Suppose $X$ and $Y$ have diagonal uniformities $\mathscr{D}$ and $\mathscr{E}$ and $f: X \to Y$ is uniformly continuous. Now if $x \in X$, a basic nhood of $f(x)$ has the form $E[f(x)]$ for some $E \in \mathscr{E}$. By uniform continuity, there is some $D \in \mathscr{D}$ such that $(x, y) \in D \Rightarrow (f(x), f(y)) \in E$. Then easily $f(D[x]) \subset E[f(x)]$, so $f$ is continuous at $x$. ∎

**35.12 Examples.** a) Let $(M, \rho)$ and $(N, \sigma)$ be metric spaces. Then $f: M \to N$ is uniformly continuous with respect to the metric uniformities $\mathscr{D}_\rho$ and $\mathscr{D}_\sigma$ iff for each $\epsilon > 0$, there is some $\delta > 0$ such that $(x, y) \in D^\rho_\delta \Rightarrow (f(x), f(y)) \in D^\sigma_\epsilon$. It is easy to see that this reduces to the usual $\epsilon$–$\delta$ requirement for uniform continuity of functions between metric spaces.

b) Any function defined on a space with the discrete uniformity to another uniform space is uniformly continuous.

c) Examples of continuous functions which are not uniformly continuous should be familiar; for instance, the function $f(x) = x^2$ from $\mathbf{R}$ to $\mathbf{R}$ (with the usual uniformity).

## Problems

### 35A. *Examples of uniformities*

Verify that each of the following is a uniformity on the set $X$ indicated:

1. The metric uniformity on a metric space $(M, \rho)$. See 35.3(b).

2. The discrete uniformity on any set $X$. See 35.3(c).

3. The trivial uniformity on any set $X$. See 35.3(d).

4. The uniformity defined on $\mathbf{R}$ in 35.3(e), having as a base the sets

$$D_a = \Delta \cup \{(x, y) \mid x > a, y > a\}, \qquad a \in \mathbf{R}.$$

## 35B. *More examples of uniformities*

Decide which of the following collections are uniformities on the sets indicated and, for each that is, give the most efficient base you can and describe the uniform topology, in familiar terms if possible.

1. On $\mathbf{R}$, let $\mathscr{D}$ be all subsets of $\mathbf{R} \times \mathbf{R}$ which contain $\Delta \cup \nabla$, where $\nabla = \{(x, -x) \mid x \in \mathbf{R}\}$.

2. On $\mathbf{I}$, let $\mathscr{D}$ be all subsets of $\mathbf{I} \times \mathbf{I}$ of the form $E_\epsilon = \{(x, y) \mid |x - y| \text{ is rational and } < \epsilon\}$, $\epsilon > 0$.

3. On $[-1, 1]$, let $\mathscr{D}$ be all subsets of $[-1, 1] \times [-1, 1]$ which contain $\Delta \cup \square$, where $\square$ is the boundary of $[-1, 1] \times [-1, 1]$.

## 35C. *Separation in uniform spaces*

Let $\mathscr{D}$ be a separating uniformity on the set $X$.

1. If $a$ and $b$ are distinct points in $X$, then for some $D \in \mathscr{D}$, $D[a] \cap D[b] = \varnothing$.

2. If $A$ is compact and $B$ is closed in the uniform topology on $X$ and if $A \cap B = \varnothing$, then for some $D \in \mathscr{D}$, $D[A] \cap D[B] = \varnothing$.

## 35D. *Bases and subbases for uniformities*

1. If $\mathscr{E}$ is a base for the uniformity $\mathscr{D}$ on $X$, then $\{E[x] \mid E \in \mathscr{E}\}$ is a nhood base at $x$, for each $x \in X$.

2. If $\mathscr{E}$ is a base (subbase) for the uniformity $\mathscr{D}$ consisting of open sets, then

$$\{E[x] \mid E \in \mathscr{E}, x \in X\}$$

is a base (subbase) for the topology of $X$.

3. Let $H$ be a subset of $X \times X$ containing $\Delta$. Then the collection of all subsets of $X \times X$ which contain $H$ is a uniformity for $X$ iff $H$ is symmetric and $H \circ H = H$.

## 35E. *Union and intersection of uniformities*

1. The intersection of two uniformities on $X$ need not be a uniformity on $X$. [Let $X = \mathbf{I}$ and for each $x \in \mathbf{I}$, let $\mathscr{D}_x =$ all subsets of $\mathbf{I} \times \mathbf{I}$ containing $\Delta \cup \{(x, 1)\} \cup \{(1, x)\}$. Then $\mathscr{D}_x$ is a uniformity on $\mathbf{I}$, but for $x \neq y$, $\mathscr{D}_x \cap \mathscr{D}_y$ is not a uniformity on $\mathbf{I}$. (See 35D.3.)]

2. If two uniformities do intersect in a uniformity, is the uniform topology of the intersection the intersected uniform topologies?

3. The union of two uniformities on $X$ need not be a uniformity on $X$. [On $\mathbf{R}$, consider the uniformity $\mathscr{D}_1$ of Example 35.3(e), construct a similar uniformity $\mathscr{D}_2$ on $\mathbf{R}$ such that for some $D_1 \in \mathscr{D}_1$ and $D_2 \in \mathscr{D}_2$, $D_1 \cap D_2 \notin \mathscr{D}_1 \cup \mathscr{D}_2$.]

Compare with 36G.

### 35F. *Uniformities on topological groups*

Let $G$ be a topological group, with $\mathscr{U}$ a base of symmetric nhoods at the identity $e$. The *right uniformity* $\mathscr{D}_R$ for $G$ has for a base all sets of the form

$$R_u = \{(x, y) \mid x \in Uy\}$$

for $U \in \mathscr{U}$, while the *left uniformity* $\mathscr{D}_L$ for $G$ has for a base all sets of the form

$$L_u = \{(x, y) \mid x \in yU\}$$

for $U \in \mathscr{U}$. If $\mathscr{D}_L = \mathscr{D}_R$, we say $G$ has *equivalent uniform structures.*

1. $\mathscr{D}_L$ and $\mathscr{D}_R$ are uniformities on $G$, whose topologies $\tau_L$ and $\tau_R$ are each the group topology on $G$.

2. If $G$ is Abelian or compact, then $G$ has equivalent uniform structures.

Since, in Section 38, we will see that every topological space which is uniformizable is completely regular, part 1 above provides the corollary: *every topological group is a Tychonoff space.* This in turn provides the result: *every linear topological space* (22C) *is a Tychonoff space.* [See 22C.5.]

### 35G. *Uniformities from different metrics*

Show that if $\rho_1$ and $\rho_2$ are metrics on $X$ and there are constants $m$ and $M$ with $0 < m < M$ such that $m\rho_1 < \rho_2 < M\rho_1$, then $\rho_1$ and $\rho_2$ generate the same uniformity on $X$.

This extends a comment made in 35.3(b).

## 36 Uniform covers

Any uniformity on $X$ can be described, without passing to $X \times X$, by giving the list of covers of $X$ each of which consists of sets "of the same size". The result is an alternative approach to the theory of uniform spaces, and one which is quite often more convenient than the approach of the previous section.

**36.1 Definition.** A cover of a uniform space $(X, \mathscr{D})$ is a *uniform cover* iff it is refined by a cover of the form $\mathscr{U}_D = \{D[x] \mid x \in X\}$ for some $D \in \mathscr{D}$.

**36.2 Theorem.** *The collection $\mu$ of all uniform covers of a uniform space $(X, \mathscr{D})$ has the properties:*

a) *if $\mathscr{U}_1, \mathscr{U}_2 \in \mu$ then for some $\mathscr{U}_3 \in \mu$, $\mathscr{U}_3 \mathrel{*<} \mathscr{U}_1$ and $\mathscr{U}_3 \mathrel{*<} \mathscr{U}_2$,*

b) *if $\mathscr{U} < \mathscr{U}'$ and $\mathscr{U} \in \mu$, then $\mathscr{U}' \in \mu$.*

*Conversely, given any family $\mu$ of covers of a set $X$ satisfying* a) *and* b), *the collection of all sets $D_{\mathscr{U}} = \bigcup \{U \times U \mid U \in \mathscr{U}\}$, for $\mathscr{U} \in \mu$, is a base for a diagonal uniformity on $X$, whose uniform covers are precisely the elements of $\mu$.*

*Proof.* a) It is sufficient to show that any two covers $\mathscr{U}_{D_1}$ and $\mathscr{U}_{D_2}$ have a common barycentric refinement. (Recall that a barycentric refinement of a barycentric refinement of $\mathscr{U}$ star-refines $\mathscr{U}$.) Pick a symmetric $D \in \mathscr{D}$ such that

$$D \circ D \subset D_1 \cap D_2.$$

Then for each $x \in X$, $\text{St}(x, \mathscr{U}_D) \subset D_1[x] \cap D_2[x]$ and it follows that $\mathscr{U}_D$ is a common barycentric refinement of $\mathscr{U}_{D_1}$ and $\mathscr{U}_{D_2}$.

b) is obvious from the definition of uniform cover.

The converse is left as a straightforward exercise (36C). ■

Thus the uniform covers describe a uniformity as well as its surroundings do. In fact, the relationship between the two should be approached in much the same spirit one approaches the open sets and closed sets in a topological space: either describes the structure equally well. Actually, there is an abundance in the literature of references to "uniform spaces" whose primary structure is a collection of covers satisfying (a) and (b) above (such a collection is often called a *covering uniformity*), so it is best to keep an open mind about the sort of structure involved when someone starts yelling "uniform space." We will find it convenient, on different occasions, to use both coverings and surroundings to describe uniformities and we emphasize this dual approach with the following convention: hereafter, a *uniformity* on $X$ will mean either a diagonal uniformity on $X$ *or* a covering uniformity on $X$.

Although, as we have said, coverings and surroundings should be used in the same way as one uses open and closed sets in a topological space, i.e., interchangeably, we should comment that the passage back and forth is not nearly as neat. The uniform covers of a uniform space translate only to a base for the surroundings and similarly, the surroundings provide, in translation, only a base (as defined below) for the uniform covers. This causes no real problems, since all the important concepts defined for uniform spaces can be defined in terms of bases for the uniformities in question.

**36.3 Definition.** A *base* for a covering uniformity $\mu$ on $X$ is any subcollection $\mu'$ of $\mu$ such that

$$\mu = \{\mathscr{U} \mid \mathscr{U} \text{ covers } X \text{ and } \mathscr{U}' < \mathscr{U} \text{ for some } \mathscr{U}' \in \mu'\}.$$

Once we are over our initial confusion, $\mu'$ will simply be called a *base for the uniformity* on $X$ (context and notation will make it clear whether the base should consist of covers or surroundings). Evidently, $\mu'$ is a base for some uniformity on $X$ iff it satisfies (a) of 36.2.

A *subbase* for the covering uniformity $\mu$ is any subcollection $\mu'$ of $\mu$ such that all finite intersections of elements of $\mu'$ form a base for $\mu$, where the *intersection* of two covers $\mathscr{U}$ and $\mathscr{V}$ of $X$ is the cover $\mathscr{U} \wedge \mathscr{V} = \{U \cap V \mid U \in \mathscr{U}, V \in \mathscr{V}\}$.

In the language of bases, we can restate (and provide a trivial strengthening of) Theorem 36.2. The proof is left as Exercise 36C. The reader is invited to think now about the corresponding result for subbases.

**36.4 Theorem.** a) *If $\mathscr{D}'$ is a base for a diagonal uniformity $\mathscr{D}$, then $\{\mathscr{U}_D \mid D \in \mathscr{D}'\}$ is a base for a covering uniformity, whose surroundings are precisely the elements of $\mathscr{D}$.*

b) *If $\mu'$ is a base for a covering uniformity $\mu$, then $\{D_{\mathscr{U}} \mid \mathscr{U} \in \mu'\}$ is a base for a diagonal uniformity whose uniform covers are precisely the elements of $\mu$.*

The importance of a self-contained theory of covering uniformities justifies restating the important properties introduced in Section 35 in terms of uniform covers. Each of the Theorems 36.5, 36.6 and 36.8 would be the definition of the property involved if you got into a discussion with someone who knew only covering uniformities.

The first theorem shows that uniform covers may correctly be interpreted as covers by sets "of the same size."

**36.5 Theorem.** *A uniformity is metrizable, generated by the metric $\rho$, iff the covers $\mathscr{U}^\rho_\epsilon = \{U^\rho_\epsilon(x) \mid x \in X\}$ of $X$ by $\epsilon$-spheres, for $\epsilon > 0$, form a base.*

*Proof.* The sets $D^\rho_\epsilon = \{(x, y) \mid \rho(x, y) < \epsilon\}$ form a base for the surroundings on $X$ so the covers consisting of the sets

$$D^\rho_\epsilon[x] = \{y \mid \rho(x, y) < \epsilon\} = U^\rho_\epsilon(x)$$

form a base for the uniform covers of $X$. ∎

**36.6 Theorem.** *If $\mu'$ is a base for a covering uniformity $\mu$ on $X$, the sets* St $(x, \mathscr{U})$, *for $\mathscr{U} \in \mu'$, form a nhood base at $x$ in the uniform topology.*

*Proof.* Let $\mathscr{D}$ be the diagonal uniformity corresponding to $\mu$. The sets $D_{\mathscr{U}}$, for $\mathscr{U} \in \mu'$, form a base for $\mathscr{D}$, so the sets $D_{\mathscr{U}}[x]$, $\mathscr{U} \in \mu'$, form a nhood base at $x$ in the uniform topology, by 35.6. But

$$D_{\mathscr{U}}[x] = \{y \mid (x, y) \in D_{\mathscr{U}}\} = \{y \mid (x, y) \in U \times U \text{ for some } U \in \mathscr{U}\} = \text{St}(x, \mathscr{U}),$$

so the theorem is established. ∎

The last theorem enables us to state the condition that a uniformity on $X$ be separated in terms of its uniform covers, as follows: a covering uniformity is separated iff whenever $x \neq y$ in $X$, then there is a uniform cover $\mathscr{U}$ of $X$ such that St $(x, \mathscr{U}) \cap$ St $(y, \mathscr{U}) = \varnothing$. This can be rephrased, in light of the existence of a star-refinement of such a cover $\mathscr{U}$, as follows: a covering uniformity is separated iff whenever $x \neq y$ in $X$, then there is some uniform cover $\mathscr{U}'$ of $X$ such that $x \notin$ St $(y, \mathscr{U}')$.

**36.7 Theorem.** *Let $\mu$ be a covering uniformity on $X$. Then the open uniform covers of $X$ form a base for $\mu$.*

*Proof.* Let $\mathscr{D}$ be the diagonal uniformity on $X$ corresponding to $\mu$. The open elements of $\mathscr{D}$ form a base for $\mathscr{D}$, so the covers $\mathscr{U}_D$, for $D$ open in $\mathscr{D}$, form a base for $\mu$. But $\mathcal{U}_D = \{D[x] \mid x \in X\}$ and (easily) if $D$ is open in $X \times X$, then $D[x]$ is open in $X$ for each $x \in X$. ■

***36.8 Theorem.*** *Let X and Y be uniform spaces. A function $f: X \to Y$ is uniformly continuous iff for each uniform cover $\mathcal{U}_1$ of Y, there is a uniform cover $\mathcal{U}_0$ of X such that $f(\mathcal{U}_0) < \mathcal{U}_1$, where $f(\mathcal{U}_0) = \{f(U) \mid U \in \mathcal{U}_0\}$. (Hence, iff for each uniform cover $\mathcal{U}$ of Y, $f^{-1}(\mathcal{U})$ is a uniform cover of X).*

*Proof.* Suppose $f: X \to Y$ is uniformly continuous and $\mathcal{U}_0$ is a uniform cover of $Y$. Let $E$ be a surrounding for $Y$ such that $\mathcal{U}_E < \mathcal{U}_0$ and let $D$ be a surrounding for $X$ such that whenever $(x, y) \in D$, then $(f(x), f(y)) \in E$. Then, easily,

$$f(\mathcal{U}_D) < \mathcal{U}_E < \mathcal{U}_0.$$

Conversely, suppose the condition of the theorem holds. Given any surrounding $E$ for $Y$, find a uniform cover $\mathcal{U}_1$ of $Y$ such that $D_{\mathcal{U}_1} \subset E$ and a uniform cover $\mathcal{U}_0$ of $X$ such that $f(\mathcal{U}_0) < \mathcal{U}_1$. Then easily, $(x, y) \in D_{\mathcal{U}_0}$ implies $(f(x), f(y)) \in D_{\mathcal{U}_1}$ implies $(f(x), f(y)) \in E$, so $f$ is uniformly continuous. ■

It is convenient at this point to include a theorem relating the most important property of uniform covers (existence of star-refinements) to the most important property of surroundings (existence of $E$ such that $E \circ E \subset D$). It says they are essentially the same.

***36.9 Theorem.*** a) *If $D \circ D^{-1} \subset E$, then $\mathcal{U}_D$ is a barycentric refinement of $\mathcal{U}_E$.*

b) *If $\mathcal{U}$ is a star-refinement of $\mathcal{U}'$, then $D_{\mathcal{U}} \circ D_{\mathcal{U}} \subset D_{\mathcal{U}'}$.*

*Proof.* a) Suppose $D \circ D^{-1} \subset E$. Let $x \in X$ and let $D[y]$ be any element of $\mathcal{U}_D$ containing $x$. It suffices to show $D[y] \subset E[x]$. But if $z \in D[y]$, then $(y, z) \in D$ and, since $(y, x) \in D$ we have $(x, y) \in D^{-1}$, so that $(x, z) \in D \circ D^{-1} \subset E$ and hence $z \in E[x]$. Thus St $(x, \mathcal{U}_D) \subset E[x]$.

b) Suppose $\mathcal{U}$ star-refines $\mathcal{U}'$. Let $(x, y) \in D_{\mathcal{U}} \circ D_{\mathcal{U}}$. Then, say, $(x, z) \in D_{\mathcal{U}}$ and $(z, y) \in D_{\mathcal{U}}$ so that there are $U_1, U_2 \in \mathcal{U}$ with $(x, z) \in U_1 \times U_1$ and $(z, y) \in U_2 \times U_2$. But then St $(U_1, \mathcal{U})$ contains both $x$ and $y$ and is contained in some $U' \in \mathcal{U}'$, and hence $(x, y) \in U' \times U' \subset D_{\mathcal{U}'}$. Thus $D_{\mathcal{U}} \circ D_{\mathcal{U}} \subset D_{\mathcal{U}'}$. ■

We end this section with an introduction to so-called fine uniformities and fine spaces. We need some preliminary material on combining uniformities. First recall that a *normal sequence* of covers is a sequence $\mathcal{U}_1, \mathcal{U}_2, \ldots$ such that $\mathcal{U}_{n+1} \overset{*}{<} \mathcal{U}_n$ for $n = 1, 2, \ldots$, and a *normal cover* is a cover which is $\mathcal{U}_1$ in some normal sequence.

**36.10 Definition.** A family $\nu$ of covers of a set $X$ is a *normal family* iff every cover in $\nu$ has a star refinement in $\nu$. Then every normal sequence is a normal family, but a sequence of covers can be a normal family without being a normal sequence (e.g., by being two normal sequences intermixed).

The proof of the following theorem is left as an easy exercise for the reader (36F).

**36.11 Theorem.** *Every normal family of covers of $X$ is a subbase for some uniformity on $X$; the converse fails.*

**36.12 Theorem.** *If $X$ is any uniformizable topological space, there is a finest uniformity on $X$ compatible with the topology of $X$.*

*Proof.* Let $\{\mu_\alpha \mid \alpha \in A\}$ be the collection of all covering uniformities compatible with the topology of $X$. Then $\mu_0 = \bigcup \{\mu_\alpha \mid \alpha \in A\}$ is obviously a normal family and hence is a subbase for a uniformity on $X$, finer than all the $\mu_\alpha$. We have finished if $\mu$ is compatible with the topology of $X$.

First, finer uniformities clearly generate finer topologies, so it is sufficient to show the uniform topology generated by $\mu$ is contained in the original topology. But a subbase for $\mu$ generates a subbase for the uniform topology of $\mu$, so this is clear. ∎

**36.13 Definition.** If $X$ is a uniformizable topological space, the uniformity constructed in 36.12 is called the *fine uniformity* on $X$, denoted $\mu_F$, and when $X$ is provided with this uniformity, it is called a *fine space.*

We can further elucidate the nature of the fine uniformity on a topological space, using the following concept.

**36.14 Definition.** An open cover $\mathscr{U}$ of a topological space $X$ is *normally open* iff $\mathscr{U} = \mathscr{U}_1$ in some normal sequence $\mathscr{U}_1, \mathscr{U}_2, \ldots$ consisting of open covers. Note that every normally open cover is an open normal cover, but the converse fails.

**36.15 Theorem.** *$\mu_F$ is the uniformity on $X$ having as a base all normally open covers of $X$.*

*Proof.* If $\mu$ is any uniformity on $X$ giving the topology of $X$ and $\mathscr{U}_1, \mathscr{U}_2, \ldots$ is any normal sequence consisting of open covers of $X$, then the collection $\mu \cup \{\mathscr{U}_1, \mathscr{U}_2, \ldots\}$ is a normal family and hence a subbase for a uniformity, which clearly still gives the same topology on $X$. It follows that every normal sequence of open covers of $X$, and hence every normally open cover of $X$, must be included in $\mu_F$.

But, conversely, the open covers in $\mu_F$ form a base for $\mu_F$, by 36.7, and each open cover in $\mu_F$ is normally open (also by 36.7). ∎

**36.16 Corollary.** *In a paracompact space, the fine uniformity is generated by all open covers.*

*Proof.* Every open cover in a paracompact space has an open star refinement, by 20.15, and thus every open cover is normally open. ∎

The diagonal analogs to the last theorems are easily described.

**36.17 Corollary.** a) *The fine uniformity $\mathscr{D}_F$ is generated by all open nhoods*

*$D$ of the diagonal such that $D = E_1$ in some sequence $E_1, E_2, \ldots$ of open sets containing the diagonal with $E_n \circ E_n \subset E_{n-1}$, for all $n > 1$.*

b) *In a paracompact space, the fine uniformity is generated by all nhoods of the diagonal.*

*Proof.* Exercise 36H. ■

**36.18 Theorem.** *Every continuous function on a fine space to some uniform space is uniformly continuous.*

*Proof.* Let $X$ have the fine uniformity $\mu_F$ and let $f: X \to Y$. If $\mathscr{U}$ is any open uniform cover of $Y$, then $\mathscr{U}$ is a normal cover and hence $f^{-1}(\mathscr{U})$ is normal and open. Then $f^{-1}(\mathscr{U}) \in \mu_F$. Thus $f$ is uniformly continuous. ■

**36.19 Theorem.** *A compact $T_2$-space has only one uniformity compatible with its topology.*

*Proof.* Let $X$ be a compact space, $\mu$ a uniformity compatible with the topology on $X$. We will show every open cover $\mathscr{U}$ of $X$ belongs to $\mu$, so that $\mu$ must be the fine uniformity $\mu_F$.

For each $x \in X$, $x \in U$ for some $U \in \mathscr{U}$. Since $U$ is open, $\mathrm{St}(x, \mathscr{U}_x) \subset U$ for some $\mathscr{U}_x \in \mu$. Find an open $\mathscr{V}_x \in \mu$ such that $\mathscr{V}_x \mathbin{*\!<} \mathscr{U}_x$ and for each $x$, pick an element $V_x$ of $\mathscr{V}_x$ containing $x$. A finite number of these, say $V_{x_1}, \ldots, V_{x_n}$, cover $X$. Let $\mathscr{W}$ be a common star refinement of the corresponding covers $\mathscr{V}_{x_1}, \ldots, \mathscr{V}_{x_n}$. Now if $W \in \mathscr{W}$, then for some $x_i$,

$$W \subset \mathrm{St}(V_{x_i}, \mathscr{V}_{x_i}) \subset \mathrm{St}(x_i, \mathscr{U}_{x_i}) \subset U_{x_i}$$

so that $\mathscr{W} < \mathscr{U}$. Thus $\mathscr{U} \in \mu$. ■

**36.20 Corollary.** *Every continuous function on a compact $T_2$-space is uniformly continuous.*

Theorem 36.19 is generalized in Exercise 41F, where the topological spaces with unique uniform structure are characterized.

## Problems

### 36A. *Exercise on refinement*

A *partition* of $X$ is a cover of $X$ whose elements are disjoint.

1. Every partition star-refines itself.

2. For any cover $\mathscr{U}$ of $X$, there is a finest partition $P(\mathscr{U})$ refined by $\mathscr{U}$. Each element $U$ of $\mathscr{U}$ is contained in a unique element $P(U)$ of $P(\mathscr{U})$.

3. $\mathscr{U}$ star-refines itself iff for some partition $P$, $\mathscr{U} < P < \mathscr{U}$. [If $\mathscr{U}$ star-refines itself, set $P = P(\mathscr{U})$; show that $\mathrm{St}(U, \mathscr{U}) \supset P(U)$ for each $U \in \mathscr{U}$.]

## 36B. *Examples of covering uniformities*

1. The collection $\mu'$ consisting of the single cover $\{X\}$ is a base for the trivial uniformity on $X$. (More accurately, $\mu'$ is a base for a covering uniformity whose associated diagonal uniformity is trivial.)

2. The collection $\mu'$ consisting of the single cover $\{\{x\} \mid x \in X\}$ is a base for the discrete uniformity on $X$.

3. In a metric space $(M, \rho)$, the covers $\mathscr{U}_\epsilon^\rho$ by $\epsilon$-spheres form a base for the metric uniformity on $M$.

4. If the single cover $\mathscr{U}$ is a base for a uniformity on $X$, the same uniformity is generated by some partition $P$ of $X$ [see 36A.3].

5. The collection $\mu'$ of all finite (countable) covers of a set $X$ is a base for a uniformity on $X$, whose uniform topology is the discrete topology. [Use partitions; see 36A.]

## 36C. *Coverings give uniformities*

1. Let $\mu'$ be a base for a covering uniformity $\mu$ on $X$. Then the collection of all sets $D_{\mathscr{U}} = \bigcup \{U \times U \mid U \in \mathscr{U}\}$, for $\mathscr{U} \in \mu$, is a base for a diagonal uniformity $\mathscr{D}$ on $X$ whose uniform covers are precisely the elements of $\mu$.

2. Let $\mathscr{D}'$ be a base for a diagonal uniformity $\mathscr{D}$ on $X$. Then the collection of all covers $\mathscr{U}_D = \{D[x] \mid x \in X\}$, for $D \in \mathscr{D}$, is a base for a covering uniformity $\mu$ on $X$ whose surroundings are precisely the elements of $\mathscr{D}$.

## 36D. *Bounded metrics*

We already know that every metric is (topologically) equivalent to a bounded metric. Prove that any metric $\rho$ is uniformly equivalent to (i.e., produces the same uniformity as) the bounded metric $\rho^* = \min(\rho, 1)$.

## 36E. *The Hyperspace*

Let $\mathscr{D}$ be a diagonal uniformity on $X$ and let $\mathscr{H}(X)$ be the collection of all closed subsets of $X$. For $A, B \in \mathscr{H}(X)$ and $D \in \mathscr{D}$, we will say $A$ and $B$ are $D$-close iff $A \subset D[B]$ and $B \subset D[A]$.

1. The sets $\{(A, B) \mid A \text{ is } D\text{-close to } B\}$, for $D \in \mathscr{D}$, form a base for a diagonal uniformity $\mathscr{D}_{\mathscr{H}}$ on $\mathscr{H}$. The resulting uniform space $(\mathscr{H}, \mathscr{D}_{\mathscr{H}})$ is called the *hyperspace* of $X$.

2. The hyperspace of a metrizable uniform space is metrizable. [Replace the metric on the space by a uniformly equivalent bounded metric (36D) and consider the resulting Hausdorff metric (2F) on $\mathscr{H}$.]

We will return to the hyperspace in 39D.

## 36F. *Normal families*

Every normal family in $X$ is a subbase for some uniformity on $X$, but the converse fails.

## 36G. *The lattice of uniformities*

We saw in 35E that the union or intersection of uniformities on a fixed set $X$ need not be a uniformity on $X$. Now show that given any family $\mathscr{A}$ of uniformities on $X$, there is a coarsest containing them all and a finest contained in all of them, so that the uniformities on $X$ form a complete lattice. [Use normal families.]

### 36H. *Fine uniformities; the diagonal case*

1. The fine (diagonal) uniformity $\mathscr{D}_F$ on a uniformizable space is the uniformity having for a base the open sets $D \supset \Delta$ such that for some sequence $D_1, D_2, \ldots$ of open sets containing $\Delta$, $D_n \circ D_n \subset D_{n-1}$ for all $n$ and $D_1 = D$.

2. In a paracompact uniformizable space, the fine (diagonal) uniformity is generated by all nhoods of the diagonal.

## 37 Uniform products and subspaces; weak uniformities

In this section, we provide the standard constructions for uniform structures on subspaces and products of uniform spaces, as well as the generalization from product structures to weak uniformities. There are no surprises.

Subspaces, in particular, can be dealt with quickly and easily.

**37.1 Definition.** If $\mathscr{D}$ is a diagonal uniformity on $X$ and $A \subset X$, the *relative uniformity* induced on $A$ by $\mathscr{D}$ is the uniformity consisting of the sets $D \cap (A \times A)$, for $D \in \mathscr{D}$. With this uniformity, $A$ is called a *(uniform) subspace* of $X$.

Verification that the relative uniformity induced on $A$ by $\mathscr{D}$ actually is a diagonal uniformity on $A$ is left to Exercise 37A.

To describe the uniform covers in a subspace $A$ of $X$, we need the following notion. If $\mathscr{U}$ is any collection of subsets of a set $X$ and $A \subset X$, the *trace* of $\mathscr{U}$ on $A$ is the collection $\{U \cap A \mid U \in \mathscr{U}\}$ of subsets of $A$.

**37.2 Theorem.** *The traces on $A$ of the uniform covers of $X$ form a base for the uniform covers of $A$.*

*Proof.* If $\mathscr{U}$ is a uniform cover of $X$, then for some $D \in \mathscr{D}$, $\{D[x] \mid x \in X\}$ refines $\mathscr{U}$. But then, obviously, $\{[D \cap (A \times A)][x] \mid x \in A\}$ refines the trace of $\mathscr{U}$ on $A$, so the latter is a uniform cover of $A$.

Conversely, if $\mathscr{U}'$ is a uniform cover of $A$, then for some $D \in \mathscr{D}$

$$\{[D \cap (A \times A)][x] \mid x \in A\}$$

refines $\mathscr{U}'$. Then $\{D[x] \mid x \in X\}$ is a uniform cover of $X$ whose trace on $A$ refines $\mathscr{U}'$. ■

**37.3 Theorem.** *The topology on a uniform subspace $A$ of $X$ is the subspace topology.*

*Proof.* It is sufficient to note that, for $D \in \mathscr{D}$ and $a \in A$,

$$[D \cap (A \times A)][a] = D[a] \cap A. \; ■$$

We turn now to the problem of defining a uniformity on the product of uniform spaces, subject to the obvious restriction that the topology of such a uniformity should be the product topology.

A definition will make life easier.

**37.4 Definition.** If $X_\alpha$ is a set for each $\alpha \in A$ and $X = \prod X_\alpha$, the $\alpha$th *biprojection* is the map $P_\alpha: X \times X \to X_\alpha \times X_\alpha$ defined by $P_\alpha(x, y) = (\pi_\alpha(x), \pi_\alpha(y))$.

***37.5 Theorem.*** *If $\mathscr{D}_\alpha$ is a diagonal uniformity on $X_\alpha$, for each $\alpha \in A$, then the sets*

$$P_{\alpha_1}^{-1}(D_{\alpha_1}) \cap \cdots \cap P_{\alpha_n}^{-1}(D_{\alpha_n}),$$

*where $D_{\alpha_i} \in \mathscr{D}_{\alpha_i}$ for $i = 1, \ldots, n$, form a base for a uniformity $\mathscr{D}$ on $\prod X_\alpha$ whose associated topology is the product topology on $\prod X_\alpha$.*

*Proof.* a) Easily, $\Delta \subset P_{\alpha_1}^{-1}(D_{\alpha_1}) \cap \cdots \cap P_{\alpha_n}^{-1}(D_{\alpha_n})$.

b) The intersection of two sets of the form $P_{\alpha_1}^{-1}(D_{\alpha_1}) \cap \cdots \cap P_{\alpha_n}^{-1}(D_{\alpha_n})$ clearly has the same form.

c) Let $D = P_{\alpha_1}^{-1}(D_{\alpha_1}) \cap \cdots \cap P_{\alpha_n}^{-1}(D_{\alpha_n})$. For $i = 1, \ldots, n$, find $E_{\alpha_i} \in \mathscr{D}_{\alpha_i}$ such that $E_{\alpha_i} \circ E_{\alpha_i} \subset D_{\alpha_i}$, and let $E = P_{\alpha_1}^{-1}(E_{\alpha_1}) \cap \cdots \cap P_{\alpha_n}^{-1}(E_{\alpha_n})$. Then if $(x, z) \in E \circ E$, we can find some $y$ such that $(x, y) \in E$ and $(y, z) \in E$. Now $(x_\alpha, y_\alpha)$ and $(y_\alpha, z_\alpha)$ each belong to $E_\alpha$ for $\alpha = \alpha_1, \ldots, \alpha_n$ and hence

$$(x_\alpha, z_\alpha) \in E_\alpha \circ E_\alpha \subset D_\alpha, \qquad \text{for} \quad \alpha = \alpha_1, \ldots, \alpha_n.$$

Thus $(x, z) \in D$, so $E \circ E \subset D$.

d) As in (c), if $E_{\alpha_i}^{-1} \subset D_{\alpha_i}$ for $i = 1, \ldots, n$, then $E^{-1} \subset D$.

Finally, to show the product uniformity gives the product topology, note that if $D = P_{\alpha_1}^{-1}(D_{\alpha_1}) \cap \cdots \cap P_{\alpha_n}^{-1}(D_{\alpha_n})$, then for $x \in \prod X_\alpha$,

$$\begin{aligned} D[x] &= \{y \mid (x, y) \in D\} \\ &= \{y \mid (x_{\alpha_i}, y_{\alpha_i}) \in D_{\alpha_i} \text{ for } i = 1, \ldots, n\} = \bigcap_{i=1}^{n} \pi_{\alpha_i}^{-1}(D_{\alpha_i}[x_{\alpha_i}]). \blacksquare \end{aligned}$$

**37.6 Definition.** The uniformity constructed in 37.5 is the *product uniformity* on $\prod X_\alpha$, and $\prod X_\alpha$ is the *product space* formed from the *factor spaces* $X_\alpha$, $\alpha \in A$.

Before describing the uniform covers on a product space, it will be convenient to introduce the uniform analog to the weak topology induced by a collection of maps from a set $X$ to topological spaces $X_\alpha$, $\alpha \in A$; namely, the weak uniformity induced by a collection of maps from a set $X$ to *uniform* spaces $X_\alpha$, $\alpha \in A$.

**37.7 Definition.** For each $\alpha \in A$, suppose $f_\alpha: X \to X_\alpha$, where $X$ is a set and $X_\alpha$ is a space with a diagonal uniformity $\mathscr{D}_\alpha$. Define $F_\alpha: X \times X \to X_\alpha \times X_\alpha$ by

$$F_\alpha(x, y) = (f_\alpha(x), f_\alpha(y)),$$

for each $\alpha \in A$. The collection of sets of the form

$$F_{\alpha_1}^{-1}(D_{\alpha_1}) \cap \cdots \cap F_{\alpha_n}^{-1}(D_{\alpha_n}),$$

where $D_{\alpha_i} \in \mathscr{D}_{\alpha_i}$ for $i = 1, \ldots, n$, is a base for a uniformity on $X$, called the *weak uniformity* generated by the maps $f_\alpha$ on $X$. Verification that this is a base for a uniformity can be obtained by a trivial rewriting of the proof of 37.5. In the same way, you can prove that the topology induced by a weak uniformity is the weak topology induced by the maps $f_\alpha$.

It is clear that the product uniformity on $\prod X_\alpha$ is the weak uniformity generated by the projection maps $\pi_\alpha$.

**37.8 Theorem.** *The weak uniformity generated by the maps $f_\alpha: X \to X_\alpha$ is the weakest uniformity making each $f_\alpha$ uniformly continuous.*

*Proof.* It is clear that each $f_\alpha$ is uniformly continuous, for if $D \in \mathscr{D}_\alpha$, then $F_\alpha^{-1}(D)$ is an element of the weak uniformity and if $(x, y) \in F_\alpha^{-1}(D)$, then $(f_\alpha(x), f_\alpha(y)) \in D$.

Suppose $\mathscr{D}$ is the weak uniformity on $X_\alpha$ and $\mathscr{D}'$ is any uniformity such that each $f_\alpha$ is uniformly continuous. We will show $F_{\alpha_1}^{-1}(D_{\alpha_1}) \cap \cdots \cap F_{\alpha_n}^{-1}(D_{\alpha_n})$ will always belong to $\mathscr{D}'$. For each $i = 1, \ldots, n$ there will be some $E_{\alpha_i} \in \mathscr{D}'$ such that if $(x, y) \in E_{\alpha_i}$, then $(f_{\alpha_i}(x), f_{\alpha_i}(y)) \in D_{\alpha_i}$. Then $E_{\alpha_i} \subset F_{\alpha_i}^{-1}(D_{\alpha_i})$ and hence

$$E_{\alpha_1} \cap \cdots \cap E_{\alpha_n} \subset F_{\alpha_1}^{-1}(D_{\alpha_1}) \cap \cdots \cap F_{\alpha_n}^{-1}(D_{\alpha_n})$$

so that the latter set belongs to $\mathscr{D}'$ as claimed. ■

**37.9 Theorem.** *If $X$ has the weak uniformity generated by maps $f_\alpha: X \to X_\alpha$, then $f: Z \to X$ is uniformly continuous iff $f_\alpha \circ f$ is uniformly continuous, for each $\alpha$.*

*Proof.* If $f$ is uniformly continuous, then $f_\alpha \circ f$ is, for each $\alpha$, because composition preserves uniform continuity.

Suppose $f_\alpha \circ f$ is uniformly continuous, for each $\alpha$. Let $D \in \mathscr{D}(X)$. Then $D$ contains a set of the form $F_{\alpha_1}^{-1}(D_{\alpha_1}) \cap \cdots \cap F_{\alpha_n}^{-1}(D_{\alpha_n})$, where $F_{\alpha_i}$ is the map associated with $f_{\alpha_i}$ (see 37.7) and $D_{\alpha_i} \in \mathscr{D}_{\alpha_i}$ (the uniformity on $X_{\alpha_i}$). But for each $i = 1, \ldots, n$, there is some $F_i \in \mathscr{D}(z)$ such that

$$(x, y) \in F_i \Rightarrow (f_{\alpha_i} \circ f(x), f_{\alpha_i} \circ f(y)) \in E_{\alpha_i}$$

since $f_{\alpha_i} \circ f$ is uniformly continuous. Thus for $(x, y) \in F_i$, $(f(x), f(y)) \in F_{\alpha_i}^{-1}(E_{\alpha_i})$, and hence, if $(x, y) \in F = \bigcap_{i=1}^n F_i$ then $(f(x), f(y)) \in \bigcap_{i=1}^n F_{\alpha_i}^{-1}(E_{\alpha_i}) \subset D$. This establishes uniform continuity of $f$. ■

It is worth the effort of rewriting the above theorems to have them set apart for the special case of product spaces.

**37.10 Corollary.** a) *The product uniformity on $\prod X_\alpha$ is the weakest making each projection $\pi_\alpha$ uniformly continuous.*

b) *A map* $f: Z \to \prod X_\alpha$ *is uniformly continuous iff* $\pi_\alpha \circ f$ *is uniformly continuous for each* $\alpha$.

Just as with weak topologies, weak uniformities generated by most collections of maps are just subspaces of product uniformities. Recall that the evaluation map $e: X \to \prod X_\alpha$ determined by a collection of maps $f_\alpha: X \to X_\alpha$ is the map $[e(x)]_\alpha = f_\alpha(x)$; that is, $e$ is defined by the relation $\pi_\alpha \circ e = f_\alpha$.

**37.11 Theorem.** *The evaluation e is a uniform isomorphism iff the maps* $f_\alpha$ *separate points in X and X has the weak uniformity given by them.*

*Proof.* Exercise 37B. Use the proof of Theorem 8.12 as a model. ■

We now turn to the problem of describing the uniform covers on a product space, or in a weak uniformity, in terms of uniform covers of the factor spaces. What we are really doing then, is developing the covering description of product, or weak uniformities.

**37.12 Theorem.** *The weak uniformity on a set X induced by maps* $f_\alpha: X \to X_\alpha$ *has as a subbase for its uniform covers the inverse images*

$$f_\alpha^{-1}(\mathscr{U}) = \{f_\alpha^{-1}(U) \mid U \in \mathscr{U}\},$$

*for* $\alpha \in A$ *and* $\mathscr{U}$ *a uniform cover of* $X_\alpha$.

*Proof.* The weak uniformity on $X$ is the coarsest making all $f_\alpha$ uniformly continuous; i.e., the coarsest making each $f_\alpha^{-1}(\mathscr{U})$ a uniform cover. Thus, if the $f_\alpha^{-1}(\mathscr{U})$ form a subbase for a uniformity on $X$, it must be the weak uniformity. But since $\mathscr{U}_1 {}^*\!< \mathscr{U}_2 \Rightarrow f_\alpha^{-1}(\mathscr{U}_1) {}^*\!< f_\alpha^{-1}(\mathscr{U}_2)$, the covers $f_\alpha^{-1}(\mathscr{U})$ form a normal family and thus, by 36.11, are a subbase for a uniformity on $X$. ■

**37.13 Corollary.** *A base for the uniform covers on a product* $\prod X_\alpha$ *of uniform spaces consists of all covers obtained as follows: Pick* $\alpha_1, \ldots, \alpha_n$ *and a uniform cover* $\mathscr{U}_i$ *of* $X_{\alpha_i}$ *for each i; then form the cover of* $\prod X_\alpha$ *by all the sets* $\prod U_\alpha$, *where* $U_{\alpha_i} \in \mathscr{U}_i$, $i = 1, \ldots, n$, *and* $U_\alpha = X_\alpha$ *otherwise.*

## Problems

### 37A. *Uniform subspaces*

1. Show that, if $\mathscr{D}$ is a diagonal uniformity on $X$ and $A \subset X$, the sets $D \cap (A \times A)$, for $D \in \mathscr{D}$, form a diagonal uniformity on $A$.

2. If $A$ is dense in the uniform space $X$ and $\mathscr{U}$ is any uniform cover of $A$ (in the relative uniformity), then $\{\overline{U} \mid U \in \mathscr{U}\}$ is a uniform cover of $X$.

### 37B. *Evaluation and the weak uniformity*

Let $X_\alpha$ be a uniform space and $f_\alpha: X \to X_\alpha$ for each $\alpha \in A$, where $X$ is a uniform space. Then the evaluation map $e: X \to \prod X_\alpha$ is a uniform isomorphism (into, not onto) iff the maps $f_\alpha$ separate points and the uniformity on $X$ is the weak uniformity given by the maps $f_\alpha$.

### 37C. *Sufficient conditions for uniform continuity*

Is there a uniform analog to 7.6? That is, if $f$ is uniformly continuous when restricted to each of two open (or closed) subsets $A$ and $B$ of $X$, whose union is all of $X$, then is $f$ uniformly continuous on $X$?

### 37D. *Metric products and subspaces*

1. A uniform product of metrizable spaces is metrizable iff the number of nontrivial factors is countable.

2. Every uniform subspace of a metrizable uniform space is metrizable.

### 37E. *Uniform quotients*

Let $X$ be a uniform space, $Y$ a set, with $f: X \to Y$ a map of $X$ onto $Y$.

1. There is a largest uniformity on $Y$ which makes $f$ uniformly continuous. It is called the *quotient uniformity* induced on $Y$ by $f$, and $Y$ with this uniformity is called a *uniform quotient* of $X$ (by $f$).

2. If $Y$ is a uniform quotient of $X$ by $f$, the uniform topology on $Y$ may differ from the quotient topology induced on $Y$ by $f$.

3. A map $f$ between uniform spaces $X$ and $Y$ is *uniformly open* iff for each surrounding $D$ for $X$, there is a surrounding $E$ for $Y$ such that $f(D[x]) \supset E[f(x)]$, whenever $x \in X$. Show that if $f: X \to Y$ is uniformly open, then $Y$ has the quotient uniformity induced by $f$.

### 37F. *Inverse limits of uniform spaces*

Construct a theory of inverse limit spectra and inverse limit spaces for uniform spaces which mimics 29C.

## 38 Uniformizability and uniform metrizability

Here we tackle two difficult, but important, questions: which topologies come from uniformities and which uniformities come from metrics?

One lemma is basic to the development of criteria both for uniformizability and for uniform metrizability. The major part of the development of this lemma has already been accomplished in Section 23. There, in 23.4, we showed that the topology generated by a normal sequence $\mathscr{U}_1 >^* \mathscr{U}_2 >^* \cdots$ is also generated by a pseudometric $\rho$, and that in fact if $\mathscr{V}_n$ is the collection of $1/2^n$ spheres measured by $\rho$, then $\mathscr{V}_n < \mathscr{U}_{n-1}$ and $\mathscr{U}_n < \mathscr{V}_{n-1}$. This easily leads to the following lemma, which says any uniform cover in a uniform space can be "approximated" by a pseudometric.

**38.1 Lemma.** *If $\mathscr{U}$ is a uniform cover on a uniform space $X$, there is a pseudometric $\rho$ on $X$ such that $\mathscr{V}_\epsilon = \{U_\rho(x, \epsilon) \mid x \in X\}$ is a uniform cover for each $\epsilon > 0$ and $\mathscr{V}_1 < \mathscr{U}$. Moreover, $\rho$ can be taken to be bounded by 1.*

*Proof.* First, using the definition of a covering uniformity, a normal sequence can be constructed "beneath" $\mathscr{U}$:

$$\cdots {}^*\!< \mathscr{U}_1 {}^*\!< \mathscr{U}_0 {}^*\!< \mathscr{U}.$$

Letting $\rho$ be the pseudometric associated with this normal sequence by 23.4, the conclusions of the lemma are easily satisfied by $\rho$.

Once $\rho$ is found, it does not hurt to replace it with $\min(\rho, 1) = \rho^*$, since $U_{\rho^*}(x, \epsilon) = U_\rho(x, \epsilon)$ for all $x$ and all $\epsilon \le 1$. ∎

The collection of uniform covers which make up a uniformity on $X$ thus gives rise to a collection of pseudometrics on $X$. What is more, this collection of pseudometrics can be used to recover the original uniformity. Thus, certain collections of pseudometrics on $X$ can lay claim to being uniform structures. In Exercise 38A, you will develop the properties such a collection of pseudometrics must satisfy.

It is also worth mentioning that, by 36.9, the diagonal analog to the normal sequence that sits under any uniform cover is the *composition sequence*

$$\cdots \subset D_2 \subset D_1 \subset D$$

contained in any $D \in \mathscr{D}(X)$ where for each $n$, $D_n \circ D_n \subset D_{n-1}$. Such a sequence can be used in much the same way to generate a pseudometric.

We are prepared now to prove our theorem on uniformizability. Our policy of using whichever description of a uniform space is most convenient is stretched to the limit here; one implication will be proved using covering uniformities, the other using diagonal uniformities.

***38.2 Theorem.*** *A topological space is uniformizable iff it is completely regular.*

*Proof.* $\Rightarrow$: Let $\mu$ be a covering uniformity on $X$ which generates the topology, and suppose $A$ is closed in $X$, $x \notin A$. For some $\mathscr{U} \in \mu$, $\mathrm{St}(x, \mathscr{U}) \cap A = \varnothing$. Let $d$ be the pseudometric (bounded by 1) associated with $\mathscr{U}$ by 38.1. Then

$$U_d(x, 1) \cap A = \varnothing.$$

Let $f: X \to [0, 1]$ be the function $f(x) = d(A, x)$. Then $f$ is easily uniformly continuous on $X$, and $f(A) = 0$, $f(x) = 1$. Thus $X$ is completely regular.

$\Leftarrow$: Suppose $X$ is completely regular. Let $S$ be the collection of continuous real-valued functions on $X$. For $f \in S$ and $\epsilon > 0$, let

$$D_{f,\epsilon} = \{(x, y) \in X \times X \mid |f(x) - f(y)| < \epsilon\},$$

and let $\mathscr{D}$ be the entourage uniformity having as base the collection of sets of the form

$$D_{f_1,\epsilon_1} \cap \cdots \cap D_{f_n,\epsilon_n}$$

where $f_1, \ldots, f_n \in S$, $\epsilon_i > 0$. In fact, $\mathscr{D}$ is apparently the weak uniformity generated on $X$ by its collection of real-valued continuous functions. We need only show $\mathscr{D}$ generates the right topology on $X$.

Note we are really proving that the topology associated with a weak uniformity is the weak topology. Suppose $A$ is closed in the original topology on $X$, and $x \notin A$. Find $f \in S$ such that $f(x) = 0, f(A) = 1$. Let $E = D_{f,1/2}$. Then if $z \in E[x]$, $(x, z) \in E$ so that $|f(x) - f(z)| < \frac{1}{2}$ and hence $|f(z)| < \frac{1}{2}$. It follows that $E[x] \cap A = \emptyset$, so $A$ is closed in the uniform topology. Thus the usual topology is smaller than the uniform topology.

For the reverse, it suffices to show $E[x]$ is open in the original topology on $X$ for each $E$ belonging to the base for the uniformity and each $x \in X$. But $E = D_{f_1,\epsilon_1} \cap \cdots \cap D_{f_n,\epsilon_n}$, and then

$$E[x] = D_{f_1,\epsilon_1}[x] \cap \cdots \cap D_{f_n,\epsilon_n}[x]$$

so we need only check $D_{f_k,\epsilon_k}[x]$ is open in $X$. But

$$D_{f_k,\epsilon_k}[x] = \{y \mid |f_k(x) - f_k(y)| < \epsilon_k\} = f_k^{-1}(f_k(x) - \epsilon_k, f_k(x) + \epsilon_k)$$

so the desired result follows from continuity of the $f_k$. ∎

A word of caution. Do not read into the complete regularity we are using to characterize uniform spaces any separation axioms which are not there. For example, every pseudometric space is uniformizable, i.e., completely regular, but the nonmetric examples of these spaces are not even $T_0$! Thus the trivial topology on any space is uniformizable.

The particular uniformity constructed in 38.2 for $X$ will not usually be the only one giving the right topology. (Topological spaces with unique uniform structure are discussed in 41F; an example of a topological space with several compatible uniformities is given in 40E.)

As might be imagined, Lemma 38.1 must serve as the stepping stone to success in any search for a uniform metrization theorem. The idea of the following proof is simple enough. According to 38.1, each element of the base of a uniformity can be described by a pseudometric. If there are only countably many pseudometrics to deal with, the result of combining them is still a pseudometric, and it will describe completely the uniformity in question.

***38.3 Theorem.*** *A uniformity $\mu$ on $X$ is pseudometrizable iff it has a countable base.*

*Proof.* If $\rho$ is a pseudometric giving the uniformity $\mu$, then $\{\mathscr{U}_1, \mathscr{U}_2, \ldots\}$ is a countable base for $\mu$, where $\mathscr{U}_n = \{U_\rho(x, 1/2^n) \mid x \in X\}$.

Conversely, suppose $\{\mathscr{U}_1, \mathscr{U}_2, \ldots\}$ is a base for $\mu$. By taking common star refinements in order, we may assume $\cdots {}^*\!< \mathscr{U}_2 {}^*\!< \mathscr{U}_1$. Let $d_n$ be the pseudometric associated with $\mathscr{U}_n$ by 38.1 and assume $d_n \leq 1$. Define

$$d(x, y) = \sum_{n=1}^{\infty} \frac{d_n(x, y)}{2^n}.$$

Then $d$ is a pseudometric on $X$, and $\{U_d(x, 1/2^n) \mid x \in X\} < \{U_{d_n}(x, 1) \mid x \in X\} < \mathscr{U}_n$, so for each $n$ the cover $\mathscr{U}_n$ belongs to the uniformity $\mu_d$ generated by $d$. Thus $\mu \subset \mu_d$.

We will be done if we show $\{U_d(x, \epsilon) \mid x \in X\} \in \mu$ for each $\epsilon > 0$. Pick $N$ large enough that $\sum_{n=N+1}^{\infty} (1/2^n) < \epsilon/2$. Now, by 38.1, the uniformity generated by $d_n$ is contained in $\mu$ for each $n$, so that $\{U_{d_n}(x, 2^n\epsilon/(4N)) \mid x \in X\}$ belongs to $\mu$, for each $n$. Let $\mathscr{U}$ be a common refinement of these $N$ covers. Then given $U \in \mathscr{U}$, for some $x_1, \ldots, x_N$ in $X$,

$$U \subset \bigcap_{n=1}^{N} U_{d_n}\left(x_n, \frac{2^n\epsilon}{4N}\right)$$

and a routine computation shows that, if $x \in U$,

$$\bigcap_{n=1}^{N} U_{d_n}\left(x_n, \frac{2^n\epsilon}{4N}\right) \subset U_d(x, \epsilon).$$

It follows that $\mathscr{U} < \{U_d(x, \epsilon) \mid x \in X\}$. ■

***38.4 Corollary.*** *A uniformity is metrizable iff it is separating and has a countable base.*

It is well at this point to correct what is a common misconception. A uniformity $\mu$ is metrizable if for some metric the covers of $X$ by $\epsilon$-spheres form a base for $\mu$. If $\mu$ is metrizable, so is the topology it generates, but metrizability of the associated topology does *not* imply metrizability of $\mu$.

**38.5 Example.** An example of the phenomenon just mentioned can be found in the countable ordinals. For each $\alpha \in \mathbf{\Omega}_0$, let

$$D_\alpha = \{(x, y) \mid x = y \text{ or } x > \alpha, y > \alpha\}$$

(see Fig. 38.1). Then $\{D_\alpha \mid \alpha \in \mathbf{\Omega}_0\}$ is a base for a uniformity which cannot be metrizable (for any countable base would lead to a collection $(D_{\alpha_n})$ with the property that $\sup_n \{\alpha_n\} = \omega_1$), but whose topology is obviously the discrete topology; indeed, $D_\alpha[\beta] = \{\beta\}$ if $\beta < \alpha$.

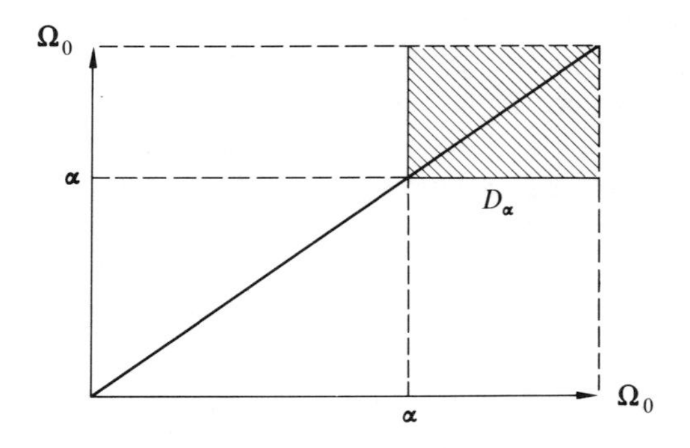


**Figure 38.1**

## Problems

### 38A. *Gage structures*

A covering uniformity $\mu$ on a space $X$ is prescribed by giving a collection of uniform covers. By 38.1, then, we could describe such a uniformity by giving the family $\{\rho_\alpha \mid \alpha \in A\}$ of pseudometrics, such that for each $\alpha$ and $\epsilon > 0$, $\mathscr{V}_\epsilon^\alpha = \{U_{\rho_\alpha}(x, \epsilon) \mid x \in X\}$ is a uniform cover; i.e., by giving the family of all pseudometrics which generate a weaker uniformity than the original uniformity (38.1 ensures that the original uniformity will be the smallest containing all these weaker uniformities).

1. The collection $\mathscr{G}$ of pseudometrics so obtained has the properties:

   a) $\rho_1, \rho_2 \in \mathscr{G} \Rightarrow \rho_1 \vee \rho_2 \in \mathscr{G}$, where $\rho_1 \vee \rho_2 = \sup(\rho_1, \rho_2)$,
   b) if $\rho$ is a pseudometric and for each $\epsilon > 0$ there is a $\delta > 0$ and a $\rho' \in \mathscr{G}$ such that $\rho'(x, y) < \delta \Rightarrow \rho(x, y) < \epsilon$, then $\rho \in \mathscr{G}$.

We call $\mathscr{G}$ the *gage* of the uniformity $\mu$.

2. Conversely, any collection $\mathscr{G}$ of pseudometrics on $X$ satisfying a) and b) of 1 is a gage for some covering uniformity $\mu$.

3. A collection $\Phi$ satisfying 1-a) only is a *base* for the gage obtained by taking all pseudometrics $\rho$ which satisfy 1-b) relative to $\Phi$.

4. Any collection $\Phi$ of pseudometrics on $X$ is contained in a smallest possible gage, called the gage *generated* by $\Phi$.

5. Since gages are in one–one correspondence with uniformities, we can treat them as uniformities. In particular, verify that if $\mathscr{G}$ is a gage for $\mu$, the uniform topology of $\mu$ on $X$ is generated by the nhoods $U_\rho(x, \epsilon)$, $\rho \in \mathscr{G}$, $\epsilon > 0$, at $x$.

6. Let $X$ be a completely regular topological space and for each $f \in C^*(X)$, define $\rho_f(x, y) = |f(x) - f(y)|$. The collection $\{\rho_f \mid f \in C^*(X)\}$ generates a gage for a uniformity compatible with the topology on $X$.

Gages do a good job of illuminating the nature of the generalization from pseudometric spaces to uniform spaces; a pseudometric space is a space with a gage generated by a single element.

### 38B. *Separation axioms in groups*

1. Every topological group (13G) is Tychonoff. [See 35F.]

2. Every locally compact topological group is normal. [Let $U$ be a nhood of $e$ such that $\bar{U}$ is compact. Let $H = \bigcup_{n=1}^{\infty} (U \cup U^{-1})^n = \bigcup_{n=1}^{\infty} (\bar{U} \cup \bar{U}^{-1})^n$. Then $H$ is an open subgroup of $G$ (hence closed) and is $\sigma$-compact, therefore Lindelöf, therefore normal. Thus $\{\alpha H \mid a \in G\}$ is a cover of $G$ by disjoint, open–closed normal subspaces. Proceed.]

### 38C. *Metrization of topological groups*

Let $G$ be a topological group (13G).

1. $G$ is metrizable iff $G$ is first countable [35F, 38.4]. (More can be shown, with some difficulty. If $G$ is first countable, then it has a *left invariant* metric $\rho$; i.e., $\rho(ax, ay) = \rho(x, y)$ for all $a, x, y \in G$.)

2. If $G$ is locally compact, $G$ is metrizable iff the identity $e$ is a $G_\delta$. (Again the metric can be taken as a left invariant.)

## 38D. *Examples on metrization*

For each of the following uniformities on the set indicated, decide whether it is metrizable, pseudometrizable or neither. If it is neither, decide whether or not the associated topology is metrizable.

1. The uniformity $\mathscr{D}$ on $\mathbf{R}$ which has for a base the sets

$$D_a = \Delta \cup \{(x, y) \mid x > a, y > a\}$$

for $a \in \mathbf{R}$.

2. The discrete uniformity on any set $X$.
3. The uniformity on $\mathbf{R}$ having for a base all countable covers of $\mathbf{R}$.

## 38E. *Uniformizability and the fine uniformity*

Give an example of a completely regular space $X$ for which the uniformity $\mathscr{D}$ generated by the continuous real-valued functions on $X$ (see 38.2) is not the fine uniformity.

## 39 Complete uniform spaces; completion

The notion of completeness can easily be carried over from metric spaces to uniform spaces. All that is needed is an appropriate generalization of the notion of a Cauchy sequence.

**39.1 Definition.** Let $X$ have a diagonal uniformity $\mathscr{D}$. A net $(x_\lambda)$ in $X$ is *$\mathscr{D}$-Cauchy* (or just *Cauchy*) iff for each $D \subset \mathscr{D}$, there is some $\lambda_0 \in \Lambda$ such that $(x_{\lambda_1}, x_{\lambda_2}) \in D$ whenever $\lambda_1, \lambda_2 \geq \lambda_0$. The corresponding covering description is as follows: $(x_\lambda)$ is *$\mu$-Cauchy*, or just *Cauchy*, iff for each uniform cover $\mathscr{U}$, there is some $\lambda_0 \in \Lambda$ such that $x_{\lambda_1}$ and $x_{\lambda_2}$ lie together in some element of $\mathscr{U}$ whenever $\lambda_1, \lambda_2 \geq \lambda_0$.

***39.2 Theorem.*** *Every convergent net is Cauchy.*

*Proof.* If $x_\lambda \to x$ and $D \in \mathscr{D}$, pick symmetric $E \in \mathscr{D}$ such that $E \circ E \subset D$. Then $(x_\lambda)$ is eventually in $E[x]$, say for $\lambda \geq \lambda_0$, and now if both $\lambda_1$ and $\lambda_2$ are $\geq \lambda_0$, then $(x_{\lambda_1}, x) \in E$ and $(x_{\lambda_2}, x) \in E$ so $(x_{\lambda_1}, x_{\lambda_2}) \in E \circ E \subset D$. ∎

**39.3 Definition.** If every Cauchy net in a uniform space $X$ converges, then $X$ is a *complete* uniform space.

The concept of completeness thus defined matches up with the definition already given for metric spaces, according to the next theorem. As preparation for reading the proof, note that if $X$ is a metrizable uniform space, generated by a metric $\rho$, then a sequence $(x_n)$ in $X$ is Cauchy in the sense of 39.1 iff for each $\epsilon > 0$, there is some $n_0$ such that $\rho(x_{n_1}, x_{n_2}) < \epsilon$ whenever $n_1, n_2 \geq n_0$; i.e., iff it is Cauchy as defined earlier for metric spaces.

**39.4 Theorem.** *A space $X$ with a uniformity $\mathscr{D}$ generated by the metric $\rho$ is complete iff $\rho$ is a complete metric.*

*Proof.* If $\mathscr{D}$ is complete, every Cauchy net, hence every Cauchy sequence, converges, so $\rho$ is complete.

Conversely, suppose $\rho$ is complete and $(x_\lambda)$ is a Cauchy net in $X$. Pick $\lambda_1 \in \Lambda$ so that whenever $\lambda, \lambda' \geq \lambda_1$, we have $\rho(x_\lambda, x_{\lambda'}) < 1$. Having picked $\lambda_1, \ldots, \lambda_{n-1}$ choose $\lambda_n$ greater than $\lambda_1, \ldots, \lambda_{n-1}$ so that whenever $\lambda, \lambda' \geq \lambda_n$ we have $\rho(x_\lambda, x_{\lambda'}) < 1/n$. Then $x_{\lambda'} \to x$ for some $x \in X$. But the terms of $(x_\lambda)$ are residually close to the terms of $(x_{\lambda_n})$ so we must then have $x_\lambda \to x$. (See 39A.) ■

There is a pitfall to be avoided here. Given a metrizable *topology* on $X$, there will always be some complete uniformity compatible with that topology, even if $X$ with this topology is not completely metrizable. See Exercise 39B for a discussion of *completely uniformizable* topological spaces.

We will develop now some of the elementary properties of complete uniform spaces. These will make complete spaces seem quite a bit like compact spaces; many of the elementary properties enjoyed by compact spaces are also found in complete spaces, in fact. The section will end with an investigation of the property, total boundedness, which is precisely the difference between completeness and compactness, and a theorem on extension of uniformly continuous functions whose range is a complete space.

**39.5 Theorem.** a) *A closed subset $A$ of a complete space $X$ is complete.*

b) *A complete subspace $A$ of a Hausdorff uniform space $X$ is closed.*

*Proof.* a) A Cauchy net $(x_\lambda)$ in $A$ is Cauchy in $X$ and hence converges. Its limit in $X$ must belong to the closed set $A$, so $A$ is complete.

b) A net $(x_\lambda)$ in $A$ which converges to $x \in X$ is Cauchy in $A$ and thus has a limit point $y \in A$. Since $X$ is Hausdorff, limits are unique, so $x = y$. Thus $A$ is closed. ■

**39.6 Theorem.** *A nonempty product of uniform spaces is complete iff each factor space is complete.*

*Proof.* Suppose $\prod X_\alpha$ is complete. Each $X_\alpha$ is homeomorphic to a closed subspace of the product and is thus complete, by 39.5.

On the other hand, suppose $X_\alpha$ is a complete space for each $\alpha \in A$. Then the projection into $X_\alpha$ of a Cauchy net in the product is Cauchy in $X_\alpha$ and thus converges to some point $x_\alpha$. The original net then converges to the point in the product whose $\alpha$th coordinate is $x_\alpha$, for each $\alpha \in A$. ■

The analogy between completeness and compactness does not extend to mapping properties. In fact, the uniformly continuous image of a complete space need not be complete. The situation is actually even worse than this; see Exercise 39C.

We turn now to a study of the difference between completeness and compactness of a uniform space.

**39.7 Definition.** A uniformity $\mathscr{D}$ on $X$ is *totally bounded* (*precompact*) iff for each $D \in \mathscr{D}$, there is a finite cover $\{U_1, \ldots, U_n\}$ of $X$ such that $U_k \times U_k \subset D$, for each $k$. Equivalently, a covering uniformity $\mu$ on $X$ is totally bounded iff $\mu$ has a base consisting of finite covers. If $X$ is equipped with a totally bounded uniformity, it is called a *totally bounded* (*precompact*) *uniform space.*

***39.8 Lemma.*** *$X$ is totally bounded iff each net in $X$ has a Cauchy subnet.*

*Proof.* Let $(x_\lambda)$ be a net in the totally bounded space $X$. Now given any $D \in \mathscr{D}$, there is a set $U_D \subset X$ such that $U_D \times U_D \subset D$ and $(x_\lambda)$ is frequently in $U_D$. Let $\Gamma = \{(\lambda, D) \mid D \in \mathscr{D} \text{ and } x_\lambda \in U_D\}$, directed by $(\lambda_1, D_1) \leq (\lambda_2, D_2)$ iff $\lambda_1 \leq \lambda_2$ and $U_{D_1} \supset U_{D_2}$. For each $(\lambda, D) \in \Gamma$ define $x_{(\lambda, D)} = x_\lambda$. Then $(x_{(\lambda,D)})$ is a subnet of $(x_\lambda)$ and, given $D_0 \in \mathscr{D}$, pick $\lambda_0 \in \Lambda$ so that $(\lambda_0, D_0) \in \Gamma$. Then

$$(\lambda, D), (\lambda', D') \geq (\lambda_0, D_0) \Rightarrow (x_\lambda, x_{\lambda'}) \in U_D \times U_{D'} \subset U_{D_0} \times U_{D_0} \subset D_0,$$

so that $(x_{(\lambda,D)})$ is a Cauchy subnet of $(x_\lambda)$.

On the other hand, if $X$ is not totally bounded, then a set $D \in \mathscr{D}$ exists such that no finite cover $\{U_1, \ldots, U_n\}$ of $X$ exists with $U_k \times U_k \subset D$ for each $k$. Then if $E \circ E \subset D$, it follows from the fact that $E[x] \times E[x] \subset D$ for each $x \in X$ that no finite sequence $E[x_1], \ldots, E[x_n]$ covers $X$. Then we can construct by induction a sequence $x_1, x_2, \ldots$ such that $x_n \notin E[x_i]$ for any $i < n$. Easily, $(x_n)$ can have no Cauchy subnet. ■

***39.9 Theorem.*** *A uniform space $X$ is compact iff it is complete and totally bounded.*

*Proof.* If $X$ is compact and $(x_\lambda)$ is a Cauchy net in $X$, $(x_\lambda)$ has a cluster point $x$, and since it is Cauchy, $(x_\lambda)$ must converge to $x$. Thus $X$ is complete. Since every net has a convergent, hence Cauchy, subnet, $X$ is also totally bounded.

To show a complete, totally bounded space is compact, note that every net has a Cauchy subnet (by total boundedness), which is a convergent subnet (by completeness). ■

We continue this section with an extension theorem. Recall that a continuous function on a subset $A$ of a metric space $X$ to a complete metric space $Y$ can be extended to a $G_\delta$-subset between $A$ and its closure by the primitive version of Lavrentieff's theorem (24.8). If the function in question is *uniformly* continuous, we can extend it to all of $\bar{A}$.

***39.10 Theorem.*** *A uniformly continuous function on a subset $A$ of a uniform space $X$ to a complete uniform space $Y$ can be extended to $\bar{A}$.*

*Proof.* For each $x \in \bar{A}$, a Cauchy net $(x_\lambda)$ in $A$ converges to $x$. The net $(f(x_\lambda))$ is easily still a Cauchy net, and hence converges in $Y$, say to $y$. We define $f(x) = y$ in this case.

To show $f$ is uniformly continuous, let $\mathscr{D}$ and $\mathscr{E}$ be the uniformities on $X$ and $Y$, respectively, and let $E' \in \mathscr{E}$. Pick symmetric $E \in \mathscr{E}$ such that $E \circ E \circ E \subset E'$. Now find an open $D \in \mathscr{D}$ such that, if $x$ and $y$ are points of $A$ and $(x, y) \in D$, then $(f(x), f(y)) \in E$. We claim if $x$ and $y$ are points of $\bar{A}$ and $(x, y) \in D$, then $(f(x), f(y)) \in E'$. This will establish uniform continuity of $f$ on $\bar{A}$.

Find nets $(x_\lambda)$ and $(y_\mu)$ in $A$ converging to $x$ and $y$ respectively. Since $D$ is open and $(x, y) \in D$, eventually $(x_\lambda, y_\mu) \in D$, and thus $(f(x_\lambda), f(y_\mu)) \in E$ eventually. But, since $f(x_\lambda)$ converges to $f(x)$, eventually $f(x_\lambda) \in E[f(x)]$ and thus $(f(x_\lambda), f(x)) \in E$. Similarly, $(f(y_\mu), f(y)) \in E$ eventually. Now it follows that $(f(x), f(y)) \in E \circ E \circ E \subset E'$, establishing the theorem. ∎

We will prove now that every uniform space can be uniformly embedded in a product of pseudometric spaces. These factor spaces all have pseudometric completions, by 24.5, so we have an obvious scheme for obtaining a completion of any uniform space, analogous to the procedure used to obtain the Stone–Čech compactification of a Tychonoff space. The idea behind the proof of the key embedding theorem is, of course, our old friend, Lemma 38.1.

***39.11 Theorem.*** *Every uniform space $X$ can be embedded in a product of pseudometric spaces, and the factors can be made metric if $X$ is separated.*

*Proof.* Let $\mu = \{\mathscr{U}_\alpha \mid \alpha \in A\}$ be a covering uniformity on $X$, which we assume is separating (the proof in the nonseparated case duplicates the proof below, with the metric identification omitted). With each uniform cover $\mathscr{U}_\alpha$, associate a pseudometric $d_\alpha$ using Lemma 38.1, and let $X_\alpha$ denote the pseudometric space $(X, d_\alpha)$. Let $X_\alpha^*$ be the metric identification of $X_\alpha$, the identification map being $h_\alpha$, and define $e: X \to \prod X_\alpha^*$ to be evaluation: $[e(x)]_\alpha = h_\alpha(x)$. Since $\mu$ is separating, any two distinct points $x$ and $y$ are at positive distance in some $d_\alpha$, and hence $e$ is one–one. The composition of $e$ with any projection map is uniformly continuous and thus $e$ is uniformly continuous. Finally, to show $e^{-1}$ is uniformly continuous, let $D_\alpha$ be an element of the diagonal uniformity on $X$ associated with $\mu$. If $d_\alpha$ is the pseudometric associated with $D_\alpha$ (by way of $\mathscr{U}_\alpha$), then note that $\{(x, y) \in X \times X \mid d_\alpha(x, y) < 1\} \subset D_\alpha$. If we let

$$T_\alpha = \{(x_\alpha, y_\alpha) \in X_\alpha^* \times X_\alpha^* \mid d_\alpha^*(x_\alpha, y_\alpha) < 1\},$$

then $T_\alpha$ belongs to the metric uniformity on $X_\alpha^*$, and thus

$$T = P_\alpha^{-1}(T_\alpha) = \{(x, y) \in \prod X_\alpha^* \times \prod X_\alpha^* \mid d_\alpha^*(x_\alpha, y_\alpha) < 1\}$$

belong to the uniformity on $\prod X_\alpha^*$. But

$$e^{-1}(T) = \{(x, y) \in X \times X \mid d_\alpha(x, y) < 1\} \subset D_\alpha.$$

This establishes uniform continuity of $e^{-1}$. ∎

**39.12 Theorem.** *Every uniform space $X$ can be uniformly embedded as a dense subspace of a complete uniform space $\hat{X}$ which is unique, in the sense that if $Y$ is any complete space containing $X$ as a dense subspace, then $\hat{X}$ and $Y$ are uniformly isomorphic, under an isomorphism leaving $X$ pointwise fixed. Moreover, $\hat{X}$ is separated iff $X$ is.*

*Proof.* $X$ can be embedded in a product of pseudometric spaces, $X \subset \prod X_\alpha$, and each $X_\alpha$ has a pseudometric completion $\hat{X}_\alpha$. Then the closure in $\prod \hat{X}_\alpha$ of $X$ is a complete uniform space containing $X$ as a dense subspace. Moreover, each $X_\alpha$ can be made metric if $X$ is separated, and then the resulting completion of $X$ will be Hausdorff and hence separated.

Uniqueness remains. But if $Y$ is any completion of $X$, then the identity map $i: X \to X$ has uniformly continuous extensions $I: \hat{X} \to Y$ and $J: Y \to \hat{X}$, by 39.10. It follows that $I$ is the required uniform isomorphism of $\hat{X}$ with $Y$. ■

**39.13 Theorem.** *The completion of a totally bounded uniform space is compact.*

*Proof.* In fact, we can show that whenever a uniform space $Y$ contains a dense totally bounded subspace $X$, then $Y$ is totally bounded.

Let $\mathscr{U}$ be any uniform cover of $Y$, $\mathscr{V}$ an open star refinement of $\mathscr{U}$, and let $\mathscr{W}'$ be a finite uniform cover of $X$ which refines the trace on $X$ of $\mathscr{V}$. It is sufficient to show $\{\overline{W} \mid W \in \mathscr{W}'\}$ is a uniform cover of $Y$ which refines $\mathscr{U}$.

First, it is a uniform cover. Let $\mathscr{W}$ be a uniform cover of $Y$ whose trace on $X$ is $\mathscr{W}'$ and find an open refinement $\mathscr{W}_0$ of $\mathscr{W}$. Since each $W_0 \in \mathscr{W}_0$ is open, $W_0 \subset \overline{W_0 \cap X}$, and each $\overline{W_0 \cap X}$ is contained in $\overline{W}$ for some $W \in \mathscr{W}'$. Thus, $\mathscr{W}_0$ refines $\{\overline{W} \mid W \in \mathscr{W}'\}$, showing the latter to be a uniform cover of $Y$.

Second, $\{\overline{W} \mid W \in \mathscr{W}'\}$ refines $\mathscr{U}$, since given $W \in \mathscr{W}$, we have $W \subset V$ and $\text{St}(V, \mathscr{V}) \subset U$ for some $V \in \mathscr{V}$ and $U \in \mathscr{U}$ and it follows that $\overline{W} \subset U$. ■

Thus the uniform completion of a totally bounded uniform space is, in fact, a compactification of $X$ (whose unique uniformity restricts to the uniformity on $X$). Conversely, given any compactification $BX$ of $X$, $BX$ has a unique uniformity (which is totally bounded), thus giving rise to a totally bounded uniformity on $X$.

The resulting one–one correspondence between totally bounded uniformities on $X$ and compactifications of $X$ will be further studied in Section 41 on proximities.

## Problems

### 39A. *Cauchy nets and Cauchy sequences*

Supply the missing details in the proof of Theorem 39.4. Specifically:

1. If $(x_{\lambda_n})$ is constructed as in 39.4 and $x_{\lambda_n} \to x$, then $x_\lambda \to x$.
2. If the requirement that $\lambda_n$ be greater than or equal to all of $\lambda_1, \ldots, \lambda_{n-1}$ is dropped, $(x_{\lambda_n})$ need not be a Cauchy sequence.
3. Every subnet of a Cauchy net is Cauchy.

## 39B. *Completely uniformizable spaces*

A topological space is *completely uniformizable* iff there is a complete uniformity which generates its topology. Thus every completely uniformizable space is completely regular.

1. A completely regular space $X$ is completely uniformizable iff the fine uniformity on $X$ is complete. [If $\mathscr{D}_1$ and $\mathscr{D}_2$ are uniformities on $X$ generating the same topology, and $\mathscr{D}_1 \subset \mathscr{D}_2$, then $\mathscr{D}_1$ complete $\Rightarrow \mathscr{D}_2$ complete.]

2. Every paracompact space is completely uniformizable.

As a consequence of 2, *every metric space is completely uniformizable.* Thus a completely uniformizable metric space need not be completely metrizable, nor can we prove a Baire theorem (25.4) for completely uniformizable spaces.

## 39C. *Mapping properties of complete spaces*

With mapping properties, the analogy between complete uniform spaces and compact spaces ends.

1. The uniformly continuous image of a complete uniform space need not be complete. (The situation is even worse. The uniformly continuous and uniformly open image of a complete space need not be complete. See the notes.)

2. If $X$ is a complete metric space, $Y$ is a separated uniform space, and $f$ is a continuous uniformly open map of $X$ onto $Y$, then $Y$ is complete. [Show that $f$ is uniformly open as a map of $X$ into the completion $\hat{Y}$ of $Y$ and conclude that $f(X) = Y$ is closed in $\hat{Y}$.]

## 39D. *Completeness of the hyperspace*

Recall that the hyperspace (36E) of a uniform space $(X, \mathscr{D})$ is obtained by forming the set $\mathscr{H}$ of all closed subsets of $X$ and taking as a base for a diagonal uniformity on $\mathscr{H}$ the collection of all sets of the form $\{(A, B) \mid A \text{ is } D\text{-close to } B\}$, for $D \in \mathscr{D}$, where $A$ and $B$ are $D$-close iff $A \subset D[B]$ and $B \subset D[A]$.

1. The hyperspace of a complete metric space is a complete metric space. [Refer to 36E.2.]

2. The hyperspace of an arbitrary complete uniform space need not be complete. [Consider a complete space $X$ of cardinality $\mathfrak{c}$ whose (covering) uniformity is the uniformity having as a base all countable covers of $X$ (36B.5).]

## 39E. *Homeomorphism does not imply uniform isomorphism*

There is an uncountable family of countable discrete metric spaces, no two of which are uniformly isomorphic. [From 17R, there are compact subsets $C_\alpha$, $\alpha \in \Omega_0$, of $\mathbf{R}$ no two of which are homeomorphic. For each $\alpha$, let $D_\alpha = \{x_{\alpha_1}, x_{\alpha_2}, \ldots\}$ be a countable dense subset of $C_\alpha$ and let $X_\alpha = \{(x_{\alpha_n}, 1/m) \mid m \geq n\}$ in $\mathbf{R}^2$. Then $X_\alpha$ is a countable discrete metric space whose completion is $X_\alpha \cup C_\alpha$. Use 39.12 to conclude that a uniform isomorphism of $X_\alpha$ onto $X_\beta$ would induce a homeomorphism of $C_\alpha$ with $C_\beta$.]

## 39F. *Filters and completeness*

Filters can be used to describe completeness in a uniform space. A filter $\mathscr{F}$ in a uniform space $X$ is *Cauchy* iff $\mathscr{F}$ contains an element of each uniform cover of $X$. (In the language of diagonal uniformities, $\mathscr{F}$ is Cauchy iff it contains a set $A$ such that $A \times A \subset D$, for each surrounding $D$.)

1. Every convergent filter is Cauchy.
2. $X$ is complete iff every Cauchy filter converges.

### 39G. *Examples on completeness and completion*

Decide which of the following uniform spaces are complete. For those that are not, try to describe their uniform completion in simple terms.

1. The uniformity $\mathscr{V}$ on $\mathbf{R}$ having as a base the sets

$$D_\alpha = \Delta \cup \{(x, y) \mid x > a, y > a\}$$

for $a \in \mathbf{R}$.

2. Any set $X$ with the discrete uniformity.

## 40 Proximity spaces

The basic purpose in this section is to introduce a new "nearness relation", called *proximity*, on a set $X$ and to establish a one–one correspondence between the proximities on $X$ and the totally bounded uniformities on $X$.

**40.1 Definition.** A *proximity space* is a pair $(X, \delta)$ where $X$ is a set and $\delta$ is a binary relation on $\mathscr{P}(X)$ satisfying, for $A$, $B$ and $C$ subsets of $X$:

$P$-1) $\varnothing \not\delta A$,

$P$-2) $a \, \delta \, a$, for each $a \in A$,

$P$-3) $A \, \delta \, B$ implies $B \, \delta \, A$,

$P$-4) $A \, \delta \, (B \cup C)$ iff $A \, \delta \, B$ or $A \, \delta \, C$,

$P$-5) $A \not\delta B$ implies there are $C, D \subset X$ such that $C \cap D = \varnothing$ and $A \not\delta (X - C)$, $B \not\delta (X - D)$.

The space is called *separated* if it satisfies the additional axiom:

$P$-6) $a \, \delta \, b$ implies $a = b$.

We speak, in practice, of "the proximity space $X$" and $\delta$ is referred to as the *proximity* on $X$. The phrase $A \, \delta \, B$ is read "$A$ is close to $B$" (or, where confusion is possible, "$A$ is $\delta$-close to $B$"). As with uniformities, it causes no real loss in generality to assume most proximities are separated.

**40.2 Examples.** a) In any set $X$ define $A \, \delta \, B$ iff both $A$ and $B$ are nonempty. This defines a proximity on $X$, called the *trivial proximity*.

b) In any set $X$, define $A \, \delta \, B$ iff $A \cap B \neq \varnothing$. This always defines a proximity on $X$, called the *discrete proximity*.

c) In any normal topological space $X$, define $A \, \delta \, B$ iff $\bar{A} \cap \bar{B} \neq \varnothing$. This provides a proximity on $X$, called the *elementary proximity*. It is separated iff $X$ is $T_0$. In a sense to be made precise in the next section, all proximities are obtainable as elementary proximities.

d) In a space $X$ with a diagonal uniformity $\mathscr{D}$, define $A \not\delta B$ iff for some $D \in \mathscr{D}$, $D[A] \cap D[B] = \varnothing$. Equivalently, from the existence of elements $E \in \mathscr{D}$ such that $E \circ E \subset D$, the definition above can be made to read, $A \not\delta B$ iff for some $D \in \mathscr{D}$, $D[A] \cap B = \varnothing$.

If a covering uniformity $\mu$ is given on $X$, the definition becomes $A \not\delta B$ iff for some $\mathscr{U} \in \mu$, $\text{St}(A, \mathscr{U}) \cap \text{St}(B, \mathscr{U}) = \varnothing$; equivalently, from the existence of elements $\mathscr{U}'$ such that $\mathscr{U}' \mathrel{*}< \mathscr{U}$, we can write the definition $A \not\delta B$ iff for some $\mathscr{U} \in \mu$, $\text{St}(A, \mathscr{U}) \cap B = \varnothing$.

A proximity which can be obtained in either of these ways (they are the same, by the translation process for uniformities) is called *uniformizable*. As we will see later in this section, all proximities are uniformizable.

In both cases, the proximity induced by a uniformity is separated iff the uniformity itself is separated.

e) As a special case of (d) (verify that it *is* a special case) in a metric space $X$, a proximity is obtained if we define $A \,\delta\, B$ iff $d(A, B) = 0$. Whenever a proximity is obtainable in this way from a metric, it is called *metrizable*. Metrizable proximities are always separated; their nonseparated counterparts are the *pseudometrizable proximities*, the definition and properties of which are obvious analogs to those of metrizable proximities.

**40.3 Definition.** In a proximity space $(X, \delta)$, we write $A \subset\subset B$ iff $A \not\delta (X - B)$. When $A \subset\subset B$, we call $B$ a *proximity nhood* (*p-nhood*, *$\delta$-nhood*) of $A$.

**40.4 Theorem.** *Proximity nhoods have the following properties, for any $A, B, C \subset X$:*

$P$-1)$'$ $\varnothing \subset\subset A$,

$P$-2)$'$ *if* $A \subset\subset B$, *then* $A \subset B$,

$P$-3)$'$ $A \subset\subset (B \cap C)$ *iff* $A \subset\subset B$ *and* $A \subset\subset C$,

$P$-4)$'$ *if* $A \subset\subset B$ *then for some* $C$, $A \subset\subset C \subset\subset B$,

*and in a separated space*

$P$-5)$'$ *if* $a \neq b$, *then* $a \subset\subset X - \{b\}$.

*Conversely, given a relation $\subset\subset$ between subsets of a set $X$ satisfying ($P$-1)$'$ through ($P$-4)$'$, we can define a proximity $\delta$ on $X$ by $A \not\delta B$ iff $A \subset\subset X - B$, and the proximity nhoods of $A$ relative to $\delta$ will be precisely those sets $B$ for which $A \subset\subset B$. Moreover, $\delta$ will be separated iff ($P$-5)$'$ holds.*

*Proof.* Left as Exercise 40B. ■

**40.5 Theorem.** *In any proximity space $(X, \delta)$:*

a) *if $A \ \delta \ B$ and $A \subset C$, $B \subset D$, then $C \ \delta \ D$,*

b) *if $A \cap B \neq \varnothing$, then $A \ \delta \ B$,*

c) *if $A \subset\subset B \subset C$, then $A \subset\subset C$.*

*Proof.* (a) follows directly from ($P$-3), while (b) follows from (a) and ($P$-2), and (c) follows from ($P$-3)′. ■

**40.6 Definition.** In a proximity space $(X, \delta)$, define $\bar{A} = \{x \mid x \ \delta \ A\}$. The result is a closure operator on $P(X)$ (Exercise 40C), thus providing a topology on $X$, called the *topology induced by* $\delta$. The topological spaces whose topologies can be derived in this way from proximities are called *proximizable.*

In the topology induced by a proximity $\delta$, the nhoods $U$ of a point $x$ are precisely the proximity nhoods of $x$; that is, those sets $U$ for which $x \subset\subset U$. It is *not* in general true, however, that the nhoods of a set $A$ are the proximity nhoods of $A$ (see Exercise 40C).

Whenever we use a topology on a proximity space $(X, \delta)$, it is assumed to be the topology induced by $\delta$.

**40.7 Theorem.** a) *$\bar{A} \ \delta \ \bar{B}$ iff $A \ \delta \ B$,*

b) *if $A \subset\subset B$, then $\bar{A} \subset B^\circ$; the converse fails.*

*Proof.* a) If $A \ \delta \ B$, then $\bar{A} \ \delta \ \bar{B}$ by Theorem 40.5. On the other hand, suppose $\bar{A} \ \delta \ \bar{B}$. Then $A \subset\subset X - B$, so for some $C \subset X$, $A \subset\subset C \subset\subset X - B$. Then $\bar{A} \subset C \subset\subset X - B$, so $\bar{A} \ \delta \ B$. Repeating the argument shows $A \ \delta \ B$.

b) If $A \subset\subset B$, then for some $C$, $A \subset\subset C \subset\subset B$ and then $\bar{A} \subset C \subset B^\circ$, so $\bar{A} \subset B^\circ$. The converse fails, even for the proximity generated by the usual metric on the plane. Let $A$ be the set $\{(x, 0) \mid x \geq 1\}$, $B$ the set of points $(x, y)$ with $x > 0$ and $|y| < 1/x$. Then $\bar{A} \subset B^\circ$, but $d(A, X - B) = 0$, so it is not true that $A \subset\subset B$. ■

**40.8 Theorem.** *The topology induced by a proximity induced by a uniformity is the uniform topology.*

*Proof.* Let $\mathscr{D}$ be a diagonal uniformity on $X$. Then $U$ is a uniform nhood of $x$ iff $D[x] \subset U$ for some $D \in \mathscr{D}$ iff $E[x] \cap E[X - U] = \varnothing$ for some $E \in \mathscr{D}$ [take $E$ symmetric so that $E \circ E \subset D$] iff $x \not\delta (X - U)$ in the proximity induced by $\mathscr{D}$ if $x \subset\subset U$ iff $U$ is a proximity nhood of $x$. ■

**40.9 Definition.** If $(X, \delta)$ and $(Y, \delta')$ are proximity spaces, a map $f: X \to Y$ is a *proximity map* (*p-map*, *$\delta$-map*) iff whenever $A \ \delta \ B$ in $X$, then $f(A) \ \delta' \ f(B)$ in $Y$.

Alternatively, $f$ is a $p$-map iff whenever $C \subset\subset' D$ in $Y$, then $f^{-1}(C) \subset\subset f^{-1}(D)$ in $X$.

**40.10 Theorem.** a) *Every p-map is continuous.*

b) *Every uniformly continuous map is a p-map.*

*Proof.* a) Let $f: X \to Y$ be a $p$-map. To show $f$ is continuous, it suffices to show $f(\bar{A}) \subset \overline{f(A)}$ for each $A \subset X$. But $x \in \bar{A}$ iff $x \,\delta\, A$, and if $x \,\delta\, A$, then $f(x) \,\delta\, f(A)$, which implies that $f(x) \in \overline{f(A)}$. Thus $f(\bar{A}) \subset \overline{f(A)}$.

b) Let $(X, \mathscr{D})$ and $(Y, \mathscr{E})$ be uniform spaces and suppose $f: X \to Y$ is uniformly continuous. Now suppose $C \sqsubset\sqsubset' B$. Find a symmetric $E \in \mathscr{E}$ such that $E[C] \subset B$ (for $C \not\delta\, Y - B$, so for some $E$, $E[C] \cap (Y - B) = \varnothing$). Let $D$ be an element of $\mathscr{D}$ such that if $(x, y) \in D$, then $(f(x), f(y)) \in E$, by uniform continuity of $f$. We assert that $D[f^{-1}(C)] \subset f^{-1}(B)$. For if $x \in D[f^{-1}(C)]$, then for some $y \in f^{-1}(C)$, $(x, y) \in D$. Then $(f(x), f(y)) \in E$ and $f(y) \in C$, so $f(x) \in E[C] \subset B$ and hence $x \in f^{-1}(B)$. Thus $D[f^{-1}(C)] \subset f^{-1}(B)$, so that $f^{-1}(C) \sqsubset\sqsubset f^{-1}(B)$. ■

**40.11 Theorem.** *If $\nu$ is totally bounded, a map $f: (X, \mu) \to (Y, \nu)$ is uniformly continuous iff it is a $p$-map (relative to the induced proximities).*

*Proof.* Let $f$ be a $p$-map. It suffices to show that $f^{-1}(\eta) \in \mu$ for each finite uniform cover $\eta$ of $Y$, since these form a base for $(Y, \nu)$.

We begin by supposing $\eta$ has two elements, say $\eta = \{N_1, N_2\}$. If

$$f^{-1}(\eta) \notin \mu,$$

then for each $\mathscr{U} \in \mu$, there is some $U \in \mathscr{U}$ such that $U$ meets both $X - f^{-1}(N_1)$ and $X - f^{-1}(N_2)$ [otherwise $\mathscr{U} < f^{-1}(\eta)$]. It follows that

$$[X - f^{-1}(N_1)] \;\delta\; [X - f^{-1}(N_2)]$$

and hence, since $f$ is a $p$-map, $(Y - N_1) \,\delta\, (Y - N_2)$. Then for every uniform cover $\mathscr{W}$ of $Y$, $\mathrm{St}\,(Y - N_1, \mathscr{W})$ meets $Y - N_2$; in particular, $\mathrm{St}\,(Y - N_1, \eta)$ meets $Y - N_2$. This is impossible. Thus $f^{-1}(\eta) \in \mu$.

We can complete the proof by showing every finite uniform cover is refined by an intersection of two-element uniform covers. Let $\mathscr{U}' = \{U_1, \ldots, U_n\}$ be a uniform cover and pick $\mathscr{U}'$ such that $\mathscr{U}' \,{}^*\!\!< \mathscr{U}$. For $i = 1, \ldots, n$ let

$$W_i = \bigcup \{U' \in \mathscr{U}' \mid \mathrm{St}\,(U', \mathscr{U}') \subset U_i\}.$$

Since each $U' \in \mathscr{U}'$ is included in some $W_i$, we have $\bigcup_{i=1}^n W_i = Y$. Also, $W_i \subset U_i$ for each $i = 1, \ldots, n$, so $\mathscr{W}_i = \{U_i, Y - W_i\}$ is a cover of $Y$ for each $i$. Moreover, each $\mathscr{W}_i$ is a uniform cover of $Y$ since (it is easily checked that) $\mathscr{U}' < \mathscr{W}_i$. But then

$$\mathscr{W} = \left\{ \bigcap_{i=1}^{n} T_i \;\middle|\; T_i \in \mathscr{W}_i \right\} = \mathscr{W}_1 \wedge \cdots \wedge \mathscr{W}_n$$

is a uniform cover of $Y$, and $\mathscr{W} < \mathscr{U}$. ■

**40.12 Definition.** A one–one, onto map $f$ such that both $f$ and $f^{-1}$ are $p$-maps is a *$p$-isomorphism*. Apparently, a one–one onto map $f: (X, \delta) \to (Y, \delta')$ is a $p$-isomorphism precisely when $A \,\delta\, B$ iff $f(A) \,\delta'\, f(B)$; i.e., precisely when $C \sqsubset\sqsubset D$

iff $f(C) \subset\subset f(D)$. Such maps are also called *proximity isomorphisms* or *$\delta$-isomorphisms.*

The next theorem follows easily from Theorem 40.10.

***40.13 Theorem.*** a) *Every p-isomorphism is a homeomorphism.*

b) *Every uniform isomorphism is a p-isomorphism.*

We now know that proximity represents a structural layer somewhere between uniformity and topology. The rest of this section will be devoted to substantiating the claim that *the theory of proximity spaces is, in a sense, a theory of totally bounded uniform spaces.*

**40.14 Definition.** We say $\mathscr{U} = \{U_\alpha \mid \alpha \in A\}$ is a *p-cover* of the proximity space $X$ iff there is a cover $\mathscr{V} = \{V_\alpha \mid \alpha \in A\}$ of $X$ such that $V_\alpha \subset\subset U_\alpha$ for each $\alpha \in A$. We call $\mathscr{V}$ a *p-refinement* of $\mathscr{U}$.

***40.15 Theorem.*** *Every proximity is induced by some totally bounded uniformity.*

*Proof.* We assert that, given a proximity $\delta$ on $X$, the collection $\mu$ of all finite $p$-covers of $X$ is a base for a uniformity on $X$, whose associated proximity is $\delta$. If so, the uniformity will easily be totally bounded, since the generating covers are finite.

First, if $\mathscr{U}_1$ and $\mathscr{U}_2$ belong to $\mu$, so easily does

$$\mathscr{U} = \mathscr{U}_1 \wedge \mathscr{U}_2 = \{U_1 \cap U_2 \mid U_1 \in U_2 \mid U_1 \in \mathscr{U}_1, U_2 \in \mathscr{U}_2\},$$

and $\mathscr{U} < \mathscr{U}_1, \mathscr{U} < \mathscr{U}_2$.

Next, given $\mathscr{U} \in \mu$, we wish to find $\mathscr{W} \in \mu$ such that $\mathscr{W} {}^*\!< \mathscr{U}$; actually, by 20B.1, it is enough to find a barycentric refinement $\mathscr{W}$ of $\mathscr{U}$. Write $\mathscr{U} = \{U_1, \ldots, U_n\}$ and let $\mathscr{V} = \{V_1, \ldots, V_n\}$ be a $p$-refinement of $\mathscr{U}$. For $i = 1, \ldots, n$ find $H_i$ and $G_i$ such that $V_i \subset\subset H_i \subset\subset G_i \subset\subset U_i$ (using 40.4), and set $A_i^1 = U_i^\circ$, $A_i^2 = X - \overline{V}_i$. Then, for each $i$, $A_i^1$ and $A_i^2$ are open and, by 40.7, $A_i^1 \cup A_i^2 = X$. Let $\mathscr{W}$ be the cover of $X$ by the $z^n$ open sets of the form $A_1^{\epsilon_1} \cap \cdots \cap A_n^{\epsilon_n}$ where each $\epsilon_i$ is 1 or 2. Then $\mathscr{W}$ is a $p$-cover, since it is $p$-refined by the cover

$$\{Y_1^{\epsilon_1} \cap \cdots \cap Y_n^{\epsilon_n} \mid \epsilon_i = 1 \text{ or } 2\},$$

where $Y_i^1 = G_i$ and $Y_i^2 = X - G_i$, and thus $\mathscr{W} \in \mu$. Moreover, $\mathscr{W}$ is a barycentric refinement of $\mathscr{U}$. In fact, given $x \in X$, $x \in V_{i_0}$ for some $i_0$ and we can show $\operatorname{St}(x, \mathscr{W}) \subset U_{i_0}$, for if $x \in W$ for some $W \in \mathscr{W}$ then the $i_0$th factor $A_{i_0}^{\epsilon_{i_0}}$ of $W$ cannot be $X - V_{i_0}$, so it must be $U_{i_0}^\circ$. Thus $W \subset U_{i_0}$.

Finally, we show that the resulting uniformity on $X$ induces the proximity $\delta$ we started with. For this purpose, it suffices to show $A \not\delta B$ iff $\operatorname{St}(A, \mathscr{U}) \cap B = \varnothing$ for some finite $p$-cover $\mathscr{U}$ of $X$.

If $A \not\delta B$, then $A \subset\subset X - B$ and hence

$$A \subset\subset C_1 \subset\subset C_2 \subset\subset C_3 \subset\subset C_4 \subset\subset X - B$$

for some $C_1, C_2, C_3, C_4 \subset X$. Set $\mathscr{U} = \{C_4, X - C_1\}$. Then since $C_3 \subset\subset C_4$ and $X - C_2 \subset\subset X - C_1$, $\mathscr{U}$ is a $p$-cover of $X$. Moreover, $\mathrm{St}(A, \mathscr{U}) \subset C_4$, so $\mathrm{St}(A, \mathscr{U}) \cap B = \varnothing$.

Conversely, suppose $\mathrm{St}(A, \mathscr{U}) \cap B = \varnothing$ for some finite $p$-cover $\mathscr{U}$. Then if $\mathscr{V}$ is a $p$-refinement of $\mathscr{U}$, $A \subset \mathrm{St}(A, \mathscr{V}) \subset\subset \mathrm{St}(A, \mathscr{U}) \subset X - B$, so $A \not\delta B$. ■

***40.16 Corollary.*** *The proximizable topological spaces are precisely the completely regular spaces.*

*Proof.* The proximizable spaces, by Theorem 40.15 together with Example 40.2(d), coincide with the uniformizable spaces. ■

**40.17 Definition.** Given a proximity $\delta$ on $X$, the totally bounded uniformity constructed in 40.15 which generates it will be denoted $\mu_\delta$ (or, in the case of the corresponding diagonal uniformity, $\mathscr{D}_\delta$).

***40.18 Theorem.*** *$\mu_\delta$ is the only totally bounded uniformity giving the proximity $\delta$.*

*Proof.* It suffices to show that the finite $p$-covers form a base for any totally bounded uniformity $\mu$ which generates $\delta$. Now the finite uniform covers form a base for $\mu$, by total boundedness. But let $\mathscr{U}$ be any finite uniform cover, $\mathscr{V}$ a finite star refinement of $\mathscr{U}$, and for each $V \in \mathscr{V}$, pick $U_V \in \mathscr{U}$ such that $\mathrm{St}(V, \mathscr{V}) \subset U_V$. Then $\mathscr{U}_0 = \{U_V \mid V \in \mathscr{V}\}$ is a finite cover which refines $\mathscr{U}$, so it suffices to show $\mathscr{U}_0$ is a $p$-cover.

It is enough to show $V \subset\subset U_V$ for each $V \in \mathscr{V}$. But if $\mathscr{W}$ is any star refinement of $\mathscr{V}$, then easily $\mathrm{St}(V, \mathscr{W}) \cap \mathrm{St}(X - U_V, \mathscr{W}) = \varnothing$; it follows that $V \not\delta (X - U_V)$, i.e., that $V \subset\subset U_V$. ■

***40.19 Theorem.*** *If $\mu$ is any uniformity inducing $\delta$, then $\mu_\delta \subset \mu$.*

*Proof.* Since both $\mu$ and $\mu_\delta$ generate the same proximity, the identity $i: (X, \mu) \to (X, \mu_\delta)$ is a $p$-map. Then, since $\mu_\delta$ is totally bounded, $i$ is uniformly continuous by 40.11. Thus $\mu_\delta \subset \mu$. ■

The study of proximities on $X$ is now revealed as merely the study of equivalence classes of uniformities on $X$, under the equivalence relation $\mu_1 \sim \mu_2$ iff $\mu_1$ and $\mu_2$ give the same proximity on $X$. Moreover, each equivalence class of uniformities contains precisely one totally bounded uniformity, the smallest uniformity in the class. As we have said, then, in certain ways the study of proximity structures reduces to the study of totally bounded uniform structures. The next section provides a good illustration.

The real reason for an interest in proximity structure lies in the fact that many of the interesting properties of uniform spaces turn out not only to be uniform invariants, but to be proximity invariants; that is, turn out to be possessed by all uniform spaces which are proximity isomorphic to any uniform space having them.

## Problems

### 40A. *Examples of proximities*

Verify that each of the following is a proximity relation on the set in question.

1. The trivial proximity (40.2a).
2. The discrete proximity (40.2b).
3. The elementary proximity on a normal space (40.2c).
4. The uniform proximity on a uniform space—both kinds. (40.2d.)
5. The metric proximity on a metric space $(X, d)$ (40.2e).

### 40B. *P-nhoods*

1. Given a proximity space $(X, \delta)$, show that the relation $\subset\!\subset$ in $(X, \delta)$ satisfies axioms $(P\text{-}1)'$ through $(P\text{-}4)'$ of 40.4, and also $(P\text{-}5)'$ if $X$ is separated.

2. Conversely, show that any relation $\subset\!\subset$ satisfying $(P\text{-}1)'$ through $(P\text{-}4)'$ will generate a proximity $\delta$, with the definition $A \not\delta B$ iff $A \subset\!\subset X - B$, for which $\subset\!\subset$ is just the $p$-nhood relation. Also, $(X, \delta)$ is separated iff $(P\text{-}5)'$ holds.

### 40C. *The proximity topology*

1. Verify that $\bar{A} = \{x \in X \mid x \,\delta\, A\}$ defines a valid closure operator on any proximity space $(X, \delta)$, making $X$ a topological space.

2. The resulting topology on $X$ is Hausdorff iff $(X, \delta)$ is separated.

3. The nhoods of $x \in X$ in this topology are precisely the sets $U$ for which $\{x\} \subset\!\subset U$.

4. More generally, if $A \subset\!\subset B$, then $B$ is a nhood of $A$ in the induced topology on $X$, but the converse fails.

### 40D. *Proximizable topologies*

Our verification that the proximizable topological spaces were precisely the completely regular ones was indirect; we showed proximizability equivalent to uniformizability. We can, however, explicitly construct a proximity on any completely regular space compatible with the topology on that space.

Given a completely regular space $X$, define $A \not\delta B$ iff for some continuous $f: X \to \mathbf{I}$, $f(A) = 0$ and $f(B) = 1$. Then $\delta$ is a proximity on $X$ compatible with the topology on $X$.

### 40E. *Subspace proximities*

Given a proximity space $(X, \delta)$ and $A \subset X$, a proximity $\delta_A$ is induced on $A$ in a natural way, namely, $B \,\delta_A\, C$ iff $B \,\delta\, C$.

1. $\delta_A$ is a proximity on $A$. It is called the *relative proximity* on $A$ and $A$ with this proximity is a *subspace* of $(X, \delta)$.

2. The topology induced by $\delta_A$ is the relative topology on $A$.

3. The proximity induced by the relative uniformity on a subset of a uniform space is the relative proximity.

Thus, subspaces of proximity spaces work well. We will see later (40F and 41C) that products and quotients do not behave as nicely.

### 40F. *Product proximities*

For each $\alpha \in Z$, let $(X_\alpha, \delta_\alpha)$ be a proximity space. We can define a product proximity on $X = \prod X_\alpha$ as follows: $A \; \delta \; B$ iff whenever $A = A_1 \cup \cdots \cup A_m$, $B = B_1 \cup \cdots \cup B_n$ then for some $A_i$ and $B_j$, $[\pi_\alpha(A_i)] \; \delta_\alpha \; [\pi_\alpha(B_j)]$ for all $\alpha$.

1. $\delta$ is a proximity on $X$, the coarsest proximity for which each projection $\pi_\alpha$ is a $p$-map.

2. The topology induced by $\delta$ is the product topology.

3. The proximity induced by the product uniformity need *not* be the product proximity, even though both produce the product topology. [Take $X = Y = \mathbf{N}$ with the usual (discrete) uniformity and proximity.]

This result can be stated: *products of p-isomorphic uniform spaces need not be p-isomorphic.*

4. The proximity induced on a product of *totally bounded* uniform spaces is the product proximity. [Use coverings.]

Apropos of 4, Dowker has noted that it is enough if all but one of the uniform spaces is totally bounded. This cannot be improved, by the example in 3.

## 41 Compactness and proximities

Here our basic purpose is to establish the one–one correspondence between the compatible proximities on a Tychonoff space $X$ and the compactifications of $X$.

***41.1 Theorem.*** *A compact Hausdorff space admits a unique proximity, given by the elementary proximity $A \; \delta \; B$ iff $\bar{A} \cap \bar{B} \neq \varnothing$.*

*Proof.* Such a space is uniformizable with a unique uniformity, so it must be proximizable with a unique proximity $\delta$.

Now if $A \not\delta B$, then $\bar{A} \not\delta \bar{B}$ by 40.7, and hence $\bar{A} \cap \bar{B} = \varnothing$.

Conversely, suppose $\bar{A} \cap \bar{B} = \varnothing$. For each $x \in \bar{A}$, $X - \bar{B}$ is open and contains $x$, so for some open set $C_x$, $x \subset\subset C_x \subset\subset X - \bar{B}$. The cover of $\bar{A}$ by the sets $C_x$, $x \in \bar{A}$, has a finite subcover, say $\bar{A} \subset C_{x_1} \cup \cdots \cup C_{x_n} = C$. Now $C_{x_i} \subset\subset X - \bar{B}$ for each $i = 1, \ldots, n$ so $C \subset\subset X - \bar{B}$ and hence $\bar{A} \subset\subset X - \bar{B}$. Thus $\bar{A} \not\delta \bar{B}$, so by 40.7, $A \not\delta B$. ∎

Now suppose $X$ is any Tychonoff space. Then $X$ can be densely embedded in various ways in compact spaces $Y$, each such $Y$ has a unique proximity, and the restriction of that proximity to $X$, call it $\delta_Y$, gives a compatible proximity on $X$ (subspace proximities are defined in the obvious way and have all the right properties; see 40E).

Conversely, given any compatible proximity $\delta$ on $X$, $\delta$ corresponds to a unique totally bounded uniformity $\mu_\delta$ on $X$ and the uniform completion $\hat{X}$ of

$(X, \mu_\delta)$ is a complete, totally bounded uniform space and thus a compactification of $X$. Moreover, since $(X, \mu_\delta)$ is a uniform subspace of $\hat{X}$ with its unique uniformity, we must have $\delta = \delta_{\hat{X}}$.

**41.2 Definition.** The unique compactification $\beta_\delta X$ of $X$ corresponding to the proximity $\delta$ is called the *Samuel compactification* of $X$, relative to $\delta$.

Thus the proximities on $X$ (and hence, the totally bounded uniformities on $X$) are in one–one correspondence with the compactifications of $X$. Moreover, the method of construction substantiates our claim in 40.2(c) that every proximity is an elementary proximity. We should state it more accurately: *every separated proximity is the restriction of an elementary proximity on a compact Hausdorff space.* With obvious modifications to the discussion so far, "separated" and "Hausdorff" can be dropped from the last sentence.

The remainder of this section is devoted to the question: when is $\beta_\delta X$ the Stone–Čech compactification $\beta X$ of $X$?

**41.3 Definition.** If $\delta_1$ and $\delta_2$ are proximities on the same set $X$, we say $\delta_1$ is *finer than* $\delta_2$ (or, $\delta_2$ is *coarser than* $\delta_1$) iff $A\ \delta_1\ B$ implies $A\ \delta_2\ B$. Hence, in finer proximities, it is harder for sets to be close together.

In the language of proximity nhoods, $\subset\subset_1$ is finer than $\subset\subset_2$ iff $A \subset\subset_2 B$ implies $A \subset\subset_1 B$.

**41.4 Theorem.** *Every family $\{\delta_\lambda \mid \lambda \in \Lambda\}$ of proximities on $X$ has a sup $\bar{\delta}$ and an inf $\underline{\delta}$.*

*Proof.* Let $\subset\subset_\lambda$ denote the $p$-nhood operation corresponding to the proximity $\delta_\lambda$. To define the inf of the proximities $\delta_\lambda$, let

a) $A \subset\subset' B$ iff $A \subset\subset_\lambda B$ for each $\lambda \in \Lambda$.

b) $A \underline{\subset\subset} B$ iff there is a set $C_s \subset X$ for each binary rational $s$ in $[0, 1]$ such that $C_0 = A$, $C_1 = B$ and $s < t$ implies $C_s \subset\subset' C_t$.

We leave the verification that $\underline{\subset\subset}$ is a $p$-nhood relation to Exercise 41A, and proceed to show it gives the finest proximity coarser than all $\delta_\lambda$.

First, if $A \underline{\subset\subset} B$, then $A \subset\subset' B$ and hence $A \subset\subset_\lambda B$ for all $\lambda$, so $\underline{\subset\subset}$ is coarser than all $\delta_\lambda$. Second, if $\subset\subset^*$ is coarser than all $\subset\subset_\lambda$, then given that $A \subset\subset^* B$, we can find $C_{1/2}$ such that

$$A \subset\subset^* C_{1/2} \subset\subset^* B$$

and then $C_{1/4}$ and $C_{3/4}$ such that

$$A \subset\subset^* C_{1/4} \subset\subset^* C_{1/2} \subset\subset^* C_{3/4} \subset\subset B.$$

By continuing in this way, we obtain for each binary rational a set $C_s$ such that $s < t$ implies $C_s \subset\subset^* C_t$, which implies $C_s \subset\subset' C_t$, where $C_0 = A$ and $C_1 = B$. It follows that $A \underline{\subset\subset} B$. Thus $\underline{\subset\subset}$ is the inf of the $\subset\subset_\lambda$ as claimed.

To obtain sups, we can take the inf of all proximities finer than the given family. (There is one such finer proximity: the discrete proximity.) An explicit construction goes as follows. Define:

a) $A \subset\subset'' B$ iff $A \subset\subset_\lambda B$ for some $\lambda \in \Lambda$

b) $A \mathrel{\overline{\subset\subset}} B$ iff there are sets $A_1, \ldots, A_m$ and $B_1, \ldots, B_n$ such that $A = \bigcup_{i=1}^m A_i$, $B = \bigcap_{j=1}^n B_j$ and $A_i \subset\subset'' B_j$ for all $i$ and $j$.

Again, the verification that $\overline{\subset\subset}$ is a $p$-nhood relation on $X$ is left to Exercise 41A. We will show it is the sup of the proximities $\delta_\lambda$.

Certainly it is finer than each $\delta_\lambda$, since if $A \subset\subset_\lambda B$ for any $\lambda$, then $A \subset\subset'' B$ and hence $A \mathrel{\overline{\subset\subset}} B$. If $\subset\subset^*$ represents any proximity finer than every $\delta_\lambda$, and if $A \mathrel{\overline{\subset\subset}} B$, then $A = \bigcup_{i=1}^m A_i$ and $B = \bigcap_{j=1}^n B_j$, where $A_i \subset\subset'' B_j$ for each $i$ and $j$; that is, $A_i \subset\subset_{\lambda_{ij}} B_j$ for some $\lambda_{ij} \in \Lambda$. Then $A_i \subset\subset^* B_j$ for each $i$ and $j$, so $A_i \subset\subset^* \bigcap_{j=1}^n B_j$, for each $i$, and hence $\bigcup_{i=1}^m A_i \subset\subset^* \bigcap_{j=1}^n B_j$; that is, $A \subset\subset^* B$. Thus $\overline{\subset\subset}$ represents the coarsest proximity finer than each $\delta_\lambda$. ■

***41.5 Theorem.*** *If $\{\delta_\lambda \mid \lambda \in \Lambda\}$ is a family of proximities on $X$, all inducing the same topology $\tau$ on $X$, then $\bar{\delta} = \sup \delta_\lambda$ also induces $\tau$.*

*Proof.* $U$ is a $\bar{\delta}$-nhood of $x$ iff $x \mathrel{\overline{\subset\subset}} U$ iff $U = U_1 \cap \cdots \cap U_n$, where $x \subset\subset_{\lambda_j} U_j$ for some $\lambda_j \in \Lambda$ iff $U = U_1 \cap \cdots \cap U_n$, where each $U_j$ is a $\tau$-nhood of $x$, iff $U$ is a $\tau$-nhood of $x$. ■

**41.6 Definition.** Given a completely regular (i.e., proximizable) topological space $X$, the finest proximity on $X$ which is compatible with the topology on $X$ is called the *fine proximity* on $X$. Its existence is guaranteed by Theorem 41.5.

***41.7 Theorem.*** *If $\delta_1$ and $\delta_2$ are compatible proximities on a Tychonoff space $X$, then $\delta_1$ is finer than $\delta_2$ iff there is a continuous $f\colon \beta_{\delta_1}X \to \beta_{\delta_2}X$ such that $f \mid X$ is the identity (i.e., iff $\beta_{\delta_1}X$ is larger than $\beta_{\delta_2}X$ in the partial order on the set of compactifications of $X$).*

*Proof.* Since $\delta_1$ is finer than $\delta_2$, the identity $i\colon (X, \delta_1) \to (X, \delta_2)$ is a $p$-map. Thus $i\colon (X, \mu_{\delta_1}) \to (X, \mu_{\delta_2})$ is uniformly continuous (see 40.11) and thus extends to the uniform completions $\beta_{\delta_1}X$ and $\beta_{\delta_2}X$, giving the required map $f$. ■

***41.8 Corollary.*** *If $\delta$ is the fine proximity on a Tychonoff space $X$, then $\beta_\delta X = \beta X$.*

## Problems

### 41A. *Supremum and infimum of proximities*

The relations $\underline{\subset\subset}$ and $\overline{\subset\subset}$ defined in 41.4 are $p$-nhood relations.

### 41B. *Freudenthal compactification*

Proximities provide us with a useful way of generating compactifications of a Tychonoff space $X$, since each proximity $\delta$ on $X$ corresponds to a unique compactification $\beta_\delta X$ of $X$.

Define $A \not\delta B$ for nonempty subsets $A$ and $B$ of $X$ iff for some compact set $K$ in $X$, $X - K = G \cup H$ where $G$ and $H$ are disjoint open sets in $X$ with $\bar{A} \subset G$, $\bar{B} \subset H$. (Thus $A \not\delta B$ iff $\bar{A}$ and $\bar{B}$ are separated in $X$ by some compact set.)

1. $\delta$ is a proximity on $X$.

2. If $X$ is rim-compact, $\delta$ is compatible with the topology on $X$. (A space is *rim-compact* iff each of its points has a base of nhoods with compact frontiers.)

3. Each point in $\beta_\delta X$ has a nhood base consisting of sets whose frontiers lie in $X$.

The compactification $\beta_\delta X$ of a rim-compact Tychonoff space thus obtained is called the *Freudenthal compactification* of $X$. From 3, it has the property that $B_\delta X - X$ is zero dimensional.

## 41C. *Quotient proximities*

If $(X, \delta)$ is a proximity space, $Y$ is a set and $f$ is a map of $X$ onto $Y$, we can, with some difficulty, provide a quotient proximity structure on $Y$; i.e., give a proximity structure on $Y$ which is the finest making $f$ a $p$-map.

We begin by defining $\delta_1$ on $Y$ by $C \,\delta_1\, D$ iff $f^{-1}(C)\,\delta\, f^{-1}(D)$; equivalently, we could define $\subset\!\subset_1$ by $C \subset\!\subset_1 D$ iff $f^{-1}(C) \subset\!\subset_1 f^{-1}(D)$. Unfortunately, $\subset\!\subset_1$ fails to satisfy Axiom $(P\text{-}4)'$ for $p$-nhoods (see 1 below). Using the idea of Theorem 41.4, we now force $(P\text{-}4)'$ by defining $\subset\!\subset_2$ by $C \subset\!\subset_2 D$ iff for each binary rational $s$ in $[0, 1]$ there is some $C_s \subset Y$ such that $C_0 = C$, $C_1 = D$ and $s < t$ implies $C_s \subset\!\subset_1 C_t$. As we will see, $\subset\!\subset_2$ is the right candidate for a quotient structure on $Y$.

1. $\subset\!\subset_1$ does not satisfy Axiom $(P\text{-}4)'$ for $p$-nhoods.

2. $\subset\!\subset_2$ is a $p$-nhood relation; i.e., $\subset\!\subset_2$ satisfies Axioms $(P\text{-}1)'$ through $(P\text{-}4)'$.

3. $\subset\!\subset_2$ makes $f$ a $p$-map and is the finest proximity on $Y$ which does so. Thus, the proximity which $\subset\!\subset_2$ represents is referred to as the *quotient proximity on $Y$ induced by $f$.*

4. The proximity induced by a quotient uniformity (37E) is the quotient proximity. (This is difficult.)

5. The topology induced by a quotient proximity need not be the quotient topology. [See 37E.2 and part 4 above.]

6. If $f^{-1}[f(H)] = H$ for each open set $H$ in $X$, then the quotient proximity is given by $\subset\!\subset_1$ and it *does* induce the quotient topology.

## 41D. *The separation identification*

There is an analog for proximity spaces to the $T_0$-identification for topological spaces. Given a nonseparated proximity space $(X, \delta)$, define $x \sim y$ iff $x \,\delta\, y$.

1. $x \sim y$ iff $\bar{x} = \bar{y}$. Thus $\sim$ is an equivalence relation.

2. Let $Y$ be the set of equivalence classes in $X$ under the equivalence relation $\sim$, with the quotient proximity induced by the projection map of $X$ onto $Y$, which takes each $x \in X$ to its equivalence class $[x]$. Then $Y$ is a separated proximity space, whose topology *is* the quotient topology given by $f$.

3. Topologically, $Y$ is the $T_0$-identification of $X$.

### 41E. *Coarsest uniformities and proximities*

The following are equivalent, for a Tychonoff space $X$:

a) $X$ has a coarsest compatible uniformity,

b) $X$ has a coarsest compatible proximity,

c) $X$ is locally compact.

[See 40.11, 41.7.]

### 41F. *Unique uniformity and proximity*

The following are equivalent, for a Tychonoff space $X$:

a) $X$ has a unique compatible uniformity,

b) $X$ has a unique compatible proximity,

c) $|\beta X - X| \leq 1$.

Chapter 10

# Function Spaces

## 42 Pointwise convergence; uniform convergence

Our overall aim in this chapter is the study of the compactness and completeness properties of subcollections $\mathscr{F}$ of the set $Y^X$ of all maps from a space $X$ to a space $Y$. To do this, a usable topology, or uniformity, must be introduced on $\mathscr{F}$ (presumably related to the structures on $X$ and $Y$), and when this has been done, $\mathscr{F}$ is a *function space.*

We have one topology for $Y^X$ and its subcollections already at hand: the product topology.

**42.1 Definition.** We say a subcollection $\mathscr{F} \subset Y^X$ has the *topology of pointwise convergence* (or, the *pointwise topology*) iff it is provided with the subspace topology induced by the Tychonoff product topology on $Y^X$.

This topology on $\mathscr{F}$ is determined solely by the topology on $Y$. The structure on $X$ plays no part. Note also that projection from $\mathscr{F} \subset Y^X$ takes the form of evaluation at a point. That is, for each $x \in X$, the projection map $\pi_x : \mathscr{F} \to Y$ is defined by $\pi_x(f) = f(x)$. The next theorem provides the reason for the name "topology of pointwise convergence" when the product topology is used in this context.

***42.2 Theorem.*** *If $\mathscr{F}$ has the pointwise topology, $(f_\lambda)$ converges to $f$ in $\mathscr{F}$ iff $(f_\lambda(x))$ converges to $f(x)$ for each $x \in X$.*

*Proof.* $(f_\lambda)$ converges to $f$ in $\mathscr{F}$ iff $(\pi_x(f_\lambda))$ converges to $\pi_x(f)$, for each $x \in X$; i.e., iff $f_\lambda(x)$ converges to $f(x)$, for each $x$. ■

We have already made a thorough investigation of the properties of product spaces. In particular, anyone who knows Tychonoff's theorem can prove the following theorem with no trouble.

***42.3 Theorem.*** *Let $Y$ be Hausdorff. A function space $\mathscr{F} \subset Y^X$, with the pointwise topology, is compact iff*

a) *$\mathscr{F}$ is pointwise closed in $Y^X$ (i.e., $\mathscr{F}$ is closed in the pointwise topology on $Y^X$),*

b) *for each $x \in X$, $\pi_x(\mathscr{F}) = \{f(x) \mid f \in \mathscr{F}\}$ has compact closure in $Y^X$.*

As we have said, one of our goals in this chapter is the discovery of conditions on $\mathscr{F}$, with various topologies, which will force compactness. The Tychonoff topology on a product space, it will be recalled, was introduced primarily for its ability to carry things like compactness from $Y$ up to $Y^X$. It is not undue pessimism, then, to predict that no interesting topology on function spaces $\mathscr{F}$ can be found for which the compactness criteria are any simpler than they are above. In fact, if we deal with Hausdorff spaces $Y$ and agree that a topology $\tau$ is "interesting" iff it is no smaller than the topology of pointwise convergence, then more is true: $\mathscr{F}$ with such a topology will be compact iff its topology reduces to the pointwise topology. For if $\tau_p$ denotes the pointwise topology on $\mathscr{F}$ and $\tau$ the larger compact topology, then $(\mathscr{F}, \tau_p)$ is Hausdorff (since $Y$ is), and the identity $i: (\mathscr{F}, \tau) \to (\mathscr{F}, \tau_p)$ is continuous and therefore a homeomorphism!

Hence, for all topologies on $Y^X$ larger than $\tau_p$ (and a good argument is made by Example 42.4 below for restricting ourselves to these), finding conditions on $\mathscr{F}$ which will force compactness must reduce to writing down the conditions of 42.3 plus additional conditions to make convergence in the new topology reduce to pointwise convergence.

**42.4 Example.** The pointwise topology on $Y^X$ is quite small. Let $X = Y = \mathbf{R}$, and for each finite subset $F \subset \mathbf{R}$, let $\chi_F$ be the characteristic function of $F$: $\chi_F(x) = 1$ if $x \in F$, 0 otherwise. The sets $F$ are directed by inclusion, and the resulting net $(\chi_F)$ converges to the function $h$ which is identically 1 on $\mathbf{R}$ although, in a natural sense, no individual term of $(\chi_F)$ seems very close to $h$. What is needed, of course, is more "small" open sets containing $h$.

Suppose now that $Y$ is equipped with a diagonal uniformity $\mathscr{D}$. Then a product uniformity is induced on $Y^X$, and our name for it should be no surprise.

**42.5 Definition.** If $Y$ is a uniform space, the product uniformity $\mathscr{D}_p$ in $Y^X$ is called the *uniformity of pointwise convergence* (or, the *pointwise uniformity*).

The topology associated with the pointwise uniformity on $Y^X$ is, of course, the pointwise topology. Another reason for calling this the uniformity of pointwise convergence is given by the next theorem.

**42.6 Theorem.** *$(f_\lambda)$ is a Cauchy net in $Y^X$ with the pointwise uniformity iff $(f_\lambda(x))$ is Cauchy in $Y$ for each $x \in X$.*

*Proof.* If $(f_\lambda)$ is Cauchy, then in particular $(f_{\lambda_1}, f_{\lambda_2})$ is eventually in each member of $\mathscr{D}_p$ of the form $P_x^{-1}(D)$, where $D \in \mathscr{D}$, and hence $(f_{\lambda_1}(x), f_{\lambda_2}(x))$ is eventually in $D$, so $(f_\lambda(x))$ is Cauchy. (See 37.4 for the definition of $P_x$.)

Conversely, if $(f_{\lambda_1}(x), f_{\lambda_2}(x))$ is eventually in each $D \in \mathscr{D}$, for each $x \in X$, then $(f_{\lambda_1}, f_{\lambda_2})$ is eventually in $P_{x_1}^{-1}(D_1) \cap \cdots \cap P_{x_n}^{-1}(D_n)$ for any $D_1, \ldots, D_n \in \mathscr{D}$ and $x_1, \ldots, x_n \in X$, so that $(f_\lambda)$ is Cauchy. ∎

Completeness in the pointwise uniformity on a function space must be dealt with somewhat differently from compactness in the pointwise topology. The main difference between the following theorem and the corresponding theorem on compactness is the fact that the conditions listed are sufficient, but not necessary. The obstacle to proving necessity is the lack of a theorem saying that uniformly continuous images of complete spaces are complete. As stated, the theorem offers no difficulty in proof.

**42.7 Theorem.** *A function space $\mathscr{F} \subset Y^X$ with the pointwise uniformity is complete if*

a) *$\mathscr{F}$ is pointwise closed in $Y^X$,*

b) *the closure of $\pi_x(\mathscr{F})$ is complete in $Y$, for each $x \in X$.*

The reader should already know that the pointwise limit of continuous functions (on the real line, say) need not be continuous, so that $C(X, Y)$ is not always complete in the uniformity of pointwise convergence.

The uniformity of pointwise convergence and its topology occupy one end of the spectrum of structures used to make function spaces out of collections of functions. At the other end sit the uniformity of uniform convergence and its topology. To introduce these, we note that the sets of the form

$$E_{F,D} = \{(f, g) \mid (f(x), g(x)) \in D \text{ for each } x \in F\},$$

for $D \in \mathscr{D}(Y)$ and $F$ a finite subset of $X$, form a base for the uniformity of pointwise convergence. Larger uniformities will be generated if we use larger sets than the finite sets in this definition (the next section provides an example of this, where the finite sets are replaced by the compact sets), and the largest uniformity of all is obtained by replacing the finite sets by $X$ itself.

**42.8 Definition.** If $Y$ has a uniformity $\mathscr{D}$, the family of sets of the form

$$E_D = \{(f, g) \mid (f(x), g(x)) \in D \text{ for each } x \in X\},$$

for $D \in \mathscr{D}$, form a base for a uniformity $\mathscr{D}_u$ on $Y^X$ called the *uniformity of uniform convergence*. Its topology, $\tau_u$, is the *topology of uniform convergence*. If $(f_\lambda)$ converges to $f$ in this topology, we say $(f_\lambda)$ *converges uniformly* to $f$. Cauchy nets $(f_\lambda)$ in the uniformity of uniform convergence are called *uniformly Cauchy*.

The next theorem provides a relationship between pointwise convergence and uniform convergence which should not be too surprising.

**42.9 Theorem.** *A net $(f_\lambda)$ converges uniformly to $f$ iff $(f_\lambda)$ is uniformly Cauchy and converges pointwise to $f$.*

*Proof.* Necessity is clear. Conversely, suppose $(f_\lambda)$ is uniformly Cauchy and pointwise convergent to $f$. For any $D \in \mathscr{D}$ we will show $(f_\lambda)$ is eventually in $E_D[f] = \{g \mid (f(x), g(x)) \in D, \text{ for each } x \in X\}$. Pick symmetric closed $T \in \mathscr{D}$ so that $T \subset D$. Now for some $\lambda_0$, if $\lambda_1, \lambda_2 \geq \lambda_0$, then $(f_{\lambda_1}, f_{\lambda_2}) \in E_T$, since $(f_\lambda)$ is uniformly Cauchy.

But then, for each $x \in X$, $f_{\lambda_2}(x) \in T[f_{\lambda_1}(x)]$ for all $\lambda_1, \lambda_2 \geq \lambda_0$. Since $f_{\lambda_2}(x)$ converges to $f(x)$ and $T$ is closed, we must have $f(x) \in T[f_{\lambda_1}(x)]$ for all $\lambda_1 \geq \lambda_0$ and $x \in X$. It follows that $(f_{\lambda_1}, f) \in E_T \subset E_D$ for all $\lambda_1 \geq \lambda_0$. Thus $(f_\lambda)$ converges uniformly to $f$. ■

Completeness in the uniformity of uniform convergence is particularly easy to describe. Part b) of the theorem below generalizes a fact the reader should certainly be aware of: if a sequence of continuous real-valued functions of a real variable converges uniformly to $f$, then $f$ is continuous.

***42.10 Theorem.*** *If* $(Y, \mathscr{D})$ *is complete, then so are*

a) $(Y^X, \mathscr{D}_u)$

b) $(C(X, Y), \mathscr{D}_u)$

*Proof.* a) Suppose $(f_\lambda)$ is uniformly Cauchy. Then $(f_\lambda(x))$ is Cauchy in $Y$ for each $x \in X$ and thus converges to a limit $f(x)$. By 42.9, the function $f$ thus defined is the uniform limit of $(f_\lambda)$. Thus $Y^X$ is uniformly complete.

b) It suffices to show $C(X, Y)$ uniformly closed in $Y^X$. Suppose $f$ is *not* continuous, say at $x$. Then for some $D \in \mathscr{D}$, $f^{-1}(D[f(x)])$ contains no nhood of $x$. If $T$ is symmetric from $\mathscr{D}$ and such that $T \circ T \subset D$, then for each $g \in E_T[f]$, a routine computation will show $g^{-1}(T[f(x)])$ is contained in $f^{-1}(D[f(x)])$ and thus contains no nhood of $x$. Thus $E_T[f]$ is a nhood of $f$ consisting of functions discontinuous at $x$. It follows that $C(X, Y)$ is uniformly closed and thus complete. ■

Conditions for compactness in the topology of uniform convergence are rare. In fact, we will limit ourselves in this direction to the comment that the compact-open topology reduces to the topology of uniform convergence when $X$ is compact, so that Ascoli's theorem (next section) applies to the topology of uniform convergence in this case (see also 43E).

For noncompact $X$, the topology of uniform convergence is simply too large to force compactness with a reasonable set of conditions. Put another way, the topology of uniform convergence will reduce to the pointwise topology for only a very limited number of subspaces of $Y^X$.

## Problems

42A. *The function space* $\mathbf{I}^\mathbf{I}$.

1. Which of the following subspaces of $\mathbf{I}^\mathbf{I}$ is compact in the pointwise topology?
   a) $\{f \in \mathbf{I}^\mathbf{I} \mid f(0) = 0\}$
   b) $\{f \in \mathbf{I}^\mathbf{I} \mid f \text{ is continuous and } f(0) = 0\}$
   c) $\{f \in \mathbf{I}^\mathbf{I} \mid f \text{ is differentiable and } |f'(x)| \leq 1 \text{ for all } x \in \mathbf{I}\}$.
2. Exhibit a countable dense subset of $\mathbf{I}^\mathbf{I}$ in the pointwise topology.
3. Is $\mathbf{I}^\mathbf{I}$ separable in the topology of uniform convergence?

## 42B. *Completeness in function spaces*

1. Let $Y = \mathbf{R} - \{0\}$. Which of the following subspaces of $Y^{\mathbf{R}}$ is complete in the pointwise uniformity?

   a) $Y^{\mathbf{R}}$
   b) $\{f \in Y^{\mathbf{R}} \mid f \text{ is continuous}\}$
   c) $\{f \in Y^{\mathbf{R}} \mid |f| \geq 1\}$.

2. Same question for the uniformity of uniform convergence.

## 42C. *Metrizability in function spaces*

Let $Y$ be metrizable, its topology generated by a metric $\rho$.

1. The uniformity of uniform convergence on the space $C(X, Y)$ of all continuous functions in $Y^X$ is metrizable by the metric $d(f, g) = \sup_{x \in X} \rho(f(x), g(x))$.

2. When is the pointwise uniformity on $C(X, Y)$ metrizable?

(See also 43G.)

## 42D. *Separability of $C^*(X)$*

For compact $X$, $C^*(X)$ is separable iff $X$ is metrizable.

## 42E. *Compact and finite-dimensional mappings*

Let $X$ be a metric space, $E$ a Banach space. A mapping $f: X \to E$ is said to be *compact* iff $\overline{f(X)}$ is compact, *finite-dimensional* iff it is compact and $f(X)$ is contained in a subspace of $E$ of finite dimension.

1. If $f_n: X \to E$ is compact for $n = 1, 2, \ldots$ and the $f_n$ converge uniformly to $f$, then $f$ is compact. [It is enough to show $f(X)$ is totally bounded.]

2. A mapping $f: X \to E$ is compact iff it can be uniformly approximated by finite-dimensional mappings. [For necessity, use the fact that $\overline{f(X)}$ is totally bounded.]

# 43 The compact–open topology and uniform convergence on compacta

It is convenient to begin our discussion of the compact–open topology by returning again to the pointwise topology. The sets $(a, U) = \{f \in \mathscr{F} \mid f(a) \in U\}$, for $a \in X$ and $U$ open in $Y$, form a subbase for the latter topology on a function space $\mathscr{F}$. A case can be made, then, for calling the pointwise topology on $\mathscr{F}$ the *point–open* topology. We can, at the same time, involve the topology of $X$ in our function space and decrease the size of the basic open sets (thus increasing the size of the topology) by replacing the points in the point–open topology by the compact subsets of $X$. The resulting topology lies somewhere between the pointwise topology and the topology of uniform convergence.

**43.1 Definition.** The *compact–open topology* (*k-topology*) on $\mathscr{F} \subset Y^X$ is the topology having for a subbase the sets

$$(K, U) = \{f \in \mathscr{F} \mid f(K) \subset U\},$$

for $K$ compact in $X$, $U$ open in $Y$. We denote this topology by $\tau_C$.

A convenient counterexample to a great many theorems is easily obtained.

**43.2 Example.** If $X$ is discrete, the compact–open topology on $Y^X$ is the pointwise topology. Thus, nothing can be carried from $Y$ to $(Y^X, \tau_C)$, in full generality, unless it can be carried to product spaces.

The following lemma, an example supporting the general rule of thumb that compact sets behaving like points, can be proved with no difficulty (or, see Exercise 17B).

***43.3 Lemma.*** *In a regular space, if $F$ is compact, $U$ open and $F \subset U$, then for some open set $V$, $F \subset V$ and $\overline{V} \subset U$.*

We denote the set of continuous functions from $X$ to $Y$ by $C(X, Y)$. Most useful examples of function spaces are spaces of continuous functions, so part b) of the next theorem is not too disappointing.

***43.4 Theorem.*** a) *If $Y$ is $T_0$, $T_1$ or $T_2$, so is $(Y^X, \tau_C)$.*

b) *If $Y$ is regular, so is $\big(C(X, Y), \tau_C\big)$.*

*Proof.* We leave to the reader the proof of part a) as well as the implied assertion in b) that $(Y^X, \tau_C)$ need not be regular for regular $Y$ (43B).

To prove b), let $K$ be compact in $X$, $U$ open in $Y$, and $f$ a continuous function in $(K, U)$. Then $f(K)$ is a compact subset of $U$ in $Y$, and by the lemma above, an open set $V$ exists with $f(K) \subset V$, $\overline{V} \subset U$. Then $f \in (K, V)$ and $(K, \overline{V}) \subset (K, U)$. Now we assert $\overline{(K, V)} \subset (K, \overline{V})$. If $g \notin (K, \overline{V})$, then for some point $a$ in $K$, $g(a) \in Y - \overline{V}$, so $g \in (a, Y - \overline{V})$. But then $(A, Y - \overline{V})$ is a nhood of $g$ not meeting $(K, V)$ and thus $g \notin \overline{(K, V)}$. Therefore, $\overline{(K, V)} \subset (K, \overline{V})$.

Now suppose $f \in \bigcap_{i=1}^n (K_i, U_i)$, where each $K_i$ is compact in $X$, each $U_i$ is open in $Y$. For $I = 1, \dots, n$ find, as above, open $V_i \subset Y$ such that $f \in (K_i, V_i)$ and $\overline{(K_i, V_i)} \subset (K_i, U_i)$. Then $f \in \bigcap_{i=1}^n (K_i, V_i)$ and $\overline{\bigcap_{i=1}^n (K_i, V_i)} \subset \bigcap_{i=1}^n (K_i, V_i)$.

Thus, $C(X, Y)$ is regular in the compact–open topology. ■

As with the pointwise topology, if $Y$ has a uniform structure, we have a uniform structure on $Y^X$ which is associated with the compact–open topology. The association is not complete, however. The uniform topology matches the compact–open topology only for spaces of continuous functions.

**43.5 Definition.** Suppose $Y$ has a uniformity $\mathscr{D}$. The *uniformity of uniform convergence on compacta*, or the *uniformity of compact convergence*, $\mathscr{D}_k$, has for a subbase the sets

$$E_{K,D} = \{(f, g) \mid f(x), g(x)) \in D, \text{ for each } x \in K\}$$

where $K$ is a compact subset of $X$ and $D \in \mathscr{D}$. The topology $\tau_k$ thus induced on $Y^X$ is the topology of *compact convergence.*

The topology and uniformity of uniform convergence on compacta derive their names from the following theorem, whose proof is obvious from the definitions involved.

***43.6 Theorem.*** a) $(f_\lambda)$ *converges to* $f$ *in the topology of uniform convergence on compacta iff for each compact subset* $K$ *of* $X$, $f_\lambda \mid K$ *converges to* $f \mid K$ *in the topology of uniform convergence on* $K$.

b) $(f_\lambda)$ *is Cauchy in the uniformity of uniform convergence on compacta iff for each compact subset* $K$ *of* $X$, $(f_\lambda \mid K)$ *is Cauchy in the uniformity of uniform convergence on* $K$.

The promised relationship between the compact–open topology and the topology of uniform convergence on compacta is given by the next theorem.

***43.7 Theorem.*** *For spaces of continuous functions the topology of compact convergence is the compact–open topology.*

*Proof.* Let $(K, U)$ be a subbasic open set in the compact–open topology, $g$ a continuous function in $(K, U)$. Then $g(K)$ is compact and $g(K) \subset U$. Find $E \in \mathscr{D}(Y)$ such that $E[g(K)] = \bigcup \{E[x] \mid x \in g(K)\} \subset U$. (This is done as follows. For each $x \in g(K)$, find $D_x \in \mathscr{D}(Y)$ such that $D_x[x] \subset U$, let $E_x \circ E_x \subset D_x$ and, by compactness, say $g(K) \subset E_{x_1}[x_1] \cup \cdots \cup E_{x_n}[x_n]$. Set $E = E_{x_1} \cap \cdots \cap E_{x_n}$ and check the required property for $E$.) Now let

$$D = \{(f, h) \mid (f(x), h(x)) \in E, \text{ for each } x \in K\}.$$

Then $D \in \mathscr{D}_k$, and if $h \in D[g]$, then $(g(x), h(x)) \in E$, so that $h(x) \in E[g(x)]$, for each $x \in K$, and hence $h(x) \in E[g(K)] \subset U$, for each $x \in K$, from which it follows that $h \in (K, U)$. Thus $g \in D[g] \subset (K, U)$. Thus, each subbasic set $(K, U)$, and hence each open set in the compact–open topology, is open in the topology of compact convergence.

Conversely, let $E_{K,D}$ be a subbasic set in $\mathscr{D}_k$. The sets $E_{K,D}[f]$, for $f$ continuous on $X$, form a subbase for the topology of compact convergence, and it thus suffices to show they are open in the compact–open topology. Pick $T$ closed and symmetric so that $T \circ T \circ T \subset D$. By compactness, $f(K) \subset T[f(x_1)] \cup \cdots \cup T[f(x_n)]$ for some $x_1, \ldots, x_n$. Set $K_i = K \cap f^{-1}(E[f(x_i)])$, $T_i = \text{Int}\,(T \circ T)[f(x_i)]$.

Then $f(K_i) \subset T_i$ for each $i$, since $f(K_i) \subset T[f(x_i)] \subset \text{Int}\,(T \circ T)[f(x_i)] = T_i$, so $f \in (K_i, T_i)$ for each $i$. Suppose $g \in (K_i, T_i)$ for each $i$. Then $g(K_i) \subset T_i$ for $i = 1, \ldots, n$, and if $x \in K$, then for some $i$, $x \in K_i$, so $f(x) \in T[f(x_i)]$ while $f(x) \in T_i \subset (T \circ T)[f(x_i)]$. Hence $(f(x), f(x_i)) \in T$ and $(g(x), f(x_i)) \in T \circ T$, so that $(f(x), g(x)) \in T \circ T \circ T \subset D$. Thus $(f(x), g(x)) \in D$, for each $x \in K$, which establishes that $g \in E_{K,D}[f]$. Thus $f \in \bigcap_{i=1}^n (K_i, T_i) \subset E_{K,D}[f]$, so the latter set is open in the compact–open topology. ■

We now introduce the concept which plays a central role in the discussion of both completeness and compactness relative to the uniformity of uniform convergence on compacta and its topology.

**43.8 Definition.** A topological space $X$ is a *k-space* (or a *compactly generated space*) iff the following condition holds:

a) $A \subset X$ is open iff $A \cap K$ is open in $K$ for each compact set $K$ in $X$.

Note that one implication in a) is trivial and never needs proving. Also, it is clear that "open" could have been replaced by "closed" in a) without harm. The $k$-spaces form a wide class of spaces, including all metric spaces, according to the next theorem, an extension of which is given in 43H.

***43.9 Theorem.*** a) *Every locally compact space is a k-space,*

b) *Every first-countable space is a k-space.*

*Proof.* a) Suppose $X$ is locally compact and $A \cap K$ is open in $K$ for each compact $K \subset X$. Let $a \in A$ and let $V$ be an open nhood of $a$ with compact closure. But then $A \cap \bar{V}$ is open in $\bar{V}$, and hence $A \cap V = (A \cap \bar{V}) \cap V$ is open in $V$ and thus in $X$. Then $a$ has a nhood contained in $A$, so $A$ is open in $X$.

b) Suppose $X$ is first countable, and $B \cap K$ is closed in $K$ for each compact $K \subset X$. If $b \in \bar{B}$, then a sequence $(b_n)$ in $B$ converges to $b$. But $(b_n) \cup \{b\}$ is a compact subset of $x$, so $B \cap [(b_n) \cup \{b\}]$ is closed and hence $b \in B$. Thus $B$ is closed. ■

The $k$-spaces are important to our discussion of convergence of continuous functions on compacta because, in these spaces, the continuous functions are precisely those which behave well on compact subsets. The proof of the following lemma, which says this more precisely, is an easy exercise in applying the definition of a $k$-space. See Exercise 43D.

***43.10 Lemma.*** *If $X$ is a k-space, $f: X \to Y$ is continuous iff $f \mid K$ is continuous for each compact $K \subset X$.*

Using this result and 43.6, which describes convergence on compacta as being precisely uniform convergence on each compact subset, the following theorem is easy.

***43.11 Theorem.*** *If $X$ is a k-space and $(Y, \mathscr{D})$ is complete, then $C(X, Y)$ is complete in the uniformity of uniform convergence on compacta.*

*Proof.* If $(f_\lambda)$ is Cauchy in the uniformity of uniform convergence on compacta, then by 43.6 $(f_\lambda \mid K)$ is uniformly Cauchy on $K$ for each compact $K \subset X$. Since $C(K, Y)$ is complete in the uniformity of uniform convergence, a continuous uniform limit $f_K: K \to Y$ exists for each $K$. It is easily seen that if $K_1 \subset K_2$, then $f_{K_2} \mid K_1 = f_{K_1}$, and from this it follows that the function $f: X \to Y$ defined by

$f(x) = f_K(x)$, for $x \in K$, is well defined. It is continuous by 43.10 above, and since $(f_\lambda)$ converges uniformly to $f$ on each compact $K \subset X$, $(f_\lambda)$ converges to $f$ in the topology of uniform convergence on compacta, by 43.6. ■

One more definition is needed before we are ready to characterize the compact function spaces in the compact–open topology.

**43.12 Definition.** Let $X$ be a topological space, $Y$ a uniform space. A family $\mathscr{F}$ of continuous functions from $X$ to $Y$ is *equicontinuous at* $x \in X$ iff for each $D \in \mathscr{D}(Y)$, there is a nhood $U$ of $x$ such that $f(U) \subset D[f(x)]$, for each $f \in \mathscr{F}$. We say $\mathscr{F}$ is *equicontinuous* provided it is equicontinuous at each point of $X$.

**43.13 Lemma.** *If $\mathscr{F}$ is an equicontinuous family of functions, so is the pointwise closure $\overline{\mathscr{F}}$ of $\mathscr{F}$.*

*Proof.* Let $f \in \overline{\mathscr{F}}$, say $(f_\lambda)$ is a net in $\mathscr{F}$ converging pointwise to $f$. Now if $D$ is any closed element of $\mathscr{D}(Y)$ and $U$ is an open set containing $x \in X$ such that $g(U) \subset D[g(x)]$ for each $g \in \mathscr{F}$, then in particular $(f_\lambda(x), f_\lambda(y)) \in D$ for each $\lambda$ and each $y \in U$. Since $D$ is closed, it follows that $(f(x), f(y)) \in D$, for each $y \in U$. Hence $f(U) \subset D[f(x)]$.

Thus $\overline{\mathscr{F}}$ is equicontinuous. ■

As we noted early in Section 42, the key to making a function space compact must be listing of enough conditions on the space to ensure that the topology involved reduces to the pointwise topology. The significance of equicontinuity is made clear, then, by the next theorem.

**43.14 Theorem.** *On an equicontinuous family $\mathscr{F}$, the compact–open topology reduces to the pointwise topology.*

*Proof.* It is enough to show that if $f_\lambda \to f$ pointwise in $\mathscr{F}$, then $f_\lambda \to f$ in the compact–open topology.

It is sufficient to consider a subbasic element $(K, U)$ of the compact–open topology which contains $f$. For each $x$ in $K$, $f_\lambda(x) \to f(x)$, so eventually, say for $\lambda \geq \lambda_x$, $f_\lambda(x)$ is in $U$. But $f(K) \subset U$ and $f(K)$ is compact, so for some $D \in \mathscr{D}(Y)$, $D[f(K)] \subset U$. By equicontinuity, each $x \in K$ has a nhood $U_\lambda$ such that $f_\lambda(U_x) \subset D[f_\lambda(x)]$ for all $\lambda$, and thus, for $\lambda \geq \lambda_x$, $f_\lambda(U_x) \subset D[f(K)] \subset U$. But the cover of $K$ by the sets $U_x$ has a finite subcover, say by $U_{x_1}, \ldots, U_{x_n}$. Pick $\lambda_0 \geq \lambda_{x_1}, \ldots, \lambda_{x_n}$. Then for any $x \in U$, $x \in U_{x_i}$ for some $i$ and hence for $\lambda \geq \lambda_0$,

$$f_\lambda(x) \in f_\lambda(U_{x_i}) \subset D[f(K)] \subset U.$$

It follows that $f_\lambda \in (K, U)$ for all $\lambda \geq \lambda_0$.

Then $f_\lambda \to f$ in the compact–open topology. ■

The last result makes the proof of Ascoli's theorem, on compactness of function spaces in the compact–open topology, almost trivial. The form of Ascoli's theorem given here is quite general; we will develop in Exercise 43E a more special form of essentially the same theorem.

In order to prove necessity of the conditions we impose, we must drop equicontinuity of $\mathscr{F}$ for the weaker condition that $\mathscr{F}$ be equicontinuous on each compact subset (more precisely, that for each compact subset $K$, the family of restrictions of members of $\mathscr{F}$ to $K$ be equicontinuous).

***43.15 Theorem*** (Ascoli). *Let $X$ be a Hausdorff, or regular, k-space, $Y$ a Hausdorff uniform space, and $\mathscr{F}$ a family of continuous functions from $X$ to $Y$. Then $\mathscr{F}$ is compact in the compact–open topology iff*

a) *$\mathscr{F}$ is pointwise closed,*

b) *for each $x \in X$, $\pi_x(\mathscr{F})$ has compact closure,*

c) *$\mathscr{F}$ is equicontinuous on each compact subset of $X$.*

*Proof.* If $\mathscr{F}$ is compact in the compact–open topology, then $\mathscr{F}$ is compact in the pointwise topology, so necessity of the first two conditions follows from 42.3. Let $K$ be any compact subset of $X$, $\mathscr{F}_K$ the family of restrictions to $K$ of members of $\mathscr{F}$. It is an easy exercise to show that $\mathscr{F}_K$ is compact in the compact–open topology on $C(K, Y)$ (which reduces to the topology of uniform convergence since $K$ is compact). We will prove this implies equicontinuity of $\mathscr{F}_K$.

Pick $x \in K$ and $E \in \mathscr{D}(Y)$. Let $D$ be a symmetric element of $\mathscr{D}(Y)$ such that $D \circ D \subset E$. Since $X$ is Hausdorff or regular and $K$ is compact, $K$ is regular. Thus a nhood $U_f$ of $x$ exists for which $f(\overline{U}_f) \subset D[f(x)]$. But $(\overline{U}_f, D[f(x)])$ is then a nhood of $f$ in the compact–open topology, and the resulting cover of $\mathscr{F}_K$ has a finite subcover, say by $(\overline{U}_{f_1}, D[f_1(x)]), \ldots, (\overline{U}_{f_n}, D[f_n(x)])$. Let

$$U = U_{f_1} \cap \cdots \cap U_{f_n}.$$

Now for $f \in \mathscr{F}$, $f \in (\overline{U}_{f_i}, D[f_i(x)])$ for some $i$ and hence $f(U) \subset f(\overline{U}_{f_i}) \subset D[f_i(x)]$, and it follows easily that $f(U) \subset (D \circ D)[f(x)] \subset E[f(x)]$, so that $\mathscr{F}_K$ is equicontinuous at $x$.

To prove sufficiency, it is enough to show that condition c) forces the compact–open topology to reduce to the pointwise topology. But by 43.14, c) does force the compact–open topology *on* $\mathscr{F}_K$ to reduce to the pointwise topology, for each compact $K \subset X$. Now let $(K, U)$ be any subbasic set in the compact–open topology on $X$. From the remarks above, $(K, U) \mid \mathscr{F}_K = \{f \mid K \mid f \in (K, U)\}$ is pointwise open in $\mathscr{F}_K$. But the map $f \to f \mid K$ is clearly pointwise continuous (pointwise convergence is preserved under restriction), and the inverse under this map of the set $(K, U) \mid \mathscr{F}_K$ is the set $(K, U)$. Thus $(K, U)$ is pointwise open. ■

## Problems

### 43A. *Sequence spaces and Ascoli's theorem*

1. The sequence space **m** (see 2H) is just $C^*(\mathbf{N})$ with the uniform metric. It is not compact. [Use Ascoli's theorem.]

2. The sequence space **c** (see 2H) is just $C^*(J)$ where $J$ is the subspace $\{0\} \cup \{1/n \mid n \in \mathbf{N}\}$ of $R$. It is not compact.

3. Is $\mathbf{c}_0$ (see 2H) compact?

## 43B. *Separation axioms*

1. Show that if $Y$ is $T_0$, $T_1$ or $T_2$ then $Y^X$ has the same property in the compact–open topology.

2. Give an example of a regular space $Y$ such that $Y^X$ with the compact–open topology is not regular.

3. If $Y$ is completely regular, so is $C(X, Y)$ with the compact–open topology [see 43.7].

## 43C. *Convergence in the uniformity of uniform convergence on compacta*

Prove (a) and (b) of Theorem 43.6.

## 43D. *Continuity on compacta*

If $X$ is a $k$-space, then $f: X \to Y$ is continuous iff $f \mid K$ is continuous for each compact $K \subset X$.

## 43E. *Arzela's theorem*

1. A subfamily of $C[a, b]$ is compact in the compact–open topology iff it is uniformly bounded and equicontinuous.

2. Let $f(s, t, u)$ be a continuous real-valued function defined for $0 \leq s \leq 1$, $0 \leq t \leq 1$, $-1 \leq u \leq 1$. For each $s \in [0, 1]$ and $x \in C[0, 1]$ with $\|x\| \leq 1$, define

$$F_x(s) = \int_0^1 f(s, t, x(t))\, dt.$$

Then the mapping $F(x) = f_x$ takes $C[0, 1]$ into $C[0, 1]$. Use Arzela's theorem to show $F$, called the Urysohn integral operator, is a compact mapping (42E).

## 43F. *Joint continuity*

A topology for a function space $\mathscr{F} \subset Y^X$ is *jointly continuous* (*admissible*) iff the map $P: \mathscr{F} \times X \to Y$ defined by $P(f, x) = f(x)$ is continuous.

1. If $\tau$ is a jointly continuous topology for $\mathscr{F}$ and $\tau \subset \tau'$, then $\tau'$ is jointly continuous. The discrete topology on $\mathscr{F}$ is jointly continuous (and hence is the largest jointly continuous topology for $\tau$).

2. Every jointly continuous topology on $\tau$ contains the compact–open topology. (So the compact–open topology is the smallest jointly continuous topology for $\mathscr{F}$ whenever it is jointly continuous.)

3. Suppose $\mathscr{F} \subset C(X, Y)$. If $X$ is a Hausdorff $k$-space, then the compact–open topology on $\mathscr{F}$ is jointly continuous. [Show that $P: \mathscr{F} \times X \to Y$ is continuous iff $P \mid (\mathscr{F} \times K)$ is continuous for each compact $K \subset X$.]

## 43G. *Metrizability of C(X)*

Let $C(X)$, the collection of continuous, real-valued functions on a Tychonoff space $X$, have the compact–open topology.

1. If $X$ is hemicompact (17I), then $C(X)$ is metrizable. [If $K_1, K_2, \ldots$ is the sequence of compact subsets of $X$ required for hemicompactness, define $\rho$ on $C(X)$ by

$$\rho(f, g) = \sum_{n=1}^{\infty} \rho_n(f, g)$$

where $\rho_n(f, g) = \min(1/2^n, \sup_{x \in K_n} |f(x) - g(x)|)$. Show that $\rho$ generates the compact–open topology on $C(X)$].

2. If $C(X)$ is first countable, then $X$ is hemicompact. [Let $f$ be the function which is identically 0 on $X$. Show that if $(K_1, W_1), (K_2, W_2), \ldots$ is a countable nhood base at $f$ in $C(X)$, then $K_1, K_2, \ldots$ is a sequence of compact sets of the kind needed to show hemicompactness.]

3. *$C(X)$ is metrizable iff $X$ is hemicompact.*

## 43H. *k-spaces*

1. A subspace of a $k$-space need not be a $k$-space.

2. The product of uncountably many copies of $\mathbf{R}$ is not a $k$-space. [Let $T$ be the subset of the product consisting of all points $x$ such that for some integer $n \geq 0$, $x_\alpha = n$ for all but at most $n$ coordinates and $x_\alpha = 0$ otherwise. Then $T \cap K$ is compact for each compact subset $K$ of the product, but $T$ is not closed.]

3. A Hausdorff space $X$ is a $k$-space iff it is a quotient of some locally compact space. [If $X$ is a $k$-space, let $T$ be the disjoint union of the compact subspaces of $X$ and find a quotient map of $T$ onto $X$.]

## 43I. *The Exponential Law*

All function spaces have the compact–open topology here. Let $X$ be a locally compact, Hausdorff space, $T$ a Hausdorff space. Then $C(X \times T, T)$ is homeomorphic to $C(T, C(X, Y))$. [If $f \in C(X \times T, Y)$, define $f_t \in C(X, Y)$ by $f_t(x) = f(x, t)$. Then the map $\varphi_f(t) = f_t$ belongs to $C(T, C(X, Y))$. The correspondence $\Phi(f) = \varphi_f$ is the desired homeomorphism.]

## 43J. *Homotopy and function spaces*

Let $C(X, Y)$ have the compact–open topology.

1. Let $X$ be a $k$-space. The path components in $C(X, Y)$ are precisely the homotopy equivalence classes.

2. Recall that $\Omega(Y, y_0)$ is the subset of $C(I, Y)$ consisting of all loops based at $y_0$. Then the path components in $\Omega(Y, y_0)$ are precisely the equivalence classes in $\Omega(Y, y_0)$ under the loop homotopy relation (i.e., the relation of homotopy relative to $\{0, 1\}$).

3. Recall that $\Omega_n(Y, y_0)$ is the subset of $C(\mathbf{I}^n, Y)$ consisting of all $n$-dimensional hyperloops based at $y_0$ (33D). Then the path components in $\Omega_n(Y, y_0)$ are precisely the equivalence classes in $\Omega(Y, y_0)$ under the relation of homotopy relative to $\partial\mathbf{I}^n$.

### 43K. *The higher homotopy groups*

All function spaces are to be given the compact–open topology. If $Y$ is a topological space and $y_0 \in Y$, let $e_n$ denote the constant loop in $\Omega_n(Y, y_0)$; that is, $e_n(x) = y_0$ for each $x \in \mathbf{I}^n$. ($\Omega_n$ is defined in 33D.)

1. $\Omega_n(Y, y_0)$ is homeomorphic to $\Omega(\Omega_{n-1}(Y, y_0), e_{n-1})$. [Use the exponential law (43I) to conclude $C(\mathbf{I}^n, Y)$ is homeomorphic to $C(\mathbf{I}, C(\mathbf{I}^{n-1}, Y))$; use the resulting homeomorphism to construct a homeomorphism from $\Omega_n(Y, y_0)$ to $\Omega(\Omega_{n-1}(Y, y_0), e_{n-1})$.]
2. $\Pi_n(Y, y_0)$ is isomorphic to $\Pi_1(\Omega_{n-1}(Y, y_0), e_{n-1})$.

Now every loop space is an $H$-space and the fundamental group of any $H$-space is Abelian. Thus, by part 2, $\Pi_n(Y, y_0)$ is Abelian for $n > 1$. (For the definition of an $H$-space and an investigation of its properties, see the book *Topology*, by Dugundji.)

## 44 The Stone–Weierstrass theorem

The few elementary ideas from algebra which are necessary to read this section will not be developed here. Consult any book on algebra.

**44.1 Definition.** $C(X)$ will denote the algebra of real-valued continuous functions on the topological space $X$, with the subalgebra of bounded functions in $C(X)$ being denoted $C^*(X)$.

Our interest does not lie in developing the algebraic properties of $C^*(X)$ and their relationship to the topological properties of $X$, although a good deal of work is currently being done in this direction. An excellent account of results of this nature can be found in the book of Gillman and Jerison.

Our look at $C^*(X)$ will be confined to some elementary topological and lattice-theoretic results, with the limited aim of developing the Stone–Weierstrass theorem (44.5). The topology we will work with on $C^*(X)$ is that induced by the metric

$$\rho(f, g) = \sup_{x \in X} |f(x) - g(x)|.$$

This is called the *uniform metric*, for the good reason that it induces on $C^*(X)$ the topology of uniform convergence (so if $X$ is compact, $C^*(X) = C(X)$ is complete in this metric, by 42.10). As is our established custom, we avoid constant reference to the fact that the background structure on $C^*(X)$ is the uniform metric by using phrases like "uniformly dense" to mean "dense in the uniform metric" and so on.

The classical Weierstrass theorem deals with the uniform approximation by polynomials of continuous functions on a closed interval. We will be deriving it as a special case of the more general Stone–Weierstrass theorem, but to prove the latter, we need a very weak form of the former.

**44.2 Lemma.** *For each $\epsilon > 0$, there is a polynomial $P_\epsilon(x)$ such that*

$$\big| |x| - P_\epsilon(x) \big| < \epsilon$$

*for each $x$ in $[-1, 1]$.*

*Proof.* From the theory of functions of a real variable (see, for example, Apostol, pp. 420 and 427), there is a binomial series $\sum_{n=0}^{\infty} a_n y^n$ which converges uniformly to $(1 - y)^{1/2}$ for $y$ in $[0, 1]$. Letting $y = 1 - x^2$ for $x$ in $[-1, 1]$, we obtain as an immediate corollary that $\sum_{n=0}^{\infty} a_n(1 - x^2)^n$ converges uniformly to $|x|$ in $[-1, 1]$. Since each partial sum of this series is a polynomial, the lemma follows. ■

We need this lemma only to establish the following fact, which is critical to the proof of the general approximation theorem (44.5).

**44.3 Lemma.** *Any uniformly closed subalgebra $\mathscr{A}$ of $C^*(X)$ is a lattice. That is, if $f$ and $g$ belong to $\mathscr{A}$, so do the functions $\min(f, g)$ and $\max(f, g)$ [defined pointwise].*

*Proof.* Since it is easily verified that

$$\min(f, g) = \tfrac{1}{2}(f + g) - \tfrac{1}{2}|f - g|,$$
$$\max(f, g) = \tfrac{1}{2}(f + g) + \tfrac{1}{2}|f - g|,$$

it evidently suffices to show that, whenever $f \in \mathscr{A}$, $|f| \in \mathscr{A}$. Suppose first that $|f| \le 1$ on $X$. Then, by 44.2, a polynomial $P_\epsilon$ exists for each $\epsilon > 0$ such that, on $X$,

$$|P_\epsilon(f) - |f|| < \epsilon,$$

and thus $|f|$ is uniformly approximated by the functions $P_\epsilon(f)$, all of which belong to $\mathscr{A}$. Thus, in this case, $|f| \in \mathscr{A}$. Now even if we do not have $|f| \le 1$ on $X$, we have $|f| \le A$ for some positive number $A$. Then applying the previous procedure, we find $|f/A| \in \mathscr{A}$, and hence $|f| \in \mathscr{A}$. ■

**44.4 Definition.** If $\mathscr{D}$ is any subcollection from $C^*(X)$, the *subalgebra $\mathscr{A}(\mathscr{D})$ generated by $\mathscr{D}$* is the smallest subalgebra of $C^*(X)$ containing $\mathscr{D}$. It always exists, since the intersection of the subalgebras containing $\mathscr{D}$ is a subalgebra. Also the uniform closure $\mathscr{B}(\mathscr{D})$ of $\mathscr{A}(\mathscr{D})$ is a subalgebra (the verification is routine), called the *uniformly closed subalgebra generated by $\mathscr{D}$.*

The Stone–Weierstrass theorem provides a set of conditions on $\mathscr{D}$ under which the uniformly closed subalgebra generated by $\mathscr{D}$ is all of $C^*(X)$. Recall that a collection of functions separates points iff whenever $x \ne y$ in $X$, for some one of the functions $f$, $f(x) \ne f(y)$.

**44.5 Theorem.** (Stone–Weierstrass). *Let $X$ be a compact, Hausdorff space. If $\mathscr{D}$ is a collection of functions in $C^*(X)$ which separates points in $X$ and contains the function identically* 1, *the uniformly closed subalgebra generated by $\mathscr{D}$ is all of $C^*(X)$.*

*Proof.* The proof bears some resemblance to our proof of Tietze's theorem earlier. We will show every function $f \in C^*(X)$ can be uniformly approximated by functions from $\mathscr{A}(\mathscr{D})$. For this purpose, no true loss of generality results in assuming $\inf_{x\in X} f(x) < \sup_{x\in X} f(x)$ (otherwise $f$ is constant and, since $\mathscr{D}$ contains 1,

$f \in \mathscr{A}(\mathscr{D})$), and then we can assume, without loss of generality, that $\inf_{x \in X} f(x) = -1$, $\sup_{x \in X} f(x) = 1$. Thus $f: X \to [-1, 1]$.

Let $A_1 = \{x \in X \mid f(x) \leq -\frac{1}{3}\}$, $B_1 = \{x \in X \mid f(x) \geq \frac{1}{3}\}$. For each $a \in A_1$ and $b \in B_1$ a function $h_{ab}$ exists with $h_{ab}(a) \neq h_{ab}(b)$. Define $g_{ab}$ on $X$ by

$$g_{ab}(x) = -\frac{4}{3}\frac{h_{ab}(x) - h_{ab}(b)}{h_{ab}(a) - h_{ab}(b)} + \frac{2}{3}.$$

Then $g_{ab}(a) = -\frac{2}{3}$, $g_{ab}(b) = \frac{2}{3}$, and $g_{ab} \in \mathscr{A}(\mathscr{D})$. Fix $a \in A_1$. For each $y \in B_1$, $g_{ay}(y) = \frac{2}{3}$, and so $g_{ay}(z) \geq \frac{1}{3}$ for $z$ in some nhood $U_y$ of $y$. A finite number of these nhoods, say $U_{y_1}, \ldots, U_{y_n}$, cover $B_1$, and a function $g_a$ can now be defined at each $x \in X$ by

$$g_a(x) = \min\{g_{ay_1}(x), \ldots, g_{ay_n}(x)\}.$$

Note that $g_a(a) = -\frac{2}{3}$ and $g_a \geq \frac{1}{3}$ on $B_1$, and $g_a \in \mathscr{B}(\mathscr{D})$ by 44.3. By repeating the procedure just used, evidently we can find a function $g \in \mathscr{B}(\mathscr{D})$ such that $g \leq -\frac{1}{3}$ on $A_1$ and $g \geq \frac{1}{3}$ on $B_1$. It follows that $|g(x) - g_a(x)| \leq \frac{2}{3}$ for $x \in A_1 \cup B_1$, and if we define

$$h(x) = \min\{g(x), \tfrac{1}{3}\}$$
$$h_1(x) = \max\{h_0(x), -\tfrac{1}{3}\}.$$

The $h \in \mathscr{B}(\mathscr{D})$ and $|h_1(x)| \leq \frac{1}{3}$ on $X - (A_1 \cup B_1)$, while also $|f(x)| \leq \frac{1}{3}$ on $X - (A_1 \cup B_1)$. This, together with the fact that $h_1(x) = g(x)$ on $A_1 \cup B_1$, yields the relation $\|f - h_1\| \leq \frac{2}{3}$.

Reapplying the process to the function $f - h_1$ and the interval $[-\frac{2}{3}, \frac{2}{3}]$, we can find a function $h_2 \in \mathscr{B}(\mathscr{D})$ such that $\|f - h_1 - h_2\| \leq (\frac{2}{3})^2$; in general, functions $h_1, \ldots, h_n \in \mathscr{B}(\mathscr{D})$ exist with $\|f - (h_1 + \cdots + h_n)\| \leq (\frac{2}{3})^n$, from which it follows that $f \in \mathscr{B}(\mathscr{D})$. ■

We can now obtain the classical Weierstrass theorem as an easy corollary to the above result.

***44.6 Theorem.*** (Weierstrass). *Every real-valued continuous function $f$ on $[a, b]$ can be approximated uniformly by polynomials.*

*Proof.* The statement is that $C^*[a, b]$ is uniform closure of the algebra $\mathscr{A}$ of all polynomials on $[a, b]$. But $\mathscr{A}$ is the algebra generated by the set $\mathscr{D}$ consisting of the functions $x$ (the identity) and 1 (the function identically one), and $\mathscr{D}$ satisfies the conditions of the Stone–Weierstrass theorem, so the uniform closure of $\mathscr{A}(\mathscr{D}) = \mathscr{A}$ is indeed all of $C^*[a, b]$. ■

By elevating the collection $\mathscr{D}$ in 44.5 to a subalgebra, we obtain the following pleasing statement of the Stone–Weierstrass theorem (generalizations of which are considered in Exercises 44A, B, C and D).

***44.7 Theorem.*** *Let $X$ be a compact Hausdorff space. A subalgebra $\mathscr{A}$ of $C(X)$ is all of $C(X)$ iff*

a) $\mathscr{A}$ *is closed (in the uniform topology),*

b) $\mathscr{A}$ *contains the constant functions,*

c) $\mathscr{A}$ *separates points in* $X$.

## Problems

### 44A. *Stone–Weierstrass theorems for noncompact X:* I

Let $X$ be an arbitrary Tychonoff space. A subset $\mathscr{A}$ of $C(X)$ is said to *separate zero sets* in $X$ iff whenever $Z_1$ and $Z_2$ are disjoint zero sets in $X$, there is some $f \in \mathscr{A}$ such that

$$\overline{f(Z_1)} \cap \overline{f(Z_2)} = \varnothing.$$

1. A subalgebra $\mathscr{A}$ of $C^*(X)$ is all of $C^*(X)$ iff

a) $\mathscr{A}$ is closed (in the uniform topology),
b) $\mathscr{A}$ contains the constant functions,
c) $\mathscr{A}$ separates zero sets in $X$. [Consider $\beta X$.]

2. The condition (c) in 1 cannot be weakened to the requirement that $\mathscr{A}$ separate points in $X$. [Consider the subalgebra of $C^*(\mathbf{N})$ consisting of all functions $f$ such that $\lim_{n\to\infty} f(n)$ exists.]

The next two problems provide theorems of the Stone–Weierstrass type for the algebra $C(X)$ if $X$ is not compact.

### 44B. *Stone–Weierstrass theorems for noncompact X:* II

If $X$ is an arbitrary Tychonoff space, the following development leads to a Stone–Weierstrass theorem for $C(X)$ with the compact–open topology.

1. If $\mathscr{A}$ is a subalgebra of $C(X)$ closed in the compact–open topology and $f \in \mathscr{A}$, then $|f| \in \mathscr{A}$. [If $(K, U)$ is a compact–open nhood of $|f|$, the methods of 44.3 can be used to produce a polynomial function $P_u(f)$ of $f$ which lies in $(K, U)$.]

2. A subalgebra of $C(X)$ closed in the compact–open topology is a sublattice.

3. A subalgebra $\mathscr{A}$ of $C(X)$ is all of $C(X)$ iff

a) $\mathscr{A}$ is closed in the compact–open topology,
b) $\mathscr{A}$ contains the constant functions,
c) $\mathscr{A}$ separates points.

Note that if $X$ is compact, the result in part 3 reduces to Theorem 44.7.

### 44C. *Stone–Weierstrass theorems for noncompact X:* III

From one point of view, the compact–open topology on $C(X)$ is unsatisfactory; unless $X$ is compact, it cannot be easily derived from the algebraic structure on $C(X)$ and thus cannot be used in any attempt to represent certain algebras as algebras of continuous functions.

To remedy this, we will consider the uniform topology on $C(X)$. Recall that a base of nhoods at $f \in C(X)$ in this topology is obtained by considering the sets

$$U(f, \epsilon) = \{g \in C(X) \mid |f(x) - g(x)| < \epsilon \text{ for all } x \in X\} \qquad \text{for} \quad \epsilon > 0.$$

We require some terminology. A subalgebra $\mathscr{A}$ of $C(X)$ is said to be *inverse closed* iff whenever $f \in \mathscr{A}$ and $Z(f) = \varnothing$, then $1/f \in \mathscr{A}$. A *star subalgebra* of $C(X)$ is any subalgebra $\mathscr{A}$ which

a) is closed in the uniform topology,

b) contains the constant functions,

c) is inverse closed.

(Note the lack so far of separation properties.) Part 1 below justifies the introduction of star algebras.

1. If $X$ is not pseudocompact, there are proper subalgebras of $C(X)$ which are uniformly closed, contain the constants and separate points.

2. A *rational function* is an element of $C(\mathbf{R})$ of the form $P/Q$, where $P$ and $Q$ are polynomials and $Z(Q) = \varnothing$. $\mathscr{R}$ will denote the algebra of bounded rational functions on $\mathbf{R}$. If $f \in C(\mathbf{R})$ and $\lim_{x\to\infty} f(x) = \lim_{x\to-\infty} f(x)$ is finite, then $f$ is in the uniform closure of $\mathscr{R}$. [Apply the Stone–Weierstrass theorem to the circle.]

3. If $\mathscr{A}$ is a star subalgebra of $C(X)$ and $f \in \mathscr{A}$, then $|f| \in \mathscr{A}$. [It is enough to show $1/(1 + |f|) \in \mathscr{A}$. But $1/(1 + |f|) = g \circ f$ where $g(t) = 1/(1 + |t|)$. Apply part 2.]

4. If $\mathscr{A}$ is a star subalgebra of $C(X)$ and $f \in \mathscr{A}$, $g \in C^*(\mathbf{R})$, then $g \circ f \in \mathscr{A}$. [It is enough to show $g \circ f/(1 + f^2) \in \mathscr{A}$. But $g \circ f/(1 + f^2) = h \circ f$, where $h(t) = g(t)/(1 + t^2)$. Apply part 2.]

5. A star subalgebra $\mathscr{A}$ of $C(X)$ is all of $C(X)$ iff $\mathscr{A} \supset C^*(X)$.

6. A star subalgebra $\mathscr{A}$ of $C(X)$ is all of $C(X)$ iff $\mathscr{A}$ separates zero sets of $X$. [Use 44A.1 to show $\mathscr{A} \cap C^*(X) = C^*(X)$ and apply 5.]

7. A star subalgebra of $C(X)$ which separates points from closed sets need not be all of $C(X)$. [Find such a subalgebra on the disjoint union of two copies of the space $\Omega_0$ of all countable ordinals.]

The next problem gives conditions under which a star subalgebra which separates points and closed sets must be all of $C(X)$.

## 44D. *Stone–Weierstrass theorems for noncompact X:* IV

Again, we consider the uniform topology on $C(X)$.

A subspace $S$ of $X$ is said to be *Z-embedded* in $X$ iff whenever $Z$ is a zero set in $S$, there is a zero set $Z_0$ in $X$ such that $Z_0 \cap S = Z$.

Let $\mathscr{A}$ be a star subalgebra (44C) of $C(X)$ which separates points from closed sets and let $\mathscr{A}^*$ denote the set of bounded functions in $\mathscr{A}$. Define an equivalence relation in $\beta X$ by $p \sim q$ iff $f^\beta(p) = f^\beta(q)$ for each $f \in \mathscr{A}^*$ (where $f^\beta\colon \beta X \to \mathbf{R}$ is the Stone extension of $f\colon X \to \mathbf{R}$). Let $T$ be the quotient of $\beta X$ thus obtained, $\mathscr{V}\colon \beta X \to T$ the identification map. For each $f \in \mathscr{A}^*$, define $\tilde{f}\colon T \to \mathbf{R}$ by $\tilde{f}([p]) = f^\beta(p)$, and let $\tilde{\mathscr{A}} = \{\tilde{f} \mid f \in \mathscr{A}^*\}$.

1. $T$ is a compactification of $X$ (i.e., for $p \in X$, $[p] = \{p\}$) and $\mathscr{V} \mid X$ is a homeomorphism.

2. $\tilde{A} = C(T)$. [By the Stone–Weierstrass theorem, it is enough to show $\tilde{A}$ separates points in $T$.]

3. A Lindelöf subspace $S$ of $Y$ is $Z$-embedded in $Y$. [Let $f \in C(S)$. For $p \in S - Z(f)$,

choose $f_p \in C(Y)$ so that $0 \le f_p \le 1$, $f_p(p) = 1$ and $f_p[Z(f)] = 0$. Let $V_p = \{x \in S \mid f_p(x) > \frac{1}{2}\}$. Then $\{V_p \mid p \in S - Z(f)\}$ is an open cover of the Lindelöf space $S - Z(f)$, so a countable subcover $\{V_{p_1}, V_{p_2}, \ldots\}$ exists. Let $g = \sum_{n=1}^{\infty} (f_{p_n}/2^n)$. Then $Z(g) \cap S = Z(f)$.]

4. Let $X$ be Lindelöf. If $\mathscr{A}$ is a star subalgebra of $C(X)$ which separates points and closed sets, then $\mathscr{A} = C(X)$. [Use 2 and 3 to show $\mathscr{A}^*$ separates zero sets in $X$.]

5. Suppose $|\beta X - X| \le 1$. If $\mathscr{A}$ is a star subalgebra of $C(X)$ which separates points and closed sets, then $\mathscr{A} = C(X)$.

## 44E. *Applications of the Weierstrass theorem*

1. If $f\colon \mathbf{I} \to \mathbf{R}$ is continuous and $\int_0^1 x^n f(x)\,dx = 0$ for each $n = 0, 1, 2, \ldots$ then $f(x) = 0$ on $\mathbf{I}$. [You have finished if you show $\int_0^1 f^2(x)\,dx = 0$.]

2. Show directly that $C(\mathbf{I})$ is separable (by exhibiting a countable dense set rather thar by appealing to 42D).

3. Show that the functions of the form $f(x) = \sum_{k=0}^{\infty} c_k e^{kx}$ are dense in $C(\mathbf{I})$.

4. Show that the functions of the form $\sum_{k=0}^{\infty} (a_k \cos kx + b_k \sin kx)$ are dense in $C([0, 2\pi])$.

# Historical Notes

## Section 1

The basis for our intuitive set theory is the Zermelo–Fraenkel set theory developed by Zermelo (*Untersuchungen über die Grundlagen der Mengenlehre* I) and strengthened by Fraenkel (*Zu den Grundlagen der Cantor–Zermeloschen Mengenlehre*). Their work rests on the researches of Cantor in the 1870's which first put mathematics firmly on a set-theoretic base. Zermelo's work, in particular, was a direct response to the Russell paradox. For an historical account of the Zermelo–Fraenkel and other axiom schemes for set theory, see Suppes (*Axiomatic Set Theory*). A list of other standard references on set theory would include Fraenkel (*Abstract Set Theory*), Hausdorff (*Set Theory*), Halmos (*Naïve Set Theory*) and Sierpinski (*Cardinal and Ordinal Numbers*). Our (postulational) approach to the ordinals in 1.19 follows that of Kelley (*General Topology*, p. 29).

Gödel (*The Consistency of the Axiom of Choice and of the Generalized Continuum Hypothesis with the Axioms of Set Theory*) proved in 1940 that addition of either the axiom of choice or the continuum hypothesis to existing set theoretic axioms would not produce a contradiction. Cohen (*Independence of the Axiom of Choice; The Independence of the Continuum Hypothesis* I, II) completed the proof of independence for each by showing neither could be deduced from the existing axioms (by showing the negation of each could consistently be added to the Zermelo–Fraenkel axiom scheme). See P. J. Cohen (*Set Theory and the Continuum Hypothesis*) for a discussion of these results and his intuition about the continuum hypothesis. Another expository reference is Cohen (*Independence Results in Set Theory*).

For additional material on lattice theory, see Birkhoff (*Lattice Theory*).

## Section 2

The study of metric spaces was initiated by Frechet in his doctoral thesis (*Sur Quelques Points du Calcul Fonctionnel*) and vigorously pursued by a host of Polish mathematicians in the 1920's. A general survey of the results obtained is contained in Sierpinski (*General Topology*) or Kuratowski (*Topology*). For placement of Frechet's work in the development of topology, see the notes to Section 3. For other comments on metric spaces, see the notes to Sections 22, 23 and 24.

The theory of metric spaces (and their topologies) is treated in Copson (*Metric Spaces*).

### Section 3

Topology owes its beginnings to a line of development which began with the first attempt to classify spaces by Riemann (*Über die Hypothesen welche der Geometrie Grunde liegen*), continued through the already mentioned work of Frechet on metric spaces in 1906, the work of Riesz (*Stetigkeit und Abstrakte Mengenlehre*) in 1909 which used a primitive version of the notion of condensation point to describe abstract spaces, the work of Weyl (*Die Idee der Riemannschen Fläche*) in 1913 who proposed studying abstract spaces in terms of neighborhood systems, and culminated in 1914 with the epic paper of Hausdorff (*Grundzüge der Mengenlehre*) who found the right axiom system for Weyl's neighborhoods, made them a suitable abstraction and thus founded modern topology. An excellent detailed account of the forces prevalent in mathematics in the 1800's which gave rise to set theory and point set topology can be found in Manheim (*The Genesis of Point Set Topology*). See also the notes to Section 42.

Weyl's paper mentioned above occupies a place in the development of the structure theory for Riemann surfaces. For modern accounts, see Springer (*Introduction to Riemann Surfaces*) or Ahlfors and Sario (*Riemann Surfaces*).

Hausdorff's axiom scheme included the $T_2$ separation axiom, which we treat in Section 13. The axiom scheme given here is due essentially to Alexandroff (*Zur Begründung der n-dimensionalen mengentheoretischen Topologie*). See also Alexandroff and Hopf (*Topologie* I). The closure operation was axiomatized by Kuratowski (*Sur l'Opération $\bar{A}$ de l'Analysis Situs*). The frontier operator also characterizes the topology. See Albuquerque (*La Notion de "Frontière" en Topologie*).

Properties of the simple extension (3A.5) of a topology are treated in Levine (*Simple Extensions of Topologies*) and Borges (*On Extensions of Topologies*). Exercise 3C is taken from Kelley (*General Topology*, p. 57). The lattice of topologies (3G) was first systematically studied by Birkhoff (*On the Combination of Topologies*). For recent developments, see Steiner (*The Lattice of Topologies; Structure and Complementation*) and van Rooij (*The Lattice of Topologies is Complemented*). The theory of Borel sets (3I) and their derivatives, the analytic sets, is developed extensively in Sierpinski (*General Topology*) and Kuratowski (*Topology*) for separable metric spaces. The extension to general metric spaces is begun in Montgomery (*Non-separable Metric Spaces*) and continued in Stone (*Non-separable Borel Sets*). For descriptions of the theory in general topological spaces, see Frolik (*On the Descriptive Theory of Sets; Baire Sets Which are Borelean Subspaces*). Their name derives from their consideration in Borel (*Leçons sur la Théorie des Fonctions*).

### Section 4

The original description of a topological space by Hausdorff (*Grundzüge der Mengenlehre*) was in terms of nhoods (paralleling our 4.2).

The Sorgenfrey line (4A) was first introduced by Sorgenfrey (*On the Topological Product of Paracompact Spaces*). The Moore plane (4B) is a classical example (see, for example, Alexandroff and Hopf (*Topologie* I, p. 31) sometimes called the *Nemitskii plane.* Exercise 4G is taken from Alexandroff and Hopf (*Topologie* I).

### Section 5

The concept of a subbase for a topology appears in Bourbaki (*General Topology, part 1*) which is translated from *Topologie Générale*, Chapters I and II, Actualités Sci. Ind. 858 (1940).

The scattered line (5C) is used in the form given by Michael (*The Product of a Normal Space and a Metric Space need not be Normal*). The process may be applied to "scatter" any subset of any topological space.

### Section 6

Ordered spaces (6D) were first studied systematically in Eilenberg (*Ordered Topological Spaces*), along lines of the questions posed in Birkhoff (*Lattice Theory*). Nachbin (*Sur les Espaces Topologiques Ordonnés; Topology and Order*) has studied ordered spaces and recent startling developments are contained in Solovay and Tennenbaum (*Iterated Cohen Extensions and Souslin's Problem*).

Subsets of ordered spaces which are ordered (6D.4) have been characterized by M. E. Rudin (*Interval Topology in Subsets of Totally Orderable Spaces*). See also Lynn (*Linearly Orderable Spaces*).

### Section 7

Many of the ideas in this section existed long before the study of topology and topological spaces became an independent discipline.

The question treated in 7B was considered by Kuratowski (*On a Topological Problem Connected with the Cantor–Bernstein Theorem*). The theory of retracts and their use in algebraic topology is covered in Spanier (*Algebraic Topology*). The material of 7L can be found in any real analysis book, for example, Royden (*Real Analysis*). The standard reference on $C(X)$ and $C^*(X)$ is Gillman and Jerison (*Rings of Continuous Functions*); the germinal reference is Hewitt (*Rings of Real-valued Continuous Functions* I). The group of homeomorphisms (7N) is considered in a fundamental paper by Whittaker (*On Isomorphic Groups and Homeomorphic Spaces*).

## Section 8

The Tychonoff topology was introduced by Tychonoff (*Über die Topologische Erweiterung von Räumen*). The box topology was considered by Tietze (*Über Analysis Situs*) and has been studied recently by Knight (*Box Topologies*). An equivalent definition of the product topology is given by Efremovic (*Invariant Definition of Topological Product*).

Weak topologies are covered in Bourbaki (*General Topology, part 1*) under the name *initial topologies.*

Theorems 8.12 and 8.16 are folk theorems of long standing used consistently in Stone–Weierstrass- and Tychonoff-type theorems (see Sections 17 and 44).

It is not universally true that projection maps fail to be closed (8A). See, for example, Noble (*Products with closed projections*). The relationship between weak topologies and the lattice of topologies (8I) is discussed in Levine (*Families of topologies on a fixed set*).

Fox (*On a problem of S. Ulam concerning Cartesian products*) provided an example of nonhomeomorphic spaces $X$ and $Y$ whose squares are homeomorphic (8J.2).

## Section 9

The quotient topology was first studied by Moore (*Concerning Upper semi-continuous Collections of Continua*) and Alexandroff (*Über stetige Abkildung kompakter Räume*). The first cohesive study of open maps is found in Aronszajn (*Über ein Urbildproblem*). A later reference is Whyburn (*Open and Closed Mappings*).

Products of quotient maps have been studied recently by Michael (*Bi-quotient Maps and Cartesian Products of Quotient Maps*). Strong topologies (9H) are covered in Bourbaki (*General Topology, part 1*) under the name *final topologies.* Covering projections (9K) are covered more fully in Spanier (*Algebraic Topology*).

## Section 10

The problem of characterizing topological spaces which can be described by sequential convergence is considered in Ponomarev (*Axioms of Countability and Continuous Mappings*), Arhangel'skii (*Some Types of Factor Mappings and the Relations between Classes of Topological Spaces*), and Franklin (*Spaces in which Sequences Suffice; Spaces in which Sequences Suffice* II). See also Dudley (*On Sequential Convergence*).

Convergence in the product topology (10.6) is considered in Tychonoff (*Über einen Funktionenräum*) and in Sections 42–44.

## Section 11

E. H. Moore (*Definition of Limit in General Integral Analysis*) and later Moore and Smith (*A General Theory of Limits*) developed the general theory of convergence motivated by the considerations in 11.4(c). It was applied to topology by Birkhoff (*Moore–Smith convergence in General Topology*) and further developed by Tukey (*Convergence and Uniformity in Topology*). Subnets were introduced by Moore (*General Analysis* I, *Part* II) and developed by Kelley (*Convergence in Topology*) who there coined the word "net."

## Section 12

The definitions of filter and ultrafilter given here are those of Bourbaki (*General Topology, part* I) and are due to Cartan (*Théorie des Filtres; Filtres et Ultrafiltres*). The idea can be found in much earlier work, e.g., Caratheodory (*Über die Begrenzung einfach zusammenhangender Gebiete*). The relationship between net and filter convergence (12.15–12.17) is developed in Bartle (*Nets and Filters in Topology*). All the fundamentals of general topology are developed using filter convergence in Kowalsky (*Topological Spaces*).

In the study of rings of functions $z$-filters and $z$-ultrafilters (12E) are important. Characterizations of many topological properties in terms of $z$-filter convergence can be found in Gillman and Jerison (*Rings of Continuous Functions*). In particular, they can be used to characterize compactness (17D) and to construct the Stone–Čech compactification of a Tychonoff space (19J). Closed filters will be used later in problems on the Wallman compactification (19K). Open filters can be used to characterize $H$-closed spaces (17K) and to construct $H$-closures (19N).

## Section 13

The $T_2$-axiom is included in the original list of axioms for a topology given by Hausdorff (*Grundzüge der Mengenlehre*). The $T_0$-axiom is usually credited to Kolmogoroff and the $T_1$-axiom to Frechet or Riesz (and spaces satisfying these axioms are sometimes called *Kolmogoroff spaces, Frechet spaces* or *Riesz spaces*, accordingly). Tietze was the first to use the term "separation axiom" (Trennungsaxiom), in 1923. The $T_0$-identification (13.2c) is due to M. H. Stone (*Application of Boolean Algebras to Topology*).

The Zariski topology (13D) crops up in algebraic geometry. See, for example, Hirzebruch (*Topological Methods in Algebraic Geometry*).

I know of no necessary and sufficient condition for the intersection of two Hausdorff topologies to be Hausdorff (13F).

Topological groups (13G) were introduced by Schreier (*Abstrakte kontinwerliche Gruppen*) and are studied intensively in Hewitt and Ross (*Abstract Harmonic Analysis* I).

Exercise 13H improves 13.9(b) and is due to Shimrat (*Decomposition Spaces and Separation Properties*).

## Section 14

Regular spaces were first introduced by Vietoris (*Stetige Mengen*). Completely regular spaces were considered by Urysohn (*Über die Machtigkeit der zusammenhängenden Mengen*) in 1925. Their importance was established with the proof of 14.13 by Tychonoff (*Über die topologische Erweiterung von Raumen*) in 1929. The name "Tychonoff space" was suggested by Tukey (*Convergence and Uniformity in Topology*, p. 84). Necessary and sufficient conditions for a quotient of a completely regular space to be completely regular are developed in Himmelberg (*Quotients of completely regular spaces*).

Properties of the double (14B) of a topological space are investigated in Engelking (*On the Double Circumference of Alexandroff*). Semiregular spaces are considered in M. H. Stone (*Applications of the Theory of Boolean rings to General Topology*) and Hewitt (*A Problem in Set-theoretic Topology*). Urysohn (*Über die Mächtigkeit der Zusammenhängenden Mengen*) introduced the separation axiom presented in 14F. Functionally Hausdorff spaces (14G) are sometimes called *Stone spaces* for reasons that can be ferreted out in Section 44.

## Section 15

The $T_4$-axiom was introduced by Tietze (*Beiträge zur allgemeinen Topologie* I) in 1923. Lemma 15.2 is attributed to F. B. Jones in Dugundji (*Topology*). Urysohn's Lemma (15.6) was proved in Urysohn (*Über die Machtigkeit der zusammenhangenden Mengen*). The Tietze extension theorem (15.8) can be found in Tietze (*Über Funktionen, die auf einer abgeschlossenen Menge stetig sind*). The theorem has been extended in several ways. See, for example, Dugundji (*An Extension of Tietze's Theorem*), Hanner (*Retraction and Extension of Mappings of Metric and Non-metric Spaces*), or Dowker (*On a Theorem of Hanner*). Theorem 15.10 appears in Lefschetz (*Algebraic Topology*).

Complete normality (15B) was added to the list of separation axioms in 1923 by Tietze (*Beiträge zur allgemeinen Topologie* I). Perfect normality was introduced by Alexandroff and Urysohn (*On Compact Topological Spaces*). Urysohn (*Über die Mächtigkeit der zusammenhangenden Mengen*) proved every perfectly normal space is completely normal (15B, C). The study of retracts, absolute retracts and ANR's (15D) began with Borsuk (*Sur les Retracts*). See also Borsuk (*Theory of Retracts*) and Hu (*Homotopy Theory*). The Urysohn extension theorem (15E) is a variant of Urysohn's lemma (see reference above). Extremally disconnected spaces (15G) were first investigated in Hewitt (*A Problem in Set-theoretic Topology*). The Hahn–Banach theorem (15H) forms a part of any course in real analysis. See, for example, Royden (*Real Analysis*).

## Section 16

Second-countable spaces were once (and occasionally still are) called *perfectly separable*. The axioms of first and second countability were defined by Hausdorff (*Grundzüge der Mengenlehre*). Separability was introduced by Frechet (*Sur Quelque Points du Calcul Fonctionnel*) in 1906. The Lindelöf property was proved for Euclidean spaces as early as 1903 by Lindelöf (*Sur Quelques Points de la Théorie des Ensembles*); the formal study of Lindelöf spaces was begun in 1921 by Kuratowski and Sierpinski (*La Théorème de Borel–Lebesgue dans la Théorie des Ensembles Abstraits*). Lindelöf spaces are called *finally compact* by authors in the Soviet Union. See, for example, Alexandroff (*Some Results in the Theory of Topological Spaces*).

Tychonoff (*Über einen Metrisationsatz von P. Urysohn*) proved sufficiency in the countable case in 16.4(c) in 1926. The result for $\mathfrak{c}$ factors is due to Pondiczerny (*Power Problems in Abstract Spaces*), Hewitt (*A Remark on Density Characters*) and Marczewski (*Séparabilité et Multiplication Cartesienne des Espaces Topologiques*). For a recent general result, see Väisälä (*The Separability of Cartesian Products*). Ross and Stone (*Products of Separable Spaces*) have written a short expository paper on separability in product spaces.

The countable chain condition (16C) plays an implicit role in the statement of a problem raised by Souslin (*Problème* 3) which is discussed at greater length in the notes on Section 21.

A variant of 16D.6 is stated by Mrowka (*Functionals on Uniformly Closed Rings of Continuous Functions*). A characterization of Lindelöf spaces in terms of $z$-filters (see 16D.4 and 16D.5) can be found in Gillman and Jerison (*Rings of Continuous Functions*, 8.2).

The exact relationship between hereditarily Lindelöf spaces (16E) and hereditarily separable spaces (those whose every subspace is separable) is unknown. According to Solovay and Tennenbaum (*Iterated Cohen Extensions and Souslin's Problem*) one can consistently assume the existence of a Souslin space (21n) and such a space is necessarily hereditarily Lindelöf but not separable.

Mapping theorems for first countable spaces are considered in Ponomarev (*Axioms of Countability and Continuous Maps*), Arhangel'skii (*Some Types of Factor Mappings and the Relations between Classes of Topological Spaces*) and Stone (*Metrizability of Decomposition Spaces*). See also Arhangel'skii (*Mappings and Spaces*).

## Section 17

Frechet (*Sur Quelques Points du Calcul Fonctionnel*) was the first to use the term "compact". He applied it to metric spaces in which every sequence of points contains a convergent subsequence or, equivalently, in which every infinite set has a limit point. Applied to general topological spaces today, these define the sequentially compact spaces (17G) and the countably compact spaces (17F), respectively. Hausdorff (*Grundzüge der Mengenlehre*) first noticed that the present-day definition, in terms of the Heine–Borel condition, is equivalent in metric spaces to the definitions given above. It was left to Alexandroff and Urysohn (*Zur theorie der topologischen Räume*) to apply this definition to general topological spaces; they called such spaces *bicompact*. Bicompactness won out over countable and sequential compactness when Tychonoff (*Über die topologische Erweiterung von Räumen; Über einen Funktionenraum*) proved it was preserved in the passage to products (17.8). This result fails for sequential compactness (see 17G.6) and for countable compactness (an example is given in Novak (*On the Cartesian Product of Two Compact Spaces*)). See also Mrowka (*Compactness and Product Spaces*) and his references for more on products of countably compact spaces.

That compactness could be described using the finite intersection property (17.4) was first noted by Riesz (*Stetigkeitsbegriff und abstrakte Mengenlehre*).

The Cantor set, (17.9c), was described by Cantor (*Über unendliche, lineare Punktmannigfaltigkeiten*, (e), p. 590).

Maximal compact spaces (17C) were studied by Balachadran (*Minimal–Bicompact Space*) and Ramanathan (*Minimal–Bicompact Spaces*).

The compactness of lattice-complete ordered spaces (17E) was proved by Frink (*Topology in Lattices*). Part (c) of 17E has been known since 1910; see Haar and König (*Über einfach geordnete Mengen*). Compactness in ordered spaces is studied in a recent monograph by Maurice (*Compact Ordered Spaces*).

17G.5 is strengthened to say that the product of $\leq \aleph_1$ sequentially compact spaces is sequentially compact, in Scarborough and Stone (*Products of Nearly Compact Spaces*). See also Kenderov (*A Certain Problem of A. Stone*).

A thorough exposition of realcompact spaces (17H) can be found in Gillman and Jerison (*Rings of Continuous Functions*). They were introduced by Hewitt (*Rings of Real-valued Continuous Functions*, I) as an aid in studying the properties of the ring $C(X)$. Recent references on realcompact spaces, including their mapping properties, are Wenjen (*Realcompact Spaces*), Isiwata (*Mappings and Spaces*), Kenderov (*On Q-spaces*) and Engelking (*Remarks on Realcompact Spaces*). Pseudocompact spaces (17J) were defined and studied by Hewitt in the paper mentioned above. Again, the book of Gillman and Jerison is a good general reference. A product theorem for pseudocompact spaces has been proved by Glicksberg (*Stone–Čech compactifications of products*).

*H*-closed spaces (17K, L) were first introduced by Alexandroff and Urysohn (*Mémoire sur les Espaces Topologiques Compacts*), and later studied by Chevalley and Frink (*Bicompactness of Cartesian Products*), M. H. Stone (*Application of the Theory of Boolean Rings to General Topology*), Katetov (*Über H-abgeschlossen und Bikompakt Räume*), Obreanu (*On a Problem of Alexandroff and Urysohn*), Scarborough and Stone (*Products of Nearly Compact Spaces*) and Liu (*Absolutely Closed Spaces*). The product theorem was proved by Chevalley and Frink.

Minimal Hausdorff spaces (17M) were first considered by Urysohn (*Über die Machtigkeit der Zusammenhangenden Mëngen*), later by Berri (*Minimal Topological Spaces; Categories of Certain Minimal Topological Spaces*), Banaschewski (*Über Hausdorffsch-minimale Erweiterung von Räumen*), Ikenaga (*Product of Minimal Topological Spaces*), Kawashima (*On the Topological Product of Minimal Hausdorff Spaces*) and Scarborough and Stone (*Products of Nearly Compact Spaces*). The product theorem mentioned in 17M is due independently to Ikenaga, Kawashima and Scarborough and Stone; the embedding theorem is Banaschewski's.

Various compactness properties are considered in a monograph by Van der Slot (*Some Properties Related to Compactness*).

Kelley (*The Tychonoff Product Theorem Implies the Axiom of Choice*) contributed 17O.2. Product theorems without the axiom of choice are discussed in Comfort (*A Theorem of the Stone–Čech Type, and a Theorem of Tychonoff Type, without the Axiom of Choice; and their Realcompact Analogues*).

The results in 17Q on projective compact spaces are due to Gleason (*Projective Topological Spaces*).

The result in 17R can be strengthened. Reichbach (*The Power of Topological Types of some Classes of 0-dimensional Spaces*) proves that there are $\mathfrak{c}$ non-homeomorphic compact subsets of **R**. The exercise is taken from Isbell (*Uniform Spaces*).

For a discussion of applications of compactness in analysis, see Hewitt (*The Role of Compactness in Analysis*).

## Section 18

Tietze (*Beiträge zur allgemeinen Topologie* I) and Alexandroff (*Über die Metrisation der im Kleinen kompakten topologische Räume*) defined local compactness independently. The concept is indispensable now in the theory of integration and the study of topological groups. See, for example, Hewitt and Ross (*Abstract Harmonic Analysis*, I).

The treatment of manifolds in this book is far from rich. A serious study of manifolds should include a reading of Bishop and Crittenden (*Geometry of Manifolds*), Wilder (*Topology of Manifolds*), Fort (*Topology of 3-manifolds*) and M. Curtis' rumored book on manifolds, when it appears. There are several books available on differentiable manifolds (18.3c), among them Auslander and McKenzie (*Introduction to Differentiable Manifolds*), Milnor (*Topology from the Differentiable Viewpoint*), Poenaru (*On the Geometry of Differentiable Manifolds*) and Hu (*Differentiable Manifolds*).

Brouwer's theorem on invariance of domain (mentioned in 18B) can be found in Spanier (*Algebraic Topology*).

The example in 18G is due to Hewitt (*On two Problems of Urysohn*); in the same paper, he modified this example to produce a regular, $T_1$-space on which every continuous real-valued function is constant.

## Section 19

Caratheodory (*Über die Begrenzung einfach zusammenhangender Gebiete*) first formally considered the problem of extending a space in 1913. Work on compactification began with Tietze (*Beitrage zur allgemeinen Topologie*, II), Alexandroff (*Über die Metrisation der im Kleinen kompakten topologischen Räume*) and Alexandroff and Urysohn (*Zur theorie der topologischen Räume*), who introduced the one-point compactification. It continued with Tychonoff (*Über die topologische Erweiterung von Räumen*), who proved that every Tychonoff space can be embedded in a compact Hausdorff space. Čech (*On Bicompact Spaces*) and M. H. Stone (*Applications of the Theory of Boolean Rings to General Topology*) gave their names to the compactification constructed by Tychonoff by proving its maximality in the sense of 19.5, 19.10 and 19.12. Construction of $\beta X - X$ relies on a form of the axiom of choice and any knowledge of its structure, or of the structure of $\beta X - X$, seems to involve the continuum hypothesis. See, for example, Gillman (*The Space $\beta N$ and the Continuum Hypothesis*).

Compactification of ordered spaces (19D) has been studied recently by Kaufman (*Ordered Sets and Compact Spaces*).

The result mentioned at the end of 19I on the Stone–Čech compactification of products is due to Glicksberg (*Stone–Čech Compactifications of Products*).

The filter description of $\beta X$ and compactifications in general can be traced from M. H. Stone (*Applications of the Theory of Boolean Rings to General Topology*), through Wallman (*Lattices and Topological Spaces*) and Alexandroff (*Bikompakte Erweiterungen topologische Räume*). A current account can be found in Gillman and Jerison (*Rings of Continuous Functions*).

The Wallman compactification (19K) was constructed in the above named paper by Wallman. Frink (*Compactifications and Semi-normal Spaces*) introduced Wallman bases (calling them normal bases) and raised the question (19L) of whether every compactification of a normal space $X$ is obtainable from some Wallman base for $X$. Answers are known in certain special cases: $\beta X$ is so obtainable (19K), the closed interval **I** is so obtainable from the open unit interval, and the property of being so obtainable is finitely productive according to Hager (*Some Remarks on the Tensor Product of Function Rings*), so that, for example, the closed unit disk is so obtainable from the open unit disk. Other work on Wallman compactifications has been done by Fan and Gottesman (*On Compactifications of Freudenthal and Wallman*), Banaschewski (*On Wallman's Method of Compactification; Normal Systems of Sets*), Njastad (*On Wallman-type Compactifications*), Steiner and Steiner (*Precompact Uniformities and Wallman Compactifications; Wallman and Z-compactifications*), Brooks (*On Wallman Compactifications*), E. F. Steiner (*Wallman Spaces and Compactifications*) and Alo and Shapiro (*A Note on Compactifications and Semi-normal spaces*).

The results in 19M are in the paper of Frink mentioned above.

The concept of $H$-closure appears as early as 1924, in Tietze (*Beiträge zur allgemeinen Topologie*, II) and 1929, in Alexandroff and Urysohn (*Mémoire sur les Espaces Topologiques Compacts*). $H$-closures were explicitly constructed by Katetov (*Über H-abgeschlossen und bikompakt Räume*) and Obreanu (*On a problem of Alexandroff and Urysohn*).

The Hewitt realcompactification (19O) was introduced by Hewitt (*Rings of Real-valued Continuous Functions* I). It is sometimes called the *Nachbin completion*. There are open questions concerning realcompactifications of product spaces; see Comfort (*On the Hewitt Realcompactification of a Product Space*).

### Section 20

Paracompactness was defined in 1944 by Dieudonné (*Une Généralisation des Espaces Compacts*) and elevated to its present high stature by A. H. Stone (*Paracompactness and Product Spaces*), who proved that every metric space is paracompact, and Bing, Nagata and Smirnov, who used Stone's result to obtain a general metrization theorem (see notes, Section 23). Sorgenfrey (*On the Topological Product of Paracompact Spaces*) showed in 1947 that products of paracompact spaces need not be paracompact. In a series of three papers, Michael (*A Note on Paracompact Spaces; Another Note on Paracompact Spaces; Yet Another Note on Paracompact Spaces*) intensively investigated the structure of paracompact spaces. Theorem 20.7 and most of Theorem 20.12, in particular, can be found in the first of these. Cedar (*Some Generalizations of Metric Spaces*) used some of the properties considered by Michael to define and study classes of spaces intermediate between the paracompact and the metrizable spaces.

For additional comments, see notes, Section 21.

The notions of barycentric and star refinement (20B) have been around since Tukey (*Convergence and Uniformity in Topology*). The results in 20C are due to Michael (*A Note on Paracompact Spaces*). Metacompactness (20D) is defined and studied in Arens and Dugundji (*Remark on the Concept of Compactness*). The example in 20F was provided by Michael (*The Product of a Normal Space and a Metric Space need not be Normal*). The example in 20H was produced by M. E. Rudin (*A Separable Normal, Non-paracompact Space*).

### Section 21

The example (20F) cited early in this section is produced in Michael (*The Product of a Normal Space and a Metric Space need not be Normal*). The normal spaces whose product with every metric space is normal have been characterized by Morita (*On the Product of a Normal Space with a Metric Space; Products of Normal Spaces with Metric Spaces; Products of Normal Spaces with Metric Spaces* II). See also Ishii (*On Closed Mappings and M-spaces* I, II).

Theorem 21.1 is proved in Tamano (*On Paracompactness*); he relies on earlier work of Corson (*The Determination of Paracompactness by Uniformities*). The conjecture of Kelley's mentioned after 21.1 can be found on p. 208 of his book (*General Topology*). The counterexample was provided by Corson (*Normality in Subsets of Product Spaces*), who also (*The Determination of Paracompactness by Uniformities*) provided the first correct result.

Several theorems relating properties of $X$ to properties of $X \times \beta X$ have been proved by Tamano (*On Compactifications*). A general survey of theorems on normality and paracompactness of products $X \times Y$ can be found in Tamano (*Normality and Product Spaces*).

The results 21.3, 21.4 and 21B are all due to Dowker (*On Countably Paracompact Spaces*). The result connecting Dowker's conjecture and the Souslin hypothesis can be found in M. E. Rudin (*Countable Paracompactness and Souslin's Problem*). Souslin's hypothesis resulted from a long standing problem posed by Souslin (*Problème* 3). A discussion of the Souslin hypothesis and its independence of the usual axioms of set theory can be found in Solovay and Tennenbaum (*Iterated Cohen Extensions and Souslin's Problem*). See also Jech (*Non-provability of Souslin's Hypothesis*) and Tennenbaum (*Souslin's Problem*). It should be repeated that although Dowker's conjecture cannot be proved within the existing set theory through the choice axiom it is still possible that a counterexample can be constructed with these axioms. According to Morita and Tamano, such a counterexample would result if there were a paracompact space $X$ and a normal space $Y$ whose product $X \times Y$ was normal but not paracompact; see Tamano (*Normality and Product Spaces*).

For recent results on independence of certain topological questions, see Solovay (*A Model of Set Theory in which Every Set of Reals is Iebesgue Measurable; Real-valued Measurable Cardinals*).

Exercise 21C is due to A. H. Stone (*Paracompactness and Product Spaces*). Other work on normality in infinite products has been done by Corson (*Normality in Subsets of Product Spaces; Examples Relating to Normality in Topological Spaces*).

## Section 22

The result in 22.4 attributed to R. D. Anderson can be found in Anderson (*Hilbert Space is Homeomorphic to the Countable Infinite Product of Lines*). The facts cited about quotients and continuous images of metrizable spaces are discussed in the notes on Section 23.

The result in 22B can be found in Hocking and Young (*Topology*). Linear topological spaces (22C) are coherently studied in Hu (*Introduction to General Topology*), Wilansky (*Functional Analysis*) and extensively in Kelley, Namioka *et al.* (*Linear Topological Spaces*). The result in 22D.4 is due to Dugundji (*An Extension of Tientze's Theorem*). The extension theorem (part 4) given in 22E is due originally to Hausdorff (*Erweiterung einer stetigen Abbildung*) in 1938. See also Kuratowski (*Remarques sur les Transformations Continues des Espaces*

*Métriques*). The proof given here, based on Dugundji's result (22D.4), is due to Arens (*Extension of Functions on Fully Normal Spaces*) in 1952. Metric extensions have been considered elsewhere by Hausdorff (*Erweiterung einer Homöomorphie*) in 1930, by Bing (*Extending a Metric*) in 1947, who rediscovered Hausdorff's result 22E.4, and by Lavrentieff (see notes, Section 24). Shapiro (*Extensions of Pseudometrics*) has done recent work on extending pseudometrics. See also his references and Willard (*Absolute Borel Sets in their Stone–Čech Compactifications*) and Gantner (*Extensions of Uniformly Continuous Pseudometrics*).

The result given in 22F can be found in Levine (*A Characterization of Compact Metric Spaces*).

## Section 23

Urysohn's metrization theorem (23.1) was proved in 1925 by Urysohn (*Zum Metrisation problem*). Theorem 23.4 is essentially the uniform metrization theorem; see the notes to Section 38. The nhood metrization theorem (23.5) is a slight alteration of a result found in Nagata (*A Contribution to the Theory of Metrization*). A recent interesting metrization theorem for compact spaces has been provided by Mardešić (*Images of Ordered Compacta are Locally Peripherally Metric*). For some other metrization theorems, see Alexandroff (*Some Results in the Theory of Topological Spaces, Obtained within the Last Twenty-five Years; On Some Basic Directions in General Topology*).

A Moore space (23.6) is usually defined in the literature to be a space satisfying the first three parts of axiom 1 in Moore (*Foundations of Point Set Theory*). Our definition is equivalent and often used. An excellent account of how the normal Moore space conjecture arises can be found in an article by Jones (*Remarks on the Normal Moore Space Metrization Problem*). It presently occupies the time of a great many good mathematicians; see, for example, Bing (*A Translation of the Normal Moore Space Conjecture*). It is a classical result of Jones that, with the continuum hypothesis (actually, with $2^{\aleph_0} < 2^{\aleph_1}$) every *separable* normal Moore space is metrizable. It is a recent result of Heath (*Screenability, Pointwise Paracompactness, and Metrization of Moore Spaces*) and Silver (unpublished as yet) that it is consistent with the axioms of set theory through the axiom of choice (but not, of course, the continuum hypothesis) to assume the existence of a non-metrizable separable normal Moore space.

The metrization theorem given in 23.7 is due to Alexandroff and Urysohn (*Une Condition Nécessaire et Suffisante pour qu'une Classe* ($\mathscr{L}$) *Soit une Classe* ($\mathscr{D}$)).

The general metrization theorem (23.9) was proved independently by Nagata (*On a Necessary and Sufficient Condition of Metrizability*), Smirnov (*A Necessary and Sufficient Condition for Metrizability of a Topological Space*) and, in somewhat different form, Bing (*Metrization of Topological Spaces*).

The theorem on metrizability of the one-point compactification of a locally compact space (23C) was essentially proved in Alexandroff and Urysohn (*Mémoire sur les Espaces Topologiques Compacts*). Metrizability of $\beta X$ (23D) is discussed in Gillman and Jerison (*Rings of Continuous Functions*).

The result 23G.3 is due independently to Nagata (*On a Necessary and Sufficient Condition of Metrizability*) and Smirnov (*On Metrization of Topological Spaces*). The result 23G.4 was first mentioned in Alexandroff and Urysohn (*Mémoire sur les Espaces Topologiques Compacts*). It can be strengthened. An example of Bing (*Metrization of Topological Spaces*, Example 3) shows that a nonmetrizable space can be the union of two *open* metrizable subsets. Necessary and sufficient conditions for a finite union of open metrizable spaces to be metrizable are given by A. H. Stone (*Metrizability of Unions of Spaces*). See also Corson and Michael (*Metrizability of Certain Countable Unions*). The addition theorem (23G.5) was actually proved in the following form: if $X$ is compact and the union of *countably* many compact metrizable subsets, then $X$ is metrizable. See Smirnov (*The Metrizability of Bicompacta Decomposable into the Sum of Sets with a Countable Base*). A general discussion of this and other addition theorems can be found in Arhangel'skii (*Mappings and Spaces*).

The metrization theorem in 23J is due to Mrs. A. H. Frink (*Distance Functions and the Metrization Problem*).

The result 23K.3 is due to A. H. Stone (*Metrizability of Decomposition Spaces*) and Morita and Hanai (*Closed Mappings and Metric Spaces*). The results are based on earlier work of Vainstein (*On Closed Mappings of Metric Spaces*). For open maps, or general quotient maps, the situation is not so nice. See the references above. A general review of what is known about quotients of metric spaces, including a great deal of Russian work not cited here, can be found in Arhangel'skii (*Mappings and Spaces*). The result in 23L can be found in Willard (*Metric Spaces all of whose Decompositions are Metric*).

For a result related to the metrization problem, see Sion and Zelmer (*On Quasi-metrizability*).

### Section 24

Complete metric spaces were introduced along with the definition of metric spaces by Frechet (*Sur Quelques Points du Calcul Fonctionnel*). Separable completely metrizable spaces are called *Polish spaces* in Bourbaki (*General Topology, Part* 2). The proof that every metric space has a completion (24.4) is based on the familiar method of defining the irrational numbers by means of Cauchy sequences and is due to Hausdorff (*Grundzüge der Mengenlehre*). Theorem 24.9 was proved by Lavrentieff (*Contribution à la Théorie des Ensembles Homéomorphes*). Without much difficulty it can be proved equivalent to a metric extension theorem (24M) complementing Hausdorff's theorem (22E.4).

The first part of 24.12 is due to Alexandroff (*Sur les Ensembles de la Première Classe et les Espaces Abstraits*). For the second part, see Mazurkiewicz (*Über

*Borelsche Mengen*), Sierpinski (*Sur l'Invariance Topologique des Ensembles $G_\delta$; sur les Ensembles Complêts d'un Espace (D)*) and Lavrentieff (*Contribution à la Théorie des Ensembles Homéomorphes*). Parts (d) and (e) in Theorem 24.13 are the work of Čech (*On Bicompact Spaces*).

The fixed-point theorem given in 24.16 is due to Banach (*Sur les Opérations dans les Ensembles Abstraits et leurs Applications aux Équations Intégrales*); see also Banach (*Théorie des Opérations Linéaires*). Other important fixed-point theorems include the Schauder theorem: every continuous map of a closed convex subset $A$ of a Banach space onto a compact subset of $A$ has a fixed point [see Schauder (*Der Fixpunktsatz in Funktionalräumen*) and Tychonoff (*Ein Fixpunktsatz*)] and the Brouwer theorem, about which more in Section 34. For a general discussion of fixed-point theorems, see Bing (*The Elusive Fixed-point Property*), Cronin (*Fixed Points and Topological Degree in Nonlinear Analysis*), or Van der Walt (*Fixed and Almost Fixed Points*). See also McKnight (*Brown's Method of Extending Fixed Point Theorems*) and Ward (*A Theorem of Fixed-point Type for Non-compact Locally Connected Spaces*). The Brouwer, Schauder and Tychonoff fixed-point theorems can be see in action in Hartman (*Ordinary Differential Equations*).

Totally bounded metric spaces (24B) were introduced by Hausdorff (*Grundzüge der Mengenlehre*, p. 108). The equivalent condition (b) for completeness in 24C is due to Cantor and appears in its present form in Hausdorff (*Grundzüge der Mengenlehre*).

For some authors, *topologically complete* means *completely metrizable*. It seems better to allow the former term to apply to arbitrary Tychonoff spaces, reserving the latter for metrizable spaces.

Banach spaces are treated in most modern books on real analysis; see, for example, Royden (*Real Analysis*, p. 181). An important extension of the notion of a Banach space, the notion of a Banach algebra, is obtained by adding further algebraic structure. Banach algebras are studied in Loomis (*An Introduction to Abstract Harmonic Analysis*) and are the primary objects of study in Naimark (*Normed Rings*).

Problem 24K is proved in Sierpinski (*General Topology*, p. 143) using continued fractions.

Picard's theorem (24L) can be found in Hartman (*Ordinary Differential Equations*) or in any book on differential equations. It was proved by Picard (*Mémoire sur la Théorie des Équations aux Derivées Partielles et la Méthode des Approximations Successives*) and Lindelöf (*Sur l'Application des Méthodes des Approximations Successives à l'Étude des Intégrals Réelles des Équations Différentielles Ordinaires*).

### Section 25

First and second category spaces were defined by Baire (*Sur les Fonctions des Variables Réelles*) and Theorem 25.4(b) was proved in the same paper. Part (a) of 25.4 is due essentially to Moore (*An Extension of the Theorem that no Countable Point Set is Perfect*). The result 25.5 was proved by Banach (*Über die Baire'sche Kategorie gewisses Funktionenmengen*).

The open mapping theorem, the uniform boundedness principle and the closed graph theorem (25D) can be found in any book of functional analysis. See, for example, Wilansky (*Functional Analysis*). Hilbert spaces (25E) are discussed in most books on real analysis, for example, Heider and Simpson (*Theoretical Analysis*), Hewitt and Stromberg (*Real and Abstract Analysis*) or Rudin (*Real and Complex Analysis*). For a more extensive treatment, see Halmos (*A Hilbert Space Problem Book*).

## Section 26

The modern notion of connectedness was proposed by Jordan (*Cours d'Analyse*) in 1893 and Schoenfliesz (*Beiträge zur Theorie des Punktmengen*), and put on firm footing by Riesz (*Die Gensis des Raumbegriffs*) with the use of subspace topologies. Before Jordan, Cantor (*Über unendliche, lineare Punktmannigfaltigkeiten* e), p. 576) used the following notion of connectedness: a set (subspace of Euclidean space) is connected iff for any two points $a$ and $b$ of the set and any $\epsilon > 0$, a finite set of points $a = x_0, x_1, \ldots, x_n = b$ can be found such that $d(x_i, x_{i+1}) < \epsilon$ for $i = 0, \ldots, n-1$. Such a finite sequence is called an *$\epsilon$-net* and sets with arbitrarily fine $\epsilon$-nets between any pair of points are now called *well-chained.* Hausdorff (*Grundzüge der Mengenlehre*) gave the first systematic account of the properties of connected sets. The notion of arcwise connectedness (Section 27) was around long before Cantor's introduction of connectedness. The separation characterization of connectedness (26.5) is due to Mazurkiewicz (*Sur un Ensemble $G_\delta$ Punctiforme, qui n'est pas Homéomorphe avec aucun Ensemble Linéaire*). Components (26.11) were introduced by Hausdorff (*Grundzüge der Mengenlehre*). Theorem 26.15 says essentially that every connected set is connected in the $\epsilon$-net sense of Cantor (as above).

Quasicomponents (26B) were introduced in 1914 by Hausdorff (*Grundzüge der Mengenlehre*). The countable connected Hausdorff space in 26C was produced by Bing (*A Countable Connected Hausdorff Space*).

## Section 27

Arcwise connectedness is much older than connectedness, having been used explicitly as early as the 1880's by Weierstrass. Locally connected spaces were introduced by Hahn (*Über die allgemeinste ebene Punktmenge, die stetiges Bild einer Strecke ist*) in 1914 and developed by Tietze (*Über stetige Kurven, Jordansche Kurvenbögen und geschlossene Jordansche Kurven*), Kuratowski (*Une définition topologique de la ligne de Jordan*) and Hahn (*Über die Komponenten offenen Mengen*) around 1920. Theorem 27.9 can be found in the latter paper. Moore

(*Concerning Connectedness im kleinen and a Related Property*) studied connectedness *im kleinen* in 1922. See also Knaster and Kuratowski (*A Connected and Connected im kleinen Point Set which Contains no Perfect Set*) and Whyburn (*On the Structure of Connected and Connected im kleinen Point Sets*).

Property $S$ was introduced in 1920 by Sierpinski (*Sur une Condition pour qu'un Continu Soit une Courbe Jordanienne*). Spaces with property $S$ are investigated in Whyburn (*Analytic Topology*) and the references cited there.

For more material on local connectedness and related properties, see Whyburn's book (just mentioned) and Hocking and Young (*Topology*).

## Section 28

Although the word was in use earlier (e.g., "the number continuum," Bolzano), the general notion of continua as connected sets with certain properties was introduced by Cantor (*Über unendliche, lineare Punktmannigfaltigkeiten*), where he regarded them as subsets of Euclidean spaces which are both closed and connected (in his $\epsilon$-net sense, see the notes on Section 26). The definition was modified to use the modern concept of connectedness by Jordan (*Cours d'Analyse*) in 1893. Compactness seems not to have become an accepted part of the definition until the 1930's although many of the results in earlier papers (e.g., Janiszewski's, see below) are proved for "bounded plane continua."

Continua irreducible between two points were considered first by Zoretti (*La Notion de Ligne*) in 1909 and developed extensively by Janiszewski (*Sur les Continus Irréductibles entre Deux Points*) in his 1911 thesis. The latter gave a proof of Theorem 28.4 based on the reduction theorem of Brouwer (*Over de Structur der perfecte Punktnerzamelingen*), a statement of which can be found in Kelley (*General Topology*). More early references are given by Sierpinski (*Théorie des Continus Irréductibles entre Deux Points*, I).

Cut points became an important part of investigations into the properties of continua in the 1920's. See, for example, Moore (*Concerning the Cut Points of Continuous Curves and of other Closed and connected Sets*) and Whyburn (*Concerning the Cut Points of Continua*). The separation order (although not the order topology) on $E(a, b)$ seems to be first mentioned in the latter paper. Theorems 28.13 and 28.14 on the characterization of the unit interval and the unit circle were essentially proved in 1920 by Moore (*Concerning Simple Continuous Curves*). A more exhaustive study of cut points and continua can be found in Whyburn (*Analytic Topology*). Recently, a bibliography on analytic topology which includes several references to results on cut points and continua has been compiled by McAllister (*Cyclic Elements in Topology, a History*).

Indecomposable continua (28A) were studied in the 1920's by Mazurkiewicz (*Un Théorème sur les Continus Indécomposables*), Janiszewski and Kuratowski (*Sur les Continus Indécomposables*) and Knaster (*Un Continu dont tout Souscontinu est Indécomposable*). The notion can be found also in Brouwer (*Zur

*Analysis situs*). A problem posed by Knaster and Kuratowski (*Problème* 2) in 1920 was not solved until 1948 by Bing (*A Homogeneous Indecomposable Plane Continuum*) although a continuum constructed in 1922 by Knaster himself (*Un Continu dont tout Sous-continu est Indécomposable*) was discovered in 1951 to be a counterexample. See also Bing and Jones (*Another Homogeneous Plane Continuum*) and Hocking and Young (*Topology*).

The result in 28E.2 is due to Sierpinski (*Un Théorème sur les Ensembles Fermés*).

Before leaving continua, we should mention one famous unsolved problem. The *plane-continuum problem* asks whether every continuum $K$ in the plane which has connected complement has the fixed-point property (that every continuous $f: K \to K$ has a fixed point). Borsuk (*Sur un Continu Acyclique qui se laisse Transformer Topologiquement et lui même sans Points Invariants*) produced a chain of three-cells (homeomorphs of $\mathbf{I} \times \mathbf{I} \times \mathbf{I}$) whose intersection fails to have the fixed-point property. Finding a counterexample for the plane-continuum problem is equivalent to finding an example similar to Borsuk's using two-cells in the plane. A recent reference on the problem is Bell (*On Fixed-point Properties of Plane Continua*).

## Section 29

Totally disconnected spaces were considered as early as 1921 by Knaster and Kuratowski (*Sur les Ensembles Connexes*), in a paper in which example 29.2 appears (on p. 241), and by Sierpinski (*Sur les Ensembles Connexes et Non Connexes*), where they are called "dispersed."

The 0-dimensional spaces occupy an important place in modern dimension theory. The inductive definition of the dimension ind $X$ of $X$ goes as follows: ind $\varnothing = -1$ and ind $X \leq n$ iff each point of $x$ has a base of nhoods $U$ with ind $[\text{Fr}(U)] \leq n - 1$. This definition provides a satisfying theory for separable metric spaces, as detailed in the classic book of Hurewicz and Wallman (*Dimension Theory*), but has serious drawbacks in more general settings. In the general case, use of one of the following dimension functions seems more appropriate: (1) Ind $\varnothing = -1$ and Ind $X \leq n$ iff given disjoint closed sets $A$ and $B$ of $X$, there is an open set $U$ with $A \subset U$, $U \cap B = \varnothing$ and Ind $[\text{Fr}(U)] \leq n - 1$, or (2) dim $X \leq n$ iff any finite open cover of $X$ has a refinement by an open cover of order $\leq n + 1$, where the order of a cover is the largest number of sets from the cover having some point in common. The three dimension functions are called the *weak inductive dimension* (ind $X$), *the strong inductive dimension* (Ind $X$) and the (Lebesque) *covering dimension* (dim $X$). All three are equal for separable metric spaces. Katetov (*On the Dimension of Non-separable Spaces*) and Morita (*Normal Families and Dimension Theory for Metric Spaces*) extended dimension theory to general metric spaces and showed Ind $X$ = dim $X$ for such spaces. Roy (*Failure of Equivalence of Dimension Concepts for Metric Spaces*) gave an example of a metric space $X$ for which ind $X \neq$ dim $X$. A comprehensive account of the status of dimension theory up to 1965 can be found in Nagata (*Modern Dimension Theory*).

Inverse limits (29.9 and 29C) were introduced in topology by Alexandroff (*Untersuchungen über Gestalt und Lage abgeschlossenes Mengen*). They are useful in the theory of compact topological groups. See, for example, Weil (*L'intégration dans les Groupes Topologiques et ses Applications*) or Pontrjagin (*Topological Groups*). In all these sources and other references from earlier periods, they are called *projective limits* (of inverse systems). That Theorem 29.13(b) fails to hold for noncompact spaces was demonstrated by Henkin (*A Problem on Inverse Mapping Systems*).

The results in 29E can be found developed in Hewitt and Ross (*Abstract Harmonic Analysis* I).

### Section 30

Perfect sets were introduced by Cantor (*Über unendliche, lineare Punktmannigfaltigkeiten* (f)). The theorem characterizing the Cantor set (30.4) is proved in Hausdorff (*Grundzüge der Mengenlehre*); see also Hausdorff (*Set Theory*). The theorem that every compact metric space is a continuous image of $C$ (30.7) is due to Alexandroff and Urysohn (*Mémoire sur les Espaces Topologiques Compacts*).

A recent reference on the Cantor set is Nunnally (*There is no Universal-projecting Homeomorphism of the Cantor Set*).

A detailed study of scattered sets (30E) can be found in Semadeni (*Sur les Ensembles Clairsemés*).

### Section 31

Peano spaces are so called because Peano (*Sur une Courbe qui remplit toute une Plane*) in 1890 shattered many ideas about dimension and continuity prevalent at the time by producing a "space-filling curve," that is, a continuous map $f$ carrying $\mathbf{I}$ onto $\mathbf{I} \times \mathbf{I}$. Twenty-five years later Hahn (*Mengentheoretische Characterisierung der stetigen Kurven*) and Mazurkiewicz (*Sur les Lignes de Jordan*) characterized the continuous images of $\mathbf{I}$ by proving Theorem 31.5. There is considerable present-day interest in finding a Hahn–Mazarkiewicz theorem for nonmetric spaces. For progress up until 1966, see Mardešić (*On the Hahn–Mazurkiewicz Problem in Non-metric Spaces*).

A different approach to the proof of the Hahn–Mazurkiewicz theorem is found in Hall and Spencer (*Elementary Topology*).

Excellent accounts of the development of the notion of *continuous curve* can be found in the book by Hurewicz and Wallman (*Dimension Theory*) and the article by Whyburn (*What is a Curve?*).

### Sections 32, 33

The fundamental group was introduced by Poincaré (*Analysis Situs; Cinquième Complément a l'Analysis Situs*) around 1900. Hurewicz (*Beiträge zur Topologie *der Deformation* I–IV) studied the fundamental group and introduced the higher homotopy group in a series of four papers in the 1930's. For more extensive accounts of homotopy theory than we are able to give, see Hilton (*An Introduction to Homotopy Theory*), Hu (*Homotopy Theory*), Massey (*Algebraic Topology: An Introduction*) or Spanier (*Algebraic Topology*). For applications of homotopy theory to other branches of topology, see Steenrod (*The Topology of Fibre Bundles*), Hurewicz and Wallman (*Dimension Theory*) or Buseman (*The Geometry of Geodesics*).

### Section 34

That $\Pi_1(\mathbf{S}^1)$ is infinite cyclic has been known since the fundamental group was introduced (see notes, Section 32). The Brouwer fixed-point theorem was first proved in 1910 by Brouwer (*Beweis des Jordanschen Kurvensatzes*). The proof given here is based on the no-retraction theorem (34.5) due to Borsuk (*Sur les Retracts*). Brouwer's theorem can be proved without the use of algebraic methods. See, for example, Whyburn (*Analytic Topology*).

### Section 35

Uniform continuity was first defined for real-valued functions defined on Euclidean spaces by Heine (*Über trigonometrische Reihen*). Uniform convergence had been defined earlier by Weierstrass (in unpublished lectures); see the notes to Section 42. Uniform continuity was defined for metric spaces by Frechet (*Sur Quelques Points du Calcul Fonctionnel*) and Hausdorff (*Grundzüge der Mengenlehre*) and the fact that continuous functions on compact spaces are uniformly continuous was implicit in the work of many who shared in the early development of topology in the 1920's. It was not until 1937 that Weil (*Sur les Espaces à Structure Uniforme et sur la Topologie Générale*) introduced the general notion of a uniform space. The approach via surroundings (as well as another approach, see the notes to Section 38) was used by Weil and developed by Bourbaki (*General Topology, Part* 1). Another approach, distinct from Weil's and Bourbaki's and much more convenient from a topological point of view, was developed by Tukey (*Convergence and Uniformity in Topology*) and is presented in the next section.

A comprehensive overview of the theory of uniform spaces, including bibliographical notes (in the preface and at the end of each chapter), can be found in Isbell (*Uniform Spaces*).

### Section 36

Uniform covers were used as the basis for defining uniform structures by Tukey (*Convergence and Uniformity in Topology*). His approach is the "best" in the eyes of many, in a sense made precise by the following quote taken from the preface of Isbell's book on uniform spaces: "... in this book, each system is used where it is most convenient, with the result that Tukey's system of uniform coverings is used nine-tenths of the time."

According to 36.16, every open cover of a paracompact space is a uniform cover (in the fine uniformity). This does not characterize paracompact spaces; the spaces it does characterize are described by Cohen (*Sur un Problème de M. Dieudonné*) and studied by Mansfield (*Some Generalizations of Full Normality*). See also Corson (*The Determination of Paracompactness by Uniformities*) and the notes to Section 39.

**Section 37**

All the results in this section are variants of basic results contained in any of the three standard monographs on uniform spaces by Weil (*Sur les Espaces à Structure Uniforme et la Topologie Générale*), Tukey (*Convergence and Uniformity in Topology*) and Bourbaki (*General Topology, part* 1).

Uniform quotients (37E) are discussed at greater length in Isbell (*Uniform Spaces*), together with their generalization, the *strong uniformities* (dual to the weak uniformities discussed in the text). Inverse limits of uniform spaces (37F) are discussed in Bourbaki (*General Topology, part* 1).

**Section 38**

The uniform metrization theorem (38.4) was stated in 1923, not in the language of uniform spaces, by Alexandroff and Urysohn (*Une Condition Nécessaire et Suffisante pour qu'une Classe* ($\mathscr{L}$) *soit une Classe* ($\mathscr{D}$)). Its statement and proof were simplified by Chittenden (*On the Metrization Problem and Related Problems in the Theory of Abstract Sets*) and A. H. Frink (*Distance Functions and the Metrization Problem*); Weil, in his 1937 monograph, gives the theorem in the language of uniform spaces. The approach via uniform covers is, of course, due to Tukey.

The uniformization theorem (38.2) is due to Weil.

Gage structures (38A) were introduced in Weil's original approach to uniform spaces.

The argument used to prove that every locally compact group is normal (38B.2) can be modified to show paracompactness without trouble, using Michael's theorem (20.7).

The results given in 38C are the cornerstone results on metrization of topological groups. 38C.1 is due to Kakutani (*Über die Metrisation der topologischen Gruppen*) and Birkhoff (*A Note on Topological Groups*). Many other results and examples can be found in Hewitt and Ross (*Abstract Harmonic Analysis* I). One fascinating result, due to Hulanicki (*On Locally Compact Topological Groups of Power of Continuum*) and Jones (*On the First Countability Axiom for Locally

*Compact Hausdorff Spaces*), is as follows: a locally compact topological group containing an open set with $\aleph_1$ elements is metrizable.

### Section 39

Complete uniform spaces were defined by both Weil and Tukey in their germinal monographs.

Uniform spaces can be completed in a manner analagous to the process used to construct metric completions in Section 24 by turning to Cauchy filters; see Bourbaki (*General Topology, Part* 1) and Exercise 39F. Our approach is that of Kelley (*General Topology*) and Isbell (*Uniform Spaces*).

Kelley (*General Topology*) raised the question of whether a space $X$ which is completely uniformized (39B) by the family of all open covers (or the family of all nhoods of the diagonal, if you wish) is necessarily paracompact. Corson (*Normality of Subsets in Product Spaces*) provided a counterexample and, by adding another condition (*The Determination of Paracompactness by Uniformities*) achieved a characterization of paracompactness. See also the notes to Section 36. Later Tamano showed Corson's extra condition was by itself necessary and sufficient for paracompactness. See Section 21 and the notes.

The example alluded to in the first part of 39C, of a uniformly open, uniformly continuous map of a complete space onto a noncomplete space, can be found in Köthe (*Die Quotientenräume eines linearen vollkommenen Räumes*, p. 33). The proof of the result in the second part is difficult with the information available; it is developed in Kelley (*General Topology*).

For further discussion of the hyperspace (39D) of a uniform space, see Isbell (*Uniform Spaces*). The filter approach to completeness and completion (39F) is employed in Bourbaki (*General Topology, Part* 1).

One of the deepest theorems available to date in uniform spaces would appear in this section if there were space to develop it. It is due to Shirota (*A Class of Topological Spaces*) and says: a Tychonoff space in which every closed discrete subspace has nonmeasurable cardinal is completely uniformizable iff it is realcompact (17H). So far, the theorem can be used without the cardinality restriction, since every cardinal number known to man is nonmeasurable. For a discussion of measurable cardinals and realcompactness and a development of Shirota's theorem, see Gillman and Jerison (*Rings of Continuous Functions*).

### Section 40

Although it had been suggested as early as 1908 by Riesz (*Stetigkeitsbegriff und abstrakte Mengenlehre*), and the idea was revived in 1941 by Wallace (*Separation Spaces*), the theory of proximity had its real beginning with Efremovic (*The Geometry of Proximity* I) in 1952, and was developed thereafter by several authors (largely in the Soviet Union), notably Smirnov (*On Proximity Spaces; On Proximity*

*Spaces in the Sense of V. A. Efremovic; On Completeness of Proximity Spaces* I, II; *On the Dimension of Proximity Spaces*). Expository accounts appear in books by Thron (*Topological Structures*) and Csaszar (*Fondements de la Topologie Générale*); the latter book is concerned with providing a common axiomatic scheme for topology, uniformity and proximity structures.

Our account of proximity structure here has been heavily influenced by a reading of notes compiled by A. H. Stone for lectures given at the University of Rochester in 1967.

For comments on the relationship between proximity structures and totally bounded uniform structures, see the notes to Section 41.

The reference in the last paragraph of 40F is to Dowker (*Mappings of Proximity Spaces*). For results of metrization of proximity spaces, see Leader (*Metrization of Proximity Spaces*).

### Section 41

The equivalence of proximity structures on $X$, totally bounded uniform structures on $X$ and compactifications of $X$ was developed by Smirnov (*On Proximity Spaces in the Sense of V. A. Efremovic; On Proximity Spaces*). A recent account of the direct relationship between totally bounded uniformities and proximities is found in Gál (*Proximity Relations and Precompact Structures*). See also Alfsen and Fenstad (*On the Equivalence between Proximity Structures and Totally Bounded Uniform Structures*). The direct relationship between proximity structures and compactifications (without the intermediate use of uniform structures) is developed in Leader (*On Clusters in Proximity Spaces; On Completion of Proximity Spaces by Local Clusters*).

Other recent references include Alfsen and Njastad (*Proximity and Generalized Uniformity*) and Smirnov (*Proximity and Construction of Compactifications with Given Properties*).

The Freudenthal compactification (40B) is treated in Isbell (*Uniform Spaces*).

Spaces with a unique uniformity (41F) were studied by Doss (*On Uniform Spaces with a Unique Structure*).

### Section 42

The study of pointwise convergence of (sequences of) functions is as old as the calculus. The study of uniform convergence began hard on the heels of the formalization of the notion of limit by Cauchy (*Cours d'Analyse de l'École Royale Polytechnique*) in 1821. It was motivated by an example of Abel (*Untersuchen über die Reihe* $1 + [m/1]x + [m(m-1)/2]x^2 + [m(m-1)(m-2)/(2 \cdot 3)]x^3 + \cdots$ *u.s.w.*) showing that pointwise convergence of continuous functions to a function $f$ was not enough to ensure continuity of $f$, as Cauchy had assumed in 1821. Seidel (*Über eine Eigenschaft der Reihen welche Discontinunliche Functionen*

*Darstellen*), in an 1848 paper, first showed (without naming it) that uniform convergence was what had been lacking. In the last half of the 19th century, in the hands of Heine, Weierstrass, Riemann and others, uniform convergence came into its own in applications to integration theory and Fourier series.

The study of sets, or spaces, of functions began with the work of Ascoli (*Le Curve Limite di una Varieta Data di Curve*), Arzela (*Funzioni di Linee*) and Hadamard (*Sur Certaines Applications Possibles de la Théorie des Ensembles*). These papers mark the beginning, not only of function space theory, but of general topology itself, for it was the questions which they raised that men like Frechet, Riesz, Weyl and finally Hausdorff (see notes, Section 3) were trying to answer.

Coherent attempts to study topologies on spaces of functions in their own right began in 1935 with Tychonoff (*Über einen Funktionenraum*), who pointed out that his product topology (see notes, Section 8) on $Y^X$ is just the topology of pointwise convergence. The term *function space* is used much earlier in connection with questions of a topological nature about sets of functions; see, for example, Birkhoff and Kellog (*Invariant Points in Function Spaces*), but I can find no earlier work which explicitly refers to a function space as a set of functions with a given topology. The uniformities of pointwise and uniform convergence were first explicitly defined by Tukey (*Convergence and Uniformity in Topology*).

The result of 42D is due essentially to M. and S. Krein (*On an Inner Characteristic of the Set of All Continuous Functions Defined on a Bicompact Hausdorff Space*).

### Section 43

The compact–open topology was first systematically defined and studied by Fox (*On Topologies for Function Spaces*) and Arens (*A Topology for Spaces of Transformations*). Most of the basic results in this section, including the description of the compact–open topology as the topology of compact convergence, can be found in these papers.

Ascoli's theorem (43.15) is a generalization of Arzela's theorem (43E): the latter was proved by Ascoli (*Le Curve Limite di una Varieta Data di Curve*) in 1883 and Arzela (*Sull' Integrabilita delle Equazioni Differenziali Ordinarie*) in 1895. Custom recognizes the priority of Ascoli by assigning his name to the general version of the theorem. The form of Ascoli's theorem we give here is due to Gale (*Compact Sets of Functions and Function Rings*); he improved an earlier general version of Ascoli's theorem given by Myers (*Equicontinuous Sets of Mappings*), which involves spaces of functions from a locally compact or first-countable space $X$ to a metric space $Y$.

Joint continuity (43F) is studied in the papers of Fox and Arens cited above; Arens' paper also contains the theorem (43G.3) on metrizability of $C(X)$. Other results relating on properties of $C(X)$ with the compact–open topology to properties of $X$ can be found in Nachbin (*Topological Vector Spaces of Continuous*

*Functions*), Shirota (*On Locally Convex Vector Spaces of Continuous Functions*), and Warner (*The Topology of Compact Convergence on Continuous Function Spaces*).

Spaces with compactly generated topologies (43.8, 43H) were considered in 1939 by Whitehead (*Simplicial Spaces, Nuclei and m-groups*). They were used in the study of function spaces by Gale in the paper cited above, where the name *k-space* first appears (he attributes it without reference to Hurewicz). Results on products of $k$-spaces (43H.2) can be found in Dowker (*Topology of Metric Complexes*) where two $k$-spaces are produced whose product is not a $k$-space, Cohen (*Spaces with Weak Topology*) where it is shown that a product of a locally compact Hausdorff space and a $k$-space is a $k$-space, and Michael (*A Note on the Product of k-spaces*) where it is shown that Cohen's result is the best possible. The result 43H.3 is due to Cohen (*Spaces with Weak Topology*).

Other references on $k$-spaces and function spaces include Mrowka (*On Function Spaces*), Brown (*Function Spaces and Product Topologies*), Weston (*A Generalization of Ascoli's Theorem*), Bagley and Yang (*On k-spaces and Function Spaces*), Poppe (*Stetige Konvergenz und der Satz von Ascoli und Arzela*) who gives a survey of known Ascoli-type theorems, and Noble (*k-Spaces*) who surveys and extends many of the results on $k$-spaces known up until 1967.

The exponential law (43I) and its application to the higher homotopy groups of Hurewicz was developed by Jackson (*Spaces of Mappings on Topological Products with Applications to Homotopy Theory*).

## Section 44

Theorem 44.6 was proved in 1885 by Weierstrass (*Über die analytische Darstellbarkeit sogenannter willkürlicher Functionen reeller Argumente*). For comments on his method of proof, see Hewitt and Ross (*Abstract Harmonic Analysis* I, p. 281). Another popular method of proof relies on Fejer's theorem on summation of Fourier series; see Sz.-Nagy (*Introduction to Real Functions and Orthogonal Expansions*, p. 430), or Apostol (*Mathematical Analysis*, p. 481). Yet another popular proof uses an interpolation formula due to Bernstein (*Demonstration du Théorème de Weierstrass Fondée sur le Calcul des Probabilités*), and can be found in McShane and Botts (*Real Analysis*). For other proofs, see Fejer (*Über Weierstrassche Approximation, besonders durch Hermitesche Interpolation*), Royden (*Real Analysis*, p. 313) and De Branges (*The Stone–Weierstrass Theorem*).

The Stone–Weierstrass theorem (44.5, 44.7) was first proved by M. H. Stone (*Applications of the Theory of Boolean Rings to General Topology*). See also Stone (*The Generalized Weierstrass Approximation Theorem*).

The approximation theorem in 44A was published by Hewitt (*Certain Generalizations of the Weierstrass Approximation Theorem*). The approximation theorem 44C.6 is due to Mrowka (*Some Approximation Theorems for Rings of Unbounded Functions*). M. Jerison has been credited with 44D.3 by Henriksen and Johnson (in the paper cited below). The result in 44D.4 was proved by Mrowka in the paper above, after Isbell (*Algebras of Uniformly Functions*) had proved it for $X$ locally compact and $\sigma$-compact. It is interesting to note that Hager and Johnson (*A Note on Certain Subalgebras of* $C(X)$) have shown that parts 4 and 5 of 44D cover the ground completely. That is, if $C(X)$ is the only star subalgebra of $C(X)$ which separates points and closed sets, then either $X$ is Lindelöf or $|\beta X - X| \leq 1$.

Stone–Weierstrass methods were introduced as a means of characterizing certain algebras as algebras of continuous functions by Anderson and Blair (*Characterizations of the Algebra of all Real-valued Continuous Functions on a Completely Regular Space; Characterizations of Certain Lattices of Functions*). Later work has been done by Henriksen and Johnson (*On the Structure of a Class of Archimedean Lattice-ordered Algebras*), Anderson (*Approximation in Systems of Real-valued Continuous Functions*) and Jensen (*A note on Complete Separation in the Stone Topology*). For a somewhat different point of view on function algebras (including the complex-valued case) see Browder (*Introduction to Function Algebras*) or the survey article by Royden (*Function algebras*).

44E.1 is stolen from Rudin (*Principles of Mathematical Analysis*).

# Bibliography

## Elementary books

Blackett, D. W., *Elementary Topology: A Combinatorial and Algebraic Approach*, Academic Press, New York (1967) [MR 35 #951].

Burgess, D. C. J., *Analytical Topology*, Van Nostrand, Princeton (1966) [MR 36 #3305].

Bushaw, D., *Elements of General Topology*, John Wiley and Sons, New York (1963) [MR 28 #2515].

Copson, E. T., *Metric Spaces*, Cambridge Tracts in Math. and Math. Physics No. 57, Cambridge University Press, London (1968) [MR 37 #877].

Hall, D. W. and G. L. Spencer, *Elementary Topology*, John Wiley and Sons, New York (1955) [MR 17, p. 649].

Hausdorff, F., *Set Theory*, 3rd Ed. (transl.), Chelsea, New York (1957) [MR 19, p. 111].

Gemignani, M. C., *Elementary Topology*, Addison-Wesley, Reading (1967) [MR 36 #838].

McCarty, G., *Topology*, McGraw-Hill, New York (1967).

Moore, T. O., *Elementary General Topology*, Prentice-Hall, Englewood Cliffs (1964) [MR 29 #4018].

Pervin, W. J., *Foundations of General Topology*, Academic Press, New York (1964) [MR 29 #2759].

Singer, I. M. and J. A. Thorpe, *Lecture Notes on Elementary Topology and Geometry*, Scott, Foresman, Glenview, Illinois (1967) [MR 35 #4834].

## Advanced books

Alexandroff, P. and H. Hopf, *Topologie* I, Springer-Verlag, Berlin (1935) [Z 13, p. 79].

Berge, C., *Topological Spaces* (transl.), Macmillan, New York (1963) [MR 21 #4401].

Bourbaki, N., *General Topology*, Part 1 (transl.), Addison-Wesley, Reading (1966) [MR 34 #5044a].

Bourbaki, N., *General Topology*, Part 2 (transl.), Addison-Wesley, Reading (1966) [MR 34 #5044b].

Čech, E., *Topological Spaces*, Rev. Ed. (transl.), Interscience, New York (1966) [MR 35 #2254].

Choquet, G., *Topology* (transl.), Academic Press, New York (1966) [MR 33 #1823].

Cronin, J., *Fixed Points and Topological Degree in Nonlinear Analysis*, Math. Survey No. 11, Amer. Math. Soc., Providence (1964) [MR 29 #1400].

Császár, A., *Fondements de la Topologie Générale*, Akadémiai Kiadó, Budapest (1960) [MR 22 #4043].

Cullen, H. F., *Introduction to General Topology*, D. C. Heath, Boston (1968) [MR 36 #4507].

Dugundji, J., *Topology*, Allyn and Bacon, Boston (1966) [MR 33 #1824].

Gillman, L. and M. Jerison, *Rings of Continuous Functions*, Van Nostrand, Princeton (1960) [MR 22 #6994].

Greever, J., *Theory and Examples of Point-set Topology*, Brooks/Cole, Belmont, Calif. (1967) [MR 36 #5876].

Hocking, J. G. and G. S. Young, *Topology*, Addison-Wesley, Reading (1961) [MR 23 #A2857].

Hu, S. T., *Elements of General Topology*, Holden-Day, San Francisco (1964) [MR 31 #1643].

Hu, S. T., *Introduction to General Topology*, Holden-Day, San Francisco (1966) [MR 33 #4876].

Hurewicz, W. and H. Wallman, *Dimension Theory*, Princeton Univ. Press, Princeton (1941) [MR 3, p. 312].

Isbell, J. R., *Uniform Spaces*, Amer. Math. Soc. Survey No. 12, Amer. Math. Soc., Providence (1964) [MR 30 #561].

Kelley, J. L., *General Topology*, Van Nostrand, Princeton (1955) [MR 16, p. 1136].

Kowalsky, H. J., *Topological Spaces* (transl.), Academic Press, New York (1965) [MR 22 #12502].

Kuratowski, C., *Topology*, Vol. I (transl.), Academic Press, New York (1966) [MR 36 #839].

Manheim, J. H., *The Genesis of Point Set Topology*, Pergamon Press, Oxford (1964).

Moore, R. L., *Foundations of Point Set Theory*, Amer. Math. Soc. Colloq. Publ. 13 (1932) [Z 5, p. 54].

Nachbin, L., *Topology and Order* (transl.), Van Nostrand, Princeton (1965) [MR 36 #2125].

Nagata, J., *Modern Dimension Theory*, Interscience–John Wiley and Sons, New York (1965) [MR 34 #8380].

Nagata, J., *Modern General Topology*, Noordhoff, Groningen (1968).

Schubert, H., *Topology* (transl.), Allyn and Bacon, Boston (1968) [MR 37 #2160].

Sierpinski, W., *General Topology*, 2nd Ed. (transl.), University of Toronto Press, Toronto (1956).

Thron, W. J., *Topological Structures*, Holt, Rinehart and Winston, New York (1966) [MR 34 #778].

Whyburn, G. T., *Analytic Topology*, 1963 Edition, Amer. Math. Soc. Colloq. Publ. 28 (1963) [MR 32 #425].

Wilder, R. L., *Topology of Manifolds*, Amer. Math. Soc. Colloq. Publ. 32 (1949) [MR 10, p. 614].

### Expository articles


Alexandroff, P., "Some Results in the Theory of Topological Spaces, Obtained within the Last Twenty-Five Years," *Russian Math. Surveys* **15,** No. 2, 23–83 (1960). [MR 22 #9947].

Alexandroff, P., "On Some Basic Directions in General Topology," *Russian Math. Surveys* **19,** No. 5, 1–39 (1964). [MR 30 #2450].

Arhangel'skii, A., "Mappings and Spaces," *Russian Math. Surveys* **21,** No. 4, 115–162 (1966). [MR 37 #3534].

Bing, R. H., *Elementary Point Set Topology*, Herbert Ellsworth Slaught Memorial Paper No. 8, *Amer. Math. Monthly* **67,** No. 7, 51 pp. (1960) [MR 23 #A614].

Bing, R. H., "The Elusive Fixed-point Property," *Amer. Math. Monthly* **76,** 119–132 (1969).

Borsuk, K., *Theory of Retracts*, Monografie Matematyczne 44, Panstwowe Wydawnictwo Naukowe, Warsaw (1967) [MR 35 #7306].

Gottschalk, W. H., *Bibliography for Topological Dynamics*, 3rd Ed., Wesleyan Univ. mimeographed notes (1967) [MR 36 #7126].

Hewitt, E., "The Role of Compactness in Analysis," *Amer. Math. Monthly* **67,** 499–516 (1960) [MR 22 #11367].

Jones, F. B., "Metrization," *Amer. Math. Monthly* **73,** 571–576 (1966).

Maurice, M. A., *Compact Ordered Spaces*, Mathematical Centre Tract No. 6, Mathematisch Centrum, Amsterdam (1964) [MR 36 #3318].

McAllister, B. L., "Cyclic Elements in Topology, a History," *Amer. Math. Monthly* **73,** 337–350 (1966) [MR 34 #780].

Ross, K. A. and A. H. Stone, "Products of Separable Spaces," *Amer. Math. Monthly* **71,** 398–403 (1964) [MR 29 #1611].

Stone, M. H., "The Generalized Weierstrass Approximation Theorem," *Math. Mag.* **21,** 167–184, 237–254 (1948) [MR 10, p. 255].

Van der Slot, J., *Some Properties Related to Compactness*, Mathematical Centre Tract No. 19, Mathematisch Centrum, Amsterdam (1968).

Van der Walt, T., *Fixed and Almost Fixed Points*, Mathematical Centre Tract No. 1, Mathematisch Centrum, Amsterdam (1963) [MR 34 #5079].

Whyburn, G. T., *What Is a Curve?*, M.A.A. Studies in Mathematics, Vol. 5, Van Nostrand, Princeton (1968) [MR 37 #895].


### Collection of papers


Bing, R. H. and R. J. Bean, *Wisconsin Topology Seminar, 1965*, Annals of Math. Studies 60, Princeton Univ. Press, Princeton (1966).

Fort, M. K., *Topology of 3-manifolds*, Prentice-Hall, Englewood Cliffs, (1962) [MR 25 #4508].

Grace, E. E., *Topology Conference, Arizona State University, 1967*, Tempe, Arizona (1967).

*General Topology and Its Relations to Modern Analysis and Algebra* (Proc. Sympos., Prague, 1961), Academic Press, New York (1962).

*General Topology and Its Relations to Modern Analysis and Algebra* II (Proc. Sympos., Prague, 1966), Academic Press, New York (1967).

## References

Abel, N. H., "Undersuchen über die Reihe

$$1 + \frac{m}{1} x + \frac{m(m-1)}{2} x^2 + \frac{m(m-1)(m-2)}{(2.3)} x^3 + \cdots \text{ u.s.w.,"}$$

*J. für die Reine und Angewandte Math.* **1,** 311–339 (1826).

Ahlfors, L. and L. Sario, *Riemann Surfaces*, Princeton Univ. Press, Princeton (1960) [MR 22 #5729].

Albuquerque, J., "La Notion de 'Frontière' en Topologie," *Portug. Math.* **2,** 280–289 (1941) [MR 4, p. 87].

Alexandroff, P., "Sur les Ensembles de la Première Classe et les Espaces Abstraits," *C. R. Paris* **178,** p. 185 (1924).

Alexandroff, P., "Über die Metrisation der im kleinen kompakten topologische Räume, *Math. Ann.* **92,** 294–301 (1924).

Alexandroff, P., "Zur Begründung der *n*-dimensionalen mengentheoretischen Topologie," *Math. Ann.* **94,** 296–308 (1925).

Alexandroff, P., "Über stetige Abbildung kompakter Räume," *Math. Ann.* **96,** 555–571 (1926).

Alexandroff, P., "Untersuchungen über Gestalt und Lage abgeschlossenes Mengen," *Ann. of Math.* **30,** 101 (1928).

Alexandroff, P., "Bikompakte Erweiterungen topologische Räume," *Mat. Sb.* **5,** 403–423 (1939) (Russian; German summary) [MR 1, p. 318].

Alexandroff, P., "Some Results in the Theory of Topological Spaces, Obtained within the last Twenty-five Years," *Russian Math. Surveys* **15,** No. 2, 23–83 (1960) [MR 22 #9947].

Alexandroff, P., "On Some Basic Directions in General Topology," *Russian Math. Surveys* **19,** No. 5, 1–39 (1964) [MR 30 #2450].

Alexandroff, P. and H. Hopf, *Topologie* I, Springer-Verlag, Berlin (1935) [Z 13, p. 79].

Alexandroff, P. and P. Urysohn, "Une condition nécessaire et suffisante pour qu'une classe ($\mathscr{L}$) soit une classe ($\mathscr{D}$)," *C. R. Acad. Sci. Paris* **177,** 1274–1277 (1923).

Alexandroff, P. and P. Urysohn, "Zur theorie der topologischen Räume," *Math. Ann.* **92,** 258–266 (1924).

Alexandroff, P. and P. Urysohn, "Mémoire sur les Espaces Topologiques Compacts," Verh. Nederl. Akad. Wetensch. Afd. Naturk. Sect. I, **14,** 1–96 (1929).

Alexandroff, P. and P. Urysohn, "On Compact Topological Spaces," *Trudy Math. Inst. Steklov* **31,** 95 pp. (1950) (Russian) [MR 13, p. 264].

Alfsen, E. M. and J. E. Fenstad, "On the Equivalence between Proximity Structures and Totally Bounded Uniform Structures," *Math. Scand.* **7,** 353–360 (1959) [Corrections appear in *Math. Scand.* **9,** 258 (1961).] [MR 22 #5958].

Alfsen, E. M. and O. Njåstad, "Proximity and Generalized Uniformity," *Fund. Math.* **52,** 235–252 (1963) [MR 27 #4207a].

Alo, R. A. and Shapiro, H. L., "A Note on Compactifications and Semi-normal Spaces," *J. Austral. Math. Soc.* **8,** 102–108 (1968) [MR 37 #3527].

Anderson, F. W. and R. L. Blair, "Characterizations of Certain Lattices of Functions," *Pac. J. Math.* **9,** 335–364 (1959) [MR 21 #4349].

Anderson, F. W. and R. L. Blair, "Characterizations of the Algebra of all Real-valued Continuous Functions on a Completely Regular Space," *Ill. J. Math.* **3,** 121–133 (1959) [MR 20 #7214].

Anderson, F. W., "Approximation in Systems of Real-valued Continuous Functions," *Trans. Amer. Math. Soc.*, **103,** 249–271 (1962) [MR 25 #436].

Anderson, R. D., "Hilbert Space is Homeomorphic to the Countable Infinite Product of Lines, *Bull. Amer. Math. Soc.* **72,** 515–519 (1966) [MR 32 #8298].

Apostol, T., *Mathematical Analysis*, Addison-Wesley, Reading (1957) [MR 19, p. 398].

Arens, R., "A Topology for Spaces of Transformations," *Ann. of Math.* (2) **47,** 480–495 (1946) [MR 8, p. 479].

Arens, R., "Extension of Functions on Fully Normal Spaces," *Pac. J. Math.* **2,** 11–22 (1952) [MR 14, p. 191].

Arens, R. and J. Dugundji, "Remark on the Concept of Compactness," *Portugal Math.* **9,** 141–143 (1950) [MR 13, p. 264].

Arens, R. and J. Dugundji, "Topologies for Function Spaces," *Pac. J. Math.* **1,** 5–31 (1951) [MR 13, p. 264].

Arhangel'skii, A., "Some Types of Factor Mappings and the Relations between Classes of Topological Spaces," *Soviet Math. Dokl.* **4,** 1726–1729 (1963) [MR 28 #1587].

Arhangel'skii, A., "Mappings and Spaces," *Russian Math. Surveys* **21,** No. 4, 115–162 (1966) [MR 37 #3534].

Aronszajn, N., "Über ein Urbildproblem," *Fund. Math.* **17,** 92–121 (1931) [Z 3, p. 27].

Arzela, C., "Sull' Integrabilita Delle Equazioni Differenziali Ordinarie," *Mem. Accad. Bologna* (5) **5,** 257–270 (1895).

Arzela, C., "Funzioni di Linee," *Atti della Reale Accademia dei Lincei, Rendiconti* **5,** 342–348 (1889).

Ascoli, G., "Le Curve Limite di una Varietà Data di Curve," *Mem. Accad. Lincei* (3) **18,** 521–586 (1883).

Auslander, L. and R. E. McKenzie, *Introduction to Differentiable Manifolds*, McGraw-Hill, New York, 1963 [MR 28 #4462].

Bagley, R. W. and J. S. Yang, "On $k$-spaces and Function Spaces," *Proc. Amer. Math. Soc.* **17,** 703–705 (1966) [MR 33 #693].

Baire, R., "Sur les Fonctions de Variables Réelles," *Ann. di Mat.* **3,** 1–123 (1899).

Balachadran, V. K., "Minimal-bicompact Space," *J. Indian Math. Soc.* [N.S.] **12,** 47–48 (1948) [MR 10, p. 390].

Banach, S., "Sur les Opérations dans les Ensembles Abstraits et leurs Applications aux Equations Intégrales," *Fund. Math.* **3,** 7–33 (1922).

Banach, S., "Über die Baire'sche Kategorie gewisser Funktionenmengen," *Studia Math.* **3,** 174–179 (1931) [Z 3, p. 297].

Banach, S., *Théorie des Opérations Linéaires*, Monogr. Mat. Warsaw-Lwów (1932) [Z 5, p. 209].

Banaschewski, B., "Über Hausdorffsch-minimale Erweiterung von Räumen," *Arch. Math.* **12,** 355–365 (1961) [MR 25 #5490].

Banaschewski, B., "Normal Systems of Sets," *Math. Nachrichten* **24,** 53–75 (1962) [MR 30 #2458].

Banaschewski, B., "On Wallman's Method of Compactification," *Math. Nachrichten* **27,** 105–114 (1963) [MR 28 #3400].

Bartle, R. G., "Nets and Filters in Topology," *Amer. Math. Monthly* **62,** 551–557 (1955).

Bell, H., "On Fixed Point Properties of Plane Continua," *Trans. Amer. Math. Soc.* **128,** 539–548 (1967) [MR 35 #4888].

Bernstein, S., "Démonstration du Théorème de Weierstrass Fondée sur le Calcul des Probabilités," *Conm. Soc. Math. Kharkoff* (2) **13,** 1–2 (1912).

Berri, M., "Minimal Topological Spaces," *Trans. Amer. Math. Soc.* **108,** 97–105 (1963) [MR 27 #711].

Berri, M., "Categories of Certain Minimal Topological Spaces, *J. Austral. Math. Soc.* **4,** 78–82 (1964) [MR 29 #579].

Bing, R. H., "Extending a Metric," *Duke Math. J.* **14,** 511–519 (1947) [MR 9, p. 521].

Bing, R. H., "A Homogeneous Indecomposable Plane Continuum," *Duke Math. J.* **15,** 729–742 (1948) [MR 10, p. 261].

Bing, R. H., "Metrization of Topological Spaces," *Can. J. Math.* **3,** 175–186 (1951) [MR 13, p. 264].

Bing, R. H., "A Countable Connected Hausdorff Space," *Proc. Amer. Math. Soc.* **4,** 474 (1953) [MR 15, p. 729].

Bing, R. H., "A Translation of the Normal Moore Space Conjecture," *Proc. Amer. Math. Soc.* **16,** 612–619 (1965) [MR 31 #6201].

Bing, R. H., "The Elusive Fixed-point Property," *Amer. Math. Monthly* **76,** 119–132 (1969).

Bing, R. H. and F. B. Jones, "Another Homogeneous Plane Continuum," *Trans. Amer. Math. Soc.* **90,** 171–192 (1959) [MR 20 #7251].

Birkhoff, G., "A Note on Topological Groups," *Compositio Math.* **3,** 427–430 (1936) [Z 15, p. 7].

Birkhoff, G., "On the Combination of Topologies," *Fund. Math.* **26,** 156–166 (1936) [Z 14, p. 280].

Birkhoff, G., "Moore–Smith Convergence in General Topology," *Ann. Math.* **38,** 39–56 (1937) [Z 16, p. 85].

Birkhoff, G., *Lattice Theory* (Revised Ed.), Amer. Math. Soc. Colloq. Publ. 25, Revised Ed., New York (1948) [MR 10, p. 673].

Birkhoff, G. D. and O. D. Kellog, "Invariant Points in Function Spaces," *Trans. Amer. Math. Soc.* **23,** 96–115 (1922).

Bishop, R. L. and R. J. Crittenden, *Geometry of Manifolds*, Academic Press, New York (1964) [MR 29 #6401].

Borel, E., *Leçons sur la Théorie des Fonctions*, Paris (1898).

Borges, C. J. R., "On Extensions of Topologies," *Can. J. Math.* **19,** 474–487 (1967) [MR 35 #3621].

Borsuk, K., "Sur les Retracts," *Fund Math.* **17,** 152–170 (1931) [Z 3, p. 27].

Borsuk, K., "Sur un Continu Acyclique qui se laisse Transformer Topologiquement et lui même sans Points Invariants," *Fund. Math.* **24,** 51–58 (1934) [Z 10, p. 134].

Borsuk, K., *Theory of Retracts*, Monografie Matematyczne 44, Panstwowe Wydawnictwo Naukowe, Warsaw (1967) [MR 35 #7306].

Bourbaki, N., *General Topology*, Part 1 (transl.), Addison-Wesley, Reading (1966) [MR 34 #5044a].

Bourbaki, N., *General Topology*, Part 2 (transl.), Addison-Wesley, Reading (1966) [MR 34 #5044b].

Brooks, R. M., "On Wallman Compactifications," *Fund. Math.* **60,** 157–173 (1957) [MR 35 #964].

Brouwer, L. E. J., "Zur Analysis Situs," *Math. Ann.* **68,** p. 422–434 (1910).

Brouwer, L. E. J., "Beweis des Jordanschen Kurvensatz," *Math. Ann.* **69,** 169–175 (1910).

Brouwer, L. E. J., "Over de Structur der perfecte Punktnerzamelingen," *Akad. Versammlungen, Amsterdam,* **18,** 833–842 (1910), and **19,** 1416–1426 (1911).

Browder, A., *Introduction to Function Algebras,* W. A. Benjamin, New York (1969).

Brown, R., "Functions, Spaces and Product Topologies," *Quart. J. Math. Oxford* (2) **15,** 238–250 (1964) [MR 29 #2779].

Buseman, H., *The Geometry of Geodesics,* Academic Press, New York (1955) [MR 17, p. 779].

Cauchy, A. L., *Cours d'Analyse de l'École Royale Polytechnique*, Chez Debure frères, Paris (1821).

Cantor, G., "Über unendliche, lineare Punktmannigfaltigkeiten," *Math. Ann.* (a) **15,** 1–7 (1879); (b) **17,** 355–388 (1880); (c) **20,** 113–121 (1882); (d), (e) **21,** 51–58, 545–591 (1883); (f) **22,** 453–488 (1884).

Caratheodory, C., "Über die Begrenzung einfach zusammenhängender Gebiete," *Math. Ann.* **73,** 323–370 (1913).

Cartan, H., "Théorie des Filtres," *C. R. Acad. Sci. Paris* **205,** 595–598 (1937) [Z 17, p. 243].

Cartan, H., "Filtres et Ultrafiltres," *C. R. Acad. Sci. Paris* **205,** 777–779 (1937) [Z 18, p. 3].

Čech, E., "On Bicompact Spaces," *Ann. Math.* (2) **38,** 823–844 (1937) [Z 17, p. 428].

Cedar, J., "Some Generalizations of Metric Spaces," *Pac. J. Math.* **11,** 105–125 (1961) [MR 24 #A1707].

Chevalley, C. and O. Frink, "Bicompactness of Cartesian Products," *Bull. Amer. Math. Soc.* **47,** 612–614 (1941) [MR 3, p. 57].

Chittenden, E. W., "On the Metrization Problem and Related Problems in the Theory of Abstract Sets," *Bull. Amer. Math. Soc.* **33,** 13–34 (1927).

Cohen, D. E., "Spaces with Weak Topology," *Quart. J. Math. Oxford* (2) **5,** 77–80 (1954) [MR 16, p. 62].

Cohen, H. J., "Sur un Problème de M. Dieudonné," *C. R. Acad. Sci. Paris* **234,** 290–292 (1952) [MR 13, p. 763].

Cohen, H., "Fixed Points in Products of Ordered Spaces, *Proc. Amer. Math. Soc.* **7,** 703–706 (1956) [MR 17, p. 1232].

Cohen, P. J., *Independence of the Axiom of Choice*, Stanford Univ. (1963).

Cohen, P. J., "The Independence of the Continuum Hypothesis," *Proc. Nat. Acad. Sci.* (I), **50,** 1143–1148 (1963) [MR 28 #1118]; (II), **51,** 105–110 (1964) [MR 28 #2962].

Cohen, P. J., *Independence Results in Set Theory*, Theory of Models (Proc. Sympos., Berkeley, 1963), North-Holland, Amsterdam (1965) pp. 39–54 [MR 33 #3908].

Cohen, P. J., *Set Theory and the Continuum Hypothesis*, W. A. Benjamin, New York (1966).

Comfort, W. W., "On the Hewitt Realcompactification of a Product Space, *Trans. Amer. Math. Soc.* **131,** 107–118 (1968) [MR 36 #5896].

Comfort, W. W., "A Theorem of Stone–Čech Type, and a Theorem of Tychonoff Type, without the Axiom of Choice; and their Realcompact Analogues," *Fund. Math.* **43,** 97–110 (1968).

Copson, E. T., *Metric Spaces*, Cambridge Tracts in Math. and Math. Physics No. 57, Cambridge University Press, London (1968) [MR 37 #877].

Corson, H. H., "The Determination of Paracompactness by Uniformities," *Amer. J. Math.* **80,** 185–190 (1958) [MR 20 #1292].

Corson, H. H., "Normality in Subsets of Product Spaces," *Amer. J. Math.* **81,** 785–796 (1959) [MR 21 #5947].

Corson, H. H., "Examples Relating to Normality in Topological Spaces," (transl.) *Amer. Math. Soc.* **99,** 205–211 (1961) [MR 23 #A1344].

Corson, H. H. and E. Michael, "Metrizability of Certain Countable Unions," *Ill. J. Math.* **8,** 351–360 (1964) [MR 30 #562].

Cronin, J., *Fixed Points and Topological Degree in Nonlinear Analysis*, Math. Survey No. 11, Amer. Math. Soc., Providence (1964) [MR 29 #1400].

Császár, A., *Fondements de la Topologie Générale*, Akadémiai Kiadó, Budapest (1960) [MR 22 #4043].

DeBranges, L., "The Stone–Weierstrass Theorem," *Proc. Amer. Math. Soc.* **10,** 822–824 (1959) [MR 22 #3970].

Dieudonné, J. A., "Une Généralisation des Espaces Compacts," *J. Math. Pures Appl.* **23,** 65–76 (1944) [MR 7, p. 134].

Doss, R., "On Uniform Spaces with a Unique Structure," *Amer. J. Math.* **71,** 19–23 (1949) [MR 10, p. 557].

Dowker, C. H., "On Countably Paracompact Spaces," *Can. J. Math.* **3,** 219–224 (1951) [MR 13, p. 264].

Dowker, C. H., "Topology of Metric Complexes," *Amer. J. Math.* **74,** 555–577 (1952) [MR 13, p. 965].

Dowker, C. H., "On a Theorem of Hanner," *Ark Mat.* **2,** 307–313 (1952) [MR 14, p. 396].

Dowker, C. H., "Mappings of Proximity Structures," *General Topology and its Relations to*

*Modern Analysis and Algebra* (Proc. Sympos., Prague, 1961), Academic Press, New York (1962), pp. 139–141. [MR 26 #4312].

Dudley, R. M., "On Sequential Convergence," *Trans. Amer. Math. Soc.* **112,** 483–507 (1964) [MR 30 #5266].

Dugundji, J., "An Extension of Tietze's Theorem," *Pac. J. Math.* **1,** 353–367 (1951) [MR 13, p. 373].

Dugundji, J., *Topology*, Allyn and Bacon, Boston (1966) [MR 33 #1824].

Efremovic, V. A., "Invariant Definition of Topological Product," *Uspehi Mat. Nauk* (*N.S.*) **7,** 159–161 (1952) (Russian) [MR 13, p. 964].

Efremovic, V. A., "The Geometry of Proximity I," *Mat. Sb.* (*N.S.*) **31,** 189–200 (1952) (Russian) [MR 14, p. 1106].

Eilenberg, S., "Ordered Topological Spaces," *Amer. J. Math.* **63,** 39–45 (1941) [MR 2, p. 179].

Engelking, R., "Remarks on Realcompact Spaces." *Fund. Math.* **55,** 303–308 (1964) [MR 31 #4000].

Engelking, R., "On the Double Circumference of Alexandroff," to appear.

Fan, K. and N. Gottesman, "On Compactifications of Freudenthal and Wallman," *Indagat. Math.* **14,** 504–510 (1952) [MR 14, p. 669].

Fejer, L., "Über Weierstrassche Approximation, besonders durch Hermitesche Interpolation," *Math. Ann.* **102,** 707–725 (1930).

Fort, M. K., *Topology of 3-manifolds*, Prentice-Hall, Englewood Cliffs, (1962) [MR 25 #4508].

Fox, R. H., "On Topologies for Function Spaces," *Bull. Amer. Math. Soc.* **51,** 429–432 (1945) [MR 6, p. 278].

Fox, R. H., "On a Problem of S. Ulam Concerning Cartesian Products," *Fund. Math.* **34,** 278–287 (1947) [MR 10, p. 316].

Fränkel, A., "Zu den Grundlagen der Cantor–Zermeloschen Mengenlehre," *Math. Ann.* **86,** 230–237 (1922).

Fränkel, A., *Abstract Set Theory*, North-Holland Publishing Co., Amsterdam (1953) [MR 15, p. 108].

Franklin, S. P., "Spaces in which Sequences Suffice," *Fund. Math.* **57,** 107–115 (1965) [MR 31 #5184].

Franklin, S. P., "Spaces in which Sequences Suffice II," *Fund. Math.* **61,** 51–56 (1967) [MR 36 #5882].

Fréchet, M., "Sur Quelques Points du Calcul Fonctionnel," *Rendiconti di Palermo* **22,** 1–74 (1906).

Frink, A. H., "Distance Functions and the Metrization Problem," *Bull. Amer. Math. Soc.* **43,** 133–142 (1937) [Z 16, p. 82].

Frink, O., "Topology in Lattices," *Trans. Amer. Math. Soc.* **51,** 567–582 (1942) [MR 3, p. 313].

Frink, O., "Compactifications and Semi-normal Spaces," *Amer. J. Math.* **86,** 602–607 (1964) [MR 29 #4028].

Frolik, Z., "On the Descriptive Theory of Sets," *Czech. Math. J.* **13,** 335–359 (1963) [MR 28 #5414].

Frolik, Z., "Baire Sets which are Borelian Subspaces," *General Topology and Its Relations to Modern Analysis and Algebra* II (Proc. Sympos., Prague, 1966), Academic Press, New York (1967).

Gál, I. S., "Proximity Relations and Precompact Structures," *Nederl. Akad. Wetensch. Proc. Ser. A.* **62,** 304–326 (1959) [MR 21 #5944].

Gale, D., "Compact Sets of Functions and Function Rings," *Proc. Amer. Math. Soc.* **1,** 303–308 (1950) [MR 12, p. 119].

Gantner, T. E., "Extensions of Uniformly Continuous Pseudometrics," *Trans. Amer. Math. Soc.* **132,** 147–157 (1968) [MR 36 #5886].

Gillman, L. and M. Jerison, *Rings of Continuous Functions*, Van Nostrand, Princeton (1960) [MR 22 #6994].

Gillman, L., "The Space $\beta$**N** and the Continuum Hypothesis," *General Topology and Its Relations to Modern Analysis and Algebra* II (Proc. Sympos., Prague, 1966), Academic Press, New York (1967) pp. 144–146.

Gleason, A., "Projective Topological Spaces," *Ill. J. Math.* **2,** 482–489 (1958) [MR 22 #12509].

Glicksberg, I., "Stone–Čech Compactifications of Products," *Trans. Amer. Math. Soc.* **90,** 369–382 (1959) [MR 21 #4405].

Gödel, K., "The Consistency of the Axiom of Choice and the Generalized Continuum Hypothesis with the Axioms of Set Theory," *Uspehi Mat. Nauk* (*N.S.*) **3,** 96–149 (1948) [MR 9, p. 559]. See also *The Consistency of the Continuum Hypothesis*, Annals of Math Studies No. 3, Princeton University Press, Princeton (1940) [MR 2, p. 66].

Gottschalk, W. H., *Bibliography for Topological Dynamics*, 3rd Ed., Wesleyan Univ. mimeographed notes (1967) [MR 36 #7126].

Haar, A. and D. König, "Über einfach geordnete Mengen," *Crelle's J.* **139,** 16–28 (1910).

Hadamard, J., "Sur Certaines Applications Possibles de la Théorie des Ensembles," *Verhandlungen des Ersten Internationalen Mathematiker-Kongresses*, B. G. Teubner, Leipzig (1898).

Hager, A. W., "Some Remarks on the Tensor Product of Function Rings," *Math. Zeitschr.* **92,** 210–224 (1966) [MR 33 #1831].

Hager, A. W. and D. Johnson, "A Note on Certain Subalgebras of $C(X)$," *Can. J. Math.* **20,** 389–393 (1968).

Hahn, H., "Über die allgemeinste ebene Punktmenge, die stetiges Bild einer Strecke ist," *Jahresber. Deut. Math. Ver.* **23,** 318–322 (1914).

Hahn, H., "Mengentheoretische Characterisierung der stetigen Kurven," *Sitzungsberichte, Akad. der Wissenschaften* **123,** p. 2433 (1914).

Hahn, H., "Über die Komponenten offenen Mengen," *Fund. Math.* **2,** 189–192 (1921).

Hall, D. W. and G. L. Spencer, *Elementary Topology*, John Wiley and Sons, New York (1955) [MR 17, p. 649]

Halmos, P. R., *Naïve Set Theory*, Van Nostrand, Princeton (1960) [MR 22 #5575].

Halmos, P. R., *A Hilbert Space Problem Book*, Van Nostrand, Princeton (1967) [MR 34 #8178].

Hanner, O., "Retraction and Extension of Mappings of Metric and Non-metric Spaces," *Ark. Math.* **2,** 315–360 (1952) [MR 14, p. 396].

Hartman, P., *Ordinary Differential Equations*, John Wiley and Sons, New York (1964) [MR 30 #1270].

Hausdorff, F., *Grundzüge der Mengenlehre*, Leipzig (1914). Reprinted by Chelsea, New York (1949) [MR 11, p. 88].

Hausdorff, F., "Erweiterung einer Homöomorphie," *Fund. Math.* **16,** 353–360 (1930).

Hausdorff, F., "Erweiterung einer stetigen Abbildung," *Fund. Math.* **30,** 40–47 (1938) [Z 18, p. 277].

Hausdorff, F., *Set Theory*, 3rd Ed. (transl.), Chelsea, New York (1957) [MR 19, p. 111].

Heath, R. W., "Screenability, Pointwise Paracompactness, and Metrization of Moore Spaces," *Can. J. Math.* **16,** 763–770 (1964) [MR 29 #4033].

Heider, L. J. and J. E. Simpson, *Theoretical Analysis*, W. B. Saunders, Philadelphia (1967) [MR 36 #1246].

Heine, E., "Über trigonometrische Reihen," *J. für die R. und Ang. Math.* **71,** 353–365 (1870).

Helgason, S., *Differential Geometry and Symmetric Spaces*, Academic Press, New York (1962) [MR 26 #2986].

Henkin, L., "A Problem on Inverse Mapping Systems," *Proc. Amer. Math. Soc.* **1,** 224–225 (1950) [MR 11, p. 675].

Henriksen, M. and J. R. Isbell, "Some Properties of Compactifications," *Duke Math. J.* **25,** 83–105 (1967) [MR 20 #2689].

Henriksen, M. and D. Johnson, "On the Structure of a Class of Archimedean Lattice-ordered Algebras," *Fund. Math.* **50,** 73–94 (1961) [MR 24 #A3524].

Hewitt, E., "A Problem in Set-theoretic Topology," *Duke Math. J.* **10,** 309–333 (1943) [MR 5, p. 46].

Hewitt, E., "A Remark on Density Characters," *Bull. Amer. Math. Soc.* **52,** 641–643 (1946) [MR 8, p. 139].

Hewitt, E., "On Two Problems of Urysohn," *Ann. of Math.* (2) **47,** 503–509 (1946) [MR 8, p. 165].

Hewitt, E., "Certain Generalizations of the Weierstrass Approximation Theorem," *Duke Math. J.* **14,** 419–427 (1947) [MR 9, p. 95].

Hewitt, E., "Rings of Real-valued Continuous Functions I," *Trans. Amer. Math. Soc.* **64,** 45–99 (1948) [MR 10, p. 126].

Hewitt, E., "A Class of Topological Spaces," *Bull. Amer. Math. Soc.* **55,** 421–426 (1949) [MR 10, p. 616].

Hewitt, E., "The Role of Compactness in Analysis," *Amer. Math. Monthly* **67,** 499–516 (1960) [MR 22 #11367].

Hewitt, E. and K. A. Ross, *Abstract Harmonic Analysis* I, Springer-Verlag, Berlin (1963) [MR 28 #158].

Hewitt, E. and K. Stromberg, *Real and Abstract Analysis*, Springer-Verlag, New York (1965) [MR 32 #5826].

Hilbert, D., "Über die Grundlagen der Geometrie," *Math. Ann.* **56,** 381–422 (1903).

Hilton, P., *An Introduction to Homotopy Theory*, Cambridge Tracts in Mathematics and Physics No. 43, Cambridge Univ. Press, Cambridge (1953) [MR 15, p. 52].

Himmelberg, C. J., "Quotients of Completely Regular Spaces, *Proc. Amer. Math. Soc.* **19,** 864–866 (1968) [MR 37 #3510].

Hirzebruch, *Topological Methods in Algebraic Geometry*, 3rd Ed., Springer, Berlin (1966) [MR 34 #2573].

Hocking, J. G. and G. S. Young, *Topology*, Addison-Wesley, Reading (1961) [MR 23 #A2857].

Hu, S. T., *Homotopy Theory*, Academic Press, New York (1959) [MR 21 #5186].

Hu, S. T., *Introduction to General Topology*, Holden-Day, San Francisco (1966) [MR 33 #4876].

Hu, S. T., *Differentiable Manifolds*, Holt, Rinehart and Winston, New York (1969).

Hulanicki, A., "On Locally Compact Groups of Power of Continuum," *Fund. Math.* **44,** 156–158 (1957) [MR 19, p. 1063].

Hurewicz, W., "Beiträge zur Topologie der Deformationen I," *Proc. K. Akad. van Wet. Amst.* **38,** 112–119 (1935) [Z 10, p. 378].

Hurewicz, W., "Beiträge zur Topologie der Deformationen II," *Proc. K. Akad. van Wet. Amst.* **38,** 521–528 (1935) [Z 11, p. 371].

Hurewicz, W., "Beiträge zur Topologie der Deformationen III," *Proc. K. Akad. van Wet. Amst.* **39,** 117–126 (1936) [Z 13, p. 229].

Hurewicz, W., "Beiträge zur Topologie der Deformationen IV," *Proc. K. Akad. van Wet. Amst.* **39,** 215–224 (1936) [Z 13, p. 283].

Hurewicz, W. and H. Wallman, *Dimension Theory*, Princeton Univ. Press, Princeton (1941) [MR 3, p. 312].

Ikenaga, S., "Product of Minimal Topological Spaces," *Proc. Japan Acad.* **40,** 329–331 (1964) [MR 29 #5213].

Isbell, J., "Algebras of Uniformly Continuous Functions," *Ann. of Math.* (2) **68,** 96–125 (1958) [MR 21 #2177].

Isbell, J. R., *Uniform Spaces*, Amer. Math. Soc. Survey No. 12, Amer. Math. Soc., Providence (1964) [MR 30 #561].

Ishii, Tadashi, "On Closed Mappings and *M*-spaces I, II," *Proc. Japan Acad.* **43,** 752–756, 757–761 (1967) [MR 36 #5904].

Isiwata, T., "Mappings and Spaces," *Pac. J. Math.* **20,** 455–480 (1967) [MR 36 #2127].

Jackson, J. R., "Spaces of Mappings on Topological Products with Applications to Homotopy Theory," *Proc. Amer. Math. Soc.* **3,** 327–333 (1952). [MR 13, p. 859].

Janiszewski, S., "Sur les Continus Irréductibles entre Deux Points," *J. de l'École Polytech.* **16,** 79–170 (1912).

Janiszewski, S. and C. Kuratowski, "Sur les Continus Indécomposables," *Fund. Math.* **1,** p. 210 (1920).

Jech, T., "Non-provability of Souslin's Hypothesis," *Comm. Math. Univ. Carol.* **8,** 291–305 (1967) [MR 35 #6564].

Jensen, G. A., "A Note on Complete Separation in the Stone Topology," *Proc. Amer. Math. Soc.* **21,** 113–116 (1969).

Jones, F. B., "On the First Countability Axiom for Locally Compact Hausdorff Spaces," *Coll. Math.* **7,** 33–34 (1959) [MR 22 #2972].

Jones, F. B., "Remarks on the Normal Moore Space Metrization Problem," *Proc. of the 1965 Wisconsin Summer Topology Seminar*, Annals of Math. Studies 60, Princeton (1966).

Jordan, C., *Cours d'Analyse de l'École Polytechnique*, Vol. 1, 2nd Ed., Paris (1893).

Kakutani, S., "Über die Metrisation der topologischen Gruppen," *Proc. Imp. Acad. Tokyo* **12,** 82–84 (1936) [Z 15, p. 7].

Katětov, M., "Über *H*-abgeschlossen und bikompakt Räume," *Časopis pro. Mat. Fys.* **69,** 36–49 (1940) (German; Czech Summary) [MR 1, p. 317].

Katětov, M., "On the Dimension of Non-separable Spaces I," *Czech. Math. J.* **2,** 333–368 (1952) (Russian; English summary) [MR 15, p. 815].

Katětov, M., "On the Dimension of Non-separable Spaces II," *Czech. Math. J.* **6,** 485–516 (1956) (Russian; English summary) [MR 19, p. 874].

Kaufman, R., "Ordered Sets and Compact Spaces," *Coll. Math.* **17,** 35–39 (1967) [MR 35 #3634].

Kawashima, Hiroshi, "On the Topological Product of Minimal Hausdorff Spaces," *TRU Math.* **1,** 62–64 (1965) [MR 36 #7103].

Kelley, J. L., "Convergence in Topology," *Duke Math. J.* **17,** 277–283 (1950) [MR 12, p. 194].

Kelley, J. L., "The Tychonoff Product Theorem Implies the Axiom of Choice," *Fund. Math.* **37,** 75–76 (1950) [MR 12, p. 626].

Kelley, J. L., *General Topology*, Van Nostrand, Princeton (1955) [MR 16, p. 1136].

Kelley, J. L., I Namioka, *et al.*, *Linear Topological Spaces*, Van Nostrand, Princeton (1963) [MR 29 #3851].

Kenderov, P., "On Q-spaces," *Dokl. Akad. Nauk SSSR* **175,** 288–291 (1967) [*Soviet Math. Dokl.* **8,** 849–852 (1967)] [MR 36 #4511].

Kenderov, P., "A Certain Problem of A. Stone," *Vestnik. Moskov. Univ. Ser. I Mat. Meh.* **23,** 5–7 (1968) [MR 37 #3511].

Knaster, B., "Un Continu dont Tout Sous-continu est Indécomposable," *Fund. Math.* **3,** 247–286 (1922).

Knaster, B. and C. Kuratowski, "Problème 2," *Fund. Math.* **1,** 223 (1920).

Knaster, B. and C. Kuratowski, "Sur les Ensembles Connexes," *Fund. Math.* **2,** 206–255 (1921).

Knaster, B. and C. Kuratowski, "A Connected and Connected *im kleinen* Point Set which Contains no Perfect Set," *Bull Amer. Math. Soc.* **33,** 106–109 (1927).

Knight, C. J., "Box Topologies," *Quart. J. Math. Oxford,* (2) **15,** 41–54 (1964) [MR 28 #3398].

Kolmogorov, A. N. and S. V. Fomin, *Functional Analysis*, Vol. 1 (transl.), Graylock Press, Rochester (1957) [MR 19, p. 44].

Köthe, G., "Die Quotientenräume eines Linearen Vollkommenen Räumes," *Math. Zeit.* **51,** 17–35 (1947) [MR 9, p. 358, p. 735].

Kowalsky, H. J., *Topological Spaces* (transl.), Academic Press, New York (1965) [MR 22 #12502].

Krein, M. and S., "On an Inner Characteristic of the Set of All Continuous Functions Defined on a Bicompact Hausdorff Space, *C.R.* (*Dokl.*) *Acad. Sci. URSS* (*N.S.*) **27,** 427–430 (1940) [MR 2, p. 222].

Kuratowski, C., "Une Définition Topologique de la Ligne de Jordan," *Fund. Math.* **1,** 40–43 (1920).

Kuratowski, C., "Sur l'Opération $\bar{A}$ de l'Analysis Situs," *Fund. Math.* **3,** 182–199 (1922).

Kuratowski, C., "Remarques sur les Transformations Continues des Espaces Métriques," *Fund. Math.* **30,** 48–49 (1938) [Z 18, p. 277].

Kuratowski, C., "On a Topological Problem Connected with the Cantor–Bernstein Theorem," *Fund. Math.* **37,** 213–216 (1950) [MR 12, p. 729].

Kuratowski, C., *Topology*, Vol. I (transl.), Academic Press, New York (1966) [MR 36 #839].

Kuratowski, C. and W. Sierpinski, "Le théorème de Borel–Lebesgue dans la Théorie des Ensembles Abstraits," *Fund. Math.* **2,** 172–178 (1921).

Lavrentieff, M., "Contribution à la Théorie des Ensembles Homéomorphes," *Fund. Math.* **6,** 149–160 (1924).

Leader, S., "On Clusters in Proximity Spaces," *Fund. Math.* **47,** 205–213 (1959) [MR 22 #2978].

Leader, S., "On Completion of Proximity Spaces by Local Clusters," *Fund. Math.* **48,** 201–216 (1960) [MR 22 #4047].

Leader, S., "Metrization of Proximity Spaces," *Proc. Amer. Math. Soc.* **18,** 1084–1088 (1967) [MR 36 #846].

Lefschetz, S., *Algebraic Topology*, Amer. Math. Soc. Coll. Publ. 27, New York (1942) [MR 4, p. 84].

Levine, N. L., "A Characterization of Compact Metric Spaces," *Amer. Math. Monthly* **68,** 657–658 (1961).

Levine, N. L., "Simple Extensions of Topologies," *Amer. Math. Monthly* **71,** 22–25 (1964) [MR 29 #580].

Levine, N. L., "On Families of Topologies for a Set," *Amer. Math. Monthly* **73,** 358–361 (1966) [MR 34 #1976].

Lindelöf, E., "Sur l'Application des Méthodes des Approximations Successives à l'Étude des Intégrales Réelles des Équations Différentielles Ordinaires," *J. Math. Pures Appl.* **10,** 117–128 (1894).

Lindelöf, E., "Sur Quelques Points de la Théorie des Ensembles," *C. R. Acad. Sci. Paris* **137,** 697–700 (1903).

Liu, Chen-Tung, "Absolutely Closed Spaces," *Trans. Amer. Math. Soc.* **130,** 86–104 (1968) [MR 36 #2107].

Loomis, L., *An Introduction to Abstract Harmonic Analysis*, Van Nostrand, New York (1953).

Lynn, I. L., "Linearly orderable spaces," *Proc. Amer. Math. Soc.* **13,** 454–456 (1962) [MR 25 #1536].

Manheim, J. H., *The Genesis of Point Set Topology*, Pergamon Press, Oxford (1964).

Mansfield, M. J., "Some Generalizations of Full Normality," *Trans. Amer. Math. Soc.* **86,** 489–505 (1957) [MR 20 #273].

Marczewski, E., "Separabilité et Multiplication Cartésienne des Espaces Topologiques," *Fund. Math.* **34,** 127–143 (1947) [MR 9, p. 98].

Mardešić, S., "Images of Ordered Compacta are Locally Peripherally Metric," *Pac. J. Math.* **23,** 557–568 (1967) [MR 36 #4530].

Mardešić, S., "On the Hahn-Mazurkiewicz Problem in Non-metric Spaces," *General Topology and its Relations to Modern Analysis and Algebra* II (Proc. Sympos., Prague, 1966), Academic Press, New York (1967), pp. 248–255.

Massey, W. S., *Algebraic Topology: An Introduction*, Harcourt, Brace and World, New York (1967) [MR 35 #2271].

Maurice, M. A., *Compact Ordered Spaces*, Mathematical Centre Tract No. 6, Mathematisch Centrum, Amsterdam (1964) [MR 36 #3318].

Mazurkiewicz, S., "Über Borelsche Mengen," *Bull. de l'Académie des Sciences*, Cracovie, 490–494 (1916).

Mazurkiewicz, S., "Un Théorème sur les Continus Indécomposables," *Fund. Math.* **1,** 35–39 (1920).

Mazurkiewicz, S., "Sur les Lignes de Jordan," *Fund. Math.* **1,** 166–209 (1920).

Mazurkiewicz, S., "Sur un Ensemble $G_\delta$, Punctiforme, qui n'est pas Homéomorphe avec aucun Ensemble Linéaire," *Fund. Math.* **1,** 61–81 (1920).

McAllister, B. L., "Cyclic Elements in Topology, a History," *Amer. Math. Monthly* **73,** 337–350 (1966) [MR 34 #780].

McKnight, J. D., "Brown's Method of Extending Fixed Point Theorems," *Amer. Math. Monthly* **72,** 152–155 (1965) [MR 31 #2718].

McShane, E. J. and T. A. Botts, *Real Analysis*, Van Nostrand, Princeton (1959) [MR 22 #84].

Michael, E., "A Note on Paracompact Spaces," *Proc. Amer. Math. Soc.* **4,** 831–838 (1953) [MR 15, p. 144].

Michael, E., "Another Note on Paracompact Spaces," *Proc. Amer. Math. Soc.* **8,** 822–828 (1957) [MR 19, p. 299].

Michael, E., "Yet Another Note on Paracompact Spaces," *Proc. Amer. Math. Soc.* **10,** 309–314 (1959) [MR 21 #4406].

Michael, E., "The Product of a Normal Space and a Metric Space need not be Normal," *Bull. Amer. Math. Soc.* **69,** 375–376 (1963) [MR 27 #2956].

Michael, E., "Bi-quotient Maps and Cartesian Products of Quotient Maps," to appear.

Michael, E., "A Note on the Product of $k$-Spaces," to appear.

Milnor, J. W., *Topology from the Differentiable Viewpoint*, Univ. Press of Virginia, Charlottesville (1965).

Montgomery, D., "Non-separable Metric Spaces," *Fund. Math.* **25,** 527–534 (1935) [Z 12, p. 321].

Moore, E. H., "Definition of Limit in General Integral Analysis," *Proc. Nat. Acad. Sci.* **1,** 628 (1915).

Moore, E. H., *General Analysis* I, Part II, Mem. Amer. Philos. Soc., Philadelphia (1939) [Z 20, p. 366].

Moore, E. H. and H. L. Smith, "A General Theory of Limits," *Amer. J. Math.* **44,** 102–121 (1922).

Moore, R. L., "Concerning Simple Continuous Curves," *Trans. Amer. Math. Soc.* **21,** 333–347 (1920).

Moore, R. L., "Concerning Connectedness *im kleinen* and a Related Property," *Fund. Math.* **3,** 232–237 (1922).

Moore, R. L., "Concerning the Cut Points of Continuous Curves and of other Closed and Connected Point Sets," *Proc. Nat. Acad. Sci.* **9,** 101–106 (1923).

Moore, R. L., "An Extension of the Theorem that no Countable Point Set is Perfect," *Proc. Nat. Acad. Sci.* **10,** 168–170 (1924).

Moore, R. L., "Concerning Upper Semi-continuous Collections of Continua," *Trans. Amer. Math. Soc.* **27,** 416–428 (1925).

Moore, R. L., "Concerning Upper Semi-continuous Collections," *Monatschefte für Mathematik und Physik* **36,** 81–88 (1929).

Moore, R. L., *Foundations of Point Set Theory*, Amer. Math. Soc. Colloq. Publ. 13 (1932) [Z 5, p. 54].

Morita, K., "Normal Families and Dimension Theory for Metric Spaces," *Math. Ann.* **128,** 350–362 (1954) [MR 16, p. 501].

Morita, K., "On the Product of a Normal Space with a Metric Space," *Proc. Japan Acad.* **39,** 148–150 (1963) [MR 27 #721].

Morita, K., "On the Product of Paracompact Spaces," *Proc. Japan Acad.* **39,** 559–563 (1963) [MR 28 #2519].

Morita, K., "Products of Normal Spaces with Metric Spaces," *Math. Ann.* **154,** 365–382 (1964) [MR 29 #2773].

Morita, K., "Products of Normal Spaces with Metric Spaces II," *Sci. Rep. Tokyo Kyoiku Daigaku, Sect. A*, **8,** 87–92 (1964) [MR 29 #4034].

Morita, K. and S. Hanai, "Closed Mappings and Metric Spaces," *Proc. Japan Acad.* **32,** 10–14 (1956) [MR 19, p. 299].

Mrowka, S., "On Function Spaces," *Fund. Math.* **45,** 273–282 (1958) [MR 20 #4773].

Mrowka, S., "Functionals on Uniformly Closed Rings of Continuous Functions," *Fund. Math.* **46,** 81–87 (1958) [MR 20 #6650].

Mrowka, S., "Compactness and Product Spaces," *Coll. Math.* **7,** 19–23 (1959) [MR 22 #8479].

Mrowka, S., "Some Approximation Theorems for Rings of Unbounded Functions," *Notices of Amer. Math. Soc.* **11,** 666 (1964).

Myers, S. B., "Equicontinuous Sets of Mappings," *Ann. of Math.* (2) **47,** 496–502 (1946) [MR 8, p. 165].

Nachbin, L., "Sur les Espaces Topologiques Ordonnés," *C. R. Acad. Sci. Paris,* **226,** 381–382 (1948) [MR 9, p. 367].

Nachbin, L., "Sur les Espaces Uniformes Ordonnés," *C. R. Acad. Sci. Paris* **226,** 774–775 (1948) [MR 9, p. 455].

Nachbin, L., "Topological Vector Spaces of Continuous Functions," *Proc. Nat. Acad. Sci.* **40,** 471–474 (1954) [MR 16, p. 156].

Nachbin, L., *Topology and Order* (transl.), Van Nostrand, Princeton (1965) [MR 36 #2125].

Nagata, J., "On a Necessary and Sufficient Condition of Metrizability," *J. Inst. Polytech., Osaka City University* **1,** 93–100 (1950) [MR 13, p. 264].

Nagata, J., "A Contribution to the Theory of Metrization," *J. Inst. Polytech., Osaka City University* **8,** 185–192 (1957) [MR 20 #4256].

Nagata, J., *Modern Dimension Theory*, Interscience–John Wiley and Sons, New York (1965) [MR 34 #8380].

Naimark, M., *Normed Rings* (transl.), Noordhoff, Groningen (1959) [MR 22 #1824].

Njastad, O., "On Wallman-type Compactifications," *Math. Zeit.* **91,** 267–276 (1966) [MR 32 #6404].

Noble, N., *k-spaces*, Doctoral dissertation, University of Rochester, (1967).

Noble, N., "Products with Closed Projections," to appear.

Novak, J., "On the Cartesian Product of Two Compact Spaces," *Fund. Math.* **40,** 106–112 (1953) [MR 15, p. 640].

Nunnally, E., "There is no Universal-projecting Homeomorphism of the Cantor Set," *Coll. Math.* **17,** 51–52 (1967) [MR 35 #4872].

Obreanu, F., "On a Problem of Alexandroff and Urysohn," *Acad. Repub. Pop. Rômane, Bul. Sti. Ser. Mat. Fiz. Chim.* **2,** 101–108 (1950) [MR 13, p. 573].

Peano, G., "Sur une Courbe qui remplit Toute une Plane," *Math. Ann.* **36,** 157–160 (1890).

Picard, E., "Mémoire sur la Théorie des Équations aux Dérivées Partielles et la Méthode des Approximations Successives," *J. Math. Pures Appl.* **6,** 423–441 (1890).

Poénaru, V., *On the Geometry of Differentiable Manifolds*, M.A.A. Studies in Mathematics, Vol. 5, Princeton (1968).

Poincaré, H., "Analysis Situs," *J. de l'École Polytech.* (2) **1,** 1–123 (1895).

Poincaré, H., "Cinquième Complément à l'Analysis Situs," *Palermo Rendic.* **18,** 45–110 (1904).

Pondiczerny, E., "Power Problems in Abstract Spaces," *Duke Math. J.* **11,** 835–837 (1944) [MR 6, p. 119].

Ponomarev, V., "Axioms of Countability and Continuous Mappings," *Bull. Acad. Polon. Sci. Sér. Sci. Math. Astr. Phys.* **8,** 127–134 (1960) (Russian. English summary.) [MR 22 #7109].

Pontrjagin, L., *Topological Groups* (transl.), Princeton Univ. Press, Princeton (1939) [MR 1, p. 44].

Poppe, H., "Stetige Konvergenz und der Satz von Ascoli und Arzela," *Math. Nach.* **30,** 87–122 (1965) [MR 32 #6400].

Ramanathan, A., "Minimal-bicompact Spaces," *J. Indian Math. Soc.* (*N.S.*) **12,** 40–46 (1948) [MR 10, p. 390, p. 856].

Reichbach, M., "The Power of Topological Types of Some Classes of 0-dimensional Sets," *Proc. Amer. Math. Soc.* **13,** 17–23 (1962) [MR 24 #A2937].

Riemann, B., "Über die Hypothesen welche der Geometrie Grunde Liegen," *Gesammelte Mathematische Werke*, B. G. Teubner, Leipzig (1892) pp. 272–287.

Riesz, F., "Die Genesis des Raumbegriffs," *Math. Naturwiss. Ber. Ungarn* **24,** 309–353 (1906).

Riesz, F., "Stetigkeitsbegriff und abstrakte Mengenlehre," *Atti IV Contr. Internat. Mat. Roma* **2,** 18–24 (1908).

Ross, K. A. and A. H. Stone, "Products of Separable Spaces," *Amer. Math. Monthly* **71,** 398–403 (1964) [MR 29 #1611].

Roy, P., "Failure of Equivalence of Dimension Concepts for Metric Spaces," *Bull. Amer. Math. Soc.* **68,** 609–613 (1962) [MR 25 #5495].

Royden, H. L., "Function Algebras," *Bull. Amer. Math. Soc.* **69,** 281–298 (1963) [MR 26 #6817].

Royden, H. L., *Real Analysis*, 2nd Ed., Macmillan, New York (1968).

Rudin, M. E., "Countable Paracompactness and Souslin's Problem," *Can. J. Math.* **7,** 543–547 (1955) [MR 17, p. 391].

Rudin, M. E., "A Separable Normal, Non-paracompact Space," *Proc. Amer. Math. Soc.* **7,** 940–941 (1956) [MR 18, p. 429].

Rudin, M. E., "Interval Topology in Subsets of Totally Orderable Spaces," *Trans. Amer. Math. Soc.* **118,** 376–389 (1965) [MR 31 #3994].

Rudin, W., *Principles of Mathematical Analysis*, 2nd Ed., McGraw-Hill, New York (1964) [MR 29 #3587].

Rudin, W., *Real and Complex Analysis*, McGraw-Hill, New York (1966) [MR 35 #1420].

Scarborough, C. T. and A. H. Stone, "Products of Nearly Compact Spaces," *Trans. Amer. Math. Soc.* **124,** 131–147 (1966) [MR 34 #3528].

Schauder, J., "Der Fixpunktsatz in Funktionalräumen," *Studia Math.* **2,** 171–180 (1930).

Schoenfliesz, A., "Beiträge zur Theorie der Punktmengen," *Math. Ann.* **58,** 195–234 (1904).

Schreier, O., "Abstrakte kontinuierliche Gruppen," *Hamb. Abh.* **4,** 15 (1926).

Seidel, A., *Über eine Eigenschaft der Reihen, welche Discontinuirliche Functionen Darstellen*, Engelmann, Leipzig (1900).

Semadeni, Z., "Sur les Ensembles Clairsemés," *Rozpr. Mat.* **19** (1959) [MR 21 #6571].

Shapiro, H. L., "Extensions of Pseudometrics," *Can. J. Math.* **18,** 981–998 (1966) [MR 34 #6719].

Shimrat, M., "Decomposition Spaces and Separation Properties," *Quart. J. Math. Oxford* (2) **7,** 128–129 (1956) [MR 20 #3519].

Shirota, T., "A Class of Topological Spaces," *Osaka Math. J.* **4,** 23–40 (1952) [MR 14, p. 395].

Shirota, T., "On Locally Convex Vector Spaces of Continuous Functions," *Proc. Japan Acad.* **30,** 294–298 (1954) [MR 16, p. 275].

Sierpinski, W., "Un Théorème sur les Ensembles Fermés," *Bull. de l'Académie des Sciences*, Cracovie, 49–51 (1918).

Sierpinski, W., "Sur une Condition pour qu'un Continu Soit une Courbe Jordanienne," *Fund. Math.* **1,** 44–60 (1920).

Sierpinski, W., "Sur les Ensembles Connexes et Non Connexes," *Fund. Math.* **2,** 81–95 (1921).

Sierpinski, W., "Théorie des Continus Irréductibles entre Deux Points I" *Fund. Math.* **3,** 200–231 (1922).

Sierpinski, W., "Sur l'Invariance Topologique des Ensembles $G_\delta$," *Fund. Math.* **8,** 135–136 (1926).

Sierpinski, W., "Sur les Ensembles Complets d'un Espace (*D*)," *Fund. Math.* **11,** 203–205 (1928).

Sierpinski, W., *General Topology*, 2nd Ed. (transl.), University of Toronto Press, Toronto (1956).

Sierpinski, W., *Cardinal and Ordinal Numbers*, Polska Akad. Nauk, Monografie Matematyczne 34, Państwowe Wydawnictwo Naukowe, Warsaw (1958) [MR 20 #2288].

Sion, M. and G. Zelmer, "On Quasi-metrizability," *Can. J. Math.* **19,** 1243–1249 (1967) [MR 36 #4522].

Smirnov, Y. M., "A Necessary and Sufficient Condition for Metrizability of a Topological Space," *Dokl. Akad. Nauk SSSR* (*N.S.*) **77,** 197–200 (1951) (Russian) [MR 12, p. 845].

Smirnov, Y. M., "On Proximity Spaces," *Mat. Sb.* (*N.S.*) **31,** 543–574 (1952) (Russian) [MR 14, p. 1107].

Smirnov, Y. M., "On Proximity Spaces in the Sense of V. A. Efremovic," *Dokl. Akad. Nauk SSSR* (*N.S.*) **84,** 895–898 (1952) (Russian) [MR 14, p. 1107].

Smirnov, Y., *On Metrization of Topological Spaces*, Amer. Math. Soc. Transl. No. 91 (1953) [MR 15, p. 50].

Smirnov, Y. M., "On Completeness of Proximity Spaces," *Trudy Mosk. Mat. Obšč.* **3,** 271–306 (1954) (Russian) [MR 16, p. 844].

Smirnov, Y. M., "On Completeness of Proximity Spaces II," *Trudy Mosk. Mat. Obšč.* **4,** 421–438 (1955) (Russian) [MR 17, p. 286].

Smirnov, Y. M., "On the Dimension of Proximity Spaces," *Mat. Sb.* (*N.S.*) **38,** 283–302 (1956) (Russian) [MR 18, p. 497].

Smirnov, Y. M., "The Metrizability of Bicompacts Decomposable into a Sum of Sets with Countable Basis," *Fund. Math.* **43,** 387–393 (1956) (Russian) [MR 18, p. 813].

Smirnov, Y. M., "Proximity and Construction of Compactifications with Given Properties," *General Topology and its Relations to Modern Analysis and Algebra* II (Proc. Sympos., Prague, 1966), Academic Press, New York (1967) 332–340.

Solovay, R., "A Model of Set Theory in which Every Set of Reals is Lebesgue Measurable," to appear.

Solovay, R., "Real-valued Measurable Cardinals," to appear.

Solovay, R. and S. Tennenbaum, "Iterated Cohen Extensions and Souslin's Problem," to appear.

Sorgenfrey, R. H., "On the Topological Product of Paracompact Spaces," *Bull. Amer. Math. Soc.* **53,** 631–632 (1947) [MR 8, p. 594].

Souslin, M., "Problème 3," *Fund. Math.* **1,** 223 (1920).

Spanier, E. H., *Algebraic Topology*, McGraw-Hill, New York (1966) [MR 35 #1007].

Springer, G., *Introduction to Riemann Surfaces*, Addison-Wesley, Reading (1957) [MR 19, p. 1169].

Steenrod, N., *The Topology of Fibre Bundles*, Princeton Univ. Press, Princeton (1951) [MR 12, p. 522].

Steiner, A. K., "The Lattice of Topologies; Structure and Complementation," *Trans. Amer. Math. Soc.* **122,** 379–398 (1966) [MR 32 #8303].

Steiner, A. K. and E. F., "Wallman and *Z*-compactifications," *Duke Math. J.* **35,** 269–276 (1968) [MR 37 #3526].

Steiner, A. K. and E. F., "Precompact Uniformities and Wallman Compactifications," *Indag. Math.* **30,** 117–118 (1968) [MR 37 #3525].

Steiner, E. F., "Wallman Spaces and Compactifications," *Fund. Math.* **61,** 295–304 (1968) [MR 36 #5899].

Stone, A. H., "Paracompactness and Product Spaces," *Bull Amer. Math. Soc.* **54,** 977–982 (1948) [MR 10, p. 204].

Stone, A. H., "Metrizability of Decomposition Spaces," *Proc. Amer. Math. Soc.* **37,** 690–700 (1956) [MR 19, p. 299].

Stone, A. H., "Metrizability of Unions of Spaces," *Proc. Amer. Math. Soc.* **10,** 361–366 (1959) [MR 21 #4410].

Stone, A. H., "Non-separable Borel Sets," *Rozprawy Matematyczne* **28** [MR 27 #2435].

Stone, M. H., "Applications of Boolean Algebras to Topology," *Mat. Sb.* **1,** 765–771 (1936) [Z 16, p. 182].

Stone, M. H., "Applications of the Theory of Boolean Rings to General Topology," *Trans. Amer. Math. Soc.* **41,** 375–481 (1937) [Z 17, p. 135].

Stone, M. H., "The Generalized Weierstrass Approximation Theorem," *Math. Mag.* **21,** 167–184, 237–254 (1948) [MR 10, p. 255].

Strecker, G. E. and E. Wattel, "On Semi-regular and Minimal Hausdorff Embeddings," *Indag. Math.* **29,** 234–237 (1967) [MR 35 #2261].

Suppes, P., *Axiomatic Set Theory*, Van Nostrand, Princeton (1960) [MR 22 #5576].

Sz.-Nagy, B., *Introduction to Real Functions and Orthogonal Expansions*, Oxford Univ. Press, New York (1965) [MR 31 #5938].

Tamano, H., "On Paracompactness," *Pac. J. Math.* **10,** 1043–1047 (1960) [MR 23 #A2186].

Tamano, H., "On Compactifications," *J. Math. Kyoto Univ.* **1,** 162–193 (1962) [MR 25 #5489].

Tamano, H., "Normality and Product Spaces," *General Topology and its Relations to Modern Analysis and Algebra* II (Proc. Sympos., Prague, 1966), Academic Press, New York (1967) 349–352.

Tennenbaum, S., "Souslin's Problem," *Proc. Nat. Acad. Sci.* **59,** 60–63 (1968).

Thron, W. J., *Topological Structures*, Holt, Rinehart and Winston, New York (1966) [MR 34 #778].

Tietze, H., "Über Funktionen, die auf einer abgeschlossenen Menge stetig sind," *J. für die R. und Ang. Math.* **145,** 9–14 (1915).

Tietze, H., "Über stetige Kurven, Jordansche Kurvenbögen und geschlossene Jordansche Kurven," *Math. Z.* **5,** 284–291 (1919).

Tietze, H., "Beiträge zur allgemeinen Topologie I," *Math. Ann.* **88,** 290–312 (1923).

Tietze, H., "Über Analysis Situs," *Abhandl. Math. Sem. Univ. Hamburg* **2,** 27–70 (1923).

Tietze, H., "Beiträge zur allgemeinen Topologie II," *Math. Ann.* **91,** 210–224 (1924).

Tukey, J. W., *Convergence and Uniformity in Topology*, Ann. of Math. Studies 2, Princeton Univ. Press, Princeton (1940) [MR 2, p. 67].

Tychonoff, A., "Über einen Metrisationsatz von P. Urysohn," *Math. Ann.* **95,** 139–142 (1926).

Tychonoff, A., "Über die topologische Erweiterung von Räumen," *Math. Ann.* **102,** 544–561 (1930).

Tychonoff, A., "Über einen Funktionenräum," *Math. Ann.* **111,** 762–766 (1935) [Z 12, p. 308].

Tychonoff, A., "Ein Fixpunktsatz," *Math. Ann.* **111,** 767–776 (1935) [Z 12, p. 308].

Urysohn, P., "Über die Mächtigkeit der zusammenhängenden Mengen," *Math. Ann.* **94,** 262–295 (1925).

Urysohn, P., "Zum Metrisation problem," *Math. Ann.* **94,** 309–315 (1925).

Vaĭnsteĭn, I. A., "On Closed Mappings of Metric Spaces," *Dokl. Akad. Nauk SSSR (N.S.)* **57,** 319–321 (1947) (Russian) [MR 9, p. 153].

Väisälä, J., "The Separability of Cartesian Products," *Colloq. Math.* **17,** 285–288 (1967) [MR 36 #4523].

Van Rooij, A. C. M., "The Lattice of All Topologies is Complemented," *Can. J. Math.* **20,** 805–807 (1968) [MR 37 #3504].

Van der Slot, J., *Some Properties Related to Compactness*, Mathematical Centre Tract No. 19; Mathematisch Centrum, Amsterdam (1968).

Van der Walt, T., *Fixed and Almost Fixed Points*, Mathematical Centre Tract No. 1, Mathematisch Centrum, Amsterdam (1963) [MR 34 #5079].

Vietoris, L., "Stetige Mengen," *Monatsh. Math.* **31,** 173–204 (1921).

Volterra, V., "Sopra le Funzioni che Dipendono da Attra Funzioni; Sopra le Funzioni Dallmee," *Atti della Reale Academia dei Lincei Rendiconti* **3,** 97–104, 141–146, 225–230, 274–281 (1887).

Wallace, A. D., "Separation Spaces," *Ann. of Math.* (2) **42,** 687–697 (1941) [MR 3, p. 57].

Wallman, H., "Lattices and Topological Spaces," *Ann. of Math.* (2) **39,** 112–126 (1938) [Z 18, p. 332].

Ward, A. J., "A Theorem of Fixed-point Type for Non-compact Locally Connected Spaces," *Colloq. Math.* **17,** 289–296 (1967) [MR 36 #5929].

Warner, S., "The Topology of Compact Convergence on Continuous Function Spaces," *Duke Math. J.* **25,** 265–282 (1958) [MR 21 #1521].

Weil, A., *Sur les Espaces à Structure Uniforme et sur la Topologie Générale*, Act. Sci. et Ind. 551, Hermann, Paris (1937) [Z 19, p. 186].

Weil, A., *L'intégration dans les Groupes Topologiques et ses Applications*, 2nd Ed., Act. Sci. et Ind. 869, 1145, Hermann et Cie, Paris (1951) [MR 3, p. 198].

Weierstrass, K., Über die analytische Darstellbarkeit sogenannter willkürlicher Functionen reeller Argumente," *S. B. Deutsch Akad. Wiss. Berlin KL. Math. Phys. Tech.* (1885), 633–639, 789–805.

Wenjen, Chien, "Realcompact Spaces," *Portugal Math.* **25,** 135–139 (1966) [MR 37 #885].

Weston, J. D., "A Generalization of Ascoli's Theorem," *Mathematika* **6,** 19–24 (1959) [MR 22 #7108].

Weyl, H., *Die Idee der Riemannschen Fläche*, Teubner, Leipzig (1913) (3rd edition translated as *The Concept of a Riemann Surface*, Addison-Wesley, Reading (1955)) [MR 16, p. 1097].

Whitehead, J. H. C., "Simplicial Spaces, Nuclei and *m*-Groups," *Proc. London Math. Soc.* (2) **45,** 243–327 (1939) [Z 22, p. 407].

Whittaker, J., "On Isomorphic Groups and Homeomorphic Spaces," *Ann. of Math.* (2) **78,** 74–91 (1963) [MR 27 #737].

Whyburn, G. T., "Concerning the Cut Points of Continua," *Trans. Amer. Math. Soc.* **30,** 597–609 (1928).

Whyburn, G. T., "On the Structure of Connected and Connected im kleinen point sets," *Trans. Amer. Math. Soc.* **32,** 926–943 (1930).

Whyburn, G. T., "Open and Closed Mappings," *Duke Math. J.* **17,** 69–74 (1950) [MR 11, p. 194].

Whyburn, G. T., *Analytic Topology*, 1963 Edition, Amer. Math. Soc. Colloq. Publ. 28 (1963) [MR 32 #425].

Whyburn, G. T., *What is a Curve?*, M.A.A. Studies in Mathematics, Vol. 5, Van Nostrand, Princeton (1968) [MR 37 #895].

Wilansky, A. W., *Functional Analysis*, Blaisdell, New York (1964) [MR 30 #425].

Wilder, R. L., *Topology of Manifolds*, Amer. Math. Soc. Colloq. Publ. 32 (1949) [MR 10, p. 614].

Willard, S., "Absolute Borel Sets in their Stone–Čech Compactifications," *Fund. Math.* **58,** 323–333 (1966) [MR 33 #4892].

Willard, S., "Metric Spaces All of Whose Decompositions are Metric," *Proc. Amer. Math. Soc.* **20,** 126–128 (1969).

Zermelo, E., "Untersuchungen über die Grundlagen der Mengenlehre I," *Math. Ann.* **65,** 261–281 (1908).

Zoretti, M., "La Notion de Ligne," *Ann. de l'École Normale Supérieure* **26** (1909).

# Index